A GEOGRAPHY OF PENNSYLVANIA

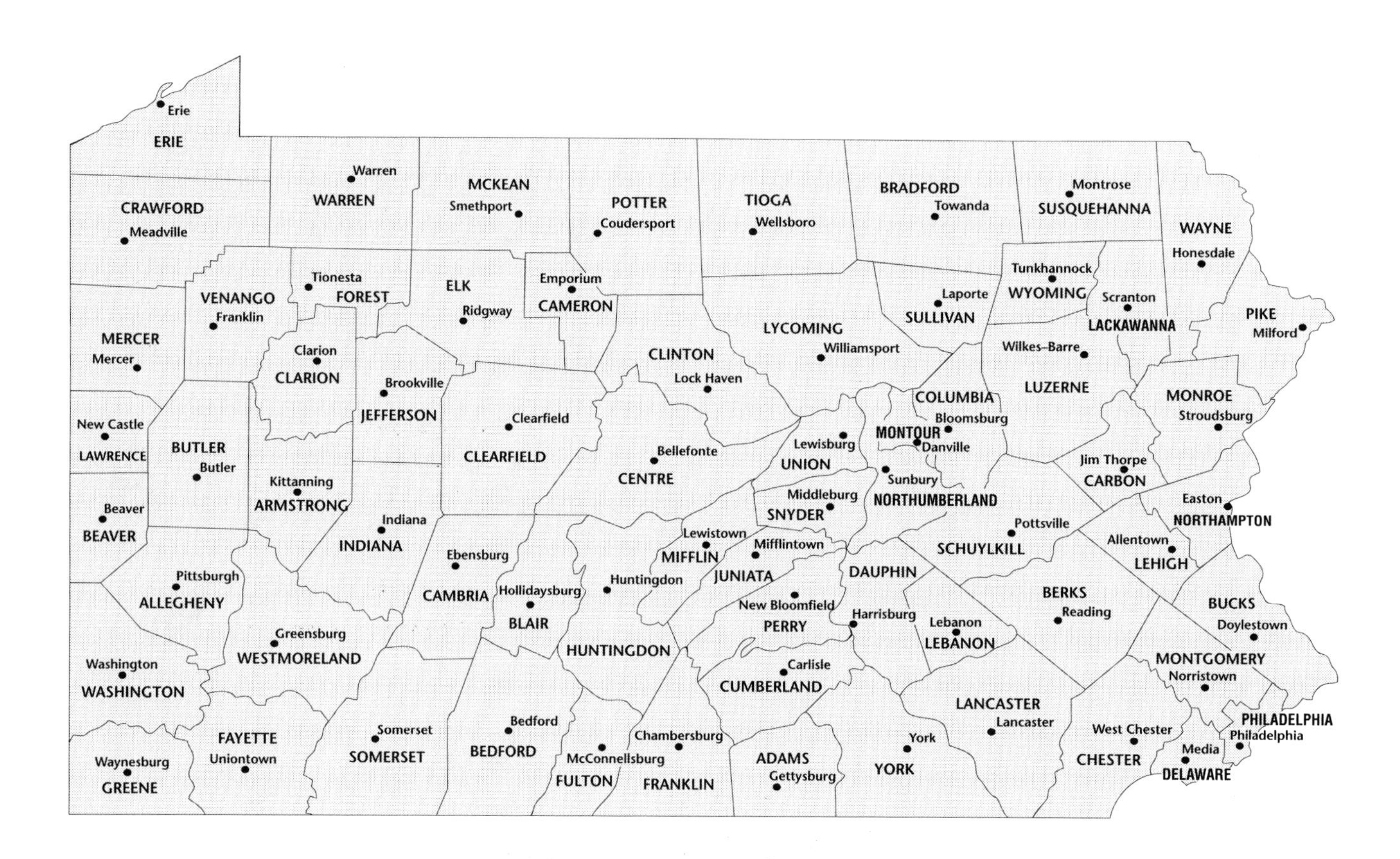

Erie
ERIE
Warren
WARREN
MCKEAN
Smethport
POTTER
Coudersport
TIOGA
Wellsboro
BRADFORD
Towanda
Montrose
SUSQUEHANNA
WAYNE
Honesdale
CRAWFORD
Meadville
Tionesta
FOREST
VENANGO
Franklin
ELK
Ridgway
Emporium
CAMERON
CLINTON
Lock Haven
LYCOMING
Williamsport
Laporte
SULLIVAN
Tunkhannock
WYOMING
Scranton
LACKAWANNA
PIKE
Milford
MERCER
Mercer
Clarion
CLARION
Brookville
JEFFERSON
Clearfield
CLEARFIELD
Wilkes-Barre
LUZERNE
MONROE
Stroudsburg
New Castle
LAWRENCE
BUTLER
Butler
Kittanning
ARMSTRONG
Bellefonte
CENTRE
Lewisburg
UNION
COLUMBIA
Bloomsburg
MONTOUR
Danville
Sunbury
NORTHUMBERLAND
Jim Thorpe
CARBON
Beaver
BEAVER
Indiana
INDIANA
Middleburg
SNYDER
Pottsville
SCHUYLKILL
Easton
NORTHAMPTON
Allentown
LEHIGH
Pittsburgh
ALLEGHENY
Ebensburg
CAMBRIA
Hollidaysburg
BLAIR
Huntingdon
HUNTINGDON
Lewistown
MIFFLIN
Mifflintown
JUNIATA
DAUPHIN
New Bloomfield
PERRY
Harrisburg
Lebanon
LEBANON
BERKS
Reading
BUCKS
Doylestown
Greensburg
WESTMORELAND
Washington
WASHINGTON
Carlisle
CUMBERLAND
MONTGOMERY
Norristown
LANCASTER
Lancaster
FAYETTE
Uniontown
Somerset
SOMERSET
Bedford
BEDFORD
McConnellsburg
FULTON
Chambersburg
FRANKLIN
ADAMS
Gettysburg
York
YORK
West Chester
CHESTER
Media
DELAWARE
PHILADELPHIA
Philadelphia
Waynesburg
GREENE

A GEOGRAPHY OF PENNSYLVANIA

E. WILLARD MILLER, EDITOR

PHILA. COLLEGE PHARMACY & SCIENCE
J. W. ENGLAND LIBRARY
RELEASED
4200 WOODLAND AVENUE
PHILADELPHIA, PA 19104-449

THE PENNSYLVANIA STATE UNIVERSITY PRESS
UNIVERSITY PARK, PENNSYLVANIA

Bib # 31744

PUBLICATION OF THIS BOOK WAS AIDED BY
A CONTRIBUTION FROM
THE PENNSYLVANIA GEOGRAPHICAL SOCIETY

Library of Congress Cataloging-in-Publication Data

A Geography of Pennsylvania / E. Willard Miller, editor.
p. cm.
Includes bibliographical references and index.
ISBN 0-271-01017-7 (alk. paper). ISBN 0-271-01342-7 (pbk.)
1. Pennsylvania—Geography. I. Miller, E. Willard (Eugene Willard), 1915– .
F149.8.G46 1944
917.48—dc20 94-575
CIP

Copyright © 1995 The Pennsylvania State University
All rights reserved
Printed in the United States of America

Published by The Pennsylvania State University Press, University Park, PA 16802-1003

It is the policy of The Pennsylvania State University Press to use acid-free paper for the first printing of all clothbound books. Publications on uncoated stock satisfy the minimum requirements of American National Standard for Information Sciences—Permanence of Paper for Printed Library Materials, ANSI Z39.48–1984.

F
149.8
G345

CONTENTS

LIST OF FIGURES

*Figs. 3.1–3.14 modified from D. J. Cuff et al., eds., *The Atlas of Pennsylvania* (Philadelphia: Temple University Press, 1989). © 1989 Temple University. By permission of Temple University Press.

*Figs. 11.1–11.12 from D. J. Cuff et al., eds., *The Atlas of Pennsylvania* (Philadelphia: Temple University Press, 1989). © 1989 Temple University. By permission of Temple University Press.

LIST OF TABLES

PREFACE

The culture and economy of Pennsylvania have been evolving for more than 300 years. Human activities have evolved from those based on primary occupations in the eighteenth century to those based on the manufactural economy of the nineteenth century and then on the service (tertiary) functions of the twentieth century. In the nineteenth century one of the world's greatest industrial societies evolved in Pennsylvania. As the state advanced into a postindustrial society, new regional patterns evolved and are still evolving.

The past is not a key to the future. With each major change in economic structure, the geographic patterns have also changed. Regions that experienced dynamic growth during the industrial era of the nineteenth and early twentieth centuries are now experiencing economic problems. Dominant economies of the past tend to resist change and repel new economic thrusts. The growth regions of today are the areas where tertiary (or service) activities, plus modern types of manufacturing, play a major role. Change will continue and Pennsylvania must recognize both the advantages and the perils of such a dynamic society.

Although each of this volume's four parts could be a book in itself, the parts and information are all related and are better understood in the context of the entire volume rather than as separate entities.

Part One presents the natural landscapes of Pennsylvania and sets the scene upon which the human geography of the state evolved.

Part Two focuses on the people of Pennsylvania. Population trends affect economic opportunity, and during the nineteenth century Pennsylvania provided the "golden streets" for waves of immigrants from Europe. It was a "melting pot" for many ethnic groups. While the culture of Pennsylvania has been influenced by many nationalities and ethnic groups, none is more important than the Germans, who provided the basis for what has come to be known as the Pennsylvania Culture Region.

Part Three is devoted to the evolution of Pennsylvania's economy, which was founded on the state's great agricultural and mineral wealth. In the nineteenth century, manufacturing became the catalyst to provide the good life for Pennsylvanians. In the twentieth century, the tertiary (service) activities have assumed dominance.

Part Four considers the urbanization of Pennsylvania. Two chapters are devoted to the historical developments and the structure of urban settlements, and then the volume concludes with major chapters on Philadelphia and Pittsburgh.

Because the broad approach makes it almost impossible for one person to write the entire book, specialists in each of the areas have prepared chapters. This is, therefore, a cooperative effort by nine faculty from the Penn State Department of Geography, plus Richard D. Schein from Penn State's Department of Plant Pathology and two former geography students, Roman A. Cybriwsky, now at Temple University, and Ben Marsh, at Bucknell University. Without their cooperation, this volume could not have been produced. Because it is a volume with multiple authors, the style of presentation varies somewhat from chapter to chapter. A selected list of references appears at the end of each chapter. More than 100 maps and graphs were produced in the Deasy GeoGraphics Laboratory under the direction of Tami Mistrick, Suzanne Peterson, and David DiBiase. We want to give special recognition to the Pennsylvania Geographical Society for their generous grant to aid the cartographic work.

1

AMERICAN ROOTS IN PENNSYLVANIA SOIL

Peirce Lewis

Pennsylvania is an old and important state, rich in historic incident and rich in geographic variety. This book is about that geography, and it has three aims:

- *To describe the places and regions of Pennsylvania*—the intricate and changing patterns of physical environment and human settlement that have merged to constitute one of America's richest geographic mosaics.
- *To explain how those patterns originated*—why people and places are located where they are, the way they are, and why maps of Pennsylvania look the way they do.
- *To explore the consequences of that rich and varied geography*—how Pennsylvania's geographic fabric has helped to shape the character and destiny of the state itself and of the United States as a whole. More than perhaps any other state, Pennsylvania has stamped the whole nation with its own distinctive personality. Some of those personality traits have proved useful and admirable, others much less so. Whether good or bad, however, Pennsylvania's impact on the rest of the country has been powerful and lasting.

This chapter seeks to explain how that happened and to put the state in a larger national context. It paints a portrait of Pennsylvania in broad strokes and primary colors. The details emerge in subsequent chapters.

The Geographic Basis of Pennsylvania's National Influence

Four elementary facts help explain why, over the years, Pennsylvania has played such an important part in the evolution of the United States as a nation.

To begin with, Pennsylvania was founded on an

exceptionally favored geographic site that got the state off to a good start and encouraged its career toward wealth and power. As chance would have it, William Penn had been granted one of the most productive bits of territory in eastern North America. For the farmers who made up the great majority of the early European population, the most alluring area was Pennsylvania's Piedmont, in the southeastern corner of the state just inland from Penn's first settlement at Philadelphia (see Fig. 2.4 in Chapter 2). It is what today's geographers refer to as Pennsylvania's *culture hearth*—where the state's character took form and which epitomizes the state in many ways. That same Piedmont region became Pennsylvania's economic backbone during the formative years before the Revolution, and indeed for much of the nineteenth and twentieth centuries too.

It was no single thing that made the Piedmont so attractive, but rather a combination: fertile soils, a benevolent climate, an absence of topographic obstacles, and easy access to main avenues of Atlantic trade. Later, as Pennsylvanians moved away from the ocean into the mountainous interior, they would discover other forms of wealth—a variety of raw materials which would make the state nationally famous—timber, oil, natural gas, but above all the most valuable coal deposits then known in the world, an almost limitless source of heat for houses and power for factories. But all of that would come later. It was on the Piedmont where Pennsylvania first took form and where Pennsylvanians developed a variety of enduring habits. It was Pennsylvania's incubator, and, for a small struggling rural colony, it was close to being ideal.

Southeastern Pennsylvania was not the kind of geographic environment that showered instant riches upon a lucky few. It had nothing resembling the silver of Mexico or the gold of Peru that had buried the Spanish Crown in treasure. Very likely that was a blessing in disguise. The enormous Mexican treasure had encouraged the Spanish government to launch into profligate spending and reckless costly military adventures, which in due course wrecked the Spanish economy and corrupted the Spanish state. In the process, the ruling classes of Mexico also developed some very bad habits, because the expectation of unending unearned wealth discouraged habits of saving and dampened respect for hard work. The geographic habitat of early Pennsylvania offered no room for cultivating those sorts of habits. Instead, it was a place that rewarded frugality and hard work, what James Lemon had called "the best poor man's country" (Lemon 1972). It was rich soil for farmers' crops, but rich soil also for the Protestant work ethic, which took root in Pennsylvania and flourished vigorously.

Partly it was Pennsylvania's *geographic variety,* and individual chapters in this book recount the main elements of that variety. From the time of earliest settlement, Pennsylvania's range of natural habitats provided a growing population with a broad variety of economic opportunity, so that over time the state became a major national force in agriculture, mining, transportation, and manufacturing. In such a rich and varied place, there was considerable room for trial and error. New ideas could be tried out in any of numerous environments, and success could yield lavish rewards. (There were plenty of errors, of course, but most of them have been conveniently forgotten.)

The people were varied too, and so were their ideas. Unlike the colonies of New England and the South, Pennsylvania's early government tolerated and invited a varied group of immigrants. Even before the Revolution, wide-open immigration policies had given Pennsylvania one of the most diverse populations in British America. It was not yet a melting pot, but rather a kind of lumpy demographic stew which contained a wider range of ethnicity than any other part of the country. In immigration, as in many other ways, Pennsylvania was a bellwether for the whole nation's later behavior.

Partly it was Pennsylvania's *geographic location* that made the state such an effective exporter of people and ideas (Fig. 1.1). Pennsylvania commands the eastern entryways to several primary routes of travel across the Appalachians (see Chapter 2). Thus, ideas that developed in southeastern Pennsylvania had a way of drifting across the mountains and out of the state by one of three great corridors: down the Great Valley[1] into the upland South, down the Ohio River toward the Kentucky Bluegrass and the Mississippi Valley, or straight overland across central Ohio into

1. The Great Valley extends almost the full length of the Appalachians from Quebec to Alabama, and sections of it have various names. In southeastern Pennsylvania, locals call it the Cumberland Valley. In Virginia, it is known as the Shenandoah Valley. Under any name, however, the Great Valley is one of the great traditional routeways in eastern North America.

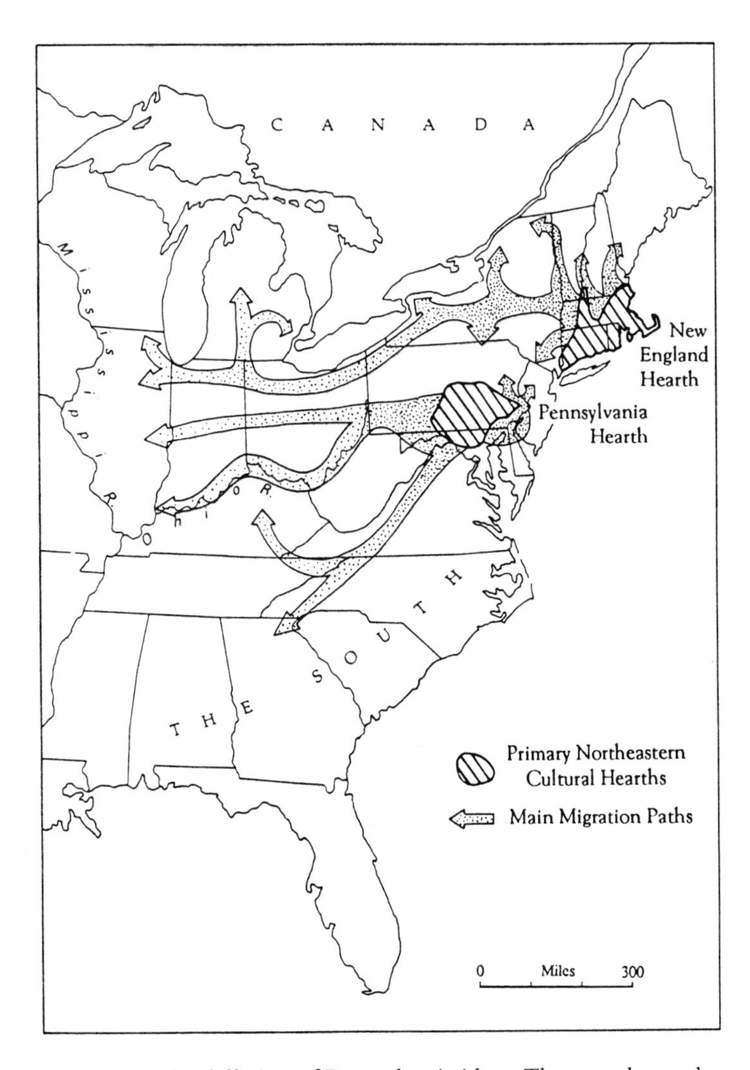

Fig. 1.1. The diffusion of Pennsylvania ideas. The map shows the spread of people and ideas westward out of Pennsylvania between the time of the Revolution and the Civil War. The Atlantic entryway to Pennsylvania was narrow—everything funneled through Philadelphia. To the west, however, Pennsylvania expatriates spread out in a vast fan, blanketing a huge area from Kentucky to the Great Lakes, and westward beyond the Mississippi River.

the center of the Midwest—roughly the present alignment of Interstate 70. When these and other trans-Appalachian routes are seen on a map, Pennsylvania's zone of influence emerges clearly, in the shape of a huge spreading fan with its narrow eastern apex near Philadelphia, but widening out westward, ultimately to cover much of the continental interior (Fig. 1.1).

Finally, it was a matter of *population,* and the power that resulted from it. From the time of the first U.S. census in 1790, for almost two centuries, Pennsylvania ranked as either the second or third most populous state in the Union.[2] Over that entire long period,

2. Pennsylvania was never first in population, although it came close several times. Most of the time, Pennsylvania was a good solid second—behind Virginia during colonial and early national times, close behind New York from 1830 to 1940. Since World War II, however, Pennsylvania's ranking has steadily

Pennsylvania's main export was people, who poured westward and southwestward into the interior of North America, carrying with them the ideas and institutions they had learned in Pennsylvania, and spreading them across the midsection of the country—that huge, vaguely defined area politicians would later call "middle America."

Images and Caricatures

Over the three centuries since Pennsylvania was founded, this old and complex place has been seen by outsiders in a wide variety of ways. Those visions of Pennsylvania are sometimes simple to the point of caricature, and it is easy to dismiss them as mere distortions of reality—as all caricatures must be. But that would be a mistake. What we think about the world is often more important than the reality of the world itself.

The best-known caricature, perhaps, is reflected in the state's official emblem, the *keystone,* an icon that dates from colonial times when Pennsylvania was seen as the central pivot of the 13 colonies, the midpoint between six northern and six southern colonies. It is no accident that both the Declaration of Independence and the Constitution were signed in Philadelphia; it was America's largest city at the time—indeed, one of the largest in the English-speaking world. Pennsylvania's leading city was an obvious central meeting place for any major assembly of important people. Both the city and the state were places to be taken seriously.

Early settlers from Britain and Germany had a different vision of Pennsylvania. They saw it as a land of fecund farms and peaceful people, where Quaker tolerance and yeoman diligence had combined with fertile soils to create a pastoral paradise—the Peaceable Kingdom of Holy Scripture implanted on the shores of North America, depicted in numerous well-known paintings by Edward Hicks (Fig. 1.2). That vision depicted Pennsylvania as a safe and comfortable haven from the miseries and privations of an older world. It was a powerful image, and responsible for making Pennsylvania one of the most desired destinations for European migrants to America. More recently, that same vision has been dusted off in modern form by the state's official tourist agency, which promotes Pennsylvania as a family vacationland. Full-color advertisements portray a fat, blissful land where New Yorkers and other urbanites can go to restore their souls and devour supercaloric meals amid scenes of bucolic tranquillity. The state's economic development board takes a similar line, advertising the state as an ideal place for enterprising corporations to invest their money, with a promise of low taxes and a contented labor force.

To others more jaundiced, Pennsylvania's image is the "rust belt"—a place of coal mines and blast furnaces, grimy cities, and obsolete factories built a century ago by labor's sweat and robber baron money—a nineteenth-century relict that the twentieth century has discarded, as wealth and people drift inexorably from the Northeast toward sunnier climes of the South and West.

To still others, Pennsylvania is "the Appalachians," but that Appalachian image itself has several faces. Planners and lawmakers have usually portrayed "Appalachia" as a land of abandoned mines and obsolete factories, declining employment, and isolated defeated people—a land destined permanently for the dole. East–west travelers, however, see the Pennsylvania Appalachians differently: as a belt of rough, wild terrain that lies inconveniently astride the main routes between the Atlantic coast and the Midwest—a permanent geographic nuisance to motorists and truck drivers of today, as to the drivers of Conestoga wagons 200 years ago. Meanwhile, to many residents of nearby states (especially young and middle-aged males), those forested Pennsylvania hills are a habitat for deer and wild turkey, patiently waiting for hunters to come each autumn and try to shoot them.

Finally, of course, to most American schoolchildren, the image of Pennsylvania is History, with a capital H, made of picturesque figures and heroic deeds: battles with Hessians along the Brandywine, Betsy Ross sewing the first American flag, the crack in the Liberty Bell, Washington's army camped in the snow at Valley Forge, Pickett's charge at Gettysburg, solemn gentlemen in powdered wigs signing sacred documents at Independence Hall.

---

declined. In 1950, with the growth of sun-belt population, the state's total fell behind California's, and then in 1980 behind Texas' total. Then, in 1990, Florida also surpassed Pennsylvania, making the state's population the fifth largest of the fifty states. Pennsylvania's population is not only stable in numbers but geographically stationary as well. As of 1990, about 85 percent of Pennsylvania's population was born in Pennsylvania. That is the highest proportion of stay-at-homes of all the 50 states.

Fig. 1.2. Edward Hicks's *The Peaceable Kingdom*. Hicks painted more than a hundred versions of this allegorical vision of Quaker Pennsylvania as heaven on earth. All of them are based on the Old Testament prophecy that with Christ's coming all God's creatures would live in harmony with one another. The painting illustrates the passage from Isaiah (11:6): "The wolf also shall dwell with the lamb, and the leopard and the calf and the young lion and the fatling together; and a little child shall lead them." In the background, Penn with God's blessing treats peacefully with the Indians, along the shores of a river that should be the Delaware at Philadelphia but looks considerably more like the Susquehanna above Harrisburg.

An Ordinary Kind of Place

But there is another image of Pennsylvania, and it reveals one of the state's most important and curious traits. A recent study has shown that Pennsylvania conveys no very clear image of regional identity; by many Americans it is seen simply as an ordinary kind of place (Zelinsky 1980; Lewis 1989). Quite unlike the South or New England or Texas, all of which possess strong regional images, much of Pennsylvania seems merely a large gray indistinct area on the average American's mental map of the United States (see Fig. 1.3). But the reason for that neutral image is very straightforward: Pennsylvania is the place where much of "middle America" originated.

The process began in colonial times. More than almost any other part of colonial America, southeastern Pennsylvania was the region where European settlers learned how to cope successfully with a dangerous, unruly country and convert it into tame, productive land—and do it not primarily as a

Fig. 1.3. Weak regional self-identity in the northeastern United States and southern Ontario. In most of the United States and Canada, people apply distinctive names to the region where they live—names like "Dixie," "Midwest," "Pacific Northwest," or "Great Lakes." In a broad swath of territory between western New England and southern Indiana, however, such regional names are not commonly used, suggesting no clear sense of regional identity. This region of weak regional identity includes much of Pennsylvania.

commercial venture, as in Virginia, and not as a member of some authoritarian religious community, as in Massachusetts, but as individuals. It was not particularly dramatic, just a long series of small experiments in which individual Americans learned little by little to design ordinary landscapes that would serve them in their daily lives—nothing fancy, just landscapes that would work. And the lessons learned in Pennsylvania were easily transferred as migrants moved westward across the mountains into the interior of North America. The idea can be phrased in slightly different terms: America formed some of its common geographic habits in Pennsylvania; small wonder the state looks familiar to common people from the American interior.

Consider just one example—a set of ideas about *towns:* how they should be laid out, and what their functions should be. Many of those urban ideas can be traced to William Penn's 1682 plan for Philadelphia, whose underlying principles were easy to imitate and apply elsewhere (Fig. 1.4). To begin with, the town's street pattern would be laid out in advance of settlement, in the form of a grid. (The grid-plan town was not a Pennsylvania invention, of course; such plans date far back into antiquity. But grid-plan towns were very rare in British North America, and they did not become common until after the founding of Philadelphia.) It was a democratic plan: all blocks were the same, at least on the map. It was utopian too: laid out in advance of settlement, it represented a fresh start, a clean break with European tradition, and a town would become as good or as bad as its residents could make it.

Beyond the platting of a grid, however, there was little thought of city planning as we know it today—no imposition of authority to restrict the way people

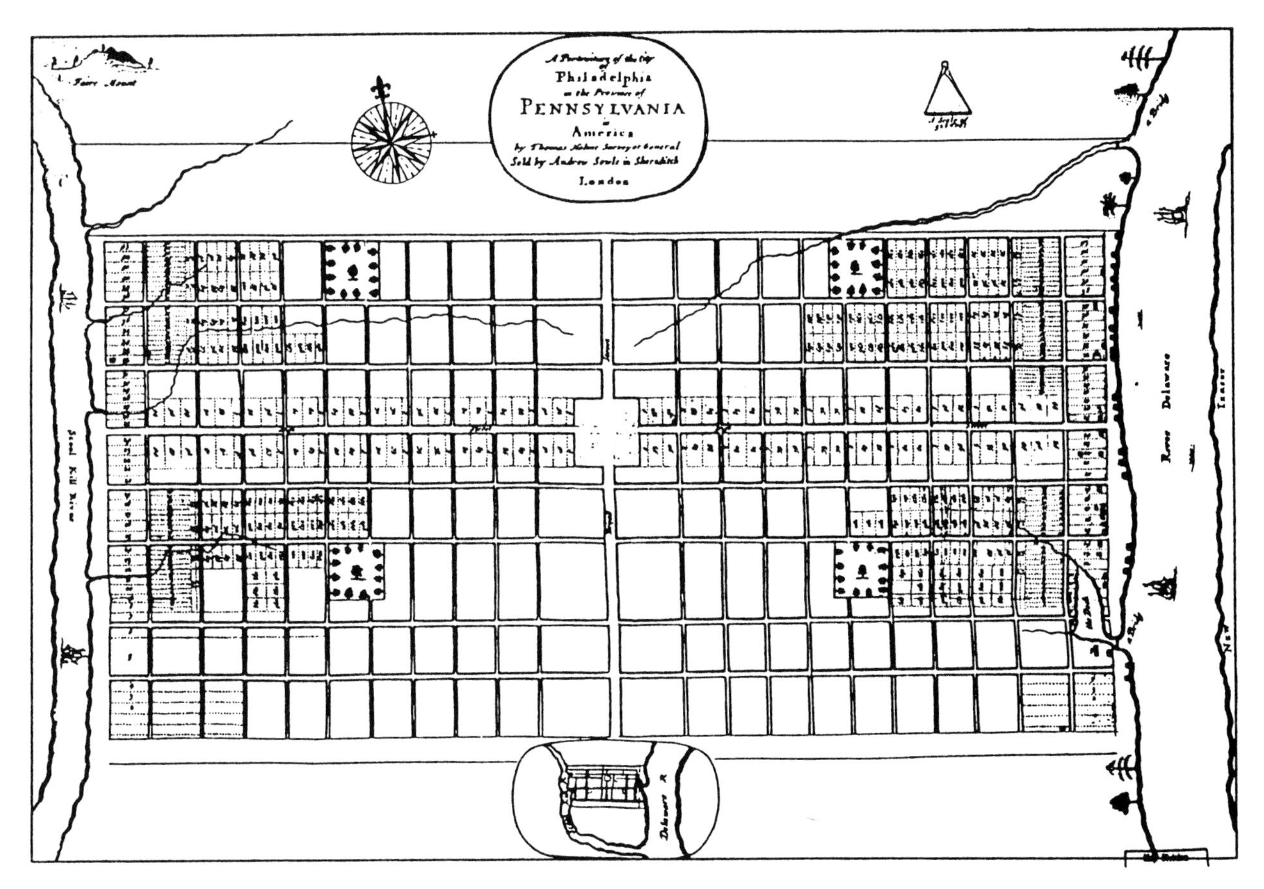

Fig. 1.4. Penn's plan for Philadelphia, London, 1682. Penn did not invent the grid plan, but Philadelphia was the first large city in British North America to use it. The layout of Philadelphia was widely imitated as Americans spread westward and planted towns in the process. Even street names were imitated: north-south streets were given egalitarian numbers; east-west streets were named after trees, not persons. More than any other kind of town plan, a grid was the easiest to lay out in advance of settlement.

used land. Lots were for sale to whoever could pay for them, and if land made a profit, so much the better. Inevitably, the center of a successful town commanded the highest prices and the main street became a marketplace, lined with shops and often named with no special flight of imagination, "Market Street." There was no nonsense about turning the center of town into a park or public green for recreation or public gatherings—no reservation of valuable downtown land to build churches. Unlike villages in New England, for example, the Pennsylvania town was not designed to create social unity or a sense of community; instead, it was expected to be economically useful. The business of a Pennsylvania town was business—to trade goods and make money. If such towns failed to elevate the spirit, they succeeded mightily in doing what most Americans wanted their landscapes to do: to promote economic prosperity. They were, in the final analysis, not just Pennsylvanian, but quintessentially American: egalitarian, optimistic, lacking in frills, and designed to be useful.

That kind of town, with its gridded streets and its downtown devoted to commerce, is almost boringly familiar to all residents of the interior and western United States. After all, most American towns are copied from it, and most Americans take them for granted. It is worth remembering, however, that the choice of a Pennsylvania model was not inevitable. Indeed, when Americans set about to design new towns for a new country, they had a considerable range of designs they might have chosen to emulate but that they quite consciously rejected.

For example, those early Pennsylvania town-builders might have consented to letting city streets develop in the winding ad hoc way of most European cities, and, for that matter, the ways of Boston and lower Manhattan—towns that predated Philadelphia and represented transplants of European forms. Or they could have chosen to control land use in their towns, to reserve choice central areas for plazas or common greens or churches, designed to give focus to the community and to elevate the spirit. That was the way things were done in places as different as New England and New Spain. Or they might have omitted towns from their settlement schemes entirely, as many Southerners did—creating community and transacting business through an elaborate system of plantations and dispersed rural communities.

But they chose none of those options. West of the Appalachians, the Pennsylvania plan won out, not because the scheme was thought to be beautiful, certainly not because it was spiritually uplifting, but because it worked. It met America's immediate needs at the crucial time that the nation was growing most rapidly and imprinting its newly formed geographic habits over the face of half a continent.

If town-planning had been Pennsylvania's only influence on the emerging fabric of America's human landscape, it might be dismissed as some sort of unimportant fluke. But it was not. As we shall see in later chapters of this book, trans-Appalachian Americans had carried with them several other basic Pennsylvania ideas about how ordinary landscapes should be organized. The state's influence was especially noticeable in three other kinds of western landscapes: those of *agriculture, transportation,* and *mining.*

With respect to *agriculture,* Pennsylvania's yeoman farmers were among the most prosperous in early America. Their huge barns, needed to store big harvests and shelter big herds, were not just functional. (Some thoughtful observers have suggested that those Pennsylvania barns, especially in the long-settled areas of the southeast, were generally much larger than they needed to be.) They were visible and self-conscious symbols of agricultural success. That symbolic vision was carried across the Appalachians, where similar barns came to dominate the horizons of the corn and wheat and dairy belts of the Midwest.

It was a similar story in the development of *transportation.* Pennsylvania's varied geography offered huge rewards to land developers and speculators, but profits could be made only if the products of town and country could be shipped cheaply to larger markets. That was no easy matter in Pennsylvania. Despite the bounty of its farmland, Pennsylvania has an abundance of rough terrain and a shortage of ocean coastline and navigable waterways. Efficient overland transportation was the only answer, so that over the course of time Pennsylvanians repeatedly led the nation in designing and engineering new fast and cheap ways of getting across difficult terrain. At various times and in quite amazing places, they created a system first of turnpikes, then canals, then railroads, and finally roads and superhighways—often in innovative ways that were quickly imitated elsewhere in the country. Since earliest times, Americans have been infatuated with mobility, with a childlike fondness for fast vehicles, whether steamboats, stagecoaches, steam locomotives, or automobiles. For better or for worse, Pennsylvania has helped nourish that attitude for almost 300 years.

Pennsylvania's contributions to ordinary American landscape design have not always been especially admirable. The landscape of *mining* is a case in point. The first really gargantuan mineral discovery in the United States was the anthracite (hard coal) of northeastern Pennsylvania. The hard-coal fields produced fabulous fortunes for a lucky few, provided work for armies of immigrant laborers, and held out a promise of riches in a hurry for those who had the right luck and who mastered a few basic techniques of corporate organization, all aimed at getting minerals from the ground as rapidly and cheaply as possible. Pennsylvanians mastered the techniques of wrenching wealth from the earth with terrible skill and effectiveness, and then taught the rest of the nation what they had learned. Later on, when gold and silver were discovered in California and Colorado and Nevada, the mining operations there were directed by people who had acquired their corporate and technical know-how in Pennsylvania. They had, unfortunately, learned some evil environmental habits in the process, and those habits were also exported to the West. There, transplanted Pennsylvanians laid waste to the landscape as nonchalantly as they had devastated the coal lands of their native state. It was the throwaway mentality on a grand scale: Pennsylvania mining companies ruthlessly plundered the land in order to make quick profits, and then when the land would yield no more, it was simply abandoned. Much of that land was so badly damaged that it was almost uninhabitable—and remains so today. That kind of throwaway land ethic was not unique to Pennsylvania, of course; it was and is widespread wherever human beings have engaged in large-scale mining. But as in so many other parts of ordinary American life, Pennsylvanians led the way—for better or for worse—and it was not always for the better.

A Practical State

So it was, over the years, that Pennsylvania made large and permanent contributions to the mainstream of American culture. As one surveys the record, however, it is remarkable how commonplace the bulk of those contributions were: material things with immediate and obvious utility to ordinary middle-class people. It is equally remarkable how little the state contributed to the abstract world of philosophy, political ideology, literature, or the arts. There are exceptions, to be sure: Benjamin Franklin, of course, Thomas Eakins (whom some critics regard as America's greatest painter), an occasional novelist such as John O'Hara, and perhaps others. But the numbers are scanty, especially when one recalls that Pennsylvania has always been one of America's most populous states and presumably should have contributed a large share to the nation's pool of intellectual and aesthetic talent. It is a plain fact, however, that Pennsylvania has produced no writers to equal Melville or Mark Twain, no architect to rival Frank Lloyd Wright, no philosopher with the stature of a John Dewey or Thoreau or Emerson, no national figure of social reform with the prestige of Jane Addams or Lincoln Steffens or Eleanor Roosevelt. In a monumental study of American leadership, sociologist E. Digby Baltzell has shown that the most creative and famous Pennsylvanians throughout history have typically made their reputations and fortunes as captains of commerce and industry, rarely as artists, abstract thinkers, or political leaders (Baltzell 1979). Pennsylvania's best-known figures have been bankers like Andrew Mellon, merchants like F. W. Woolworth and John Wanamaker, magnates like Andrew Carnegie and Henry Clay Frick, or exploiters of minerals, like Colonel Edwin Drake, who discovered oil at Titusville. Pennsylvania's most influential institutions have been powerful business corporations like U.S. Steel and the Pennsylvania Railroad, not schools of literature or transcendental philosophy. It is almost as if Pennsylvanians had bred in themselves a deliberate suspicion of abstract ideas or even actions that did not possess some immediate workaday utility—valuing common material things like turnpikes and steel ingots, skeptical of ideas that seemed too cerebral or too utopian. Pennsylvanians have made excellent engineers and barn-builders—mediocre artists and philosophers.

This avoidance and denigration of abstract or utopian thought is especially conspicuous in the field of government. At various times during the last century or so, for example, strong currents of political reform have swept across much of the country, but they have bypassed Pennsylvania with systematic regularity—and Pennsylvania politicians have a long-standing reputation for mediocrity and worse. The state's most famous political leaders have been power-brokers and spoilsmen with little trace of high-minded idealism. There were exceptions, of course, conspicuously Benjamin Franklin; Andrew Curtin, the state's civil war governor; and Gifford Pinchot, the maverick Republican governor (1922–26; 1930–34). The early twentieth-century bosses, Matthew Quay, Boies Penrose, and Joseph Grundy, are more typical of Pennsylvania's political leaders. Two perceptive observers of American politics have remarked that Pennsylvania "was a state where important people were in business, and politics was left to faintly disreputable leaders" and go on to describe Pennsylvania's typical delegation to the U.S. House of Representatives as "a collection of political hacks" (Barone and Ujifusa 1985, 1136).[3] As for executive leadership, Pennsylvania's record is dismal, reflected in what might be called "the Buchanan effect"—the remarkable fact that over a period of more than 200 years this large, rich, diverse state produced just one U.S. President, James Buchanan (1857–61), who is unanimously rated by historians as one of the most ineffectual presidents in American history. (Buchanan distinguished himself by sitting in the White House for the four years before the Civil War, fecklessly watching the nation prepare to tear

3. "Faintly" disreputable is a polite way of putting it. Pennsylvania's state legislators in the early twentieth century had a reputation for breathtaking venality, with elected representatives routinely bribed to do the bidding of the state's great wealthy and powerful corporations. According to a famous aphorism in the late nineteenth century, the Standard Oil Company "did everything to the Pennsylvania legislature except to refine it."

itself apart and doing nothing of significance to prevent it.)

The "Buchanan effect" is particularly striking when Pennsylvania's political figures are compared with the legion of distinguished statesmen from several large old eastern states nearby. With the exception of Benjamin Franklin, Pennsylvania never produced a political or intellectual leader of the same stature as Virginia's Monroe or Madison, much less that of Washington or Jefferson or Lee. No Pennsylvania political clan has ever come close to achieving the influence of New York's Roosevelts, or the Adams or Kennedys of Massachusetts. Only on rare occasions have Pennsylvania's governors or U.S. senators exerted much influence or commanded much respect beyond the borders of the state.

Pennsylvania's "Individualistic" Political Culture

It is a nagging question why a large old state, so favored by nature and founded under such auspicious conditions, should have focused its collective attention so doggedly and single-mindedly on material matters and economic gain and remain so suspicious of ideas—so remiss in providing America with political or moral or intellectual leadership.

There is no easy answer to that question, but a map by political scientist Daniel Elazar suggests that Pennsylvania's curious role may lie deeply embedded in the basic political geography of the United States (Elazar 1984) (Fig. 1.5). Elazar sees in America what he calls three "political cultures," by which he means the commonly agreed-on system of political beliefs about how any cultural group should meet its social and political needs. He calls those three groups respectively *moralistic, traditionalistic,* and *individualistic.* Pennsylvania's political behavior may well result from its dominance by an individualistic political culture. To understand what that means, it is useful to see how those political cultures fit together, and how Pennsylvania in turn fits into a larger national pattern.

The *moralistic* political culture is geographically dominant to the north of Pennsylvania and has its main roots in Puritan New England. Members of a moralistic culture tend to view public institutions (such as school, church, and government) as necessary constructive instruments by which society

Fig. 1.5. Political cultures of the United States. Pennsylvania is dominated by what Daniel Elazar calls "individualistic" political culture.

reforms its errors and perfects its ways. Thus, in a moralistic political culture educated citizens believe it is their civic duty to seek careers aimed at improving the moral condition of society—as professors, preachers, or government officials. If it happens in the process that some leaders in a moralistic society grow rich, that is seen as unimportant. Public virtue is always held in higher esteem than private wealth.

Elazar's *traditionalistic* political culture lay to the south of Pennsylvania. In that culture, largely rooted in the old slaveholding South, people achieve their political aims through connections with traditional extended families, not through formal institutions. In such a culture, loyalties to one's clan always claim precedence over loyalty to the state—or even one's own person. Blood feuds are endemic in such societies, of course, where the righting of supposed wrongs is accomplished by direct personal action—not by recourse to the courts—and where personal violence involves not just one person but his entire kith and kin.

Pennsylvania, however, is another matter. The state lies mainly within the political culture that Elazar calls *individualistic*, so called because political and social goals are effected not through institutional groups, as was customary in New England, and not through families, as in the South, but through individual persons acting alone. It is a freewheeling, laissez-faire way of looking at things, and if Elazar's map is right it dominates popular thinking in Pennsylvania.

If this analysis is correct, or even partly correct, it explains a good deal about Pennsylvania, and perhaps about the United States as a whole. It is difficult to escape the conclusion that when Pennsylvanians emigrated to the West, they carried their political philosophies with them, just as they carried their Pennsylvania ideas about planting towns and designing barns and then strewed those artifacts across the continental interior. Ideas about politics and society must have followed similar routes—and if that is so, the well-known American love of free enterprise, the well-known suspicion of politicians, and the stubborn antipathy for governmental regulation all had deep roots in Pennsylvania. In such a political climate, bright young men and women did not migrate naturally into such institutions as government or the church. They went instead into individual enterprise—most often business enterprise—and the freer the better. Social pressures did not steer them into poorly paid but morally uplifting careers as writers or preachers or teachers, nor were such people often rewarded with prestige and public admiration. Public distinction in Pennsylvania came from building practical (and well-paying things) like bridges or steel ingots. Fame was won through wealth; good deeds were less important. A career in government was not seen as an uplifting public calling, but rather as just another kind of business job.

Baltzell (1979) and others have argued that this attitude has deep religious roots. New England's Puritan tradition, for example, was strongly authoritarian. It was deeply tinged with the doctrine of Original Sin—the idea that all humans were the children of Adam and Eve, born sinners, and that it was the obligation of superior people to curb that inborn tendency to sin. Superior folk, the favored ruling classes, had an obligation to guide with a strong hand less-fortunate people, to impose virtue from above, so to speak. Pennsylvania, by contrast, was rooted in Quaker tradition, with a deep belief in personal responsibility, and distrust for political authority so profound that it sometimes verged on anarchy. In Quaker tradition, virtue resided in managing one's own private affairs—minding one's own business—and that meant staying out of politics. One's relationship with God was always personal and private, dependent on what Quakers called the "Inner Light." It needed no hierarchy of ministers or priests, much less authoritarian institutions of school or church or state. To the Pennsylvania Quakers, all persons were equal in the sight of God. Leadership—an idea that implies authority—was deeply suspect. Henry Adams, the historian and essayist, summed it up well in 1889:

> The only true democratic community then existing in the eastern States, Pennsylvania was neither picturesque or troublesome. . . . The State contained no hierarchy like that of New England; no great families like those of New York; no oligarchy like the planters of Virginia and South Carolina. Too thoroughly democratic to fear democracy, and too much nationalized to fear nationality, Pennsylvania became the ideal American State, easy, tolerant and contented. If its soil bred little genius, it bred less treason. With twenty different religious creeds, its practice could not be narrow, and a strong Quaker element made it humane. . . . To politics the Pennsylvanians did not take kindly. Perhaps their democracy was so deep an instinct that they knew not what to do with political power when they gained it; as though political power were aristocratic in its nature, and democratic power a contradiction in terms. (Adams 1889, quoted in Baltzell 1979, 8)

Such ideas, of course, did not remain peculiar to Pennsylvania. They are basic to much of middle America's popular thinking—and that is just the point. Various lines of evidence all suggest that those egalitarian ideas, so typical of middle-class thinking in the United States, first flourished in America along the shores of the Delaware in 1682, and over time spread westward across the whole country.

It is easy to denigrate a place like Pennsylvania, and just as easy to praise it, but neither course seems particularly fruitful. On the one hand, the single-minded pursuit of wealth and the systematic deprecation of authority and abstract ideas did not often

Fig. 1.6. Lithograph of Bellefonte, Centre County, 1847. While the artist makes the town look tidier than it probably was, this lithograph projects the image of an ideal community in a democratic republic: orderly, egalitarian, and economically successful. Many small market towns in Pennsylvania were like this, and thousands of American towns deliberately emulated the Pennsylvania model.

Fig. 1.7. With an abundance of rough terrain and a shortage of navigable water, Pennsylvanians necessarily went into the business of road-building on a grand scale. Here Interstate 80, one of America's main transcontinental highways, heads westward through a water gap in Bald Eagle Mountain and disappears in the distance as it ascends the Allegheny Front, the most formidable obstacle to east-west transportation in the state. (See Figs. 2.1 and 2.8.)

produce important political leaders, much less high-minded public institutions. On the other hand, Pennsylvania's persistent concern with the practical material objects of ordinary life provided America with an extremely useful set of tools at a time when much of the nation was being settled, when such tools were badly needed. In the final analysis, Pennsylvania was and is a practical place, and few Americans would see any reason to apologize for that. In that regard, as in so many others, American values and institutions are deeply rooted in Pennsylvania soil.

It is to the physical nature of that soil that we now turn our attention, to the fabric of the land—how it came to be, and how, over the years, Pennsylvanians found it useful to adjust themselves to that fabric.

Bibliography

Adams, H. 1889. *The United States in 1800.* Ithaca: Cornell University Press, Great Seal Books, 1955. (The first six chapters of vol. 1, *History of the United States of America During the First Administration of Thomas Jefferson,* first published in 1889.)

Baltzell, E. D. 1979. *Puritan Boston and Quaker Philadelphia: Two Protestant Ethics and the Spirit of Class Authority and Leadership.* New York: The Free Press.

Barone, M., and G. Ujifusa, eds. 1985. *Almanac of American Politics, 1986.* Washington, D.C.: National Journal.

Elazar, D. 1984. *American Federalism: A View from the States.* Third edition. New York: Harper & Row.

Glassie, H. 1967. *Pattern in the Material Culture of the Eastern United States.* Philadelphia: University of Pennsylvania Press.

Klein, P. S., and A. Hoogenboom. 1980. *A History of Pennsylvania.* Second edition. University Park: The Pennsylvania State University Press.

Lemon, J. T. 1972. *The Best Poor Man's Country: A Geographical Study of Early Southeastern Pennsylvania.* Baltimore: Johns Hopkins University Press.

Lewis, P. 1975. "Common Houses, Cultural Spoor." *Landscape* 19(2):1–22.

———. 1989. "The Mosaic of Pennsylvania." In *The Atlas of Pennsylvania,* ed. D. J. Cuff et al. Philadelphia: Temple University Press.

Reps, J. 1965. *The Making of Urban America: A History of City Planning in the United States.* Princeton: Princeton University Press.

Zelinsky, W. 1980. "North America's Vernacular Regions." *Annals of the Association of American Geographers* 70(1):1–16.

Cumulo-nimbus cloud over Mount Nittany

Susquehanna River in the Appalachian Plateau

PART ONE

THE NATURAL LANDSCAPES

The physical geography of Pennsylvania is a fascinating subject—and for two quite separate reasons. In the first place, Pennsylvania occupies a border zone between several quite different sorts of environments. In terms of landforms, Pennsylvania stretches from the Atlantic Coastal Plain at Philadelphia westward across the full width of the Appalachian Mountains to the margins of North America's great interior lowlands on the borderlands of Ohio. In terms of climate and natural vegetation, Pennsylvania falls within the boundary zone between sub-boreal forest climates of the North and the subtropical forest climates of the South, and between the harsh continental regimes of the Northwest and the milder warmer oceanic regimes of the Atlantic Coast. In terms of natural resources, Pennsylvania contains some of America's most valuable mineral deposits and some of its best agricultural soils, but at the same time it includes some of the most barren and formidable land in eastern North America. In sum, Pennsylvania's physical geography is astonishingly varied, and it is hardly surprising that earth scientists of many persuasions have found the state endlessly fascinating.

In the second place, that varied physical geography formed a backdrop for a rich and eventful human geography. From times of earliest European settlement, Pennsylvania was a laboratory where men and women from foreign lands learned the art of becoming Americans. As it turned out, the hugely varied physical geography of Pennsylvania was a generous and tolerant environment for many people, but a fierce and unyielding enemy for others. That story, of how people learned or failed to cope with that varied environment, is the story of Pennsylvania—and the story of America in microcosm. But none of those stories makes much sense unless one first understands the foundation that underlies that human endeavor: the physical geography of Pennsylvania.

2

LANDFORMS AND HUMAN HABITAT

Ben Marsh and Peirce Lewis

The Overwhelming Importance of Geology and Landforms

To understand the geography of Pennsylvania, the place to begin is with a study of the state's landforms (Fig. 2.1) and the geologic circumstances that produced those landforms. Geology and landforms directly affect a remarkable variety of physical conditions in Pennsylvania, which in turn have profoundly influenced the behavior and geography of Pennsylvanians. In some instances, geology affects human environment immediately and directly; in other instances, the effect is indirect and more complicated. But the chain of causation is clear and striking, if one knows where to look.

To begin with, events in geologic history determine the kinds of rocks that underlie the earth's surface, as well as the arrangement of those rocks. Certain rocks contain valuable minerals, such as iron or coal or limestone; other rocks are economically useless. Some rocks contain an abundance of groundwater and yield it up easily; from other rocks, groundwater is difficult and expensive to obtain. Some rocks produce fertile agricultural soils; others are barren.

Bedrock geology also is a major determinant of stream patterns in Pennsylvania. Over the course of geologic time, many streams have an uncanny ability to locate themselves in zones of weak rock while carefully avoiding more resistant rocks. A map of Pennsylvania's rivers and streams exhibits patterns that strongly resemble the patterns on the geologic map.

Those rivers and streams, in turn, as they erode downward, shape the form of the land, determining the slope and height and shape of mountains, hills, ridges, and plains. Frequently, in Pennsylvania, rivers also determine the location of mountain passes and help explain the eventual locations of transportation lines—canals, railroads, roads, and even superhighways.

Taken together, this web of geologic and topo-

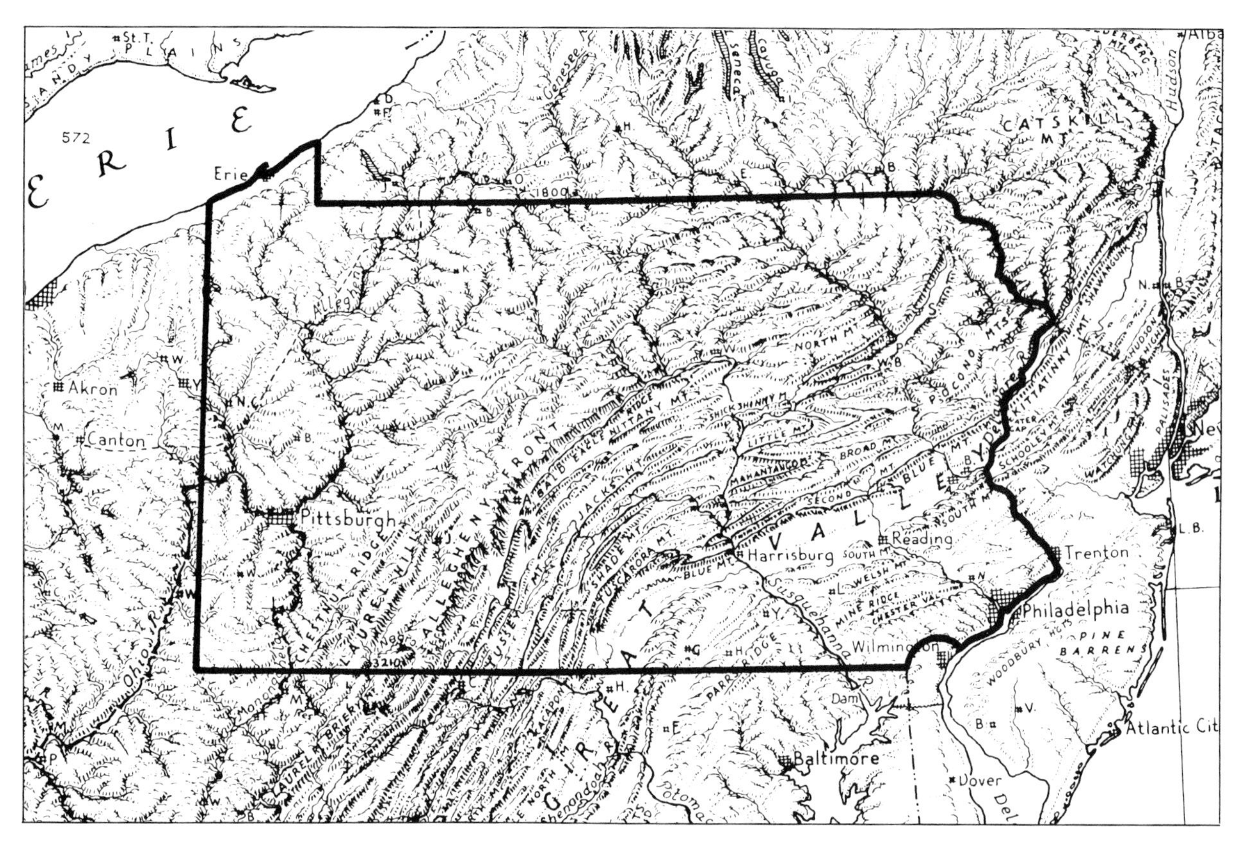

Fig. 2.1. Landforms of Pennsylvania and neighboring areas. This classic topographic map, © 1957 Erwin Raisz, illustrates several important points about the landscape of Pennsylvania: the state consists of a number of elongate regions parallel to the coast, relief is lower, closer to the ocean (where erosion has worked longer), and the texture of the landscape is finer in the southeast, where the rocks had been more deformed during mountain-building.

graphic circumstance strongly influences where people live or do not live—and how they make a living, and how well. It also goes a long way toward explaining why the contemporary landscape looks the way it does. Thus, for the sake of logical efficiency, we start at the beginning, with two patterns: Pennsylvania's landforms and the underlying arrangement of rocks.

Topographic Regions of Pennsylvania

The landforms of Pennsylvania sort themselves neatly into seven main topographic regions—that is, areas of considerable size in which topographic and geologic conditions are much the same and which differ significantly from other regions nearby (Fig. 2.2). In some parts of the United States, boundaries of topographic regions are subtle and fuzzy and one can travel from region to region and hardly notice a change. This is rarely true in Pennsylvania. Differences among the state's topographic regions are striking, and the boundaries between them are sharp and unsubtle. Travelers in Pennsylvania are likely to notice them, whether looking for them or not.

The geographic arrangement of the major regions is simple. From Philadelphia in the southeast to Erie in the northwest, there are seven regions in all, arranged roughly in a series of bands that are parallel to the trend of the Appalachian Mountains and the Atlantic Coast.

1. The easternmost region is the *Atlantic Coastal Plain,* a narrow strip of sandy, low-lying land immediately adjacent to the Delaware River. Although Pennsylvania's share of Coastal Plain is small and its topography is aggressively featureless, the region has played a major role in the state's history.

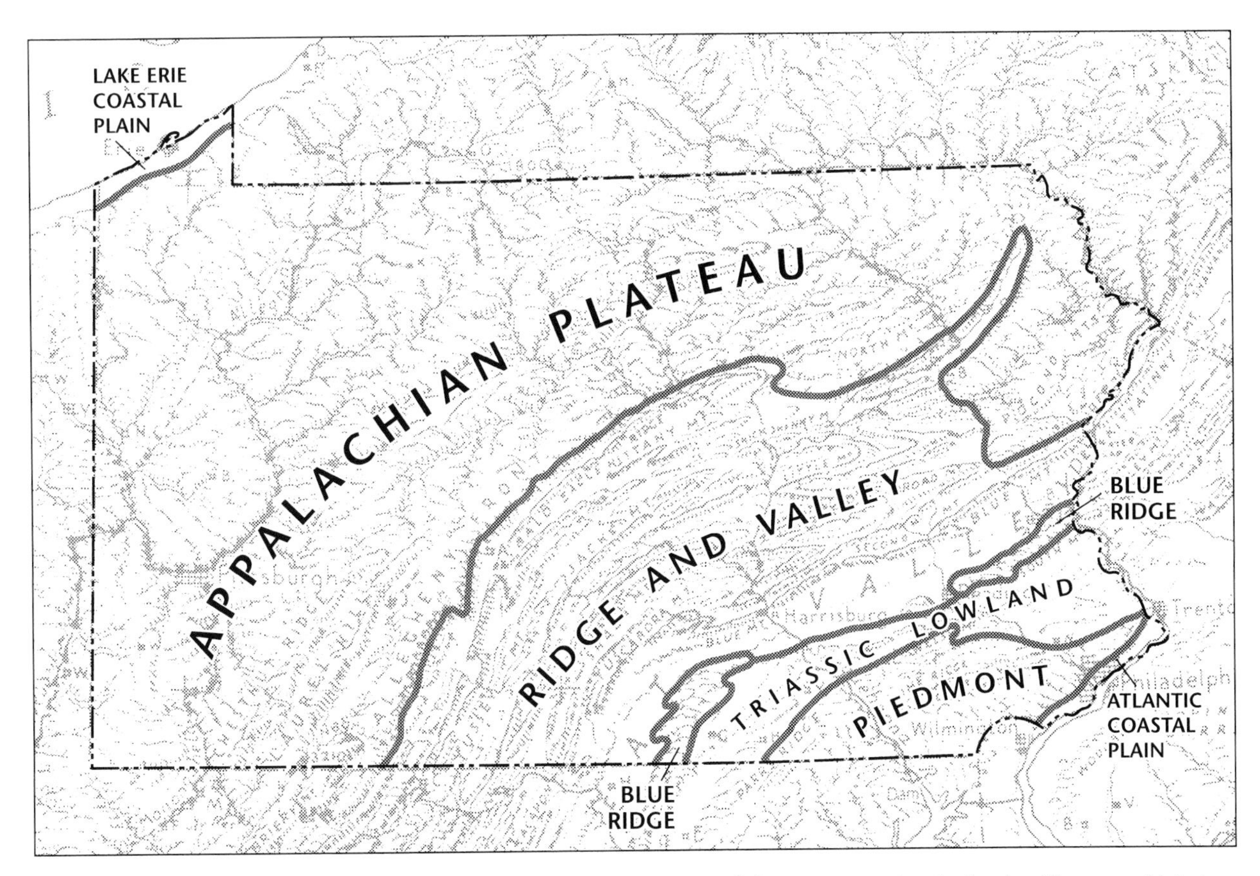

Fig. 2.2. Topographic regions of Pennsylvania. Extensive areas of the state contain similar landforms, which is a reflection of the similar processes that sculpted the earth in those regions.

William Penn landed on the Coastal Plain, and the old city of Philadelphia was built on it. It is, in fact, Pennsylvania's window on the Atlantic Ocean, and its link to the sea-lanes of the world.

2. Immediately inland from the Coastal Plain is the *Piedmont,* a gently rolling, well-drained plain rarely more than 500 feet above sea level. The boundary between the two regions is the famous "Fall Line," where elevations drop abruptly from the Piedmont to the Coastal Plain. Streams like the Schuylkill and the Lehigh cascade down across the edge of the Piedmont in a series of falls and rapids—hence the name "Fall Line." Waterpower from the falls encouraged the growth of early mill towns like Manayunk (now part of Philadelphia) and Conshohocken (an industrial suburb of Philadelphia). The falls also blocked navigation, so that Philadelphia at the foot of the falls of the Schuylkill became a major port and warehouse center. Seagoing ships could go no farther upstream, so cargo destined for the interior had to be unloaded there.

Piedmont soils vary in quality, but include the best in Pennsylvania, indeed some of the best in the eastern United States. The region is consequently spread with prosperous farms knit together with a dense network of roads and market towns. The Piedmont is the core of traditional Pennsylvania and through history has been the state's most picturesque and consistently prosperous region. All of the Piedmont is underlain by metamorphic rocks, divided into two subregions. Harder rocks, rich in quartz, support a hilly *Piedmont Upland* area in the southern parts of the Piedmont; the metamorphosed limestone of the area around Lancaster constitutes the rich soils of the *Conestoga Lowlands.*

3. The *Triassic Lowland* is an irregularly shaped belt that parallels the Piedmont to its northwest. It is made of relatively young and weak sedimentary rocks into which volcanic rocks have intruded themselves. Some of those volcanic rocks make ridges, which are topographically unimpressive but

historically important. The Battle of Gettysburg was fought for control of one of those ridges.

Although to the geologist the bedrock is very distinctive, the landscape of the Triassic region is quite similar to the adjacent Piedmont, into which it is intruded. The weak sedimentary rocks in the region have developed into fertile lowlands, like the Conestoga Lowlands, while the volcanic ridges strongly resemble the more-rugged landscape of the Piedmont Uplands.

4. Farther to the northwest lies the *Blue Ridge* region, two segments of mountain range each locally named "South Mountain." Each is a section of a larger mountain system, protruding into the state from Maryland, in one case, and from New Jersey, in the other. The southern stub, called the Carlisle Prong, is an extension of the great Blue Ridge Mountains of the southeastern United States. There the Blue Ridge is a massive bulwark, the highest and most formidable mountains east of the Rockies. In Pennsylvania, however, it is a mere shadow of its former self, reduced to a narrow ridge of low, rough country. The northern stub is the Reading Prong, technically a small section of the large *New England* topographic region. Between the two prongs is a sizable gap, one of the biggest breaks in the entire Appalachian mountain barrier and a major avenue leading to the west.

5. Inland from the Blue Ridge is one of America's most distinctive topographic regions, the so-called *Ridge-and-Valley* section of the Appalachians. The name describes it: long, narrow valleys separated by long, narrow ridges aligned parallel to one another over long distances. From an airplane or spacecraft it appears as if some great cosmic rake had been dragged along the backbone of the Appalachians from northeast to southwest. None of the ridges is very high. Typically, they rise about 1,000 feet above the valley floors, and none of the ridge crests exceeds 3,000 feet in absolute elevation. But the ridges lie athwart main lines of east-west transportation, which explains why frustrated early settlers dubbed them the "Endless Mountains." "Watergaps," where streams have cut notches through the ridges, played an enormous part in determining where trans-Appalachian canals, railroads, and roads were built.

A major subsection of the Ridge-and-Valley region is distinctive and appropriately named *the Great Valley,* the easternmost and biggest valley in the Ridge-and-Valley region. Locally in Pennsylvania it has other names: the Cumberland Valley or the Lehigh Valley, just as Virginians call their segment the Shenandoah Valley. Such localisms obscure the fact that the Great Valley of Pennsylvania is merely a small part of a trench of continental scale that extends almost without interruption nearly 2,000 miles from Quebec to Alabama. For as long as history records, the Great Valley has been one of the most important routeways in eastern North America. In early National times, Pennsylvania's portion served as a major avenue for emigrants leaving the eastern seaboard for various destinations in Tennessee, North Carolina, and Kentucky. Today it is the route of Interstate Highways 78 and 81. The Great Valley, and several other limestone-floored valleys, hold enclaves of very fertile farmland within the generally rugged and thin-soil landscape of the Ridge-and-Valley region.

6. To the west, everything changes as the corrugated Ridge-and-Valley country abruptly gives way to the dissected upland of the *Appalachian Plateau.* The southeastern edge of the Plateau is the Allegheny Front, an abrupt escarpment that is 1,500 feet high and is the most formidable obstacle to east-west transportation in Pennsylvania. West and northwest of the Allegheny Front, the Plateau's surface is underlain by warped sedimentary rocks, including thick layers of bituminous coal well-suited for making coke and steel. Among the coal beds is the fabulously rich "Pittsburgh seam," one of the most valuable mineral resources in the world. The "Pittsburgh seam" fueled America's heaviest industries while the nation was rising to world power in the late nineteenth and early twentieth centuries. Pools of oil and natural gas also occur, including the famous location near Titusville where Colonel Edwin Drake in 1859 drilled the first successful commercial oil well in the United States and in the process invented the petroleum industry.

Population densities on the Appalachian Plateau are extremely uneven, a direct result of geologic conditions. Almost everywhere, the Plateau surface has been dissected by rivers and streams into a chaos of valleys and knobby hills. This kind of country does not encourage either agriculture or easy road-building. Dense concentrations of people occur mainly where coal is found—almost exclusively in the southwest within a 100-mile radius of Pittsburgh. Indeed, a population map of western Pennsylvania exhibits patterns that look remarkably like those on a map of bituminous coal deposits. Much of the northern edge of the Appalachian Plateau has no coal and is so isolated and formidable that it remains sparsely populated. Its few scattered towns depend heavily on seasonal income from hunters and vacationers.

7. The Lake Erie Coastal Plain in extreme northwestern Pennsylvania resembles the Atlantic Coastal Plain in the southeast in several ways. It is a narrow strip of lowland along the water, and it is important not because of its size but because it is the site of a good-size city (in this case, Erie) and because of the access it provides Pennsylvania (to the Great Lakes, in this case). Unlike the plain at Philadelphia, the Lake Erie Coastal Plain is far from featureless. It rises in a series of steplike terraces from the lakeshore to a high escarpment that marks the northwest margin of the Appalachian Plateau.

Topography and location have combined to make the Lake Erie Coastal Plain a major transportation corridor. Because the plain is by far the easiest passage for east-west travel in this part of the world, it has long been the route for main-line railroads and expressways that connect the cities of upstate New York and northern New England with Cleveland, Chicago, and the Midwest. Topography has also influenced the local agricultural economy. The stepped hills above Lake Erie, and the lake itself, discourage unseasonable frosts, so the Plain has become an important producer of fruit.

The Geologic Underpinnings: Pennsylvania as the Eroded Core of the Ancient Appalachians

These regional descriptions are mere sketches. To give them substance, one must understand the geologic processes that carved the topography and formed the character and texture of the present landscape in myriad ways.

The story goes back a long way, far into geologic history. With a few exceptions, the geologic structure of Pennsylvania was created during the sequence of events that are collectively called "the Appalachian Orogeny"—that is, the mountain-building that created the ancient Appalachian range as the African tectonic plate collided with the North American plate. Some 200 or 300 million years ago, a mountain system that may have been as high as the Rockies, and that stretched from Mexico to Newfoundland, was created. The main height of that ancient range is now gone, worn away by eons of erosion. But the roots of the ancient Appalachians remain, and the landforms of present-day Pennsylvania are carved largely from them. Nearly all the basic geologic and geographic patterns of contemporary Pennsylvania directly reflect Appalachian geologic history in one way or another. The key to understanding the state's primary landforms, and much of its human geography, is understanding that history.

Appalachian history can be sketched in three stages: first, by surveying the rocks themselves as they were before the Appalachians were pushed up; second, by seeing how the Appalachian Orogeny deformed those rocks from their original condition; and third, by reconstructing the erosional events that removed the bulk of the Appalachian ranges and etched the present topography from Appalachian roots.

Appalachian Raw Materials: The Rocks of Pennsylvania

Most of Pennsylvania's rocks are sedimentary or originated as sedimentaries, which means that they were deposited in layers at the bottom of deep oceans, in shallow seas, along ancient beaches and river valleys, or in swamps—in short, wherever eroded materials were dumped.

The types of rock at the surface reflect the environments that existed long ago when the sediment was laid down. In some places, layers of clay were deposited in still, muddy water. Eventually the layers were compressed to become *shale,* the most common of all sedimentary rocks. In other places, grains of sand were laid down on beaches or in rivers. These grains were cemented together to become *sandstone*. Gravel was deposited by rapidly moving streams, and when it was turned to rock it became *conglomerate*. The limy shells and skeletons of marine organisms drifted to the bottom of shallow seas and were hardened to become *limestone*. Carbon-rich organic material in swamps became *coal* when it was compressed into rock.

Most of Pennsylvania's sedimentary rocks were laid down over hundreds of millions of years in a shallow sea bounded on one side by the interior of the North American continent and on the other side by a now-gone volcanic island arc. With the exception of a small area—the Triassic Lowlands, where the rocks are slightly younger—the rocks in Pennsylvania exactly span the geologic "Ages" of the Paleozoic Era.

The sedimentary rocks of Pennsylvania form a thick sequence; in some areas the layers stack up to a depth of eight miles. Such thicknesses of sedimentary rock are highly unusual, and Pennsylvania is a mecca for geologists who are interested in sedimentary rocks and the paleontological history of early life on earth that those Paleozoic rocks reveal. The only ocean basins that approach that depth anywhere on earth today are associated with zones where oceanic tectonic plates are being jammed beneath continental plates. Something much like that evidently happened during the Paleozoic Era, when North and South America collided twice with Europe and Africa, temporarily closing up the Atlantic Ocean each time and grossly deforming the rocks along the zone of collision. Students of North American geology surmise that there must have existed a gradually deepening downwarp in the crust at or near the zone of collision, a kind of trough that served as a trap for huge quantities of sedimentary materials, filling up nearly as fast as it lowered. Most of Pennsylvania's rocks, it is believed, originated as sediments in that basin. As the depth of the basin changed, different kinds of rocks were deposited.

The Appalachian Orogeny and Its Effect

Such an accumulation of sediments, however deep it might be, does not produce a mountain range; the sediment, lying at or near sea level, can only be the raw material for the range. The mountains themselves were created in the violent sequence of events called the Appalachian Orogeny (the word "orogeny" comes from the Greek and means "the creation of mountains"). In geologic terms, the orogeny was a period of violent plate collision and compression that caused rocks to bend, rupture, melt, and change their forms. During events such as this one, sections of the crust are uplifted to produce mountain ranges, while other sections are depressed to produce abyssal trenches in the ocean. Areas closest to the orogenic axis tend to be most deformed. Farther away, the deformation will be less extreme. This precisely reflects the pattern in Pennsylvania.

The axis of the Appalachian Orogeny was an elongate zone that evidently lay roughly beneath the contemporary Atlantic Coast. Most of the rocks in the Piedmont and Blue Ridge regions of southeastern Pennsylvania show signs of extreme deformation, whereas toward the central and western parts of the state the deformation systematically grows less and less extreme. That systematic change, as one travels inland from the coast, largely explains why Pennsylvania's geologic and topographic regions are lined up parallel to one another (see Fig. 2.3). The rocks of southeastern Pennsylvania were violently deformed, those of central Pennsylvania were only moderately deformed, and in western Pennsylvania the deformation was quite mild. The present-day geography of Pennsylvania reflects that three-part difference in many ways.

After the Appalachian Orogeny was complete and the ancient Appalachian Mountains had already begun to erode, the Atlantic Ocean started to open up—and it has been opening ever since. But at the earliest stages, during a time spanning the geologic periods known as the Jurassic Period and the Triassic Period, the ocean began to open up in a number of places along the edge of the continent during the stretching of the crust. Rifting and faulting marked the would-be ocean basins, and sediment and lava flowed into them at the fracture zones. One of the biggest of these fracture zones passes through Pennsylvania and created the complex and broken mass of rock that underlies the Triassic Lowlands. This aftershock of the Appalachian Orogeny was the last wrenching geologic change to affect the state.

The Sculpting of Pennsylvania

The landscape of Pennsylvania is no longer dominated by the great mountains of the Appalachian Orogeny, but only by their eroded roots. Every square mile of the deformed rock mass has been undergoing erosion for many millions of years, and all the original mountains are long gone. Landforms here are not the direct result of orogeny; high places are not the result of upwarps, nor low places downwarps. Instead, high places are where rocks are resistant and the work of rain and streams is incomplete, and low places are where rocks are weak and have readily washed to the sea. This process, called *differential erosion,* explains most of Pennsylvania's present-day landforms. All the mountains

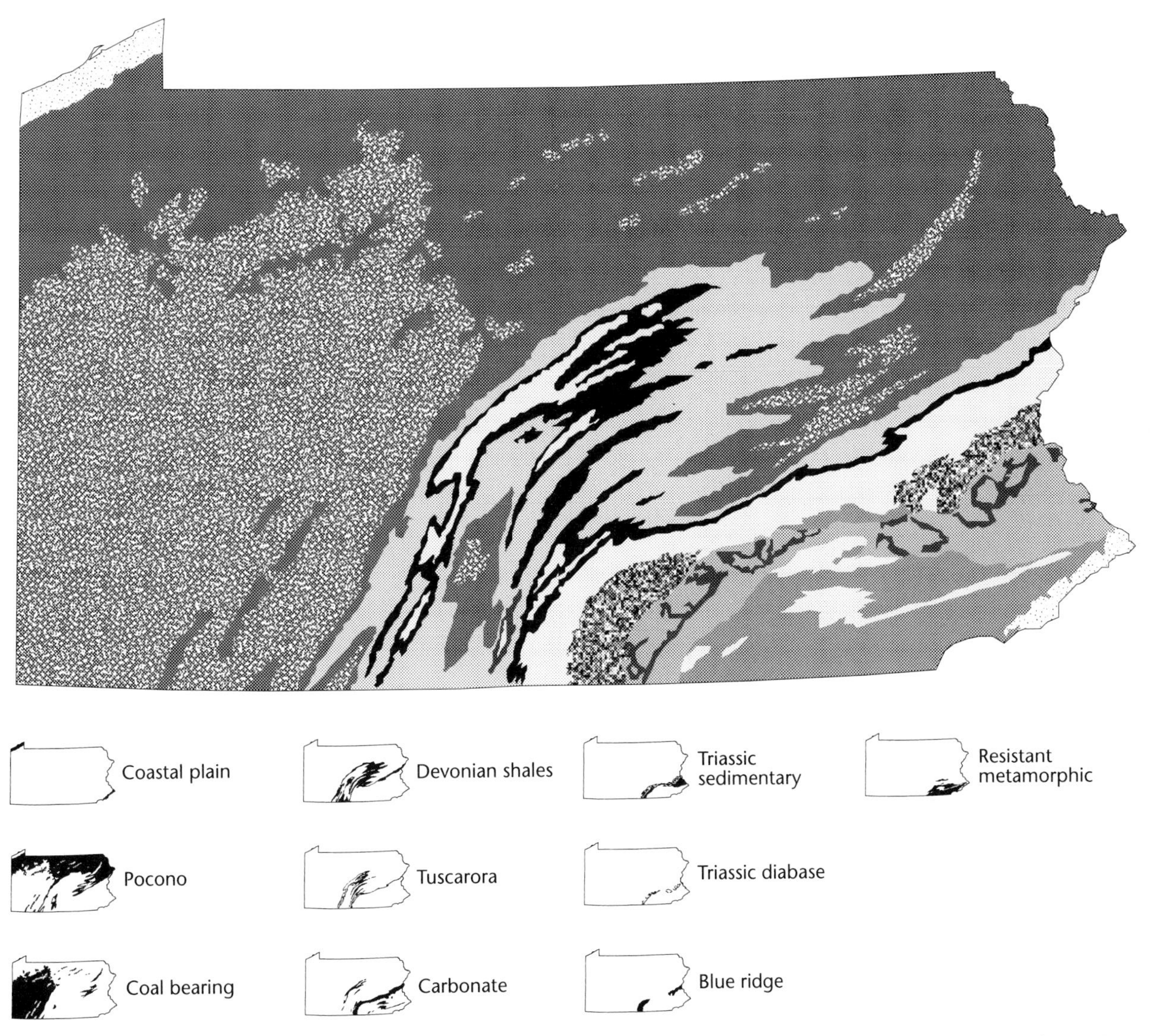

Fig. 2.3. Geologic regions of Pennsylvania. Each different rock type occurring at the surface produces a distinctive landscape. Resistant rocks, such as sandstone and granite, produce ridges and other uplands; limestone and shale create valleys. This map identifies ten general types of rocks occurring in the state, which are associated with ten areas of distinctive landscape. This map defines a number of subregions *within* the topographic regions shown in Fig. 2.2, but it also emphasizes similarities *between* parts of topographic regions.

presently in Pennsylvania are places where the orogeny brought a tough rock, such as sandstone or granite, to the surface. Valleys were formed where weaker rock, such as shale or limestone, came to the surface.

Despite Pennsylvania's stormy tectonic history and its often rough terrain, the state is completely lacking in really high elevations. Mount Davis in Somerset County, at 3,216 feet above sea level, is the highest place in the state but is hardly a peak—instead, it is merely a local high place along the edge of the Allegheny Front. The most striking thing about Pennsylvania's "mountain" topography is not lofty peaks or alpine precipices, but the remarkable consistency of high elevations. From a fire tower or scenic overlook anywhere in the rough central part of the state, one can look out and see a planar horizon that looks as if some gigantic carpenter's plane had trimmed off all the ridge crests to about the same elevation—what geomorphologists call "accordant summits." In the Appalachian Plateau, where sedimentary rocks are nearly flat-lying, accordant uplands are easy to explain. They commonly result when a single extensive sheet of resistant sandstone has been carved by streams into a hodgepodge of flat-topped knobs—what in the arid western United States would be called mesas.

Elsewhere in the state, however, especially in the Ridge-and-Valley region, where rock structure varies greatly from place to place, accordant summits cannot be explained by the effect of erosion on homogeneous rocks. Instead, it is believed that those accordant elevations were formed during an enormously long period of tectonic stability that followed the Appalachian Orogeny. In the more than 200 million years since the Appalachian Orogeny ended, the major geologic processes at work over most of Pennsylvania have been the slow, inexorable weathering of rock into loose soil particles at the surface and the erosion of that weathered material by running water. Two hundred million years was time enough and more for erosion to wear away all but the faintest traces of mountainous topography (although, as we have seen, Appalachian rocks remained immediately below the surface, as the eroded roots of the ancient mountain system). The result of such prolonged erosion was apparently a gently undulating and faintly sloping surface called a "peneplain" (from the Greek, meaning "almost a plain"). That gently rolling surface has been called the "Schooley Peneplain," named after Schooley Mountain in northern New Jersey, where it was first described. Pennsylvania has long been recognized by earth scientists as a classic area for the study of long-term stream erosion, and debates on the topic of peneplains have been ranging intermittently for almost a century.

Once that peneplain was formed, more uplift followed, completing the story of how Pennsylvania's landscape was carved. First, one must imagine that the entire Schooley Peneplain was broadly uplifted over a very large area. Compared with the violence and complexity of previous tectonic episodes, it was a gentle uplift, only 1,000 feet or so, but enough to start the process of erosion all over again. Streams that had been wandering idly across the surface of the peneplain to sea level once more began to carve valleys, chewing into the surface of the peneplain. The weakest rocks were eroded first, so that valleys developed in areas of nonresistant rock while resistant rocks were left standing at the level of the old peneplain—all at about the same elevation.

That general uplift was followed by a long quiet period during which streams were free to go about their erosional business. Once they had cut down about 1,000 feet (the amount of the Schooley uplift), however, they were so close to sea level that they were powerless to cut down further, at which time they again began to wander cross-country and begin a new peneplain—called the Harrisburg Surface, after the low-relief topography of the Great Valley, where it was first described. That second peneplain, however, was only partly completed. It is best developed on weak rocks close to streams that flow to the Atlantic. That is the reason one can drive from Philadelphia to Harrisburg and, except for occasional low ridges made of patches of harder rock, cross what appears to be the same gently rolling terrain. In the Ridge-and-Valley region, the Harrisburg Surface takes the form of flat-bottom valley floors carved in limestone or shale but isolated from other such valleys by high sandstone ridges.

The latest major episode of Pennsylvania's erosional history began much more recently. With the second peneplain half finished, there was yet another episode of general uplift—about several hundred feet. This new episode of erosion is most clearly visible where streams are most active—for example, along the lower courses of the big streams and rivers. At Philadelphia, for example, the Schuylkill River immediately reacted to the uplift by cutting down into the Piedmont. Upstream from the Fall Line, the Schuylkill has carved a deep gorge, now occupied by the Schuylkill Expressway and the fringes of Fairmount Park. The Schuylkill gorge is extending its way upstream, much as a gully erodes headward into a farmer's fresh-plowed field. It has now extended itself almost to Valley Forge. Upstream, however, from Valley Forge to Reading, the Schuylkill wanders peacefully across a gently and as-yet-undissected section of the Harrisburg Surface.

The Susquehanna system is doing much the same sort of thing, busily chewing into that surface. Because the Susquehanna is a big and powerful river, it has managed to cut downward into the surface along much of the way between Havre de Grace, Maryland, and Harrisburg, and its tributaries have followed suit. The entrenchment of the Susquehanna can be seen clearly from the U.S. 30 bridge across the river between Lancaster and York.

Much of the lower Susquehanna gorge has been flooded by water impounded behind a series of power dams.

Typically, then, an eastern or central Pennsylvania landscape exhibits three distinct levels: (1) a ridge-crest level, which is a slightly reduced version of the Schooley Peneplain; (2) a broad, rolling lowland, the remnant Harrisburg Surface of the partially completed second peneplain; and (3) the narrow valley floors of streams that are now cutting downward into the Harrisburg Surface. In much of eastern and central Pennsylvania, big streams (especially in their lower reaches) are actively cutting down into bedrock, often crossing alternating bands of weak and resistant materials. The result is that few of Pennsylvania's main rivers that flow to the Atlantic are navigable.

Geology and Erosion: How the Topographic Regions Were Formed

The geologic processes that constructed and sculpted Pennsylvania over millions of years have worked in significantly different ways in the various parts of Pennsylvania. In some places the rocks were heavily deformed, while in others they are almost in their pristine form. And near the sea, erosion has been energetic and continual, while farther inland sculpture has had less effect. Overall, however, the effect of the state's long geologic history has been to produce landforms that intimately reflect the bedrock beneath it, both in the broad sweep of the topography that mirrors the continental collision of the orogeny and in the individual valleys and hills resulting from local exposures of weaker and stronger bedrock (Fig. 2.4).

The topographic regions of Pennsylvania were created by the differences in original bedrock, in orogenic deformation, and in subsequent erosion. These differences in geologic processes have created the differences in landscape that can be seen between different parts of the state. Each region should be understood according to its particular history. The discussion of the regions will progress from the most heavily deformed and thoroughly eroded part of the state—the southeast—to the barely bent and mildly eroded northeast (Fig. 2.5).

Piedmont: The Violent Southeast

The Piedmont and the adjacent Blue Ridge Mountains contain some of the most heavily deformed rocks in Pennsylvania. The combination of compression and heat near the orogenic axis converted nearly all the sedimentary rocks into metamorphic rocks. (Again the word comes from the Greek: *meta* = change; *morphos* = form.) Thus, under great heat and pressure, shale was metamorphosed into slate or schist, limestone became marble, sandstone became quartzite. Even igneous rocks, such as granite, were sometimes metamorphosed, and in some instances metamorphic rocks were metamorphosed again.

The structure of those metamorphic rocks goes a long way toward explaining the landforms of southeastern Pennsylvania, for metamorphics are the rocks beneath almost all of the Piedmont and the Blue Ridge. The intense heat and pressure of metamorphism converted the parent rock into pliable material that flowed like hot, dense glue deep below the earth's surface. Later, as they cooled, minerals in the rock recrystallized, often concentrated in long, thin parallel sheets oriented at right angles to force and parallel to the direction of flow. It is similar to what happens when a child mixes sticky peanut butter with sticky jam in a jar and spreads it on bread with a knife. The different ingredients—peanut butter and jam in the sandwich, minerals like white quartz and dark-green hornblende in the earth—tend to stay segregated and to line up with one another in swirling, discontinuous sheets. When such rocks appear at the earth's surface, as they do in Pennsylvania's Piedmont and Blue Ridge regions, geographic patterns emerge that are highly predictable and depend on the kinds of minerals that make up the metamorphic bands. In general, topography tends to reflect at the surface the banded patterns of the metamorphic rocks beneath, simply because the minerals in some rocks permit them to erode more quickly than others.

Because of the differences in how rocks erode, the Piedmont is readily divided into two subregions. A portion of it, especially the fertile lowlands of Lancaster County, is formed from weaker rocks, such as a semi-metamorphosed limestone called "Conestoga marble." This rock contains an abundance of limy minerals that weather quickly and leave behind a thick, clay-rich residuum that is some of the best soil in the eastern United States. This rich agricultural area is called the Conestoga Lowlands.

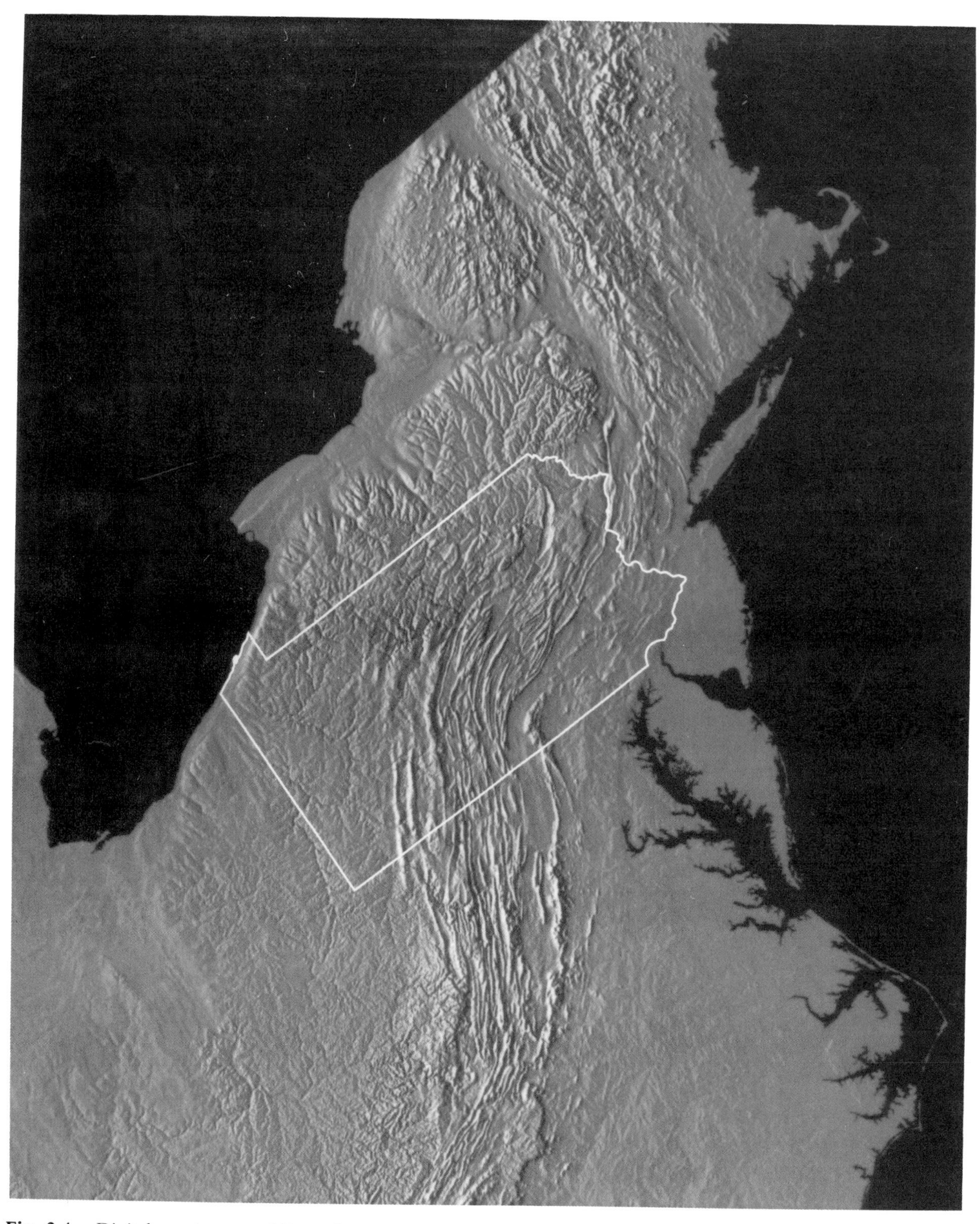

Fig. 2.4. Digital terrain map of Pennsylvania. A detail from Thelin and Pike 1991. The obvious general trend of the Appalachians—southwest to northeast, and roughly parallel to the present coastline—is directly perpendicular to the 200-million-year-old impact of the African continent from the southeast. Individual features in the region—the valleys, mountains, and even rivers—show the effect of millions of years of erosion on the deformed mass of rocks of different resistances. Every upland is underlain by sandstone or other tough rock, and most lowlands are underlain by weak rocks like shale or limestone. This shaded-relief map was produced by a computer from millions of individual elevations.

Lined up in bands south of the Conestoga Lowlands and extending beyond the Maryland border is a much more rugged area, called the Piedmont Upland. This area is underlain by harder rocks, such as the "Chickies formation"—a rock that started out as sandstone but was metamorphosed to become a very resistant quartzite. It is very resistant to weathering and slow to erode; at the surface, it produces elevated rough country. Slow weathering and a quartz-rich bedrock produce thin and sandy soils. Throughout Lancaster County, one can drive across lush, rolling farmland and know that the Conestoga formation lies immediately beneath the surface. Never far away, however, are elongate wooded ridges that occur wherever the Chickies quartzite appears at the surface.

Here and there, concentrations of valuable minerals are found in the metamorphic rock of the Pennsylvania Piedmont. Iron is the most valuable of those Piedmont minerals, occurring in widespread but usually small deposits throughout the region. In the early nineteenth century these deposits formed the basis of an embryonic iron industry, the beginnings of the Industrial Revolution in eastern North America. However, the Piedmont iron deposits were too small to justify the large expenses associated with present-day mining technology, and all were eventually abandoned. In a number of places the Conestoga formation and related marbles are quarried as a source of lime, and the harder rocks are excavated to make crushed rock.

The Blue Ridge: Remembrance of an Earlier Orogeny

Like the adjacent Piedmont, the matching small prongs of the New England and Blue Ridge regions are made of metamorphic rock. Like the Piedmont too, they were deformed and pushed westward during the Appalachian Orogeny. Because they are made of very resistant rock, the features are steep hills with thin, sandy soils, like exaggerated versions of the Piedmont Upland region. The rock of the Blue Ridge and New England regions, however, is much older than the Paleozoic rock of the rest of the Appalachians. It had already been deformed once into mountains and eroded almost flat, before being re-deformed during the Appalachian Orogeny.

These two mountains are typical of the roots of ancient mountains. Hugely complex in geology, and thoroughly eroded over the millions of years, the features are more ridges than mountains. Too steep and sandy for agriculture, and too high to be easily crossed, the ridges have served since earliest historic times to divert settlement, travel, and development along the lowlands on either side and away from direct access between the coast to the interior except at the few sizable gaps. But being sparsely occupied and relatively close to several cities, these areas are now being eyed as areas for recreation and second-home development.

The Ridge-and-Valley: The Moderate Geology of Central Pennsylvania

The landforms of the Ridge-and-Valley region are dramatic for their regularity if not for their topographic relief. Straight, wooded ridges stand 1,000 feet above fertile lowlands, or above strip-mined floors of the anthracite valleys (Fig. 2.6). This landscape is the result of the intermediate level of deformation that affected the middle part of the state. As we have seen, the intensity of Appalachian deformation decreased away from the axis of the orogeny. Inland from the Piedmont, the pressures were no longer sufficient to convert any but the softest sediments into metamorphic rocks, so that here they remain recognizable as sediments. Nevertheless, the forces were sufficient to crumple the thick layers of bedded rock into wrinkled folds—much as a rug will be wrinkled and folded if it slides across a slippery floor and collides with a wall.

Geologists logically call this section of the Appalachians the "Folded Appalachians." It is the erosion of those folded sedimentary rocks that produces the distinctive topography of the Ridge-and-Valley region. The two names are exactly parallel: "Folded Appalachians" is a geologic term that refers to underlying rocks, whereas "Ridge-and-Valley" is a geographic term that refers to topography. The two names refer to the same area. Indeed, in no part of America is the relationship between terrain and subsurface geology any closer or more elegant. (Compare Figs. 2.1 through 2.5.)

Just as in the Piedmont and the Blue Ridge, topography in the Ridge-and-Valley region was created by differential erosion—that is, weak rocks eroded more rapidly and became lowlands, while resistant rocks eroded more slowly and were left behind as uplands. Because the axis of folding runs

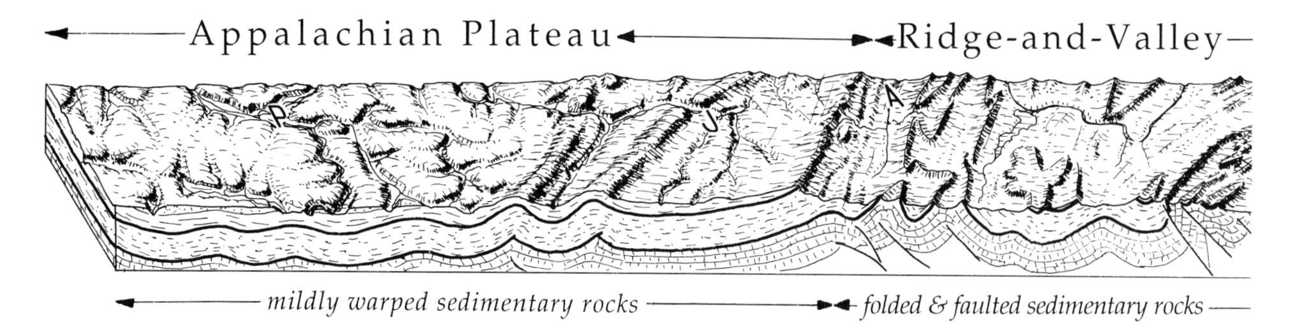

Fig. 2.5. Structure section across Pennsylvania. This east-west section across the southern part of the state shows the close relationship between the topography on the surface and the underlying geology (illustrated on the front face of the block diagram). The shaded area on the inset map shows the location, and the low-angle perspective, of this diagram. In addition to the four landform regions named on the figure, several other features are visible. Between the Great Valley, at the eastern edge of the Ridge-and-Valley, and the Triassic region, low hills of the Blue Ridge are visible. To the east (right) of the Piedmont, the narrow lowland of the Coastal Plain can be seen; the rise between the Coastal Plain and the Piedmont is the Fall Line. The cliff between the Appalachian Plateau and the Ridge-and-Valley regions is the great Allegheny Front. Cities on the map, identified by letters, are *P*ittsburgh, *J*ohnstown, *A*ltoona, *H*arrisburg, *L*ancaster, *R*eading, and *Ph*iladelphia.

northeast-southwest, the trend of the ridges and valleys runs in the same direction—ridges occurring where resistant rocks appear at the surface, and valleys occurring where the rocks are weak. It is a common error to believe that upfolds make ridges and that downfolds make valleys. The folding itself does *not* produce the ridge-and-valley topography; it is the differential erosion of alternating bands of weak and resistant rocks that does the job.

To understand any piece of Ridge-and-Valley topography, therefore, it is necessary to know only two things: (1) the particular geometry of folding and (2) the particular sequence of rocks. Both repeat themselves with great regularity. Thus, while Ridge-and-Valley topography tends to be quite fine-grained, with surface conditions likely to change abruptly within very short distances, the terrain tends to be repetitive, as the same sequences of rocks are repeatedly brought to the surface by folding, under the same kinds of structural conditions.

Three terms are used to describe various types of recurring folds, and each fold is associated with a highly characteristic kind of landscape. An upwarp (like the top of a rural mailbox) is called an *anticline*. The biggest and most fertile valleys in the region are anticlinal: Nittany Valley, in the center of the state, where the city of State College is located, as well as Kishacoquillas Valley, a sheltered Eden 20 miles to its southeast. A downwarp (like the bottom of a canoe) is called a *syncline*. The biggest and most famous syncline creates the Wyoming Valley, where the cities of Scranton and Wilkes-Barre are located, but similar synclines are responsible for burying and preserving anthracite coal wherever it is found. A third term is used to describe conditions along the flank of an anticline or syncline where rocks locally are tilted in only one direction; such rocks are said to be *homoclinal* (*homo* = same; *cline* = slope). Most of the ridges in the Ridge-and-Valley region were carved by erosion from homoclinal sedimentary beds—almost always a highly resistant sandstone or conglomerate sandwiched between two beds of weak shale. From Blue Mountain on the southeastern margin of the Ridge-and-Valley, to Bald Eagle Mountain on the northwest, most of the ridges in the region were formed that way. Correspondingly, some of the major valleys in the region are homoclinal—conspicuously the Great Valley, and Bald Eagle Valley, which runs for about 100 miles along the foot of the Allegheny Front.

Under certain conditions, however, ridges were formed where a resistant bed arched up at the surface along the crest of an anticline. Tuscarora Mountain, a 50-mile-long ridge that interposes its forbidding bulk squarely across the main corridors of trans-Appalachian roads and railroads, is an example of such an anticlinal ridge. Ridges can also be formed where a narrow bit of resistant bed is preserved in the elevated trough of a syncline—Nittany Mountain near State College is perhaps a classic example, as is Terrace Mountain above Mount Union.

A very characteristic subregion within the Ridge-

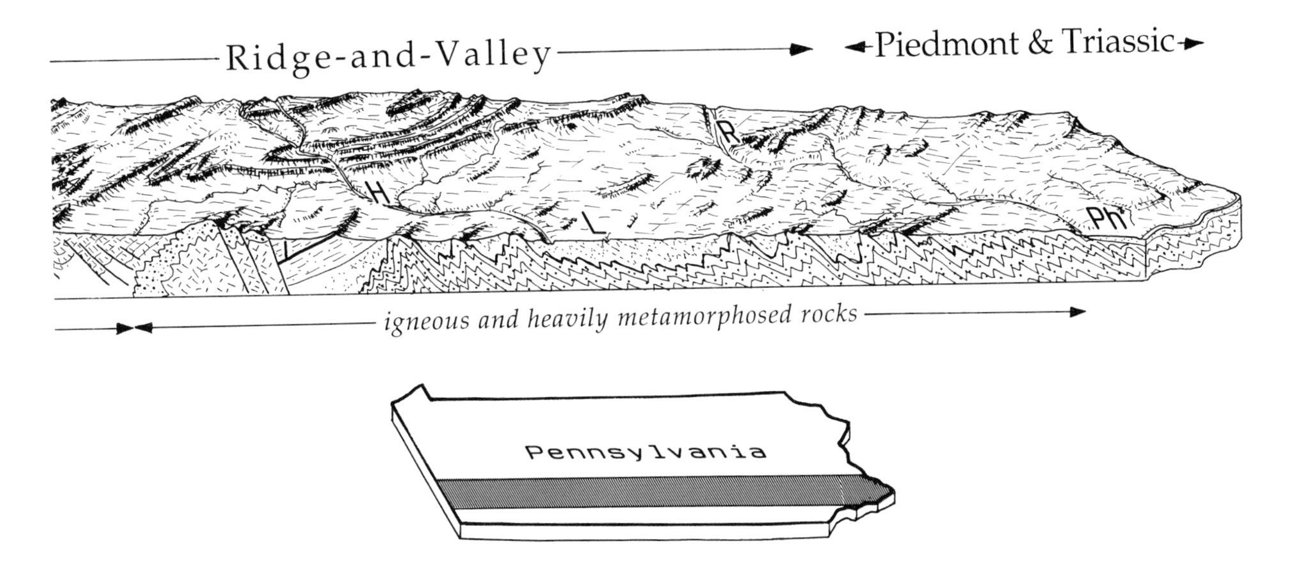

and-Valley is the set of broad *limestone valleys.* Immensely thick sequences of limestone and dolomite, collectively called the "Trenton group," are among the oldest rocks in the region and are therefore exposed chiefly in anticlines. Those Trenton limestones are overlain by a layer of weak shale, and then by an extremely resistant layer of sandstone, the Tuscarora quartzite. Everytime that sequence is arched upward in an anticline (which happens repeatedly in the Ridge-and-Valley region), a characteristic landscape results: a valley underlain by limestone and shale and flanked by Tuscarora ridges. The Great Valley is like that, as are the Nittany and Kishacoquillas valleys. The limestone typically breaks down into fertile soils, the same way the rocks in the Conestoga Lowlands do. In Kishacoquillas Valley, a fairly typical anticlinal limestone valley, the combination of fertile soils and ridge-girt isolation has attracted Amish and Mennonites who have sought to practice a traditional agricultural way of life with a minimum of interference from the outside world.

High-quality limestone is mined from these valleys to be fired into lime for cement, for use as crushed rock, or to provide metallurgical-grade flux for the iron industry. Because limestone dissolves in water, streams often disappear from the surface to produce a network of caves and underground streams. Beautiful and dramatic to behold, these cave systems create serious water-supply problems. Surface streams flowing out of the ridges disappear into sinkholes before they can be used, and the well-connected caverns rapidly export the water away from most wells. Any wells that are drilled must penetrate deeply into the hard limestone. Because the groundwater flows directly from the surface into these caves, bacteria, silt, and other pollutants are not filtered out by the soil, as would be the case under more usual circumstances. The dissolved limestone also makes the groundwater "hard"—harsh-tasting, poor at dissolving soap, and a source of calcareous scale in teapots, boilers, and plumbing. These problems combine to make the water supply in limestone valleys unpredictable in location, difficult to reach, unpleasant to use, and easily polluted.

In contrast to the anticlinal limestone valleys, the very upper layers of the geologic sequence are preserved in long, deep synclines and produce a world that is utterly different: the *anthracite valleys* (see Fig. 2.7). Instead of prosperous farms and limestone, there are numerous small mining towns in valleys underlain with shale and anthracite. Anthracite is coal that has been metamorphosed by deep burial and compression during the Appalachian Orogeny. Metamorphosis turned the coal into a shiny black rock that is clean to handle, clean to burn, and very difficult to mine out of the deep and tightly folded beds. (The Broadtop coal fields—in the south-central part of the state and also in the Ridge-and-Valley region—are equivalent to the anthracite valleys in most ways, except that the rock was somewhat less metamorphosed.)

Mining of Pennsylvania anthracite began in the early nineteenth century, first at the surface and then, as the demand for anthracite skyrocketed, in shafts that were often sunk to great depths. Culm, the

Fig. 2.6. Ridge-and-Valley landscape. The Susquehanna River flowing south toward Harrisburg (which is immediately off the right edge of the photo). Each ridge in this part of the Ridge-and-Valley region is the edge or side of a resistant layer of sandstone breaching the surface. The shaley soils of the valleys support productive agriculture, but sandy, rocky soils of the mountains are left in woods. Over the many millions of years, the Susquehanna has cut a dramatic water gap through Second Mountain, near the town of Marysville, in the middle-right of the picture.

waste material from the mines, was piled up in great heaps around the mine openings, typically cheek-by-jowl with the little wood "patch" town that sprang up next to the pit-head. Anthracite mining was a booming business, but it continued to boom only as long as Americans were burning coal in their parlor stoves and home furnaces. When that stopped happening, around the time of World War II, most mining stopped too, plunging the valleys into an acute and permanent depression. Most of the mines are now abandoned, and the underground diggings forgotten—except when they catch fire, as they sometimes do. Such fires in long-neglected mine tunnels are virtually impossible to extinguish, and at least one town, Centralia, has been largely evacuated and abandoned because of subsurface mine fires. Today it is difficult to find any part of the anthracite valleys that has not been dug up or dumped on or otherwise ravaged by mining at some time during the last 150 years. Although the worst damage is now half hidden under a growth of scrubby trees, some parts of the anthracite valleys are the closest thing to a man-made desert to be found in the United States. Even where damage is not visible at the surface, the groundwater and the streams emerging from springs are dramatically polluted by bright-

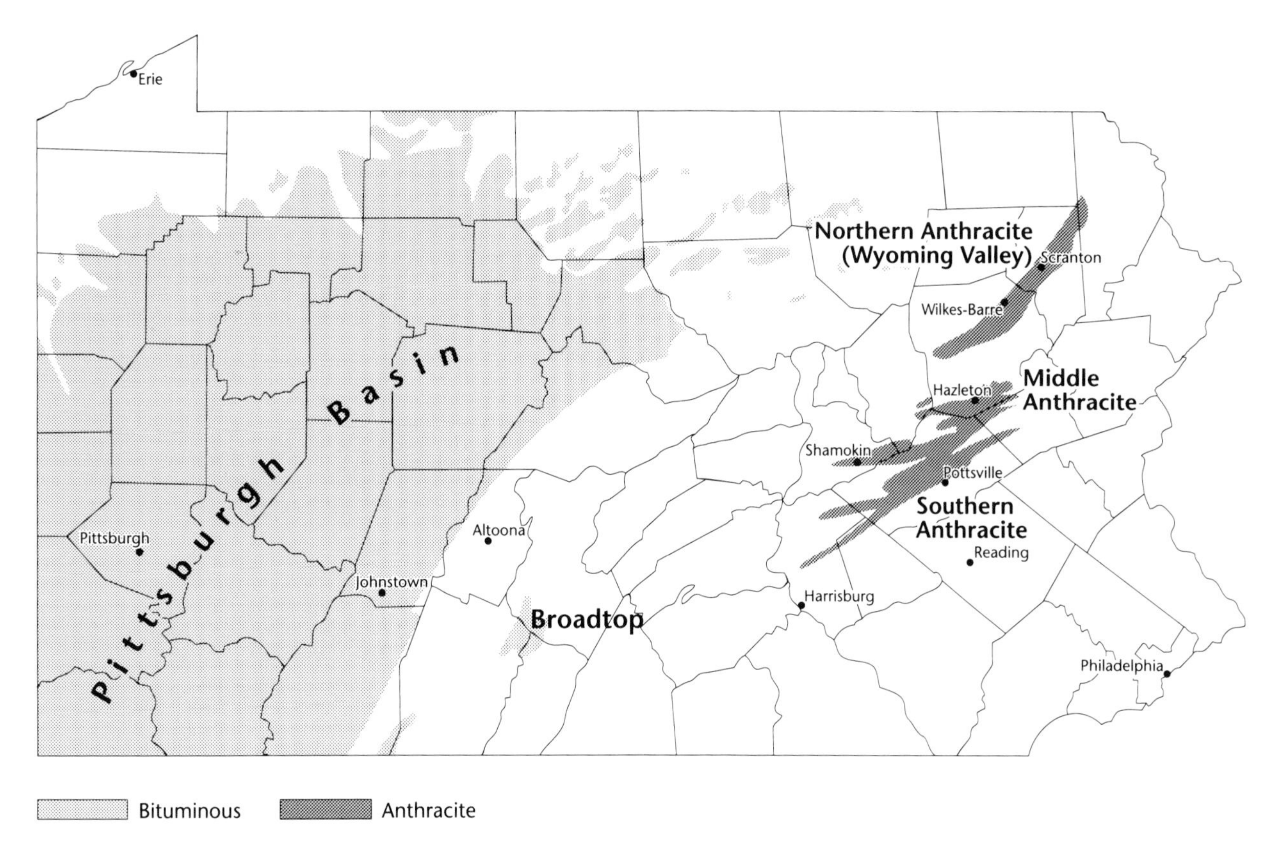

Fig. 2.7. Coal regions of Pennsylvania. The two major types of coal are both found in Pennsylvania, in two very different parts of the state. Bituminous coal, which is relatively unaltered by tectonic deformation, occurs in the broad and gently folded basins of the western part of the state, mostly within the Appalachian Plateau province. Anthracite, metamorphosed bituminous coal, has been squeezed into a few narrow synclines in the northeast, all within the Ridge-and-Valley province.

orange iron oxide and by sulfuric acid leached from the carbon-rich shale disturbed by deep-mining.

Middle Paleozoic sedimentary rocks—mostly of Devonian age—underlie the irregular topography found in irregular belts between the anticlinal limestone valleys and the synclinal anthracite valleys. The largest tract of this subregion of *Devonian shale lowland* is the wide north-south band within which the middle Susquehanna River flows from the foothills of the Allegheny Front to Blue Mountain, just north of Harrisburg. The rocks are dominated by shales of varying strengths, interbedded with thin layers of limestone and sandstone. The highly variable rock types create a miscellaneous landscape, with scrubby, rolling hills, gentle valleys with deep limestone soil, and a fair spattering of woodland. The character of any particular spot inevitably depends on slope and soil quality, and that depends largely on whether the spot is underlain by shale, limestone, or sandstone. In this middle landscape it could be any of the three. Over the years, a variety of mineral resources—including limestone, brick clay, and iron ore—have been produced from these rocks.

Visually, the most prominent and distinctive features of the Ridge-and-Valley region are the *sandstone ridges* themselves, although they occupy only a small part of the region's total area. Geologically, the ridges are carved from six different ages and types of sandstones, and the harder sandstones naturally created taller ridges. For example, the Pottsville conglomerate is a coarse, gravel-rich sandstone that makes the low ridges surrounding the anthracite valleys. By contrast, the much older Tuscarora quartzite (a massive form of quartz-rich sandstone) creates the majestic barrier along the rim of the limestone valleys. But the similarities between the ridges overwhelm the differences. The typical ridge is narrow, often knife-edged, and extends in a straight line for tens of miles before it loops abruptly back on itself to enclose a valley. Most ridges are cut here and there by watergaps, a fortunate circumstance for travelers who need to travel across the grain of the mountainous area. By contrast, travel *along* the valleys is relatively easy.

Being hard, steep, and sandy, these ridges bear some of the worst soil in the state. The rock weathers only slowly, and the slope permits water to carry away whatever soil may form. The sandstone bedrock ensures that what soil occurs is too coarse to hold water and too acid to hold nutrients. Although these ridges carry much of the most handsome forest in the southeastern half of the state, that forest is only a sparse and faint copy of the great forests that once existed on the rich lowland soils. Nonetheless, by virtue of its proximity to the growing urban population of the eastern United States, the forest remains a locally important source of timber and a significant recreational area as well.

The Appalachian Plateau: Mild-Mannered Geology of Western and Northern Pennsylvania

As elsewhere in the state, the physical character of Pennsylvania's Appalachian Plateau is the result of the kinds of rocks that are found there, the way those rocks are arranged, and, consequently, how they have been eroded to produce the contemporary landscape.

Compared with the rest of Pennsylvania, the rocks of the Appalachian Plateau reflect a tranquil geologic history. The region is underlain by a thick stack of Late Paleozoic sedimentary rocks, including a great deal of shale interbedded with sandstone and modest amounts of limestone. Near the bottom of this sedimentary stack is the very massive and very resistant Pocono sandstone—which we met earlier, forming one of the ridges that rims the anthracite valleys of northeastern Pennsylvania. At first glance, these Paleozoic rocks look as if they had been undisturbed by Appalachian orogenic upheavals and are often described as flat-lying (hence the term "plateau," from the Greek for "flat," which applies better to the geology of the region than to the rugged topography). That description is almost but not quite accurate. It is true that the tight folding of central Pennsylvania stops abruptly at Bald Eagle Valley, just below the Allegheny Front. It is also true that the rocks often appear to be flat when viewed at a local scale in any of western Pennsylvania's innumerable road-cuts. But in actuality, over the region as a whole, the rocks of the Appalachian Plateau are broadly and systematically warped. That regional warping, however slight it may seem, has important consequences.

The basic structure can be described simply: the rocks of western Pennsylvania are broadly depressed into a shallow oval basin whose deepest central portion lies in the extreme southwestern corner of the state and is called the *Pittsburgh Basin*. This colossal structure underlies not just western Pennsylvania but a good share of eastern Ohio and most of West Virginia as well. The rocks of the Pittsburgh Basin resemble a stack of shallow oval platters, the largest on the bottom, with each successive platter smaller than the one below, so that they all nest within one another with their lips all roughly at the same level. This subtle deformation of the geology makes parts of the plateau subtly different from one another.

The uppermost and smallest platters represent the strata that contain coal, and the coal-bearing parts of the plateau are found chiefly in the central section of the Basin (Fig. 2.7). Bituminous coal occurs in great abundance near the top of the sequence, including the famous "Pittsburgh seam." Most of the population of western Pennsylvania is still concentrated close to the Pittsburgh seam, because most towns developed around the mining of coal or around transportation and heavy industry dependent on that fuel resource. The richest coal resources in the state are mostly around Pittsburgh, but the coal beds of the Plateau also extend far to the northeast within several narrow synclines, to within 80 miles of the New Jersey border. In fact, those synclines extend long shovel-shaped tongues into the north-central part of Pennsylvania and are responsible for the presence of bituminous coal mines in isolated locations where they otherwise might not be expected. The Blossburg coal fields southeast of Wellsboro are perhaps the best known, not because the coal is of especially high quality but because it was the closest bituminous coal to the markets of upstate New York. Here, as often happens, strategic location pays off.

The lowest and largest platter—the bottom member of the sedimentary sequence—is the resistant Pocono sandstone. At Pittsburgh the sandstone is far underground, buried under stacks and stacks of interbedded coal and shale. On the outer edge of the

Fig. 2.8. The Allegheny Front. The eastern edge of the Appalachian Plateau, between Lock Haven and Williamsport. The photograph looks toward the northeast. The Allegheny Front divides two major landform regions in Pennsylvania. The Ridge-and-Valley region starts in earnest immediately to the right of this picture. The stream-cut, but generally flat, topography in the left part of the photograph is typical of the high, northern Appalachian Plateau. The various resistant rock layers, capped by the Pocono sandstone, are visible as steps in the abrupt cliff-face that runs through the middle of the picture. That cliff is formed where those layers, which tilt toward the left, have been truncated by erosion.

Appalachian Plateau, however, the Pocono sandstone crops out in a broad band that encircles the Plateau in Pennsylvania on its northern and eastern margins. Places where the Pocono sandstone comes to the surface in the Plateau region are the highest parts of Pennsylvania, and also the site of some of the deepest valleys and greatest cliffs in the state. Because the Pocono sandstone is underlain in most places by weak shale that erodes easily, its outer rim drops abruptly away in steep escarpments that face outward from the center of the Basin, exactly the way the edge of a platter faces outward. The most formidable of those escarpments is the Allegheny Front, where the sandstone curls to the surface on the eastern edge of the plateau, forming the highest parts and steepest parts of the topographic region (see Fig. 2.8). Similar escarpments can be found all over north-central Pennsylvania, wherever erosion has undercut the edge of the Pocono. Crossing the Front has always been a Herculean task. For example, builders of the Pennsylvania Canal, which was originally designed to link the Susquehanna River system with the Ohio, simply gave up on the idea of building a set of canal locks to lift boats up the face of the Front. Instead, one of the most amazing feats of engineering in the history of American transportation was accomplished: barges were simply hauled up the Front on a series of inclined planes. Later, in the 1850s, when the Pennsylvania Railroad's main line was built to connect Philadelphia with Pittsburgh, designers were faced with the problem of laying out tracks that would climb 1,500 feet while maintaining gradients low enough to allow steam engines to pull long trains without stalling. The solution was to build a series of enormous sinuous curves that snake their way in and out of stream valleys cut into the Front as they climb gradually up onto the Plateau above Altoona. One of these bends—the so-called Horseshoe Curve—is so spectacular that it has become a tourist attraction and is well known by rail buffs all over the nation. Where streams have managed to cut through the Pocono sandstone into the weak rock below, the results are steep-sided gorges. An especially dramatic one along Pine Creek in Tioga County is enthusiastically touted as "the Grand Canyon of Pennsylvania."

Because the Pocono sandstone is massive and resistant, it breaks down grudgingly into thin, stony soil and has therefore been left almost entirely in forest—the emptiest country in Pennsylvania, populated mainly by vacationers in the summer and by deer-hunters in the autumn. The section of the Plateau in the east that is called the Pocono Mountains shows the handsome but unproductive landscape created by Pocono sandstone. The familiar resorts and summer homes of that area illustrate the typical land-use found on this type of topography. In the northwestern part of the state, the high barren Plateau country contains the Allegheny National Forest, which owes its existence to the fact that hardly anybody except the federal government took interest in acquiring such unrewarding country.

A few ridges rise above the high surface of the Appalachian Plateau in the west-central part of the state. Severe geologic deformation does not extend west of the Allegheny Front, but the last gasps of the Appalachian Orogeny warped and faulted the southeastern margins of the Plateau into gentle, wavelike undulations—in effect, a series of broad, open anticlines and synclines. They are aligned parallel to the folds in the Ridge-and-Valley region but are much more subdued. It is as if the shallow platter of the Allegheny Basin were slightly fluted—like the wavy wrinkles in a scallop shell. There are half a dozen or so such wrinkles in the Plateau, anticlines and synclines about 20 miles wide that deform the entire thick layer of upper Paleozoic rocks. Despite the lack of geologic drama, those wrinkles have far-reaching geographic consequences.

Two of the anticlines between Pittsburgh and Johnstown are topographically conspicuous because they form the high forested ridges called Laurel Hill and Chestnut Ridge. In both structures, coal measures at the top of the sedimentary sequence have been uplifted and stripped away by erosion, exposing the Pocono sandstone below, which erodes slowly and produces high elevations. Laurel Hill is the closest mountainous country to Pittsburgh and has long been a favored place for city people to build summer homes or to go skiing in the winter.

Soils of the Appalachian Plateau are rarely very good; at worst, they are wretched—thin, sandy, rock-laden, and acidic. Farming in most parts of the region is a fairly marginal business, and in some parts conditions are so bad that nobody has even bothered to try—or if they did, they promptly gave it up. The bituminous coal regions of the Plateau are in a severe state of economic depression and have been since the introduction of large-scale strip-mining equipment, which hugely reduced the need for labor. The bituminous regions of western Pennsylvania have been losing population longer and more consistently than any other part of the state. The economy of the regions contrasts sharply with the fast-growing Piedmont region of eastern Pennsylvania.

Appalachian Aftershocks: The Triassic Lowland

For most of Pennsylvania, the end of the Appalachian Orogeny spelled the end of violent geologic deformation. In the border zone between the inner Piedmont and the Blue Ridge regions, however, deformation did not end with the orogeny. Post-Appalachian deformation has made a significant impact on the contemporary landscape here, as well as on American history.

It was a new kind of deformation, a kind of Appalachian aftershock, or rebound. During most of the Appalachian Orogeny, rocks had been subjected to compression—the result of Europe's collision with North America. (It was compression, after all, that had metamorphosed the rocks of the Piedmont and Blue Ridge regions and had crumpled the sedimentary rocks of the Ridge-and-Valley region.)

Toward the end of the Paleozoic Era, however, mostly during the Triassic Period (but starting somewhat earlier, during the Jurassic), the two continents once again began to drift apart, reopening the Atlantic Ocean and subjecting the eastern fringes of North America to wrenching strain—not compression, this time, but instead tension. The most intense Triassic Period tension was concentrated in the metamorphic belt that would eventually become the Piedmont and the Blue Ridge regions. The immediate consequence throughout much of that border zone was block-faulting, where good-size blocks of the earth's crust broke loose from one another and shifted position, up and down, rather as drifting icebergs break apart and bob up and down when a massive ice-floe comes apart. That sort of thing happened in Pennsylvania along the belt of Triassic Lowland that stretches all the way from Maryland to New Jersey and ranges in width from 30 miles where it crosses the Delaware River to a narrow 5 miles along the Lancaster-Lebanon county line. Pennsylvania's Triassic Lowland is only part of a long trench that extends from the middle of Virginia to the Palisades of the Hudson River across from New York City. Similar Triassic landscapes are scattered up and down the Atlantic Seaboard from the Carolinas to Massachusetts. All along that band, it is as though a great linear block of Piedmont and Blue Ridge metamorphic rocks slipped downward on its inland margin and left a yawning wedge-shape vacancy, something like a trapdoor that has fallen partly open. The vacant space was filled with layer upon layer of weak sedimentary rocks. Simultaneously, as the blocks were pulled apart, molten rock welled toward the surface from below, so that the Triassic sedimentaries are penetrated with diabase, a dense black volcanic rock. Some of those diabase intrusions follow fault lines running vertically through the strata and are called dikes. Others injected themselves horizontally like thin sheets between the sedimentary layers and are called sills.

Most of the Triassic sedimentaries are quite weak and have been eroded into gently rolling topography that closely resembles adjacent Piedmont and Great Valley terrain. Thus, a traveler can drive the freeway from Lancaster on the Piedmont, to Harrisburg in the Great Valley, and cross 20 miles of Triassic Lowland without noticing much change in the look of the land—except that Triassic rocks and soils are often brick-red in color. The Triassic diabase, however, is more resistant than the sediments and consequently forms knobs and low ridges. The Pennsylvania Turnpike rides along the crests of these low ridges of black rock for a significant part of its route between Philadelphia and Harrisburg. These ridges are mostly of little consequence, except that General George Meade happened to locate his Army of the Potomac atop one of them at Gettysburg on July 1, 1863. The Battle of Gettysburg that followed was Robert E. Lee's futile attempt to chase Meade off the diabase ridge called Cemetery Ridge, and off the diabase knob called Little Round Top.

Even more importantly, however, the Triassic region breaks the wall of the Blue Ridge–New England mountain system in Pennsylvania, and the consequences have reverberated through American history. In Virginia and the Carolinas, the Blue Ridge constitutes the highest, roughest, and wildest country in the eastern United States. There it is the main barrier to transportation between the Atlantic Coast and the continental interior. In Pennsylvania, however, the westbound traveler encounters no massive mountain barrier, only a pair of paltry stubs. Between those two stubs lies one of the largest gaps in the whole Appalachian chain, a beckoning gateway 80 miles wide that opens directly from the Piedmont into the Great Valley to the southwest, and toward the transmountain corridors of the Susquehanna and Juniata rivers to the northwest.

Oddly enough, this portentous gap bears no common name, probably because it is so wide that travelers pass through without noticing it. Whether noticed or not, the gap is used by a ganglion of major east-west and north-south routes—the main line of the Pennsylvania Railroad (later the Penn-Central), the Pennsylvania Turnpike, and three interstate highways, to mention only the most important ones. Even the Susquehanna River uses the gap, thus avoiding the embarrassment of having to cut through the Blue Ridge. For more than two centuries, Pennsylvanians have poured westward and southwestward through this gap to fill the midlands of America with transplanted Pennsylvanians and the institutions they carried with them. Thus, the gap became a junction of national importance—and an obvious logical place to situate the state capital. At the intersection of the Great Valley and the Susquehanna River, and commanding the main gap in the Blue Ridge, Harrisburg is one of the most accessible locations in Pennsylvania. Today, in the outskirts of Harrisburg, near Carlisle and Hershey, trucking companies have been locating some of their biggest terminals, a testimony to the fact that this remains one of the most important transportation junctions in eastern North America.

After the Orogeny: Pennsylvania Calms Down

Since the Triassic Period, Pennsylvania's geologic history has calmed down considerably. No longer at the active edge of colliding and separating continents, the gradual widening of the Atlantic Ocean left Pennsylvania in the stable interior of a huge tectonic plate. In the western United States, there were new episodes of orogeny as great mountain chains were formed, volcanoes erupted, and lava poured out across the land. Through it all, Pennsylvania remained relatively tranquil. The time was not entirely uneventful, however, and three events are worthy of special notice to round out the history of Pennsylvania's landforms:

- First, the sea lapped up over the southeastern margins of the state and left behind sediments that formed Pennsylvania's tiny piece of the Atlantic Coastal Plain.
- Second, during the entire period since the Appalachian Orogeny, the whole state was subject to prolonged erosion by running water, completely eliminating the ancient Appalachian ranges (as we saw) but meanwhile leaving behind some remarkable river patterns and some remarkable landforms as well.
- Third, and much more recent, the northern parts of the state were invaded by continental glaciers at least twice. The country under the glacier was altered, but so was a good deal of territory beyond the touch of the ice.

The first event is of minor importance only to Pennsylvania—unless one happens to live in or near Philadelphia. The third event caused major changes, but chiefly in the northern regions of the state. The second was and still is of major significance all over the state.

Atlantic Invasion: The Coastal Plain and the City of Philadelphia

Over the last 60 million years or so, the Atlantic Ocean has repeatedly slopped up over the low-lying margins of the North American continent, leaving behind thin sheets of sedimentary deposits, such as the one forming the subtle topography apparent in the small area of Coastal Plain in Pennsylvania. The continent has gradually risen a few hundred feet over the millions of years and revealed on the Coastal Plain the sediment laid down when the areas were flooded under the edge of the sea. The water was seldom very deep, so the deposits are the kind produced by shallow water and fluctuating shorelines: mud settling out from murky water, sand from river deltas and beaches—in short, the same kinds of materials that are being deposited today in the shallow Atlantic offshore from New Jersey.

Those oceanic deposits have produced the low-lying, almost flat Coastal Plain, which stretches along the Atlantic Ocean and the Gulf of Mexico from New York City to the Rio Grande. Only the extreme innermost margin of the Atlantic Coastal Plain extends into Pennsylvania, however—a thin sliver of Coastal Plain land lies along the Delaware River, between Chester and Levittown, with Philadelphia squarely in the middle; some hilltops on the Piedmont near Pennsylvania also carry a gravelly cap of Coastal Plain sediments. Outside Pennsylvania, the Coastal Plain is a very large region, big enough to include entire states, such as Florida and Louisiana, and large chunks of many others. Most of eastern and southern New Jersey, for example, from Newark to Cape May, is part of the Coastal Plain, as is nearly all of Delaware.

The Atlantic Coastal Plain is a scant five miles wide at Philadelphia—almost its widest point in the state. Philadelphia was established on the raised surface of the Plain, and consequently most of the old city is immune from normal flooding of the Delaware or Schuylkill rivers. As the city grew, however, it pressed into the swampy floodplain of the Delaware. As time went on, this area was filled on a hit-or-miss basis for use by commerce and industry—such as steel mills near Levittown, huge oil tank-farms in South Philadelphia, a tangled skein of railroads and superhighways alongside the riverbanks, and the Philadelphia International Airport, built on sediment dredged from the river bottoms of the Schuylkill and the Delaware. Although there are pockets of affluence on the Coastal Plain in Philadelphia, most of the wealthiest people have fled to the Fall Line hills of the Piedmont, in a string of suburbs that extend along the Main Line of the Pennsylvania Railroad, northwest from Fairmount Park. In the Philadelphia metropoli-

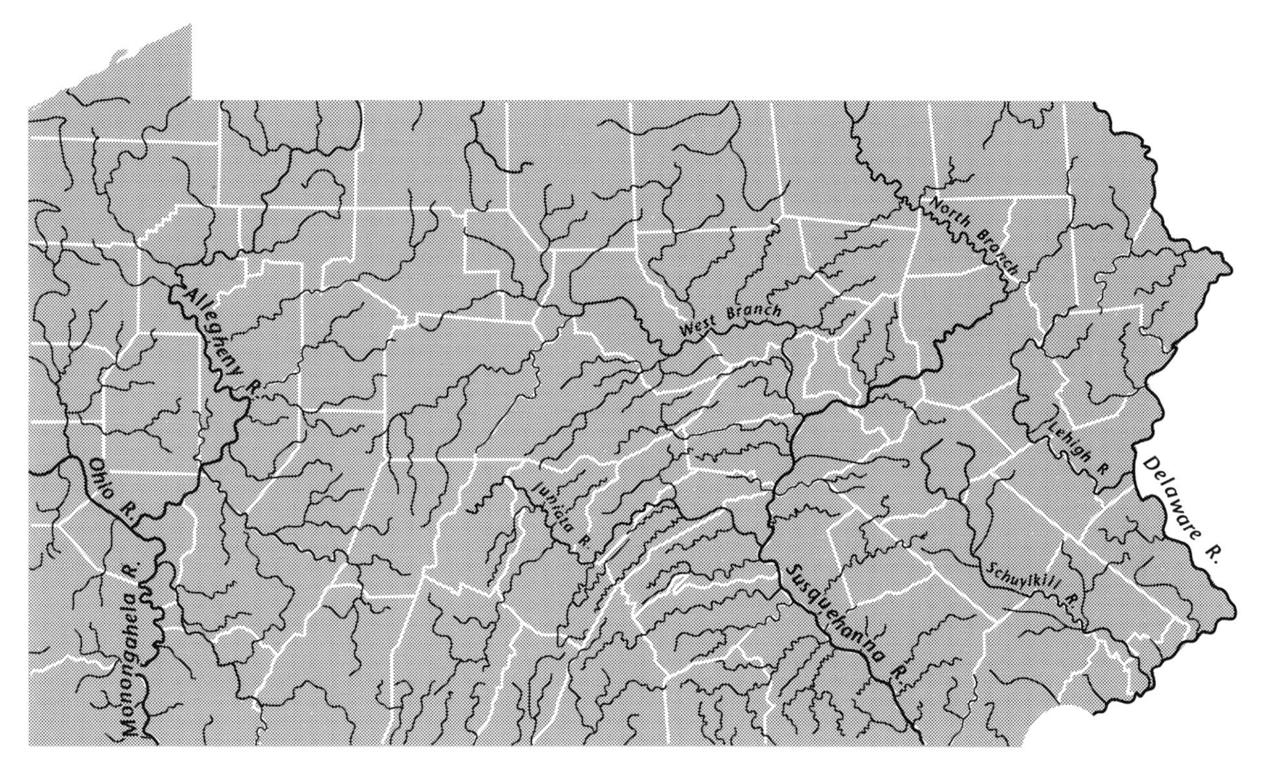

Fig. 2.9. Streams and rivers of Pennsylvania. All of the streams of the state reflect the geology around them. In the Ridge-and-Valley province, in the central part of the state, smaller streams have eroded into the softer rock that lines the long, straight valleys. In contrast, the big rivers of the east have been able to cut directly across the resistant layers of rock, cutting water gaps through the ridges in their paths as they head for the Atlantic. In the Appalachian Plateau, to the north and the west, stream patterns seem much more haphazard because the flat-lying rock has offered little guidance to their evolution.

tan area, therefore, the boundary between the Piedmont and the Coastal Plain is not simply a geologic and topographic boundary line. It is a social, economic, and racial seam that slashes across Philadelphia from northeast to southwest and divides it into two very different parts.

River Patterns

Rivers and streams, and the patterns they describe on the land, are important far out of proportion to the tiny amount of land they actually cover. All the kinds of surface transportation that have tied the state together since the earliest times have followed streams, for example, because streams are the nearest thing to flat in this complex landscape. Streams are also the primary agents in carving the landscape, so to a substantial degree stream patterns are the foundation of all geographical patterns. It is an interesting paradox in Pennsylvania that the courses of the smaller streams are intimately responsive to the influence of the underlying geology, while the larger streams have been emphatically oblivious to it (see Fig. 2.9).

Geologic Control on Smaller Streams and Rivers

In most of Pennsylvania, many stream patterns are a direct result of bedrock geology. The reason is simple. Most of the state's rivers and streams got their courses by headward erosion—that is, by the process of tiny headwater streams clawing their way into previously undissected terrain.

Three quite different kinds of stream patterns are found in Pennsylvania, each reflecting one of the three main geologic subdivisions of the ancient Appalachians. One kind of pattern is found on the nearly flat sedimentary rocks of the Appalachian

Plateau; a second occurs on the folded sedimentary rocks of central Pennsylvania; and the third has taken form on the metamorphic rocks of eastern Pennsylvania.

The western stream patterns, like the rocks beneath, are the simplest and easiest to understand. Although most streams west and north of the Allegheny Front achieved their present courses by eroding headward into gently warped sedimentary rocks, the warping is so gentle that it is not uncommon to find the same stratum extending over sizable tracts in the headwaters of any particular stream. Thus, as far as the stream's tiny headwaters are concerned, they are eroding into uniform material, and when that happens the stream's course tends to be quite irregular. When numerous tiny streams in the headwaters of any given river system are all doing the same thing at the same time—eroding headward into homogeneous material—the result is a pattern that on a map looks like the picture of a gnarled tree. Such a pattern is called dendritic (from the Greek *dendron,* "tree"), and nearly all stream systems of the Appalachian Plateau exhibit such a pattern. The ancient Monongahela system, with its intricate network of tributaries, is a fine example of a dendritic pattern—although such patterns are found elsewhere in the state where rocks of uniform resistance to erosion extend over substantial areas. In fact, dendritic stream patterns are found in parts of the Piedmont region, where certain metamorphic rocks are so uniformly weak that streams erode into them randomly.

While the dendritic stream patterns of the Appalachian Plateau make a pleasing pattern on the map (see Fig. 2.9), on the ground they have confounded travelers ever since the first path was cut across the region. The intricate network of winding streams has dissected the Plateau into a seemingly endless and random array of knobs and hills. Except on a very few (and very expensive) interstate highways, motorists traveling across much of the Plateau are confronted with a nightmarish succession of hills and blind curves that make driving dangerous and monotonous, and construction of good roads very expensive. Isolation from the outside world has been a chronic problem in much of the Appalachian Plateau, and the stream pattern is responsible for much of that separation.

Another kind of stream pattern occurs where there are extreme differences between very weak and very resistant rocks at the surface. That kind of pattern is the trademark of the Ridge-and-Valley region, where streams run for miles and miles, carving valleys into long, narrow belts of folded sedimentary rocks and leaving long, narrow ridges in between. The geometry of such stream systems is highly distinctive, and early geographers call such systems a "trellis pattern" because of the angular way streams and their tributaries are lined up with each other. Nowhere in the world is a trellis pattern more perfectly developed than in the middle Susquehanna and Juniata river system in the heart of the Ridge-and-Valley region between the Allegheny Front and Blue Mountain.

Geologic Independence of Large Streams and Rivers

The biggest rivers and streams in Pennsylvania act very differently from the smaller ones. For a long time, students of Appalachian drainage patterns have puzzled over why some of the main streams of the Ridge-and-Valley region depart now and then from the path of least resistance along bands of weak rocks and abruptly cut straight across belts of very resistant rocks, and thus slice through the ridges those resistant rocks have created. The Susquehanna River does that several times just north of Harrisburg, cutting through Blue Mountain and several other equally imposing ridges to form "water gaps" (see Fig. 2.6). The most assiduous maker of water gaps in Pennsylvania, however, is the Juniata River, which slices transversely across almost the entire width of the Ridge-and-Valley region, cutting straight through some of the most formidable ridges in the northern Appalachians. In the process, the Juniata has created one of the most important routes in North America. Because the Juniata crosses the Appalachians roughly from west to east, it has traditionally served as the main corridor for travelers heading west across the mountains. That corridor is part of a primary trans-Appalachian route, second only in importance to the Mohawk Valley of upstate New York. The Juniata Valley was the route of the Pennsylvania Canal, as well as the Pennsylvania Railroad that succeeded the canal. U.S. Highway 22 follows much the same route.

A variety of ingenious explanations for the origin of these many water gaps have been offered. The simplest is that the pattern of rivers and streams is older than the mountain through which they cut. Streams flowing roughly southeastward, across the thoroughly eroded flat surface of the Schooley

Peneplain, started cutting into the hard rock beneath themselves when renewed uplift occurred. The streams are still trapped in the same water gaps and still cutting down, after tens of millions of years. But recent images taken from spacecraft have yielded information that may shed more light on the problem. Those images from space seem to reveal fracture zones that run across the Appalachians approximately at right angles to the trend of the main ridges, and some of those fracture zones appear to be lined up with major water gaps. The rivers have cut down most rapidly in the weak zones, and streams like the Juniata found their courses by following zones of weakness—not just along bands of nonresistant shales and limestones but transversely across zones of fractured sandstone as well.

In big westward-flowing river systems, however, the situation was quite different. Most of the landscape is Appalachian Plateau, and streams are cutting into flat-lying rock. Thus, the geologic obstacles, so common along rivers like the Susquehanna, the Schuylkill, and the Delaware, are seldom encountered on the Allegheny or the Monongahela, both of which are easily navigable along much of their courses. (That was a fact of no small importance in the Appalachian Plateau, where overland transportation has always been difficult and expensive and where the export of bulky commodities such as lumber and coal demanded cheap transportation before they could be exploited.) The two rivers combine to form the Ohio, which, with the exception of an easily bypassed riffle at Louisville, Kentucky, is navigable all the way to the Mississippi and thence to the Gulf of Mexico.

That contrast between eastern and western rivers has made an enormous difference between the two sides of the state. Eastern Pennsylvania contained some of the most productive land in eastern North America, but the rivers were not usable without major navigational improvements. Very early in the state's history, therefore, there were powerful incentives to make rivers navigable by building canals alongside them, or (more rarely) overland, or, because canal-building was often impossible or ruinously expensive, to build all-weather roads. Thus it was that eastern Pennsylvanians took to road-building very early in the state's history, and from the beginning the routes of eastern Pennsylvania often converged in places far from navigable water. It was no accident that Lancaster and York, the main inland market centers of the Pennsylvania Piedmont, were among the first American cities to grow up and thrive without being located on the seacoast or a major river, and the first to depend on roads—not rivers—for commercial transportation. In western Pennsylvania, however, it was quite a different story. Because travel by land was tedious, expensive, and slow, people and towns inevitably congregated along the banks of the many navigable rivers, where transportation was easy and cheap, especially for the barges that carried the bulky loads of coal, oil, lumber, and iron ore that formed the backbone of the region's economy. Western Pennsylvania was therefore never a very appealing place for farmers, but the combination of coal and cheap water transportation made it wonderful for heavy industry.

That contrast between modes of transportation has set eastern and western Pennsylvania apart from the beginning of European settlement, and it continues to do so. Eastern Pennsylvania's elaborate and sophisticated road system ultimately led to Philadelphia—to the old Atlantic world. Western Pennsylvania's navigable rivers led to Pittsburgh and the Ohio River beyond—to the interior of North America. As with so many other things in Pennsylvania, that fundamental economic and political division springs directly from the hard facts of physical geography.

Glaciation in Pennsylvania

A map of glaciation in the northeastern United States might suggest that the gods of glaciers wanted to avoid Pennsylvania (Fig. 2.10). During the Pleistocene epoch—that is, over the last million years or so—most of the northeastern United States was heavily and repeatedly glaciated. Most of Pennsylvania was not. While ice sheets covered huge areas to the north, west, and east of Pennsylvania, they merely nibbled at the northern fringes of the state.

There were two well-documented episodes of glaciation in the territory that would become Pennsylvania, and the two were separated by a long interglacial period. (There were undoubtedly other early glacial advances, but evidence for them is obliterated by the more recent ones.) The first of the two, several hundred thousand years ago, is often called the "Illinoisan stage" because similar deposits were first described in Illinois. The Illinoisan ice sheet, covering the northeastern and northwestern

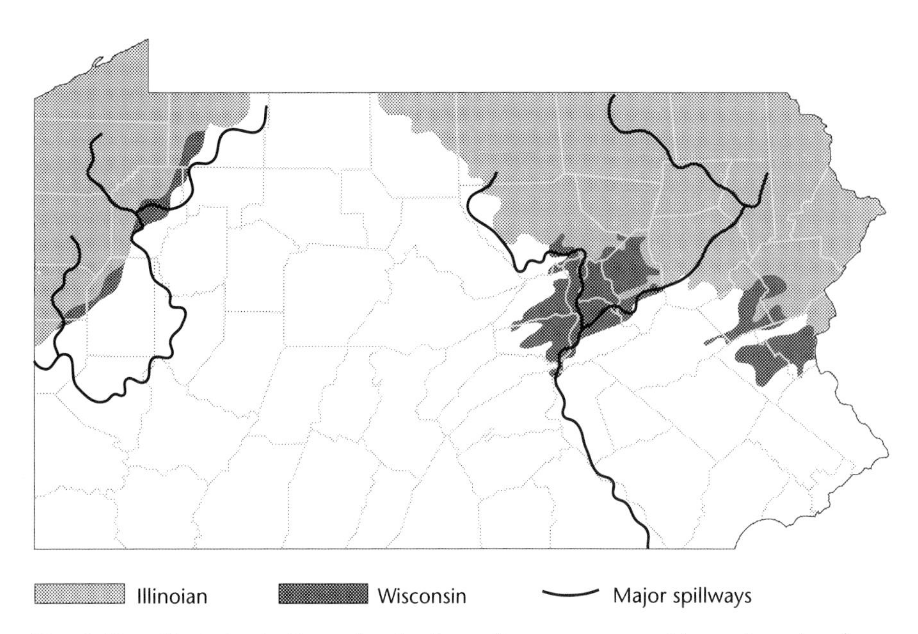

Fig. 2.10. Glaciation in Pennsylvania. According to current evidence, large ice sheets advanced into Pennsylvania twice, each time in the form of two broad lobes covering the ends of the state. The second advance traveled slightly less far than the first. Most of the major southward-flowing rivers served as spillways for the melting ice from the glaciers, which caused their valleys to be cut widely and then filled with sand and gravel to about thirty feet above present water level.

corners of the state and a wide strip across the north, was the larger of the two. In central Pennsylvania, a broad tongue of Illinoisan ice advanced down the Susquehanna River valley to a point slightly below Sunbury—in short, halfway across the Ridge-and-Valley region. A long time has passed since the Illinoisan stage, however, and because most glacial deposits tend to be fragile, most of them are heavily eroded and quite inconspicuous in the contemporary landscape.

The more recent advance was called the "Wisconsinan stage," and it was a different matter because it withdrew from Pennsylvania less than 15,000 years ago. That is the merest twinkling of an eye in geologic time—these glacial deposits in Pennsylvania lie on rocks that are 30,000 *times* older—and the imprint of Wisconsinan glaciation is still very fresh. Although Wisconsinan glaciers were less aggressive than the Illinoisan, and therefore covered slightly less of the northeastern and northwestern corners of the state, their effects are pronounced and very noticeable. Their effects, however, are quite different in the two corners of the state.

In northeastern Pennsylvania, the ice reached south as far as a line between the Delaware Water Gap and the New York border near its middle. In the Ridge-and-Valley region, the ice intruded to the extreme northern edge of the anthracite region, about midway between Scranton and Wilkes-Barre in the Wyoming Valley. Its severest effects, however, were felt on the Appalachian Plateau, especially the portions of Pike, Wayne, and Monroe counties that are loosely called the Poconos. Like so much of the higher parts of the Appalachian Plateau, the surface of the Poconos is underlain by almost flat-lying resistant sandstone. There was probably not much soil to begin with, but the advancing ice removed whatever there was and scoured the bedrock clean. Then, as the ice withdrew, it left behind a patchy carpet of sand and gravel, a random scattering of huge, isolated boulders, and shallow depressions that have subsequently filled with water and vegetation to form small lakes and swamps (almost the only natural lakes in Pennsylvania). This high, barren country is worthless for agriculture and totally lacking in valuable minerals, and its white pine forest was cut down during the lumber boom of the nineteenth century. Over the last several decades, however, the Poconos have been touted as a tourist mecca for honeymooners and, because of its proximity to New York City, a favored blue-collar resort. Recently, refugees from the city and its suburbs have been moving into this glacier-scrubbed country in search of cheap land and insulation from urban problems.

In northwestern Pennsylvania, the Wisconsinan ice sheet flowed south to a line roughly from New Castle to Warren, but glaciers here behaved very differently than in the Poconos. The ice had spilled into the area from the shallow basin now occupied by Lake Erie and previously occupied by another lake in the same place. As the ice sheet crossed that trough, therefore, it picked up huge volumes of the silts and clays that typically collect on the bottoms of lakes. Then, as the ice moved sluggishly across the edge of the Appalachian Plateau of northwestern Pennsylvania, it plastered those silts and clays over the pre-glacial terrain, filling stream-cut valleys and smearing the flat-topped surface of the Plateau with a coating of compacted silts and clays, occasionally mixed with gravel, sand, and boulders—a material glaciologists call "till." When the glacier finally melted away from northwestern Pennsylvania, the landscape was as it would be if some monstrous bricklayer had casually smeared mortar over a rough stone surface. Very little of the pre-glacial Plateau surface is visible beneath the coating of till, and the gently undulating surface is much smoother than the land that had existed before the ice arrived. Farmland in the glaciated country of northwestern Pennsylvania is much superior to the unglaciated Plateau surface to the south, so that even casual travelers are likely to notice the glacial boundary by the improved quality of farms as well as by the more subdued landforms.

So, glaciers greatly improved agriculture in northwestern Pennsylvania by importing soil from the north, while they had quite the opposite effect on the high, hard Plateau of the northeast. Pennsylvania's glacial landscapes represent a nice sampling from a much larger North American context. The Poconos, for example, resemble huge areas in Canada and New England where the ice scraped soils from the surface of resistant rocks and strewed the ravaged land with sand and boulders when it melted. The glaciers thus ruined most of Canada and New England for any practical agricultural purposes. Northwestern Pennsylvania, by contrast, is like the Midwest where ice generally improved things by laying down a nearly flat coating of silt-rich till atop a much rougher subsurface.

Glacial Effects on Lakes and Rivers

When the ice finally melted, it left behind a greatly altered drainage system. The most conspicuous glacial relict in Pennsylvania is Lake Erie, and the adjacent Lake Erie Coastal Plain on which it once lay, when it was larger. The lake occupies a long valley that was originally carved into a belt of weak shale by a pre-glacial river. In glacial times, a tongue of ice flowed westward through that valley and gouged it into a shallow basin. When the ice disappeared, it left behind not a river valley but a shallow depression that filled up with water and became Lake Erie. During the earliest stages of glacial withdrawal, however, the lake was higher and bigger than at present because its outlet by way of Niagara Falls was blocked by the glacier. This ancestor of Lake Erie backed up and rose to a level where its water could find an outlet southward across Ohio and Indiana to the Ohio River. As the ice continued to melt away, however, lower outlets were uncovered, and each time that happened the ancestral Lake Erie dropped abruptly to a new level.

This behavior explains the terrain of the Lake Erie Coastal Plain near the city of Erie. Each time the level of the lake stood still for a while, its storms and waves would cut a new shoreline, very similar to the present one—with wave-cut bluffs, sandy beaches, rows of sand dunes, and shallow muddy underwater slopes offshore. Then, abruptly, the lake would drop again and its waves once more begin the process of cutting a new shoreline. A whole series of abandoned shorelines—called terraces—line the inland margin of the Lake Erie Coastal Plain like stairsteps from the lake at the bottom to the high sandstone escarpment at the top that marks the northwestern lip of the Appalachian Plateau.

These glacial lake terraces are both picturesque and useful. Transportation engineers have found the nearly level terrace surfaces to be attractive sites for building the major railroads and roads that run parallel to the lake on their way from Cleveland to Buffalo. And they are places with an exceptionally long season in which to grow fruit because their slopes provide good drainage of freezing air, while the nearby lake discourages unseasonable frosts. The Lake Erie Coastal Plain terraces are covered with orchards and have become an important center for Pennsylvania's nascent wine industry.

The glaciers of northern Pennsylvania caused major disruptions of the state's river system well beyond the boundaries of the ice. The most obvious effect was the prodigious volumes of water dumped into south-flowing streams, such as the Delaware, the Susquehanna, and the Allegheny (see Fig. 2.10). All these rivers were temporarily turned into huge glacial "spillways" that were widened, deepened, and supplied with abundant amounts of sand and gravel. This coarse sediment accumulated in the channels of the southward-flowing streams and in many places raised their levels by 30 feet or more. When the supply of new sediment waned with the final retreat of the glaciers, the rivers cut deep channels into the gravelly surfaces of the older streambeds, leaving behind broad, flat surfaces—called "stream terraces"—beside most of the rivers in the state. The terraces are flat, fertile, easy to build on, and easy to run railroads and highways along, following the rivers. Many of the riverside cities of Pennsylvania are built on Wisconsinan-stage terraces. Much of Harrisburg, Williamsport, Easton, Pittsburgh, and other cities are in that situation. The most widespread flood damage in the state occurs during the rare storms when the rivers reclaim their floodplains, as they rise up once every few decades to the level they reached each year while glaciers were melting.

The glacial ice also blocked northbound streams. We have already seen how such an ice dam at Niagara helped create a larger glacial Lake Erie. In Pennsylvania, the most important blockage was that of the north-flowing Monongahela. That blockage, as it turned out, had a powerful influence on the course of American history by creating and refining what is now the major transportation route through the central Midwest—the Ohio River system. Today's Monongahela, together with its tributaries, is merely the headwater portion of a much larger preglacial river system that flowed north to join a still-larger river that evidently flowed off to the northeast, toward the St. Lawrence River. The Ohio River as we know it today simply did not exist. The Monongahela and other north-flowing rivers were dammed by the glacial ice, and their surplus waters spilled into a channel that skirted the edge of the ice, flowing southwestward in the general direction of the Mississippi River roughly parallel to the edge of the ice. That newly formed ice-marginal stream became the Ohio River. For a time, its volume of water was enormous, because it was fed not only by north-flowing streams like the Monongahela (whose lower sections had been amputated), but also by colossal volumes of water melting off the edge of the ice cap a few miles to its north along most of its length. The drainage map of western Pennsylvania, therefore, is a composite of very old and very young rivers. As is often the case, the subtle and obscure influence of the glaciers has had a dramatic and obvious impact on human habitation.

Learning More About the Physical Landscape

We have seen the small number of internal and external processes that have worked together—or in sequence or against each other—to sculpt the complex pattern that is the landscape of Pennsylvania. The physical landscape must be seen as the *first* component of the varied modern environment of the state. It is "first" in several senses. It was the earliest to exist, of course. It is also first in the sense of being the foundation on which all other parts of the state's environment have come into existence. And it is the first requirement to which all users of the land—from earliest settler to the most recent homeowner—must adapt.

Knowledge of landforms is useful for many purposes. It can help us understand how the history of an otherwise familiar area develops. It helps to show the local environment of a town or county. Landowners, zoning boards, and concerned citizens all benefit from a knowledge of topography when deciding how land should best be used. And the ability to read the specifics of soil, slope, and bedrock adds pleasure and appreciation to all outdoor activities and recreation, such as travel, hiking, and boating.

A fondness for landforms, and the satisfaction that comes from understanding the processes that created them, can be lifelong rewards for the careful observer. Because landform patterns tend to be so regular in Pennsylvania and to reflect the geology so closely, any careful observer can learn much about an area with relatively little information.

A good starting point is a topographic map of an area of interest. The U.S. government publishes first-quality maps of every inch of the state in several scales. The standard map series comprises the "seven-and-a-half-minute maps" (so called because that's how much latitude and longitude they cover). These maps show the streams, roads, human settlements, and political boundaries, as well as the topography, of 85-square-mile sections of the state drawn at a scale of 1 inch to 2.4 miles. These maps are available at many libraries and can be purchased from engineering supply stores or ordered directly from the U.S. Geological Survey, Map Distribution, Federal Center, Building 41, Box 25286, Denver, CO 80225.

Additional information on the particular conditions that shaped the land of Pennsylvania, and the general processes that shape topography anywhere, can be found in the references listed at the end of this chapter.

Bibliography

General Works on Landforms

Bloom, A. L. 1969. *The Surface of the Earth.* Englewood Cliffs, N.J.: Prentice Hall.

Shelton, J. S. 1966. *Geology Illustrated.* San Francisco: W. H. Freeman.

Strahler, A. H., and A. N. Strahler. 1991. *Modern Physical Geography.* Fourth edition. New York: Wiley.

Thornbury, W. D. 1965. *Principles of Geomorphology.* New York: Wiley.

Works on Landforms of the United States

Fenneman, N. M. 1938. *Physiography of the Eastern United States.* New York: McGraw-Hill.

Shimer, J. A. 1971. *Field Guide to Landforms in the United States.* New York: Macmillan.

Thornbury, W. D. 1968. *Regional Geomorphology of the United States.* New York: Wiley.

Landform Maps

Garret, W. E., ed. 1985. *Atlas of North America: Space Age Portrait of a Planet.* Washington, D.C.: National Geographic Society.

Raisz, E. 1957. *Landforms of the United States.* Boston: Erwin Raisz.

Thelin, G. P., and R. J. Pike. 1991. "Landforms of the Conterminous United States." *Miscellaneous Investigation Series,* Map I-2206. Washington, D.C.: U.S. Geological Survey.

Works on Pennsylvania Landforms and Geology

Berg, T. M. *Geologic Map of Pennsylvania.* 1984. Harrisburg: Pennsylvania Topographic and Geologic Survey.

Cuff, D. J., et al., eds. 1989. *The Atlas of Pennsylvania.* Philadelphia: Temple University Press.

Schultz, G. H., ed. Forthcoming. *The Geology of Pennsylvania.* Fourth series, Special Publication 1. Harrisburg: Pennsylvania Geologic Survey.

Van Diver, B. B. 1990. *Roadside Geology of Pennsylvania.* Missoula, Mont.: Mountain Press.

3

CLIMATE

Brent Yarnal

Climate and Its Controls

Climate has a direct impact on all Pennsylvanians, dictating such basic facts of life as how we dress, how we build our homes, and what agricultural goods we produce. Climate determines what natural vegetation presently grows in the state. It also controls the erosional and weathering processes forming the landscape. Thus, the present and past climate of Pennsylvania influences our lives profoundly, and we can safely say that the climate in the years to come will also shape the lives of succeeding generations.

Most discussions of climate focus on the elements of weather and climate—heat (temperature), moisture (precipitation and clouds), and pressure (winds). The difference between weather and climate is subtle but is essentially a matter of time and experience. *Weather* denotes the state of the atmospheric elements now and in the near future—say, the next few days. It can be sensed directly: We can feel heat, rain, and wind. On the other hand, *climate* is the long-term state of the atmospheric elements measured over time spans of months, seasons, years, decades, centuries, and even longer. It is an abstraction that cannot be felt, but only perceived intellectually.

Climate consists of all daily weather events during a specified period. It is more than just "average weather," and includes all the extreme departures from average that make the climate of Pennsylvania so interesting, and sometimes miserable! Long mid-winter cold snaps and blistering summer heat waves are important parts of Pennsylvania's climate. Therefore, when we learn that the average temperature in the state is about 50°F and that the average annual precipitation is around 40 inches, we must remem-

Figures and text in Chapter 3 are modified from D. J. Cuff et al., eds., *The Atlas of Pennsylvania* (Philadelphia: Temple University Press, 1989). © 1989 Temple University. Reprinted by permission of Temple University Press.

ber that those numbers represent a host of weather and climate possibilities.

Climate varies over both space and time, and there are important differences in the climate of various areas of Pennsylvania. For example, the southeast tends to be the warmest part of the state in all seasons. We shall also see that the climate of Pennsylvania varies from year to year, with some years or even decades being warmer or cooler, or wetter or drier, than others.

The elements of climate are controlled by four factors. The first, and perhaps most important, factor is the relationship of Pennsylvania to the angle of the sun in the sky. The latitude of Pennsylvania and the season determine whether the sun's rays are strong, as with the high sun of summer, or weak, as in the case of the low winter sun. A second controlling factor is Pennsylvania's relationship to the circulation of the atmosphere. Great rivers of air flow from the west over the state, steering storms into the region and pulling air from the northern interior of the continent or from the Gulf of Mexico (Fig. 3.1). If the main axis of these westerly winds (called the "jet stream") is to the south, the weather is relatively cool and dry. If the jet stream is to the north, the weather is relatively warm and humid. When the jet stream passes nearly overhead, we are located on the seam, or "weather front," between these two contrasting air masses and the weather tends to be stormy. A third control on Pennsylvania's climate is the proximity to large bodies of water. Although the state is quite close to the Atlantic Ocean, its climate is quite continental, rather than maritime, in its nature because the westerlies blow from the interior of the continent. However, Lake Erie does act as a moderating influence on the climate of the areas immediately adjacent to the lake. The final control on climate is the variable terrain of the state (Chapter 2). Both regional temperature and precipitation patterns are strongly influenced by the lay of the land. For instance, the rugged upland areas of the Appalachian Plateau are much cooler and snowier than the smooth, lower areas of the southeast.

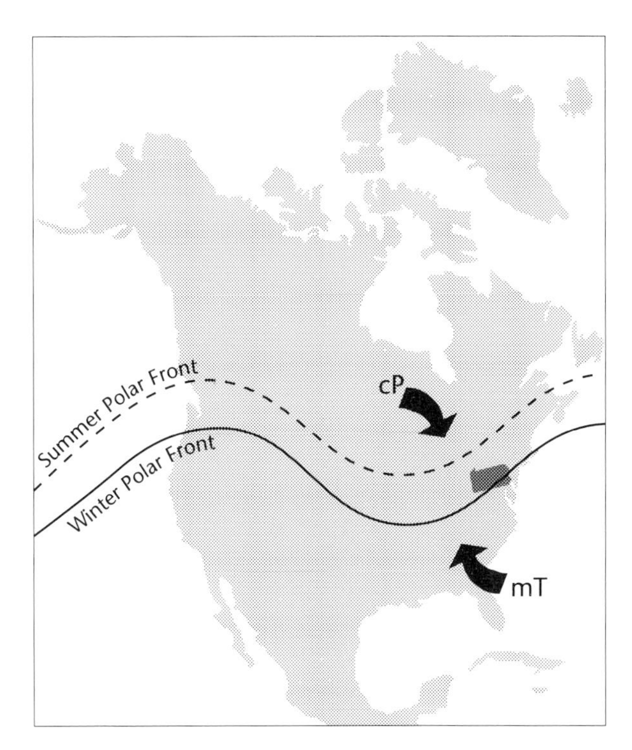

Fig. 3.1. Average position of the polar front and associated air masses in winter and summer. Broad arrows denote general direction of air-mass flow.

The weather fronts between the relatively cool, dry air coming from the northern interior of the continent, and the warm, humid air coming from the Atlantic Ocean and Gulf of Mexico, produce the stormy weather and associated precipitation so characteristic of Pennsylvania's climate. The passage of these migratory weather systems explains the day-to-day variability of Pennsylvania climate: one day being cool, cloudy, and misty; the following day being warm, muggy, and hazy; the next day being violently stormy followed by clearing skies and cool and breezy conditions. In fact, Pennsylvanians *expect* the climate to have these day-to-day changes; it is abnormal when they do not occur. Nevertheless, when climate information is combined to produce the long-term, average statistics for the state, these daily weather experiences are filtered out and sometimes forgotten in the process. No wonder the studies of weather and climate (meteorology and climatology, respectively) are often treated as separate and distinct fields of science. However, one cannot understand the climate of Pennsylvania without considering its weather too.

The maps and diagrams in the next section portray the climate of Pennsylvania. However, the climate values given are long-term (about 30-year) averages and hide a great deal of information about the day-to-day and even year-to-year variability of Pennsylvania's climate. Therefore, in the last section the year-to-year variability of Pennsylvania's climate is presented.

Variability of Pennsylvania's Climate over Space

The distribution of the elements of climate over the state is controlled primarily by the terrain. In Fig. 3.2 we do see the influence of latitude: The southern part of the state is warmer than the northern part by 4° to 6°F. But the association between the orientation of the isotherms (lines of equal temperature) in this figure and the topographic regions of Pennsylvania (see Fig. 2.4 in Chapter 2) is obvious. The warmest areas of Pennsylvania correspond to the Atlantic Coastal Plain and the Piedmont regions, with progressive cooling into the higher elevations of the Blue Ridge and Ridge-and-Valley regions. The Appalachian Plateau is the coolest area of the state, although it is much warmer in the southwest, south of Pittsburgh and west of Chestnut Ridge, and much cooler in the vicinity of McKean County and the town of Bradford. A slight moderating of temperature by Lake Erie is experienced along the narrow margin along the lake but is too local to show in Fig. 3.2.

Two factors combine to create this topographic control on the temperatures of Pennsylvania. First, temperature normally decreases with height in the atmosphere, so the highland regions of the state tend to be cooler than the lower areas. Second, topography influences low-level airflow. The southeastern areas are in a natural corridor for the direct flow of relatively warm air off the Atlantic from areas to the south. The mountain fronts tend to cut off these low-level flows, preventing them from reaching the mountainous interior of the state. However, there is clear evidence for this warm air penetrating the water gaps and following the valleys of the Ridge-and-Valley region and the Susquehanna River. Within the higher, cooler Appalachian Plateau, warm southerly airflows from the Gulf of Mexico first reach the southwest but are progressively inhibited from traveling northward. Lake Erie moderates temperature in its immediate vicinity, but the area around Bradford, which is removed from the moderating effect of the lake and is relatively cut off from all southerly flows, is the coolest region of the state.

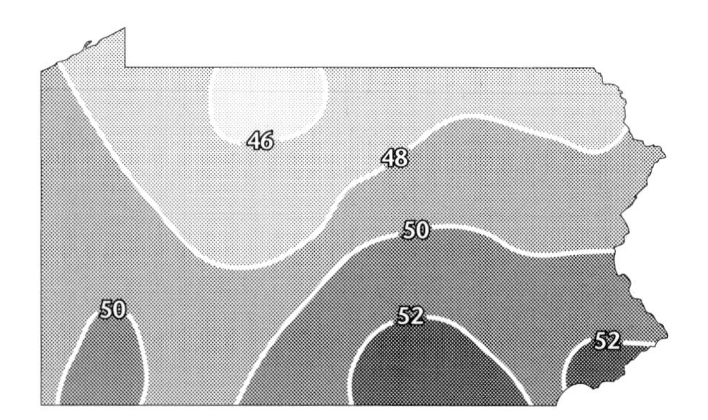

Fig. 3.2. Average annual temperature, Pennsylvania (°F).

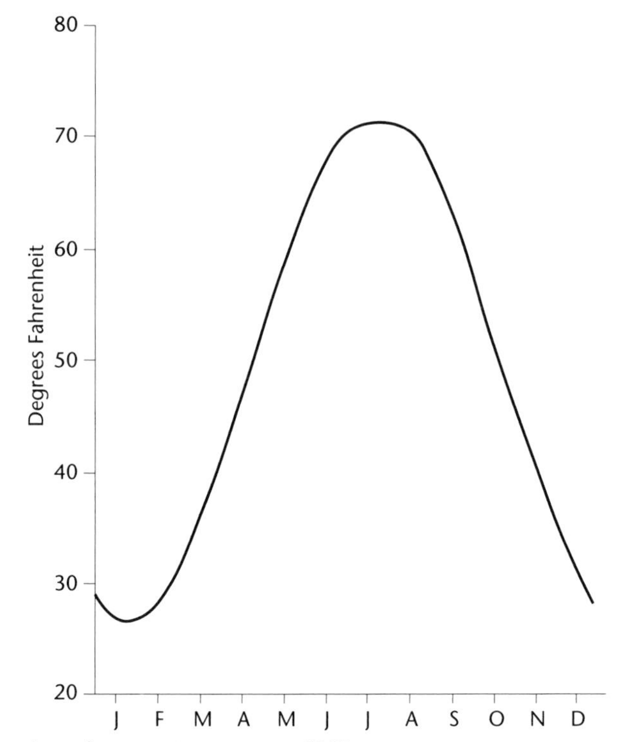

Fig. 3.3. Average march of temperature, Pennsylvania (°F).

The temperature values observed in Fig. 3.2 can be collapsed into one average temperature value for the entire state: 48°F. Furthermore, that all-Pennsylvania temperature value can be broken into its monthly values to produce a curve representing the annual march of temperature (Fig. 3.3). The annual march is controlled by the elevation of the noon sun in the sky and by the length of day. The low noon sun and small number of daylight hours in January produce the lowest temperatures. As the sun migrates to a higher position and day length increases, temperatures rise,

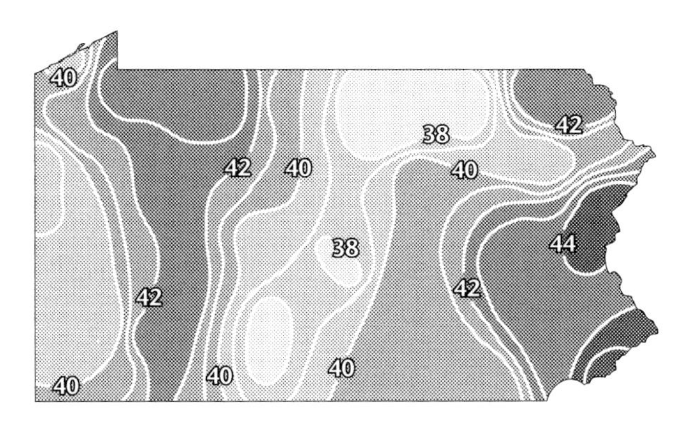

Fig. 3.4. Average annual precipitation, Pennsylvania (inches).

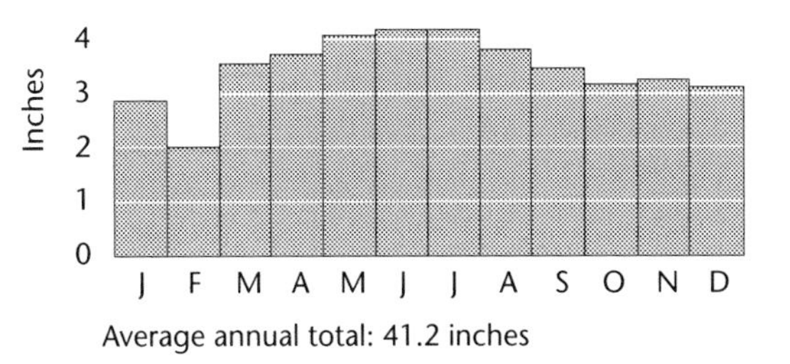

Fig. 3.5. Annual march of precipitation, Pennsylvania (inches).

reaching their maximum in July. Temperatures fall again as solar energy decreases with the coming of autumn. The distribution of annual temperature illustrated in Fig. 3.2 looks much the same in all seasons: The southeast is always the warmest and the Appalachian Plateau is always cool, with the south being warmer and the north cooler.

The distribution of Pennsylvania's precipitation (Fig. 3.4) is much more complex than temperature but is also greatly influenced by the terrain. Whereas heat and moisture flow into the state from the south, most of the precipitation-producing weather fronts move from west to east. As they enter the state in the west, these systems and their clouds are lifted by the intervening Appalachian Plateau, with the most moisture squeezed out at the highest elevations. The decreasing elevations encountered as the systems continue east result in decreasing precipitation totals in the Ridge-and-Valley region. Because these eastward-moving storms are most frequent and strongest in spring, this effect is best observed in that season.

The precipitation maximum around Warren in the northwest is a combination of the effect of uplift by the highlands on storm systems and the effect that Lake Erie has on those airflows traveling from the west and northwest over the lake. The lake adds moisture to these flows, which supplements their already high precipitation totals. This "lake effect" is best seen in late fall and winter when the forces adding moisture to the air from the lake are at a maximum.

The precipitation maximum in the southeast results from two factors, one of which made this region the warmest part of the state. The southerly airflows coming directly off the Atlantic bring in considerable moisture, as well as warmth, in all seasons, resulting in higher precipitation totals. Furthermore, occasional strong storms moving along the Atlantic coastline pass over the southeast, bringing heavy precipitation that does not reach the rest of the state. The highest totals are found in the area extending southwest from the New Jersey border along the approximate axis of the Reading Prong of the Blue Ridge, again illustrating the impact that forced uplift has on precipitation totals. The extreme northeast of the state is shielded by this mountain front and receives less precipitation than the region directly to its south.

Again, the data used to produce the precipitation values shown in Fig. 3.4 can be reworked to create an all-Pennsylvania average precipitation total (41.2 inches) and an annual march of precipitation (Fig. 3.5). Ample precipitation is witnessed in all seasons, with a slight maximum occurring in the summer season. Put simply, warm air holds more moisture than cold air, so the humid conditions of summer provide more moisture for precipitation than do the much stormier but drier conditions of winter. Most of the summer rainfall is produced by thunderstorms, with most stations observing 30 to 35 storms a year over the state. The annual march of precipitation varies from region to region across the state.

When temperatures are low enough, winter precipitation falls as snow. Annual snowfall ranges between wide limits from year to year and from place to place. Some years have only a few inches total; in other years several feet accumulate. However, long-term annual average snowfall over the state (Fig. 3.6) does conform to the terrain and temperature patterns noted earlier. Annual snowfall averages less than 30 inches in the southeast, where the higher temperatures mean that less precipitation falls as snow, while the lower temperatures, lake effect, and uplift by the mountains combine to

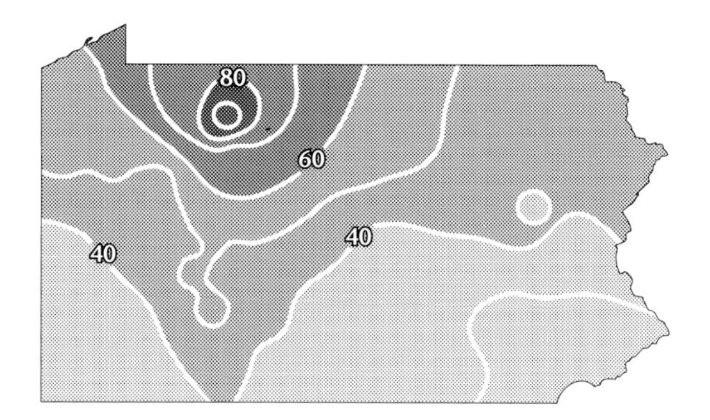

Fig. 3.6. Average annual snow fall, Pennsylvania (inches).

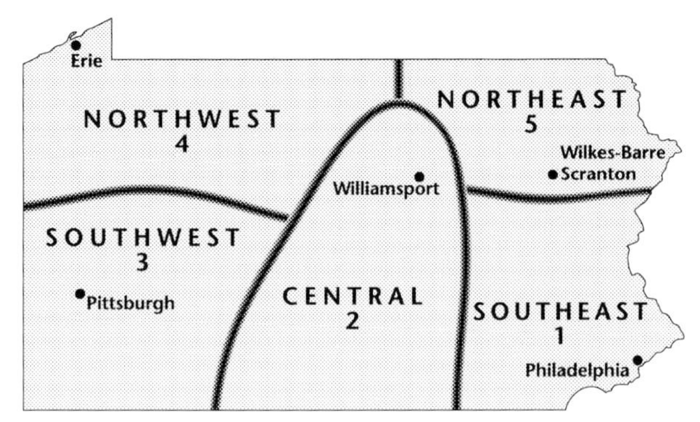

Fig. 3.7. Climate regions of Pennsylvania.

produce nearly 100 inches a year in portions of McKean County. Measurable accumulations of snow generally occur between late November and mid-March, although snow has been observed as early as the beginning of October and as late as May in the higher elevations of the northern counties. Greatest monthly accumulations occur in the coldest months (December, January, and February), but the highest totals for a single storm occur in March and early April, when the moisture supply increases with the warming air of spring.

The Climate Regions of Pennsylvania

Other elements of climate besides temperature and precipitation are only occasionally collected by weather observers. As a result, it is impossible to map the distributions of such elements as surface wind or cloud cover over the state of Pennsylvania. However, it is possible to use wind and cloud data from a few stations to represent the broad climatic regions of the state.

The climate regions of Pennsylvania (Fig. 3.7) have been determined by a sophisticated computer-assisted statistical procedure applied to the precipitation data for the state. Although a computer chose these categories, the regions are identifiable in terms of the topography of the state. The very wet area of the Atlantic Coastal Plain, the Piedmont, and the Reading Prong is represented by Region 1, while the relatively dry Ridge-and-Valley area is approximated by Region 2. The Appalachian Plateau contains Regions 3, 4, and 5, which represent the moderately wet southwest, the wet area influenced by Lake Erie in the northwest, and the drier areas of the northeast, respectively. In the following paragraphs, one key weather station is used to represent the climate of each of these regions. These weather stations are not sited so as to capture the climate of their respective regions perfectly; they were chosen simply because the data were available.

Climate Region 1: The Southeast

Data from Philadelphia are used to represent the climate of the southeast (Fig. 3.8). The annual solar cycle is evident in the temperature curves (Fig. 3.8a), with average temperatures ranging from about 32°F in January to more than 77°F in July, for an annual average range of about 45°F. The average annual temperature in Philadelphia is approximately 55°F.

Within each month, the range between average maximum and average minimum temperature varies somewhat, with the November through February period averaging around a 16°F range and the rest of the year averaging around a 21°F range. In other words, the difference between the daytime high and the nighttime low is not as great in the winter as it is the rest of the year. The weak sun and short period of daylight during the winter months prevents the region from heating up as much during the day. This phenomenon is present in all of Pennsylvania's climate regions.

In this region, the summers are longer and hotter than in the rest of the state. Heat waves of a few days

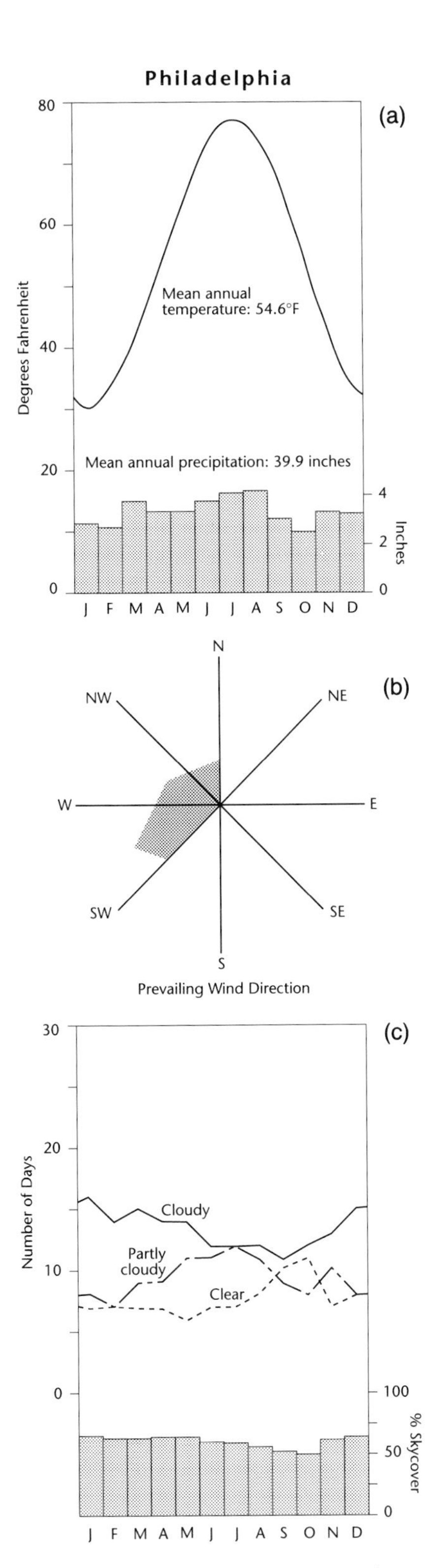

Fig. 3.8. Climate Region 1: Southeast (Philadelphia data). The annual prevailing wind direction in (b) is based on average monthly wind direction.

to a week in length are common from the beginning of July to mid-September. During these periods, high humidities and light winds combine with the heat to make conditions oppressive. On the other hand, winters are comparatively mild, with fewer subfreezing days than in all other regions of Pennsylvania. As a result, relatively long growing seasons exist in the key agricultural lands of the Piedmont.

Precipitation averages about 40 inches a year in Philadelphia, which is considerably less than the expected average for the southeastern part of the state. There is ample precipitation in all months, with distinct month-to-month variation (Fig. 3.8a). The highest monthly total is about 4.1 inches in both July and August, while the lowest occurs only two months later, in October, with just over 2.5 inches. June, July, and August, the months when warmest winds blow from the south-southeast (Fig. 3.8b) and bring the maximum amount of moisture from the Atlantic Ocean into the region, experience the most rainfall. The minimum in October corresponds to a period of relatively clear skies and high pressure (Fig. 3.8c). Comparatively low precipitation totals for December through February correspond to the period when the coldest and driest northwesterly winds blow from the interior of the continent. The average annual snowfall in this region is less than 40 inches, with a minimum of around 25 inches in the coastal plain. Snow cover persists about one-third of the time during the winter season.

Pennsylvania is a cloudy state, as the data for the southeastern region demonstrate (Fig. 3.8c). As cloudy as it is, however, the data show the least cloud cover in the state. In Philadelphia, about three days a week are considered "cloudy" (more than six-tenths of the sky is covered by cloud), more than two days a week are "partly cloudy" (from three- to six-tenths cloud cover), and considerably less than two days a week are "clear" (less than three-tenths cloud cover). Clear skies are least likely in late winter, spring, and early summer and most likely in October, when an average of about 11 clear days can be expected. The heaviest cloud cover comes in winter and early spring. However, all months are far from clear, with the smallest percentage of cloud cover coming in October (55 percent) and the largest in January (66 percent), for an annual average of 62 percent. Converted to tenths sky cover, this is greater than six-tenths (6.2/10.0), which means that the southeast's average falls in the cloudy category for the entire year.

Climate Region 2: Central

The central Ridge-and-Valley region is represented by the climate of Williamsport (Fig. 3.9). This region is transitional between the more continental plateau area and the relatively more maritime southeast. The regular ridge-and-trough pattern of the central region makes its climate unique; the climates of the ridge tops are quite different from those of the adjacent valley bottoms. Because most towns are located in valleys, however, the description here will be of the climate of the valley areas.

The annual cycle of temperature in Williamsport (Fig. 3.9a) is quite similar to that of Philadelphia, but temperatures in Williamsport are about 3 to 4°F lower throughout the year, with an annual average of less than 51°F. This results in shorter, less oppressive heat waves than in the southeast, but also longer, colder winters.

The annual precipitation total of 40 inches in Williamsport is virtually the same as Philadelphia's and fairly typical of the totals usually measured in the valley bottoms of the central region (Fig. 3.4). Nevertheless, the distribution of precipitation throughout the year (Fig. 3.9a) is quite different from that experienced in the southeast and is representative of the central region. Neither the August maximum nor the October minimum seen in the Philadelphia precipitation data are observed in Williamsport. Instead, two peaks in precipitation are present, one coming in mid-spring and the other in mid-summer, with July getting more rain than August. The least precipitation comes in winter, reaching a minimum in January. However, much of the winter precipitation falls as snow, averaging about 50 inches a year in the region.

The relationship between surface wind and precipitation is not clear in Williamsport, as it is in Philadelphia. In the southeast, the land is flat and the winds at the surface are representative of large-scale regional airflow. In Williamsport, the local arrangement of ridges and valleys steers the winds so they always come from the west (Fig. 3.9b). This is typical in the ridge-and-valley system of the central climate region and is confirmed by wind data from other towns in the region, such as Harrisburg and State College, where local topography also completely controls windflow patterns. In all cases, however, the wind always comes from either a westerly or a southerly direction, suggesting that the large-scale atmospheric flow above the ridge tops comes from those directions.

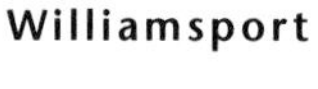

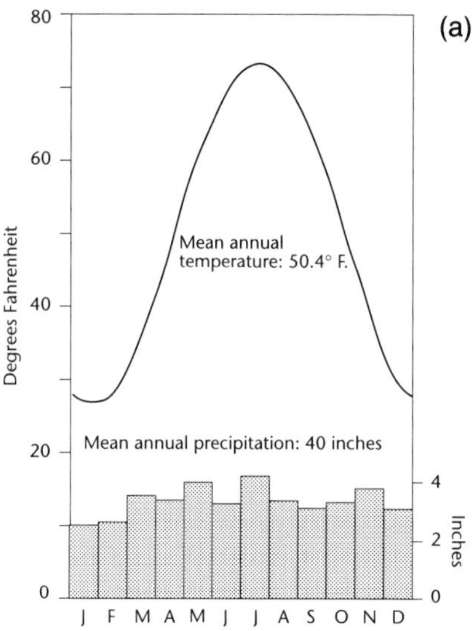

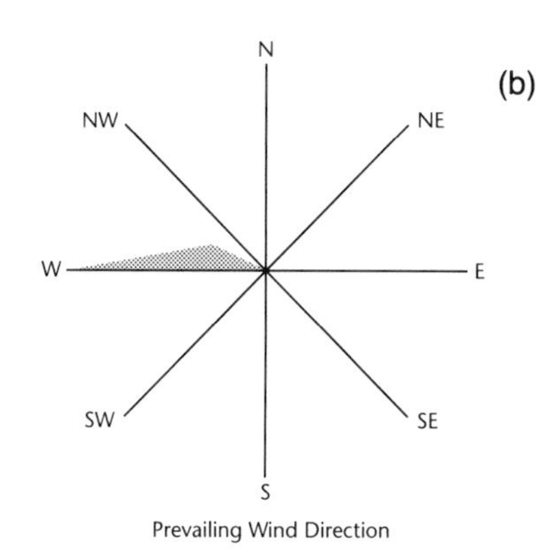

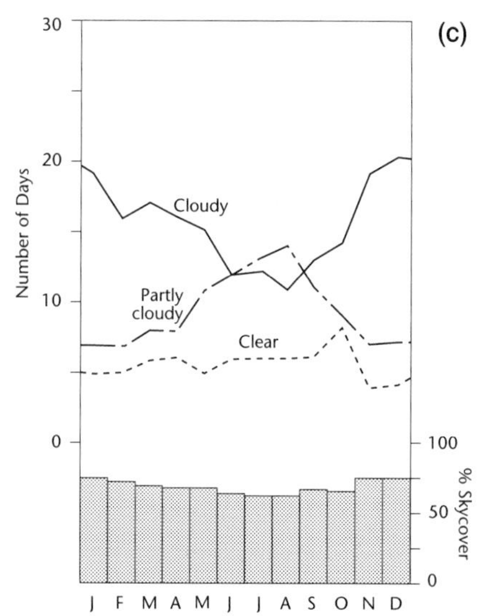

Fig. 3.9. Climate Region 2: Central (Williamsport data).

The central part of the state is even cloudier than the southeast, averaging 68 percent sky cover annually (Fig. 3.9c). Like the southeast, clearest skies are experienced in October and heaviest cloud cover comes in winter. On average, there are 25 fewer clear days and 24 more cloudy days in Williamsport than in Philadelphia.

Climate Region 3: The Southwest

The climate of the southern area of the Appalachian Plateau is represented by Pittsburgh's data (Fig. 3.10). This region is warmer than the more northerly areas of the plateau (Regions 4 and 5), but cooler and less humid than the southeast (Region 1). The annual march of temperature (Fig. 3.10a) does not differ much from the temperatures in the southeastern and central regions of the state, and monthly and annual ranges are similar too. The annual average temperature of about 50°F is considerably cooler than the southeast but about the same as the central part of the state.

The precipitation characteristics of the southwest (Fig. 3.10a) set this region apart. The spring and summer months from March to August receive around 3.5 inches a month, with a slight peak in July but no overwhelming dominance by any one month in this six-month period. In contrast, the six fall and winter months average about 2.5 inches, again with little difference between months. In all seasons, precipitation events are more frequent in the southwest, but precipitation totals are smaller than in other regions, producing roughly comparable average annual totals from 38 to 44 inches, depending on elevation.

The wind data shown in Fig. 3.10b are poor, showing local topographic control. Nevertheless, in this region winds do come primarily from the southwest in most seasons, implying that the dominant directions of wind flow are southerly and westerly.

The more frequent storms in this region result in fewer clear days and greater cloud cover (Fig. 3.10c). In fact, with an average daytime sky cover of 71 percent, less than 60 clear days a year, and more than 200 cloudy days, the southwest ranks as the cloudiest area of the state and as cloudy as any area of the 48 contiguous United States—a dubious distinction indeed.

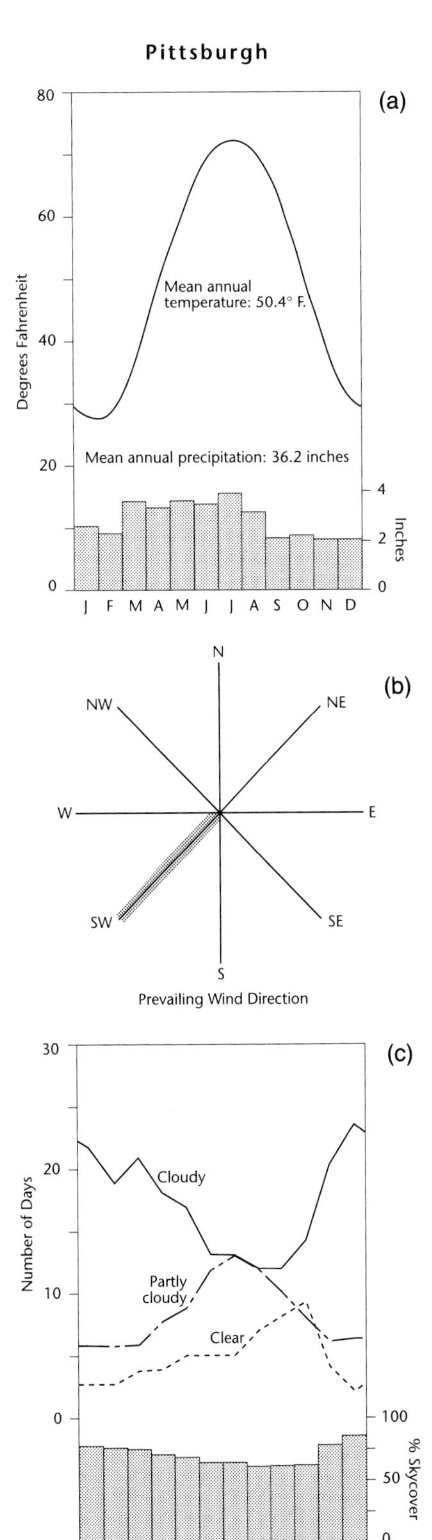

Fig. 3.10. Climate Region 3: Southwest (Pittsburgh data).

Climate Region 4: The Northwest

The northwestern area of the Appalachian Plateau is a region of sharp climatic contrasts. Therefore, selection of Erie to represent the climate of the region (Fig. 3.11) is based on the availability of data rather than on the station's representativeness.

The northwest is cooler than all other regions of Pennsylvania. Even with the moderating effect of the lake on temperature, winter temperatures are still lower in Erie (Fig. 3.11a) than at the other representative stations. The lake's influence on temperature is more noticeable in summer, when monthly maxima are suppressed by a few degrees Fahrenheit along its shore. Wintertime temperatures in the northwest are considerably lower away from the lake. Surprisingly, so are summertime temperatures, especially around McKean County. An important result of the lake's effect on temperature is to extend the length of the growing season along its margin compared with the surrounding countryside by about a month and a half. The low average annual temperature for the region as a whole (47°F in Erie, lower in other areas) suggests a relatively short growing season, especially away from the lake.

The lake's moisture input to westerly airstreams, when coupled with the forced uplift of those airstreams by the plateau, causes enhanced precipitation totals in the northwest. This is especially true for the fall months of October and November, which in other regions are often the driest months. Otherwise, the annual precipitation curve (Fig. 3.11a) is similar to the Pittsburgh curve representing the southwest, with an extended "wet" season of eight months, averaging around 3.5 inches, and a shortened "dry" season of four months, averaging roughly 2.5 inches. Note the very low February precipitation total, indicating that in some years Lake Erie freezes, cutting off the moisture source from the lake and causing lake-enhanced snows to be curtailed.

Wind data from the flatland city of Erie (Fig. 3.11b) suggest that, except for the odd month of March with its northerly winds, the region is dominated by southerly winds. Although there is probably some local topographic effect muddling the pattern, these data suggest that the same planetary-scale physical factors influencing the winds of low-lying Philadelphia, where southerly winds were also dominant, control the wind in Erie. The big difference between the two areas is that low-

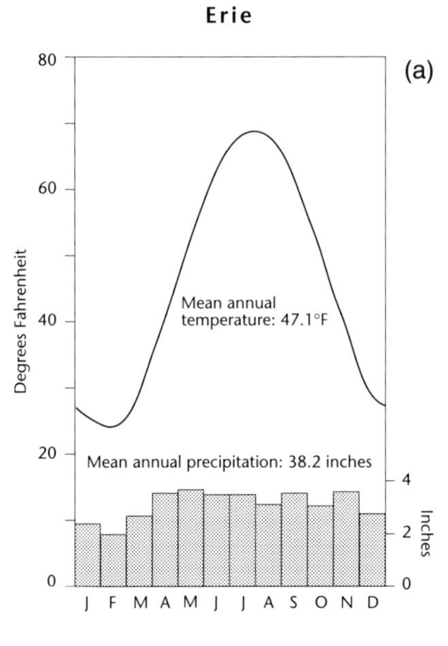

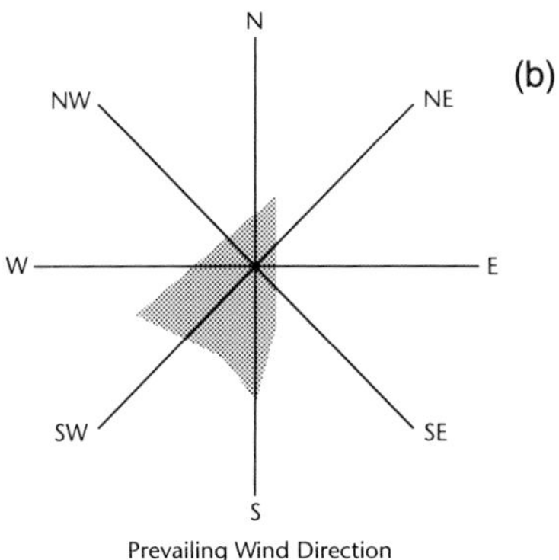

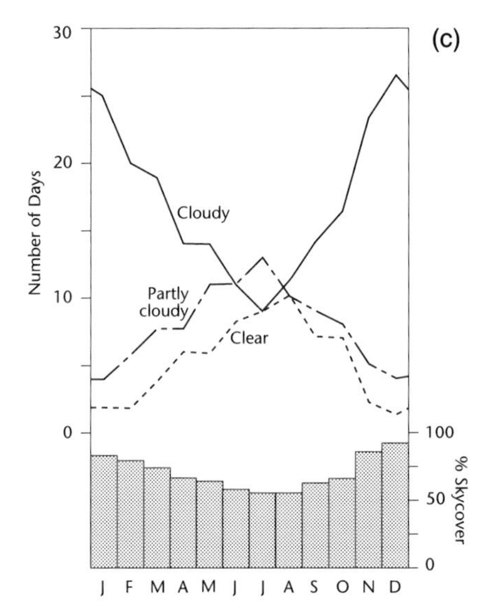

Fig. 3.11. Climate Region 4: Northwest (Erie data).

level southerly winds reaching Erie are effectively cut off from the warm, moist air of the Atlantic. As important as westerly flow coming off the lake is to the climate of the northwest, it is not the dominant direction of wind flow in the area.

The northwest is only slightly less cloudy than the southwest, with 70 percent of the sky covered during daylight hours, only 64 clear days, and more than 200 cloudy days a year on average (Fig. 3.11c). Unlike all other parts of the state, the summer months have the clearest skies, with 10 clear days in August. Winter months are extreme opposites in which, because of the lake effect on cloud formation, only 1 or 2 clear days a month are experienced from November through February. In addition, those winter months average more than 23 cloudy days a month.

Climate Region 5: The Northeast

Northeastern Pennsylvania climate is represented by the Wilkes-Barre/Scranton airport at Avoca (Fig. 3.12). The region is characterized by both the greatest annual temperature range and the largest annual precipitation range in the state (Fig. 3.12a). The July mean temperature of over 72°F and January mean of 26°F produce an average annual range of over 46°F, making this the least maritime and most continental region, despite its proximity to the Atlantic Ocean. Winter precipitation, especially in January and February, is lower here than in all other parts of Pennsylvania. Coupled with summertime precipitation totals comparable to those of other regions, including a July average of more than 4 inches, an annual range in precipitation of more than 2 inches is observed. Furthermore, the low winter totals give this region the lowest annual precipitation in the state—less than 35 inches. The average annual temperature of 49°F is comparable to other Appalachian Plateau regions, with the exception of the cool northwest.

Like the average wind conditions at Williamsport, the wind conditions at Avoca are controlled by the local topography, with the southwest-to-northeast orientation of the Lehigh Valley channeling the winds (Fig. 3.12b). Nevertheless, the average southwesterly direction suggests a dominance by southerly and westerly winds in all seasons.

Cloud conditions (Fig. 3.12c) are similar to those of Williamsport, located just upwind of Avoca. The daylight sky averages a 68 percent cloud cover, with

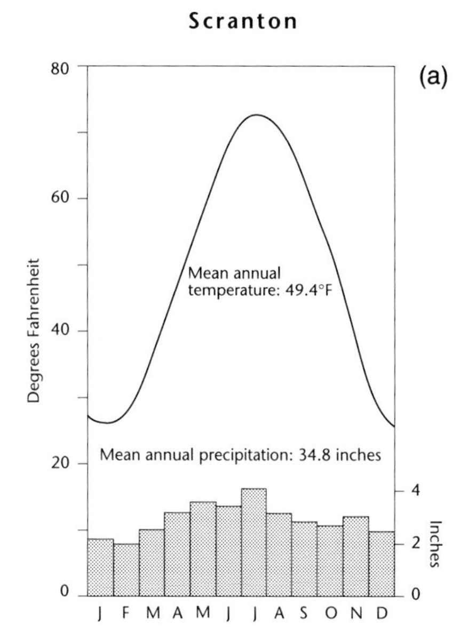

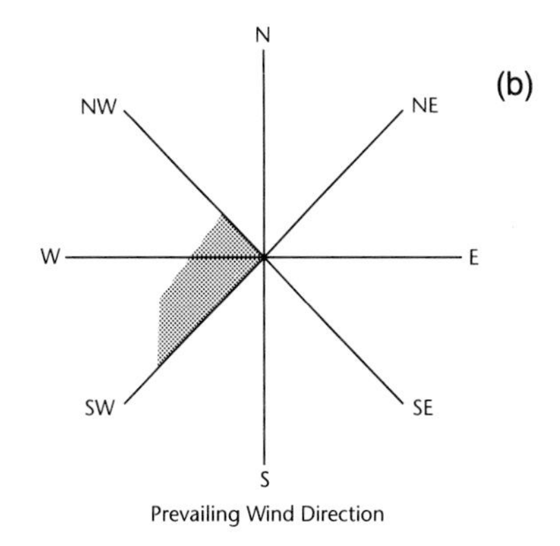

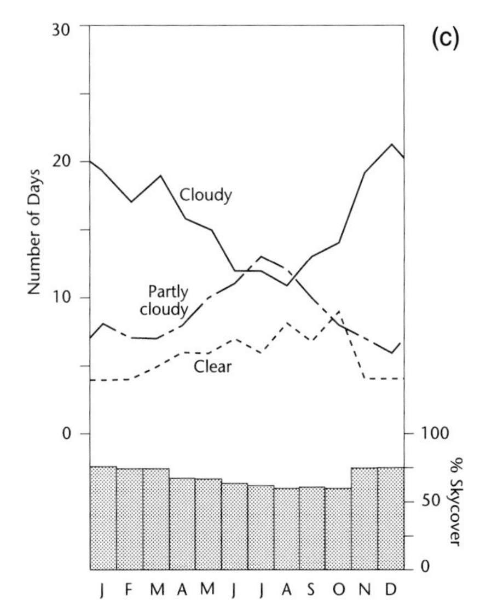

Fig. 3.12. Climate Region 5: Northeast (Scranton data).

an average of 70 clear days and 188 cloudy days a year. Summer and early fall have the clearest skies, while winter skies (November through March) are excessively cloudy.

Variability of Pennsylvania's Climate over Time

Climate varies not only over space but also over time, with one year, or even a decade, being different from the next. For instance, it is not at all unusual in Pennsylvania to have a mild, wet winter followed by a bitter-cold, dry winter. Fig. 3.13 shows the annual variations in Pennsylvania's temperature for the period 1931–1984. Data from the entire state were combined to create an all-Pennsylvania average temperature for each year. The long-term average temperature for the state for the fifty-five-year period is 48°F and is represented by the horizontal line cutting across the graph. The thin, jagged line shows the year-to-year variation. The broad line results from the application of a mathematical technique to the annual data to "smooth out" the jagged line in order to see more clearly any long-term trends that might be present.

Several clear patterns emerge. Temperatures from the beginning of the period (and presumably before 1931 as well) were above the average up to the mid-1950s. From that point, conditions cooled and stayed below normal throughout the 1960s and early 1970s. Near the end of the period, temperatures began to rise again.

The explanation for these periods of warm and cool conditions rests with the westerly winds and the position of the polar front (see Fig. 3.1). During the warmer years, ending in the mid-1950s, the jet

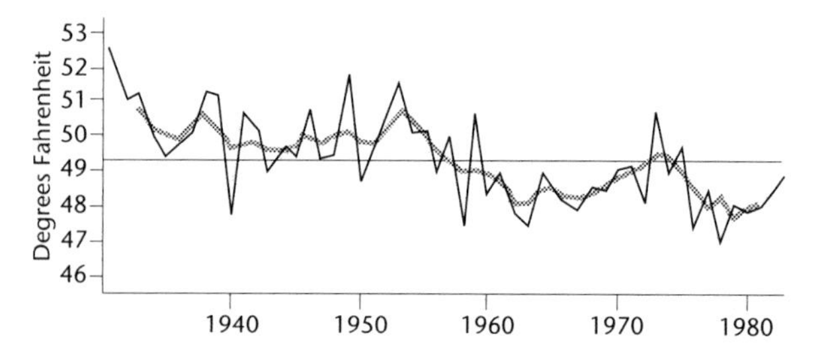

Fig. 3.13. Variability of all-Pennsylvania average annual temperature (°F), 1931–1984. Narrow line denotes annual values; broad line denotes smoothed values.

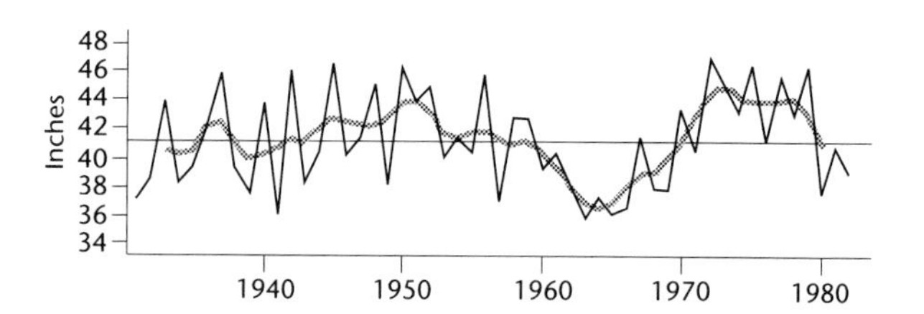

Fig. 3.14. Variability of all-Pennsylvania average annual precipitation (inches), 1931–1984. Narrow line denotes annual values; broad line denotes smoothed values.

stream winds over Pennsylvania were quite strong and flowing very nearly from due west. This placed the average position of the polar front farther to the north, meaning that less polar air and more tropical air would be over Pennsylvania, thus resulting in higher temperatures. During the two decades following the mid-1950s, the westerlies weakened and the jet stream flowed in a slightly more north-south direction. The average position of the polar front moved south, and polar air became a more frequent visitor to Pennsylvania, with accompanying lower temperatures. The short record of warming at the end of the period is too short for generalizations to be made but probably denotes further meanderings of the westerly winds.

The precipitation record for the period 1931–1984 shows a different pattern (Fig. 3.14). Precipitation to the late 1950s was nearly average (41.2 inches). Each year of the 1960s was decidedly below normal, whereas almost every year in the 1970s was above normal. Again, the position of the polar front and the orientation of the westerly winds explain this pattern. The strong westerlies of the first part of the period did not generate either extraordinarily high or low precipitation totals. During the 1960s, the average position of the polar front was depressed south and east of its earlier position, cutting off much of the subtropical moisture usually flowing into Pennsylvania and resulting in persistent drought. With the coming of the 1970s, the polar front maintained its more southerly position in the eastern United States but migrated to the west a few hundred miles. This change in position brought more moisture into the area from the south and placed the average position of the storm track over the state, resulting in above-normal precipitation for Pennsylvania.

In sum, Pennsylvania's climate is controlled by its mid-latitude, continental position and its relationship to the westerly winds and associated polar front. Variations in the climate over the surface of the state are largely a result of the terrain, while variations in the climate over time are related to meanderings of the westerlies and the polar front.

Bibliography

NOAA. 1982. "Climate of Pennsylvania." *Climatography of the United States No. 60*. Asheville, N.C.: National Oceanic and Atmospheric Administration.

Trewartha, G. T., and L. H. Horn. 1980. *An Introduction to Climate*. Fifth edition. New York: McGraw-Hill.

White, D., M. Richman, and B. Yarnal. 1991. "Climate Regionalization and Rotation of Principal Components." *International Journal of Climatology* 11:1–25.

Yarnal, B. 1989. "Climate." In *The Atlas of Pennsylvania*, ed. D. J. Cuff et al. Philadelphia: Temple University Press.

Yarnal, B. 1993. *Syntoptic Climatology in Environmental Analysis: A Primer.* London: Belhaven Press.

Yarnal, B., and D. J. Leathers. 1988. "Relationships Between Interdecadal and Interannual Climatic Variations and Their Effect on Pennsylvania Climate." *Annals of the Association of American Geographers* 78:624–641.

4

WATER RESOURCES

E. Willard Miller

Water is the most common and easily recognized substance on the earth. Pennsylvanians are accustomed to having unlimited supplies of inexpensive, clean water at their immediate disposal. Setting aside the fact that it is necessary for life, water remains an extremely pleasurable, gratifying substance. Of all drinks, cold, clear, tasteless water is still the most satisfying drink.

Water is abundant in Pennsylvania, and at the present time the supply is adequate for all needs. Because there is such an abundance of water, it may be difficult to understand that a crisis can be developing. The signs of this are the water restrictions widely imposed in towns across the state during drought periods, and possibly even more important is the number of systems that have been contaminated, forcing families to boil water for extended periods. If the signs of deteriorating water quality are recognized, there is still time to be sure there is an adequate supply in the future. If the initial signs of impending water problems are ignored, our style of living can be directly affected. We must recognize that water, in the quantity and quality desired, will never again be free to take and use without a thought.

Water Availability

The water available in Pennsylvania comes from two sources. The major source of water is precipitation, which provides about 78 percent of the water in the state. The second source is streams flowing into Pennsylvania from neighboring states.

The precipitation in the state ranges from about 32 to 60 inches annually, but if these extremes are averaged the entire state receives about 41 inches on each of its 45,309 square miles. In an average year the precipitation totals about 33,000 billion gallons. In a drought year the precipitation available falls to about

22,600 billion gallons, and in a wet year the amount may exceed 40,000 billion gallons.

In an average year, this amount of water provides about 728 billion gallons of water for each square mile of land. In a drought year, the amount falls to about 500 billion gallons, and in a wet year it will exceed 880 billion gallons a square mile.

Of the water from precipitation, about only half is returned immediately to the atmosphere by evaporation or transpiration, about one-third infiltrates the soil and bedrock, and less than one-fifth runs off the land into streams and rivers.

Groundwater Resources

Groundwater, or underground water, is the water that issues from springs or can be pumped from wells. In Pennsylvania, the availability of groundwater is directly related to the geology of the area. The type of rock—sedimentary, metamorphic, or igneous—and the geologic structure are the controlling factors. In addition, bedding, schistosity, cleavage, and fracture planes exert some control on the occurrence of groundwater. For example, in southeastern Pennsylvania, in the Ridge-and-Valley and Piedmont regions, the geologic structure is extremely complex and the yield of groundwater varies greatly from one place to another.

The porosity of the different classes of rocks in Pennsylvania is important in groundwater yield. Igneous and most metamorphic rocks are extremely dense and are usually poor aquifers. Springs from these rocks are rare, and wells drilled into them do not yield large quantities of water. Of the sedimentary rocks, sandstones are normally major sources of groundwater. Some wells in sandstone may be more than 1,000 feet deep. Limestones and shales, because of their low porosity, are usually poor aquifers. The unconsolidated sand and gravels of the Atlantic Coastal Plain are excellent aquifers.

The glacial drift deposits of northeastern and northwestern Pennsylvania are major aquifers. In these areas the water table is an undulating surface that generally stands higher beneath upland areas than beneath the adjacent valley areas, and slopes gradually downward to the level of the lakes and streams. Consequently, a depression in the land surface that intersects the water table may produce springs known as "depression springs." Small springs of this type are numerous in northeastern and northwestern Pennsylvania.

Water Consumption

In Pennsylvania, the annual consumption of water is about 17,000 million gallons, or approximately half the annual precipitation. Of the water used, about 95 percent comes from surface water withdrawal, and only about 5 percent from groundwater. Of the water uses, thermoelectrical generation takes nearly 60 percent, manufacturing about 30 percent, public water usage 9 to 10 percent, and all other uses, including mineral industries, domestic wells, irrigation, livestock, golf courses, and others, less than 3 percent.

River Systems

There are five river basins in the state, of which the Susquehanna, with an area of 21,038 square miles (followed by the Ohio with 15,614 square miles), is largest. Of the three smaller basins, the Delaware has 6,460 square miles, the Potomac 1,584, and the St. Lawrence 608 square miles. The magnitude of a river is shown best by its discharge (see Fig. 4.1), which is the volume of flow past a given point, expressed in cubic feet per second.

Within Pennsylvania, the U.S. Geological Survey has established about 240 gauging stations to measure flow. At each gauging station, the height, or stage, of the river is measured by an automatic gauge. This, combined with a measured profile of the cross section at that point, gives the cross-sectional area of the river. Periodic measurements of stream velocity are made so the discharge can be calculated by multiplying area in square feet by velocity in feet per second to yield cubic feet per second. An experienced hydrologist at a gauging station can estimate the discharge from the height alone, without measuring velocity. The water-height recorder is the key to discharge estimates.

The Susquehanna Basin has annual precipitation per square mile varying from 520 to 700 billion gallons. The discharge varies from 40,000 to about 70,000 cubic feet per second. Somewhat smaller, the Ohio Basin has an annual discharge of 30,000 to

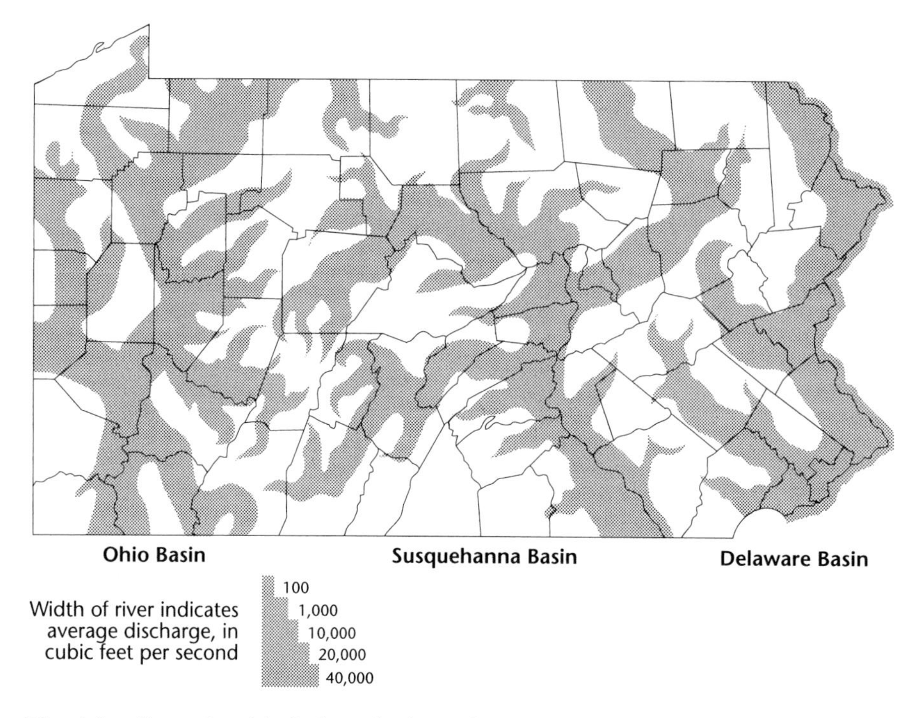

Fig. 4.1. Pennsylvania's drainage basins and average flow of water.

60,000 cubic feet per second. The other smaller basins have lower discharges.

It must be emphasized that the total supply of water in any basin varies greatly from year to year. In drought years the amount may be only 60 to 75 percent of an average year, while wet years will have 120 to 130 percent more water. If there are extended periods of drought over several years, there are cumulative results that will affect the discharge.

Water Activities

The water resources of the state provide the basis for a wide variety of recreational activities, hydroelectric power, and transportation of commodities.

Water-Based Recreation

Water offers a wide range of recreational opportunities. Pennsylvania's 45,000 miles of streams range from tiny trout streams to mile-wide rivers. Several hundred small lakes are concentrated mostly in the northeast and northwest corners of the state, plus about 4,000 man-made impoundments. In all, there are nearly 200,000 acres of flatwater surface in the state.

Nearly half the land of Pennsylvania is owned by federal, state, and local government agencies. At the federal level, a number of agencies control the development of the state's water resources. The National Park Service has jurisdiction over two districts on the Delaware River: the Delaware Scenic and Recreational River Area and the Delaware Water Gap National Recreation Area. In 1973, a 108-mile stretch of the Delaware was incorporated into the national Wildlife and Scenic River System. This portion of the Delaware is regarded as one of the best canoeing rivers in the northeastern United States. Although the river has many calm stretches, there are some Class I rapids every 8 to 10 miles. While canoeing is the dominant activity, waterskiing is allowed in the Delaware Water Gap area, and there is some excellent cold-water fishing as well.

The U.S. Fish and Wildlife Service operates facilities related to the conservation of fish and wildlife resources in Pennsylvania. The Allegheny National Fish Hatchery and the Lamar Northeast Fishing Center provide trout for federal management programs. The Erie National Wildlife Refuge protects nearly 8,000 acres of waterfowl habitat in the northwestern corner of Pennsylvania. In southeastern Pennsylvania, the Tinicum National Environmental Center provides an opportunity to observe wildlife in its natural freshwater tidal marsh habitat.

Flatwater recreational opportunities are provided by three federal agencies. The U.S. Army Corps of Engineers administers twenty-four multipurpose reservoirs. Most of these reservoirs—designed to allow low flow conditions and also control floodwaters—provide such recreational activities as swimming, boating, fishing, and picnicking. Likewise, the Soil Conservation Service operates twenty-four projects in sixteen counties. These projects have a water surface area of 5,463 acres available for water-based recreational activities. The Federal Energy Regulatory Commission regulates eleven hydroelectric projects collectively, providing more than 28,000 acres for such recreational activities as marina facilities, swimming, and picnic areas.

In addition, the U.S. Forest Service manages the Allegheny National Forest, which covers more than 500,000 acres of land in northwest Pennsylvania. Water resources are plentiful within the National Forest, which includes the Kinzua Dam on the Allegheny River. Developed recreation includes four beaches, six boat launches, and 17 campgrounds, 10 of which are on the shores of the 7,634 acre-Allegheny Reservoir. More than 500 miles of streams provide a bounty of fishing.

About 38 percent of the land of Pennsylvania is owned by the state. Much of this land is for multiple-use purposes such as hiking and picnicking. While the Pennsylvania Fish Commission owns less than 1 percent of the state-owned land, it is perhaps the most important state agency affecting water-based recreation. The Fish Commission owns or controls seventy-four lakes, dams, and reservoirs and owns or leases 39 miles of streams. The commission operates sixteen fish culture streams and provides small fish and guidance to 175 cooperative nurseries operated by sportsmen's groups. In addition, the commission is responsible for almost all of Pennsylvania's fishing resources. About 25,000 of the state's 45,000 miles of streams provide significant fishing resources. About 15,000 miles of streams are managed partially or entirely as cold-water fisheries, and 10,000 miles are managed for warm/cool-water fisheries.

The Pennsylvania Fish Commission is also responsible for the enforcement of state boating regulations. One of the most prominent and important programs involves maintaining and operating its 253 boating and fishing access areas. Recreational and boating opportunities are available on almost all public flatwater.

The use of water facilities has grown rapidly in Pennsylvania. For example, more than 120,000 boaters use the scenic river area of the Delaware annually. But the growth has brought a number of problems. Overuse of streams and rivers can lead to physical and ecological damages, such as water pollution and shoreline erosion. Studies suggest limiting the number of boaters on a stream each day, redistributing use among different sections of the river, limiting the size of the river trips, and improving scheduling at facilities. Competition for the limited quality water resources continues to increase. There is a need for a state plan to provide adequate water recreation in the future.

Hydroelectric Development

In 1992, about 16 percent of Pennsylvania's hydroelectric potential had been developed by the state's utility companies. There are eight facilities in operation, of which five are in the Susquehanna Basin, one is on a tributary of the Delaware River, and two are in the Ohio Basin. These eight facilities have a capacity of 3,510,000 megawatts per hour (MWH), of which 3,060,000 MWH is located on the lower Susquehanna.

It is evident that the hydroelectric power potential has not been developed. In a recent survey by the U.S. Army Corps of Engineers, 263 additional potential sites in Pennsylvania were selected. The development of these sites would add approximately 4,367 MWH of generating capacity to Pennsylvania's electric grid. At 175 existing dams, 1,476 megawatts of electric generating capacity could be installed, and almost twice that amount—some 2,890 megawatts—could be developed at 80 other sites where dams would have to be built.

In general, this hydro potential is greatest on the major tributaries of the state's three major river systems: the Susquehanna, the Ohio, and the Delaware. In central Pennsylvania, most of the hydro potential is located on the Juniata River and on the West Branch and main body of the Susquehanna, although most of the potential exists at sites without dams. The southwestern counties have by far the greatest amount of hydro potential at existing dams. Most of these are located on the Monongahela and Allegheny rivers. Other concentrations of sites with potential occur on the Conemaugh River (a tributary of the Allegheny River) and along the Schuylkill River, which flows into the Delaware.

The amount of energy produced by a hydro plant depends on a number of factors. Of greatest significance are the amount of water available, the seasonal variations, and the river level. These variations can be great on small streams or where dams provide only temporary storage. In Pennsylvania, spring has the highest flow when there are heavy rains and snowmelt. In contrast, late fall and winter are the low periods of water availability. However, the extremes of flow are mostly not great, and hydro facilities could be expected to maintain a fairly even production throughout the year.

Water Transportation

Over time, the importance of waterways for transportation of people and goods has varied greatly in Pennsylvania. The Delaware River provided the route for William Penn to establish his colony at Philadelphia in 1682. For decades, the waterways of southeastern Pennsylvania provided passageways for settlers moving westward. In the middle of the nineteenth century, Pennsylvania was a leader in the development of canals, but the canal era lasted barely a quarter of a century. Of the rivers of Pennsylvania, only the Delaware and Ohio river systems have remained important water routes from the time of the earliest settlements to the present. The Susquehanna cuts across the Ridge-and-Valley region, which limits its potential as a water route because of its shallow depth and rock obstructions.

The Delaware River has been a major route for Pennsylvanians since the state was first settled. Although the river has a total length of 326 miles, it is navigable only from its mouth to Trenton, New Jersey, the head of tidal flow. The navigable portion of the Delaware does not require a system of locks and dams. The U.S. Army Corps of Engineers maintains the navigability of the river with an appropriation from Congress. The work entails removal of sediments and shoals, maintenance of the banks, environmental assessments, and biological studies. The depth of the river is maintained at about 40 feet.

The U.S. Army Corps of Engineers reports that there are 238 piers, wharves, and docks between Trenton, New Jersey, and the mouth of the Delaware River. The facilities vary from modern to antiquated terminals. In 1965, the Philadelphia Port Corporation was established to revitalize the port facilities and to manage competitive port activities. A basic goal was to increase the share of cargo handled by Philadelphia. The corporation now owns a terminal with six berths, an area of 104 acres, a 90,100-square-foot roll-on, roll-off vehicle transit shed, two 100,000-square-foot warehouses, and a 1,000,000-cubic-foot refrigerated warehouse. It also leases a number of piers. Most of the piers in Philadelphia are privately owned and are designed to handle a specific type of commodity. Comparable facilities are found in New Jersey.

The Delaware River ranks as one of the great rivers of commerce in the United States. The ports of the river rank first in maritime import tonnage, with about 62 to 67 million tons annually. Of the imports, crude oil provides the greatest tonnage to supply the refineries on its banks. Other important import commodities include meat, sugar, lumber, motor vehicles, foodstuffs, and machinery. The Delaware River ports have maintained their market share in the highly competitive North Atlantic trade. The New York metropolitan area has long had about 55 percent of the total market share, the Delaware River ports have had 18 to 20 percent of the total, Baltimore about 15 percent, and the Norfolk complex 11 to 12 percent.

The operations of the Delaware River ports have a significant effect on the region's economy. In financial reports from the Delaware River Port Authority, the activities of the ports create 100,000 jobs with an annual payroll of $825 million and generating $92 million in state and local taxes.

The navigable portion of the Ohio River system in Pennsylvania consists of the Ohio, Allegheny, and Monongahela rivers, which have been improved by a series of dams and locks. Before improvements began in the 1830s, traffic was largely limited to the Ohio River. Only vessels with little draft could navigate the Monongahela and Allegheny rivers. The development of the system by the U.S. Army Corps of Engineers for navigation consisted of construction of a series of locks and dams, the removal of snags, and channel dredgings. In addition, there have been many changes in the dams and locks, the design of river vessels, and terminal facilities.

Eight locks and dams on the Allegheny River provide slack water for 72 miles from Pittsburgh to East Brady. Traffic north of Kittanning, however, has diminished so greatly that the four dams and locks between Kittanning and East Brady are used very little. Nine locks and dams on the Monongahela provide slack-water navigation from Fairmont, West Virginia, for 128.7 miles to Pittsburgh, and the Ohio River has three locks and dams from Pittsburgh to the Ohio border. Lock capacities are an extremely critical factor for the traffic on the waterways. The lock chambers vary in size from 56 by 360 feet to 110 by 720 feet. The smallest locks are at the headwaters, the larger locks are downstream. Many types of facilities on the riverbanks handle different types of commodities. The city of Pittsburgh has constructed a modern wharf on the Monongahela to handle freight.

The Ohio River system of Pennsylvania provided the means for transporting people and commodities from the time of early settlement in the eighteenth century. Pittsburgh became the focal point for settlers moving westward on the Ohio River. With the development of mining and manufacturing by the middle of the nineteenth century, coal and manufactured products emerged as the major commodities of river traffic. Coal and coke had the greatest traffic, reaching a maximum in the 1940s with 12 to 16 million tons annually. With the decline of the coal and iron and steel industries in western Pennsylvania, these commodities have declined in the river trade. Sand and gravel have grown in importance and now constitute about 75 percent of the total tonnage. Movement of petroleum products has increased too, as well as movement of chemicals (such as sulfuric acid), cement, machinery, and agricultural products.

Water Pollution

With the exploitation of the state's resources and the development of an industrial society, the pollution of both surface water and groundwater has increased. Because streams contain natural sediments, and groundwater dissolves many types of minerals, it has been difficult to determine at what point water can be considered polluted. In March 1961, an international conference in Geneva, Switzerland, provided a definition of water pollution that is now widely accepted: "Water is considered polluted when its composition or state is directly or indirectly modified by human activity to an extent such that it is less suitable for purposes it could have served in its natural state."

The water resources of Pennsylvania are subject to pollution in varying degrees from a number of sources. The exploitation of the coal resources of the state has created a major water pollution problem. Many coal seams, primarily bituminous, have sulfur deposits. When these iron sulfide deposits come in contact with water, a weak sulfuric acid results. The acid drainage from both surface and underground mines has acidified several thousands of miles of Pennsylvania's streams. The acid mine drainage has affected the ecology of the streams as well as the use of the water for many industrial purposes. Although techniques to purify the streams are available, the processes are costly and the problem of acid mine drainage persists.

Exploitation of the petroleum resources in western Pennsylvania has presented a problem of pollution from both groundwater and surface water. In the production of a single barrel of oil, as much as twenty barrels or more of saltwater are produced. Disposal of the saltwater presents a major problem to oil producers. In addition, since oil was first discovered in 1859, thousands of wells have been depleted. For decades, when the casing and tubing were pulled out of the well, the saltwater from the oil horizon was free to move into the freshwater horizons near the surface polluting them. While all oil wells must now be plugged to prevent pollution of the clean water aquifers, the plugging of the well is not always satisfactory and pollution of clean water may occur.

While it has long been recognized that raw sewage presents a health problem, development of sewage treatment systems still lags in parts of the state. In addition, many of the older sewage treatment plants are inadequate and must be improved to meet established health standards. This process of modernization will take many years. Recently, impure water resulting from improper treatment of waste materials has been discovered in a few spots in the state. When this occurs, water for human consumption must be obtained from nearby communities or the water boiled before consumption.

In the past, many industrial plants disposed of hazardous waste materials by dumping their waste materials on the ground, burying the wastes in containers that did not provide permanent storage, or dumping the wastes in nearby streams. Many of these waste sites are now known to have materials containing carcinogens. In an initial survey by the U.S. Environmental Protection Agency in 1981, at least 528 disposal sites varying from a few months to decades old were located in the state. The distribution of these sites reflects population distribution, with the most prominent concentrations in the southeast around Philadelphia and Delaware counties, and a secondary concentration in the southwest in Allegheny County. Lesser concentrations are around Reading, Lancaster, York, Allentown, Bethlehem, Harrisburg, and Erie.

The pollution of water from these waste sites has received much public attention in recent years. People living near the affected places are vitally interested in conditions that may affect their living circumstances. In addition, authorities at various levels—from local to national—often become adversaries as the problems evolve and solutions are attempted. Legislation to control the disposal of hazardous waste materials is frequently prohibitive. As a result, the removal of hazardous sites has been amazingly slow. As the problems of removal linger on, there are important ramifications for land use within the area, for economic development, and for a range of political-social problems.

Another type of water pollution comes from normal household functions. Household refuse is disposed of either by burning or by controlled disposal in landfills. The latter requires a detailed knowledge of the geological nature of the subsoil and the proximity of groundwater. The potential for hazardous wastes entering the water system from either disposal method is high. The use of synthetic detergents has become a common practice in households in Pennsylvania. These commercial detergent powders contain a surface-active agent that simulates soap but gives an unpleasant taste to water when it exceeds 1 pound per million gallons of water. Further large quantities of detergents in rivers will kill fish and often produce sufficient foam to prevent navigation. Because detergents are resistant to bacterial decomposition, biological treatment plants cannot purify water containing detergents. New treatment processes are needed.

Disposal of such waste materials as lubricating oil from a motor vehicle by simply dumping them on the ground is a common practice. While a single disposal of such waste is of minor importance, the potential of 12 million people introducing waste products into the water system becomes a major concern. Many of these practices can be altered with little effort on the part of individuals. To illustrate, old motor oil can be saved, cleaned, and used again. This simple procedure not only prevents pollution of the water system but also is a way to conserve a major resource.

Highly dangerous chemicals used in agriculture as weed killers and insecticides can also pollute the water system. As much as 600 pounds of insecticides are sprayed each year on each acre of agricultural land, and consequently streams contain about 2.5 pounds of insecticide per million gallons of water. In addition, the effects of these insecticides on humans and fish have largely been ignored. Mineral fertilizers rich in phosphorous and nitrogen (phosphates and nitrates) are also carried into the water network and infiltrate into the groundwater. These pollutants have a number of effects on animals and humans—for example, water containing about 1 pound of nitrate in 600 gallons of water produces "blue disease" (methaemoglobmaemia) in babies, and excessive nitrates that come in contact with decaying organic matter will kill aquatic fauna.

Domestic animals are also a source of water pollution. The washing of barns and other buildings, and water flowing from manure piles, produces a waste that enters the groundwater system as well as surface streams. This animal waste pollutes the streams, reducing their value for other uses.

In order to guarantee adequate clean water in Pennsylvania, certain steps must be taken. First, surveys of water resources must be continued and techniques for discovering new sources of water improved. Development of the field of hydrogeology is fundamental to the success of these programs. Second, there must be a greater effort to solve the problems of water pollution that have evolved during the twentieth century. Most critical in this undertaking will be continued development of more effective sewage treatment plants and the removal of the hazardous waste sites. Finally, a public education program is needed to make the average citizen aware that water problems are on the horizon if proper care is not taken to solve existing pollution. A major aspect of this program must be the development of conservation. In the past, water has been consumed lavishly. If this continues, it will become even more difficult to secure an adequate water supply.

There is no water shortage in Pennsylvania at the present time, but if clean water in adequate quantities is to be available in the future, past practices of polluting the streams and groundwater must be changed. There is time to solve the present problems of local shortages and contamination before they become acute. The more fundamental question is whether the present political and societal structure will be adequate to avert a future catastrophe.

Water Control and Management

The general public now understands that the waters in the state must be controlled and managed. This process can take a number of forms, including flood control and quality control. It has also been recognized that river basins located in more than one state require interstate cooperation, such as the Delaware River Basin Commission, a unique institutional framework for collaboration of the federal government and the states of Delaware, New Jersey, New York, and Pennsylvania.

Flood Control

Because many of Pennsylvania's streams and rivers are situated in V-shaped valleys, flooding has traditionally been a common phenomenon. There have even been spectacular floods—such as the Johnstown flood of 1889 when a dam broke, and the flood following Hurricane Agnes in 1972. More important in the long-run, however, were the periodic floods occurring after heavy precipitation. In the period from 1955 through 1975, Pennsylvania ranked third in monetary damage from floods, with California ranking first and Connecticut virtually tied with Pennsylvania in second place. Direct monetary losses were in the tens of millions of dollars, and businesses and industries endured lengthy recovery periods following each flood.

In order to control floods, more than 30 flood-control dams have been built since 1941 (Fig. 4.2).

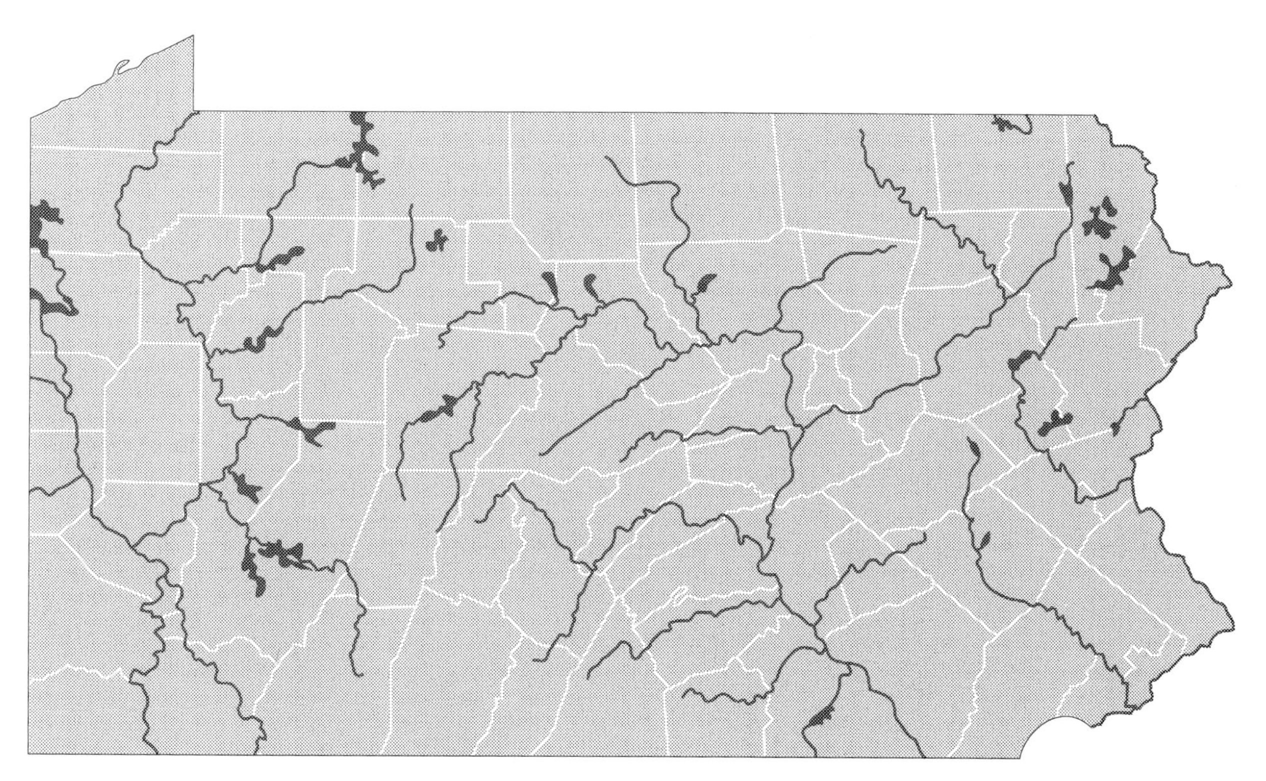

Fig. 4.2. Pennsylvania's major flood control dams.

The greatest efforts were in the Ohio Basin on the Allegheny and Monongahela river tributaries because of the great industrial development on these major rivers. Floods devastated the industries located on the floodplains of this region. The second major effort was in the upper Susquehanna River tributaries, to control floods in the heavily populated Scranton and Wilkes-Barre areas.

A number of other flood-control projects have been completed, consisting of more than 13 levees and flood-control walls and including a 26,000-foot cement wall levee at Sunbury, a 25,200-foot earth levee at Wilkes-Barre, and a 71,000-foot earth levee at Williamsport. This work continues, with the most recent new levee, at Lock Haven, started in 1992. In addition, the U.S. Army Corps of Engineers has undertaken channel improvements to improve the flow of water on more than 30 streams.

The flood-control development on Pennsylvania's rivers and streams has reduced flood damage downstream. For example, no major floods have occurred at Pittsburgh since 1936. When there is a massive amount of precipitation, such as occurred from Hurricane Agnes in 1972, there is strong evidence that no system of flood controls can prevent disastrous flooding. Heavy precipitation in a short period of time will cause an overflow in any reasonable levee system and in a short period of time fill the flood-control dams. Consequently, use of floodplains must be controlled in order to prevent economic disaster.

Quality Control

Similar to water allocation and use, legal principles governing the quality of Pennsylvania's waters are derived from common law. Several common-law doctrines form the framework for protection of waters from pollution, including nuisance, trespass, and strict liability for ultrahazardous activities. These doctrines were developed over the centuries in the United Kingdom and the United States and still play a key role in the protection of Pennsylvania's waters. In the twentieth century, the traditional doctrines have been codified as statutory and regulatory controls of pollution have developed.

The first water pollution legislation in Pennsylvania was the 1905 Purity of Waters Act, which prohibited disposal of sewage in "waters of the state." Specifically excluded from the act was sewage discharge from existing sewer systems, waste from coal mines, and waste from tanneries. In 1923, the General Assembly enacted additional legislation empowering the State Department of Health to issue orders and regulations to protect water supplies from pollution and other contamination.

The modern era of water pollution control began in Pennsylvania with the Clean Streams Law in 1937. Although it has been amended several times, the basic parameters of the law have remained the same. The Clean Streams Law, which established that discharges of sewage and industrial waste into Pennsylvania's waters were not natural uses of the waters, and which set a water standard below which courts applying common law could not dip, also declared and clarified once and for all that pollution was a public nuisance.

As a regulatory mechanism, the Clean Streams Law established a permit system for discharging treated sewage and industrial wastes into state waters. It also required that polluted water be treated before it was discharged into clean water. Sewage and waste could be discharged into waters that were not clean, or into waters that were contaminated with "acid coal mine drainage." The law explicitly exempted from its regulations acid mine drainage and silt from coal mines until "practical means for the removal of the polluting properties of such drainage shall become known." To carry out the terms of the law, the Sanitary Water Board was established and authorized to adopt rules and regulations "as may be deemed necessary for the protection of the purity of the waters and the Commonwealth, or part thereof, and to purify those now polluted, and to assure the proper and practical operation and maintenance of treatment works approved by it." All plans for construction of new sewer systems and treatment works are required to be submitted to and approved by the Sanitary Water Board.

In 1945, the Clean Streams Law was amended, increasing Pennsylvania's regulation over pollution of all waters in the state. The amendment included additional regulation of discharges from coal mines, specifically prohibiting discharge of acid mine drainage into the "clean waters" of the state. All coal mines are now required to submit plans controlling not only acid waters but also silt and other solids.

Finally, in 1965, the General Assembly amended the Clean Streams Law to fully regulate all discharges into state waters. A "finding" of the 1965 amendment recognized that the earlier versions of the law "failed to prevent an increase in the miles of polluted water in Pennsylvania" and the amendment specified that "mine drainage is the major cause of stream pollution in Pennsylvania and is doing immense damage to the waters of the Commonwealth." This established the policy to "reclaim and restore to a clean, unpolluted condition every stream in Pennsylvania that is presently polluted." Extensive requirements for obtaining permits for coal mines were added, and no permit could be issued if a discharge would "become inimical or injurious" to the waters of the state. All penalty provisions were updated in 1980, and the State Department of Environmental Resources was given clear authority to issue orders, assess civil penalties, and proceed with summary changes. In the same year, a citizen-suit provision was added, allowing any person who is adversely affected by polluted water to bring an action against the Department of Environmental Resources for failure to take action under the Clean Streams Law.

An important adjunct to the Clean Streams Law is the Pennsylvania Sewage Facilities Act. By requiring municipalities to adopt and periodically update an "Official Sewage Facilities Plan," the act provides a comprehensive basis for planning and siting both community and individual systems for management of sewage wastes.

Until 1972, federal law did not play a significant role in protection of water quality. From 1948 until its revisions in 1972, the federal Water Pollution Control Act left primary responsibility for pollution control to the states. The 1972 act focuses largely on technology-based effluent standards and "control at the source." Central to the act is the National Pollution Discharge Elimination System, which establishes discharge limitations and compliance deadlines. As modified in 1977, the Water Pollution Control Act requires the Environmental Protection Agency to set "best conventional pollution control technology limits." The federal law uses many legal tools to spur improvement of the nation's water, including planning and financial assistance, the building of treatment works, and special programs for clean lakes.

The development of Pennsylvania's water laws has been based on the challenges and problems of a changing economy from the colonial period to the present postindustrial age. Although the state has some of the most sophisticated laws and regulatory programs addressing water supply and quality, the legal foundation remains hindered by separate and too often uncoordinated doctrines and rules. Progress in the evolution of a "State Water Plan" and a "Comprehensive Water Quality Management Plan" has not been coordinated with complimentary laws capable of addressing water resources in a holistic fashion. Water quantity and quality must be husbanded and managed on the basis of the fundamental concept of the total water system.

Delaware River Basin Commission

The members of the Delaware River Basin Commission (DRBC), created in 1961, are the governors from the Basin states and the U.S. Secretary of the Interior, who are supported by a technical and support staff of about fifty people who are under the supervision of an executive director.

The DRBC has focused on several major management issues in the Delaware River Basin. Concerning water quality, the DRBC compact states: "The Commission may assume jurisdiction to control future pollution and abate existing pollution in the waters of the Basin, . . . and the Commission, after a public hearing may classify the water of the Basin and establish standards of treatment of sewage, industrial or other waste, . . . including protection of both surface and ground waters." Concerning water resources, the DRBC has focused on planning to ensure an adequate water supply. The compact directs the DRBC, "in accordance with the doctrine of equitable apportionment," to "allocate the water of the Basin to and among the states." The commission has the power to acquire, operate, and control projects and facilities for storing and releasing water, for regulating flows and supplies of surface water and groundwater of the Basin, and for protecting public health. It also is responsible for stream quality control, economic development, improvement of fisheries, recreation, dilution and abatement of pollution, prevention of undue salinity, and for assessing the environmental impacts of any projects.

During the 1960s, the attention of the DRBC was directed specifically to the water quality of the Basin, which had progressively worsened. The U.S. Public Health Service initiated the Delaware Estuary Comprehensive Study as a means for controlling water pollution in the Basin. This study developed one of the first comprehensive estuary water-quality models and led the DRBC to adopt new and higher water-quality standards for the estuary in 1967.

When the Environmental Protection Agency (EPA) was created in the 1970s, it displaced, to a significant extent, the DRBC's role as the environmental regulator of the Basin. As a result, the DRBC now works in conjunction with the states and the EPA on environmental issues. The Basin has responded to the environmental programs, and water quality has improved greatly.

Bibliography

ASCS. 1983. *Conestoga Headwaters: Rural Clean Water, Progress Report.* Harrisburg, Pa.: U.S. Department of Agriculture.

Becker, A. E. 1978. *Ground Water in Pennsylvania.* Educational Series No. 3. Harrisburg: Pennsylvania Topographic and Geologic Survey.

Brezina, E. R. 1988. "Water Quality Issues in the Delaware River Basin." In *Ecology and Restoration of the Delaware River Basin,* ed. S. K. Majumdar, E. W. Miller, and L. E. Sage. Easton: Pennsylvania Academy of Science.

Department of Environmental Resources. 1991. *1990 Pennsylvania Water Quality Assessment.* Harrisburg, Pa.: Division of Water Quality, Bureau of Water Quality Management.

Frey, R. F. 1990. "Overview of Surface and Ground-Water Quality in Pennsylvania." In *Water Resources in Pennsylvania,* ed. S. K. Majumdar, E. W. Miller, and R. R. Parizek. Easton: Pennsylvania Academy of Science.

Graefe, A. R. 1990. "Water-Based Recreation in Pennsylvania." In *Water Resources in Pennsylvania,* ed. S. K. Majumdar, E. W. Miller, and R. R. Parizek. Easton: Pennsylvania Academy of Science.

Helwig, O. J. 1985. *Water Resources: Planning and Management.* New York: Wiley.

Lynch, J. A., E. S. Corbett, and D. W. Aurand. 1975. "Effects of Management Practices in Water Quality and Quantity." In *Proc. Minimal Watershed Management Symposium.* USDA Forest Service Gen. Tech. Rep. NE-13.

Lynch, J. A., E. S. Corbett, and R. J. Hutnik. 1975. "Effects of Clear-Cutting Water Resources in Pennsylvania." In *Clearcutting in Pennsylvania.* University Park: The Pennsylvania State University.

Majumdar, S. K., E. W. Miller, and R. R. Parizek, eds. 1990. *Water Resources in Pennsylvania: Availability, Quality, and Management.* Easton: Pennsylvania Academy of Science.

Majumdar, S. K., E. W. Miller, and L. E. Sage, eds. 1988. *Ecology and Restoration of the Delaware River Basin.* Easton: Pennsylvania Academy of Science.

Miller, E. W. 1990a. "Water Quantity, Quality, and the Future." In *Water Resources in Pennsylvania,* ed. S. K. Majumdar, E. W. Miller, and R. R. Parizek. Easton: Pennsylvania Academy of Science.

———. 1990b. "Water Transportation in Pennsylvania." In *Water Resources in Pennsylvania,* ed. S. K. Majumdar, E. W. Miller, and R. R. Parizek. Easton: Pennsylvania Academy of Science.

Powledge, F. 1982. *Water: The Nature, Uses, and Future of Our Most Precious and Abused Resource.* New York: Straus Giroux.

Shaw, L. C. 1984. *Pennsylvania Gazetteer of Streams.* Part 2. Harrisburg, Pa.: Department of Environmental Resources, Water Resources Bulletin No. 16.

Weston, R. T., and J. R. Burcat. 1990. "Legal Aspects of Pennsylvania Water Management." In *Water Resources in Pennsylvania,* ed. S. K. Majumdar, E. W. Miller, and R. R. Parizek. Easton: Pennsylvania Academy of Science.

5

SOIL RESOURCES

E. Willard Miller

Although large areas of the state have a similar climate and similar vegetation, there is great variability of soils. The soils of Pennsylvania vary from the highly productive soils of the Lancaster lowlands to the soils on the Appalachian Plateau that are too poor to justify cultivation.

The Nature of Soils

Soils are the basis of food production, necessary for human survival. The importance of soil has long been recognized by soil scientists. As early as 1891, N. S. Shaler of the U.S. Geological Survey wrote: "This slight and superficial and inconsistent covering of the earth should receive a measure of care which is rarely devoted to it." And fifty years ago, G. N. Coffey of the U.S. Department of Agriculture stated: "It [soil] is one of the great formations in which the organic and inorganic kingdom meet and derives its distinctive character from this union."

Soils are natural, dynamic, three-dimensional bodies on the earth's outer crust and are composed of minerals, organic matter, and living organisms. They thus have chemical, physical, and mineralogical properties resulting from the interaction of climate and organisms acting on parent material conditioned by relief over periods of time. Soils are usually the major consideration in the quality of land resources.

Factors of Soil Formation

The five major environmental factors in soil formation are climate (*c*), vegetation (*v*), organisms (*o*), relief (*r*), and parent material (*p*). Soil (*s*) is the dependent variable relying on the interaction of five partially independent variables with specified time

(t) limits. The development of a soil at a place can be expressed as:

$$s = f(c,v,o,p,r)t$$

Climate

In early studies of soil formation, climate was considered the dominant soil-forming factor. Climate is still recognized as important, but other factors are now known to be significant too. All aspects of climate play a role in the *pedogenic,* or soil-forming, processes. Climate influences the development of soils in Pennsylvania in at least three ways. It is involved in the weathering of parent material, in internal soil processes, and in the transport of parent and weathered materials, causing erosion of the soil bodies.

Within the soil body, climate has a marked influence on pedogenic processes. Climate provides water for the soil, for the leaching of soil particles and removal of the dissolved chemical compounds. Because precipitation is fairly uniformly distributed in Pennsylvania throughout the year, the leaching process stops only when the soil freezes in winter. Soils are moist during the entire year, and during heavy precipitation periods soils may be saturated. In the north-central section of the state, the short warm season reduces the effectiveness of the pedogenic processes. In many areas, raw humus collects to a depth of several inches on the surface.

Relief

The topography of an area influences soil formation most strongly by controlling surface runoff and erosion, by determining the quantity of precipitation retained in the soil, and by directing the movement of soluble materials from one area to another. In addition, the effectiveness of solar radiation depends on the direction and degree of the slope of the land. By altering soil temperatures and water availability, topography influences the effectiveness of weathering.

The availability of water in the soil is greatly influenced by the relief of the area. Low-lying soils in relatively flat areas usually retain a larger proportion of water than soils on adjacent slope areas. On the Appalachian Plateau, some soils are designated as upland or lowland to reflect their topographic position. If the parent material is uniform on a slope, a gradual increase in water may develop downslope, with the possibility of water-bogged soil at the lower level.

Vegetation and Organisms

The soils of Pennsylvania developed under a forest covering. The organic material collects on the surface of the soil, and the decaying action of the bacteria destroys most of the organic matter before it can become part of the soil. Because the decaying process is relatively slow, particularly in the colder area of northern Pennsylvania, the raw humus is not decomposed and collects on the surface.

In the farming section of the state, when the forest cover has been removed, organic matter must be added to the soil by growing grass and other plants. Because organic matter in soils is soluble and is continuously being lost, it must be replaced. It is estimated that in Pennsylvania organic matter decomposes in the soil at a rate of about 1 to 4 percent annually. Consequently, most soils are light in color.

The soil's texture is an important factor in the maintenance of organic matter. The finer-textured soils absorb the decomposing organic matter more effectively than coarse soils. Further, organic molecules absorbed on clays are partially protected from decomposition by microorganisms, adding to the organic content of the soil.

Parent Material

The original material from which soil is derived may be local bedrock or transported material (Fig. 5.1). A number of processes are required in the development of a soil. The first process, weathering, causes the rock material to disintegrate and decompose. The two major types of weathering in soil development are physical weathering, which produces smaller particles by mechanical action from the parent material, and chemical weathering, which changes the chemical composition of the parent material. Physical and chemical weathering occur simultaneously.

The origin of parent material may be classified into two major groups. The residual parent material consists of sedimentary, metamorphic, and/or igneous bedrock. The transported material is parent material that has been moved from one location to

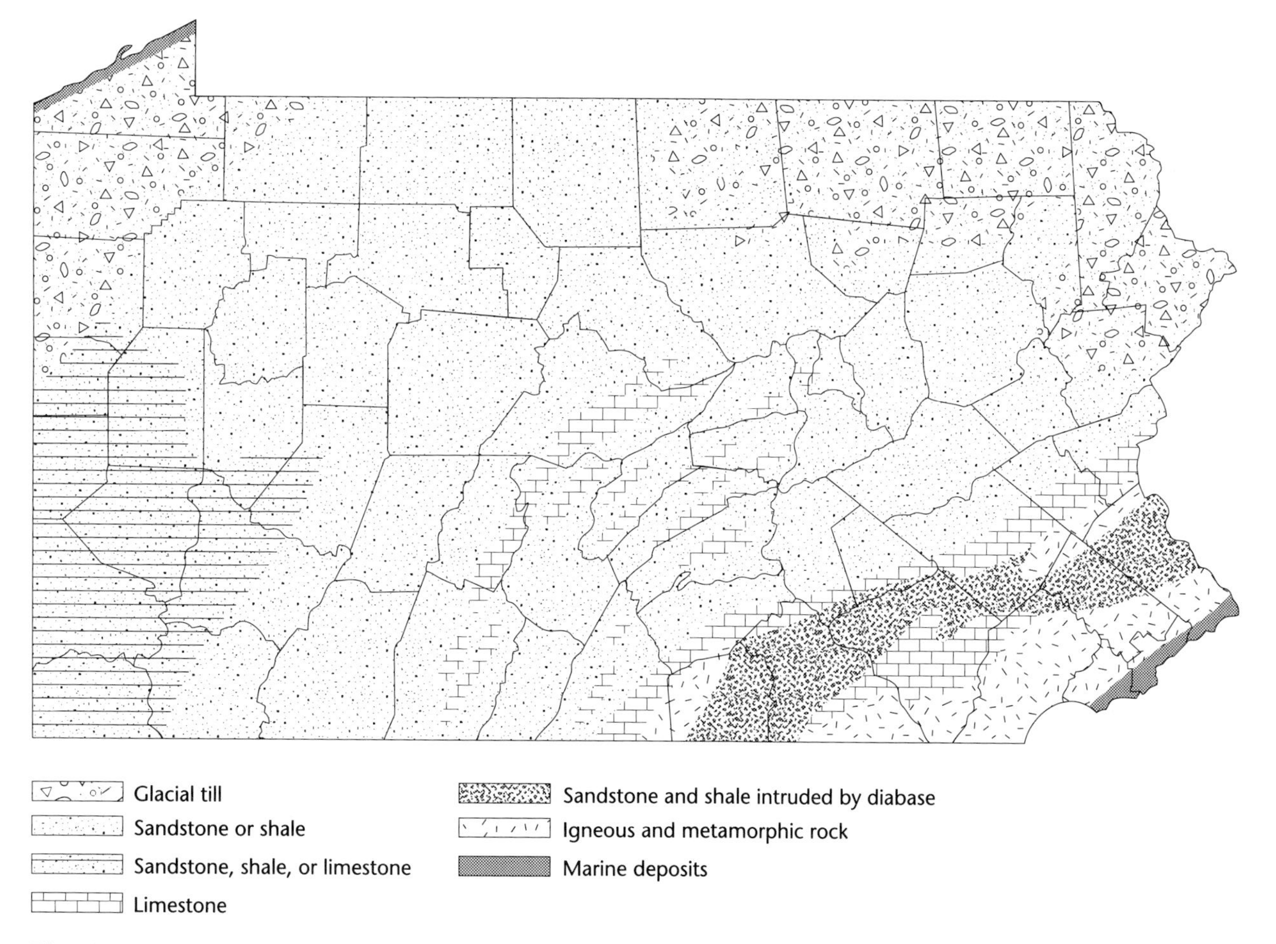

Fig. 5.1. Parent material of Pennsylvania's soils.

another. In Pennsylvania the sedimentary rocks—sandstone, shale, and limestone—are the parent rocks for most of the soils. Sandstone weathers to coarse-textured (sandy) soils, and shale and limestone weather to fine textures (clay).

Transported parent material differs from residual materials both in origin and in form. Transported material usually occurs in an unconsolidated state. In Pennsylvania the glacial deposits of northeastern and northwestern Pennsylvania provide the greatest areas of transported material. Transported material also occurs on floodplains of rivers and streams and on lake sediments, particularly on the Lake Erie plain.

Soil Horizons

As soils evolve from parent rock materials to mature soils, distinct layers known as soil horizons develop. Every mature soil has three distinct soil horizons. The horizons differ from one another in one or more properties, such as color, texture, structure, and porosity. They may vary in thickness from a few inches to several feet. Generally, the horizons merge into one another.

The A-horizon, the uppermost layer of soil, is usually called the surface soil. This is the layer where organic material accumulates. The organisms, when they die, form the humus material of the soil. When the humus content of a soil is high, the color is dark; when the humus content is low, the soil is light-colored. More leaching occurs in the A-horizon than at deeper horizons. Thus, the A-horizon is known as one of eluviation, or downward movement. As a consequence, the A-horizon has lost some of its soluable minerals to the B-horizon.

The B-horizon lies directly below the A-horizon and is frequently called the subsoil. Together the A- and B-horizons make up the "solum." The dominant feature of the B-horizon is the concentration of the oxides, silicates, clays, and humus materials leached out of the A-horizon. This is known as the zone of illuviation (or deposition). The B-horizon normally has more clay than the A-horizon because this substance is also removed from the overlying horizon.

The C-horizon provides the parent material. It has been little affected by the pedogenic processes.

Time Relationships

Time plays a major role in the soil-forming process. The effect of time is judged by the extent to which the parent material has been changed into soil. This process of change has several stages. The initial stage is characterized by accumulation of organic matter in the upper layer of the soil and by limited removal of the soluble materials by the weathering processes. At this stage, only the upper layer (A-horizon) and the bedrock (C-horizon) are developed. The soil is mature when a layer (the B-horizon) develops above the bedrock. When the three layers become highly differentiated, the final stage has been reached. This is known as a climax soil.

In Pennsylvania, all three stages of development take place. The glaciated region and alluvial floodplains tend to have minimal horizonal evolution, which is indicative of a short development time. Most of the soils on the Appalachian Plateau are also immature, with poorly developed horizons. On the steep slopes of the hill lands, erosion is rapid and the weathered material is continually removed, and in places fresh unweathered bedrock is exposed. Erosion, along with the addition of coarse fragments to the soil from the bedrock, keeps the majority of Pennsylvania soils from becoming mature soils. Only on the Piedmont of southeastern Pennsylvania and in the Lancaster and ridge-and-valley limestone lowlands have mature soils developed.

Soil Classification Systems

The soils of different areas have distinctive characteristics. Therefore, soil classification systems must consider the relationships between soils. A synthesis of knowledge about the different soils is essential, as is an understanding of how soils respond to human cultivation. Through the centuries, general classification systems have been developed for a wide variety of soil uses.

U.S. Comprehensive Soil Classification System

By the twentieth century, more scientific information was becoming available, and a number of classification systems have been developed. In 1960, the U.S. Department of Agriculture introduced a new system: The U.S. Comprehensive Soil Classification System. This system, now widely accepted, was based on the current properties of the soil rather than on the soil's genesis, environment, or virgin conditions.

Certain assumptions guided the development of this system. Soils were classified according to their measurable properties. The properties selected were those affecting or resulting from soil genesis. If more than one property met this requirement, the property with the greatest significance for plant growth was selected. The defining of soil classes was to accommodate all soils. Finally, the classification was to be flexible, so that modifications could be made when new scientific information evolved.

An entirely new terminology in which Greek and Roman word roots are combined to form precise descriptions of soil properties was devised. The Comprehensive Soil Classification System consists of ten "orders" of soils (of which only four are found in Pennsylvania). Each order is divided into "suborders" reflecting the influence of climate and drainage conditions. Suborders are divided into "great groups" on the basis of horizon development. There are further subdivisions into "subgroups," "families," and finally into "soil series," each of which is named after a place where that type of soil exists on the earth's surface. An example of how a soil is classified is given in Table 5.1 using the Doylestown series. A soil is classified beginning with the "order," so the Doylestown series is ultimately described as Typic-Fragiaqualf, fine-silty, mixed-mesic.

Pennsylvania Soils

There are three soil orders in Pennsylvania—Alfisols, ultisols, and inceptisols (Fig. 5.2)—but these can be subdivided into 300 soil series, of which 93 predominate.

Table 5.1. U.S. Comprehensive Soil Classification System, Alfisol Order to Doylestown Series

Category	Category Name and Meaning
Order	Alfisol (*sol* is the root syllable of the order name and means "soil"). High basic minerals and clay accumulation.
Suborder	Aqu (Aqualf). Characteristics associated with wetness.
Great Group	Fragi (Fragiaqualf). Presence of fragipan.
Subgroup	Typic (Typic-Fragiaqualf). Within normal range of Fragiaqualf.
Family	Fine-silty, mixed mesic (Typic-Fragiaqualf, Fine-silty, Mixed Mesic). Medium-textured with mixed mineralogy in mesic soil temperature belt.
Series	Doylestown (member of the Typic-Fragiaqualf, Fine-silty, Mixed Mesic family). Specific arrangement of characteristics.

Source: Cunningham et al. 1977.

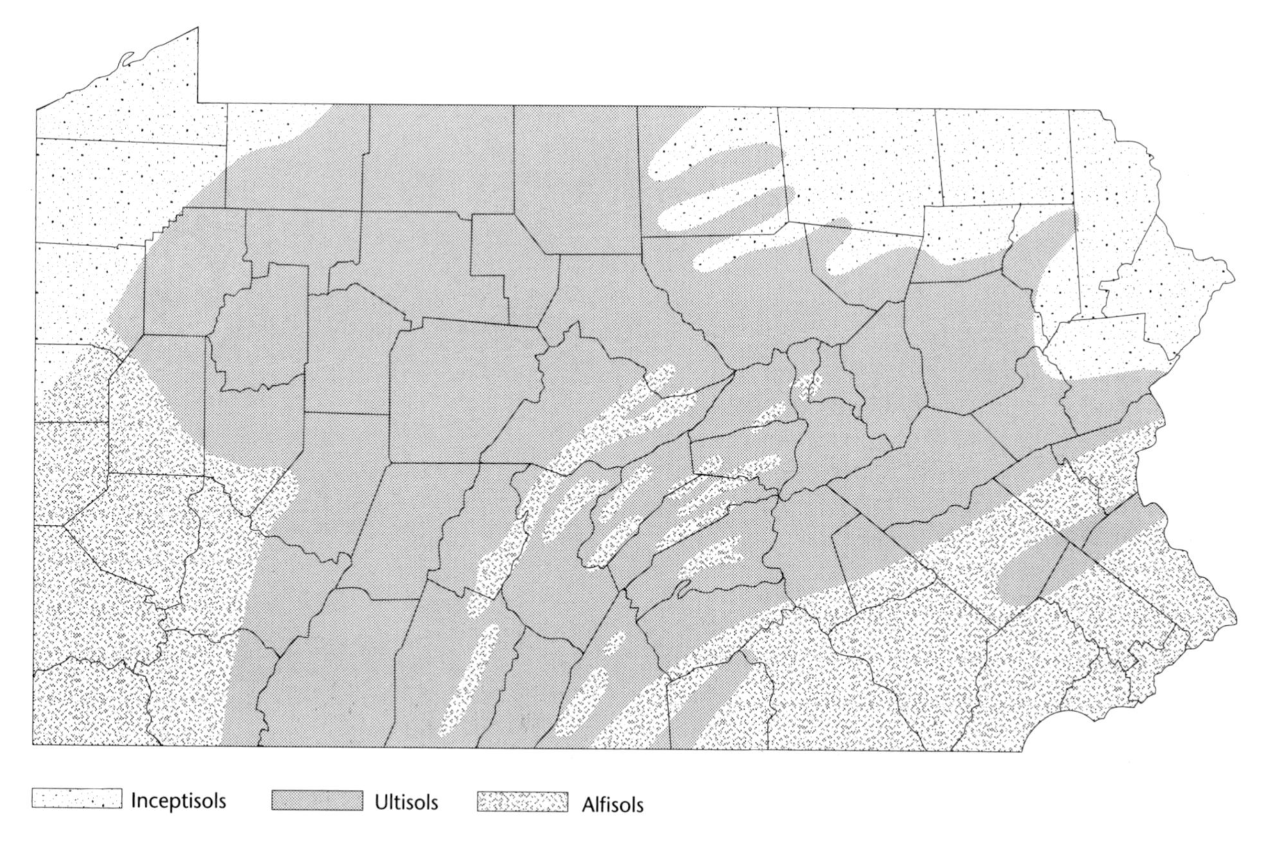

Fig. 5.2. U.S. Comprehensive Soil Classification.

Alfisols

The Alfisols are widely distributed in Pennsylvania. They are found on limestone and dolomite areas of the Appalachian Plateau, the Ridge-and-Valley region, and the Piedmont of the southeast. Some of the weathered glacial areas possess an Alfisol too. The Alfisol soils extend over a wide range of temperature and moisture conditions that vary from humid to seasonally dry. They have a yellowish-to-brown A-horizon, indicating that considerable leaching has occurred. The soil horizons have developed over a long period of time because these soils have originated on stable landscapes in Pennsylvania. Most of these soils have a depth of 20 to 40 inches to bedrock with about one-third having a fragipan of 16 to 40 inches. Most have a good to moderately good drainage. Although the Alfisol soils have experienced some leaching, moderate-to-high reserves of basic materials are still preserved. They maintain a fine platelike-to-granular structure with a surface texture of loam to silt loam. These are the most productive soils in the state.

Ultisols

Ultisols are found throughout western, northern, and eastern Pennsylvania in areas without significant amounts of limestone. In central Pennsylvania they are associated with sedimentary sandstone and shale and metamorphic rocks. They are found in the southeast on the sedimentary and igneous rocks of the Piedmont, the Reading Prong, and South Mountain. The structure of the ultisols is quite similar to that of the Alfisols, but they are more thoroughly leached of their soluble basic materials. These are the most weathered of all mid-latitude soils. Ultisols commonly have developed over long periods of time but may occur on young surfaces if the parent material weathers rapidly.

The A-horizon is usually gray or red, depending on the amount of aluminum or iron oxides in the weathered material. This horizon is normally highly acidic and poor in humus matter. The small quantity of humus materials present in the A-horizon has usually been deposited there by tree roots. Once the forest cover is removed, the meager store of nutrients is soon consumed and crop yields decline rapidly. These soils are found in Pennsylvania where there has been a significant abandonment of farms. Only with intensive fertilization and soil conservation can permanent agriculture be accomplished.

Inceptisols

Inceptisols are primarily located on the glacial deposits of northeastern and northwestern Pennsylvania. These are young soils with one or more poorly developed horizons. Most cannot be classified as mature soils. Although some of the soluble bases have been removed by weathering from the A-horizon, these soils do not possess a well-defined illuvial horizon. They are usually moist, with a light surface horizon. In Pennsylvania they developed under forest covering and are poor in humus matter. These soils can be made productive with fertilization and conservational practices.

Entisols

Entisols are youthful soils that have not had sufficient time to develop genetic horizons. The parent materials of these soils are normally recent deposits of alluvium in river floodplains. Entisols may be of any color, depending on the origin of the parent material. The quantity of organic matter varies from very high to low. To date, entisols have not been mapped in Pennsylvania according to the new classification.

Soil Erosion

Soil erosion in Pennsylvania is almost entirely a response to running water. The greatest amount of erosion occurs on nonagricultural rural land, including mines and quarries, farmland where there is no cropland or pasture, and sparse woodland. Within this area, however, there are great contrasts of erosion. For example, in the woodland areas the erosion rates are modest, providing only about 15 percent of the state's total soil erosion. In contrast, on strip-mined land the erosion rates are very high. This class of land provides about 45 percent of the annual eroded soil tonnage. The strip-mined areas in the bituminous and anthracite fields provide about two-thirds of the state's total soil removal.

On the agriculturally productive land, the erosion rate is highest on the cropland, which accounts for about 40 percent of the state's total soil erosion. The erosion on the cropland varies considerably in the state. The amount of erosion is greatest in the Appalachian region, the Piedmont, and the Ridge and Valley region. In the Appalachians, about 67 percent of the cropland is affected by excessive erosion; on the Piedmont, about 55 percent; and in the Ridge-and-Valley section the proportion is around 45 percent. In the state, the average rate of erosion on cropland varies from about 2 tons of soil an acre on the Erie Lake Plain to 9.8 tons an acre in the fruit belt of the South Mountain area of Adams and Franklin counties.

Bibliography

Cuff, D. J. 1989. "Soils." In *The Atlas of Pennsylvania,* ed. D. J. Cuff et al. Philadelphia: Temple University Press.

Cunningham, R. L., et al. 1977. *Soils of Pennsylvania.* Progress Report 365. University Park: The Pennsylvania State University, College of Agriculture.

Simonson, R. W. 1962. "Soil Classification in the United States." *Science* 137:10727–11034.

U.S. Soil Conservation Service. 1960. *Soil Classification: A Comprehensive System.* Washington, D.C.: Soil Survey Staff, 7th Approximation.

6

FOREST RESOURCES

Richard D. Schein and E. Willard Miller

A society's activities are to a large extent influenced by vegetation, as well as by land, soil, water, and climate. A great deal of a society's prosperity will depend on its domestic resources, and its foreign activities may be largely determined by its efforts to obtain that which it does not have at home. Our Pennsylvanian sense of place will be enlarged if the basic vegetation at the time the first Europeans arrived, and how the vegetation affected Pennsylvania's development and affects our present lives, are understood.

The character of vegetation depends on landforms, soils, water, and climate. The associated fauna is determined by the vegetation as well as by the basic physical resources. A biologist might view human activities as a part of animal activities—their nature and character partially but strongly affected by vegetation.

What Is Vegetation?

"Vegetation" is the word used for plants *collectively,* or as a whole. Specialists divide vegetation into communities—that is, into recognizable associations of plants. This kind of stable association of plants depends on a certain environment or habitat. Ecologists have divided the living surface of Earth into major communities, called *biomes.* A biome is a complex interassociation of all plants and animals in a region. Some biomes have names based on region's plants because the plants give the region its visible character. Chapter 2 describes Pennsylvania as made of different rocks of different ages, variously folded, tilted, and eroded. That is a correct geological view of a land mass. But as one travels across the state, only rarely can one see a bit of that Pennsylvania exposed—perhaps high on a ridge, or in a road cut.

Indeed, a good bit of the beauty of Pennsylvania is in its plants. How beautiful would landforms be judged if they were denuded of vegetation, as they are in many areas of the anthracite region of northeastern Pennsylvania or in the strip-mined areas of the Appalachian Plateau? The most visible kinds of plants in Pennsylvania are trees. Pennsylvania is part of the *deciduous forest biome,* a fact so overwhelmingly obvious 300 years ago that it gave the state its name. William Penn combined his name with the Latin word *silva* to create the name "Pennsylvania."

Some other biomes of the world are the deserts, the taiga (northern coniferous forest), the tundra, grasslands, and tropical rain forests. The deserts and the taiga are the largest areas. Tropical rain forest, or jungle, is not as common as common knowledge makes it. The deciduous forest is somewhere inbetween in size but does not make up a very large part of the earth's surface.

Within a biome are smaller communities, such as the oak-hickory forest of the Ridge-and-Valley region of central Pennsylvania. Each species within the association has its own environmental requirements—wide or narrow. As the environment changes—as temperature, for instance, decreases with elevation—the relative frequency of a species' occurrence may change, or it may disappear. Pennsylvania is large enough that some environmentally related shifts occur—that is, the character of the vegetation changes somewhat as one moves far enough east or west, north or south. These shifts in the composition of Pennsylvania's forests can be seen as the oak-hickory-yellow-poplar forest of the south changes to a white-pine-hemlock-hardwood forest, and the latter, as we move north, becomes a beech-birch-maple forest. But Pennsylvania is too small—in the sense of not having within its borders great variations in climate—to encompass the whole area of these important plant communities. The northern hardwood forest extends into New York and beyond, and Pennsylvania's oaks and hickories are an extension of the vegetation of the Appalachian ridges to the south (Fig. 6.1).

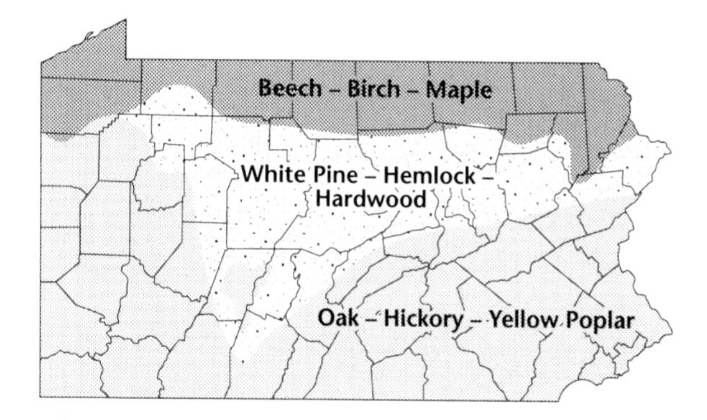

Fig. 6.1. Major forest regions of Pennsylvania.

The Woods of Penn's Woods

Is the vegetation in Pennsylvania unique to the state? Not when compared with the rest of the eastern United States and western Europe. The state's vegetation is very much like that of neighboring states with similar climates. Similar types of forests are found in western Europe, where the climate is similar to Pennsylvania's.

The deciduous forest, which has trees that lose their leaves as the days shorten and cool, is a major forest region in Pennsylvania. This forest is adapted to moderate, seasonally changing temperatures. The other major requirement is moisture. For optimal growth and development, these trees generally require 20 or more inches of rain well-distributed throughout the year. Soils are much less important. Many nondeciduous trees, particularly conifers or cone-bearers (the needle-leaf trees, such as pine) have similar requirements, so they are found in much the same area, often mixed in with the hardwood forest. These areas comprise the *Eastern Deciduous Forest.*

Proceeding northward, one passes from an oak-hickory-poplar forest through a transitional forest of white pine, hemlock, and hardwood and then into the beech-birch-maple forest. There are thus three forest regions in Pennsylvania. All three of these forest types have their larger proportions in other states extending into New York, New England, and the South.

1. *The oak-hickory-yellow-poplar forest* is sometimes divided into two sections:

- The white oak, northern red oak, hickory type in northern, shallow-soil areas
- The yellow poplar, white oak, northern red oak type at lower elevations to the south

The yellow poplar is actually not common in the northern reaches of the oak-hickory-yellow-poplar forest. This indicates that the forest types do not have tight boundaries or a single-proportioned distribution of species throughout their ranges. Great differences in elevation (from the Atlantic Coastal Plain to the Allegheny Plateau, a range of over 3,000 feet), soils, and such things as the direction a slope faces cause stands to vary greatly. The species of these forests are of very wide adaptation.

As a rule, as one progresses northward and westward, oaks (black, white, scarlet, and northern red) become by far the most common trees within this forest type. Tulip tree is more common to the south and east in Pennsylvania (there are some almost pure stands in Maryland and New Jersey). The hemlock intrusion occurs in the northern part of this forest where it covers part of the Ridge-and-Valley region. Northern ridge slopes in the northern and western sections often have other intrusions from the north, particularly of beech and maple. White pine is common in the northern portions and was once extensively logged there.

2. *The white pine, hemlock, hardwood forest* is not a distinct forest type, but only a large transitional zone between the oak-hickory-yellow-poplar forest to the south and the northern hardwoods. Almost all the species of the first type of forest are present, often in similar proportions. However, the very large white pines and hemlocks that once stood in this area formed the basis of the enormous Pennsylvania logging industry of the last half of the nineteenth century. In the northern and higher portions of this forest, the northern hardwoods are common. Yellow poplar occurs too, but usually near streams and at lower elevations.

3. *The beech-birch-maple,* or *northern hardwood, forest* includes beech, yellow birch, and sugar maple as the major components, but also good proportions of black cherry, hemlock, black birch, red maple, white ash, basswood, and northern red oak. Because oaks are definitely less common in this area, the woods have a quite different character. Aspen and paper birch are common after fire and logging. The northern hardwood forest occupies the counties in the Northern Tier and can be enjoyed in a spectacular way by traversing the state on U.S. Route 6. Note that the types of trees make no distinction between glacial and nonglacial soils.

Forest Structure

A north temperate zone woodland has a six-layer stratification: a canopy of largest trees (which name the forest), an understory of spreading, shorter trees (such as dogwoods and witch hazels), a shrub and sapling layer, an herb layer, a moss or ground layer, and a subterranean layer of roots and microflora. None of these layers is continuous, except perhaps the last. The canopy is broken by death and windfalls. The understory trees, the shrubs, and the herb layer are favored in areas where light can get through and competition is diminished.

Early travelers in eastern Pennsylvania commented in superlatives on the height and density of the canopy trees and the thickness of the shrub and herb layers. Settlers and visitors saw this density as a positive indication of soil fertility. Travelers saw it as an impedance of nearly insurmountable proportions.

Throughout the state, prominent understory trees are dogwoods, witch hazels, and striped maples. The region has a rich shrub vegetation, commonly rhododendrons, mountain laurel (the state flower), spicebush, deciduous holly, and various viburnums. The state shares a herbaceous flora with its neighbors, the most commonly observed species being black mustard, chicory, various daisies, brown-eyed susans, wild orchids, asters, goldenrod, cattail, fireweed, pokeweed, sedges, and various legumes, including the vetches. The richness of the herbaceous flora is greatest in the southeastern Piedmont because of the superior soils and milder winters.

What Our Ancestors Found

A virgin forest existed in the state when settlement began in the seventeenth century. Indeed, all the Piedmont agricultural land, the valleys of the Ridge-and-Valley region, and the highland pastures of the Appalachian Plateau had to be cleared of the forest. There was almost no flat, fertile, unforested land, and where it did occur it was often not trusted by the early settlers who bypassed even Indian-cleared land to spend years of backbreaking work clearing their own. Best estimates are that only about 2 or 3 percent of Pennsylvania was not forest. Most of this was in fewer than a dozen grassland areas of a few hundred acres each (much like the prairie in the Midwest) scattered through the state. The Ridge-and-Valley region was not settled for a century, and experience soon taught that fertile valley soils extended only minimal distances up the ridge slopes. Much land cleared earlier for crops could be used later only for pasture; later still, it was abandoned and reverted to forest. Extensive clearing was done in the Plateau, but within a few decades these soils would not support crops well, although they served adequately as permanent pasture. The high-water mark of cleared land was reached about a century ago. Now, extensive clearings have again become forest, where deer and turkey hunters stumble across stone walls and segments of barbed wire miles from roads or neighbors.

Uses of Wood in Colonial Pennsylvania

How did the Pennsylvania forest affect the development of Pennsylvania and the lives of its settlers? Very strongly, both positively and negatively. Of the negative aspects, consider the necessity and difficulty of clearing land, making trails and roads, and establishing communities in dense forests, making establishing and extending settlement a slow process.

But the positive aspects are still with us. In primitive societies, the culture and the environment become integrated, and a balance develops and is sustained. Often the hardship and material poverty of a culture is directly related to a meagerness of resources. What the people eat, the nature of their houses and tools, their clothing, and their rituals are all reflections of what and how much the environment provides.

So it was with colonists in the days of European expansion. In early stages of colonization, quality of life depended on what the land provided. Pennsylvania provided a great deal. Consider wood. A colonial farmer's first task was to clear land. From those trees a lean-to or cabin was built and heated. With skill, simple log houses could be strong, dry, and warm. Some endure to this day, often as the core structures of much larger houses, hidden by two centuries of additions and siding. Barns and other outbuildings were of wood too, or wood and stone. The basic building materials were plentiful and free for the considerable labor of gathering them. In the temperate climate that was so much like northwestern Europe, colonists grew what they knew (wheat, rye, flax, hemp, European vegetables and fruits) and augmented those with products of the woods (chestnuts, acorns, mulberries, hickory nuts, walnuts, persimmon, crabapples, pawpaws) and other American plants (maize, tobacco, beans, squash, melons).

The colonist's tools were in large part made of wood. Planes, augers, axes, and scythes used metal working edges, but the handles and bodies were wood. Shovels, forks, scoops, stalls, rakes—and boxes, baskets, barrels, and bins—and dozens of tools no longer common, were all made of wood. Furniture was wood, not only the pine, oak, and fruitwood of Europe, but also new woods of superior quality. Tulip tree was an easily worked hardwood that gave planks more than 3 feet across for doors, panels, and cabinet tops. Much good country furniture was and is made of tulipwood. Two great new woods were black cherry and black walnut. By 1800, American furniture-grade hardwood was beginning to be in short supply and European artisans turned increasingly to tropical woods: teak, mahogany, and rosewood. Cherry and walnut were quickly recognized as furniture woods of the highest workability and beauty, and by the time of the Revolution, Philadelphia craft workers were known for their fine pieces made from these woods. Owning such fine furniture was beyond the hopes of common people, but in America the rural artisans could build with fine woods, and their products—chests, tables, chairs, doughtrays, dower chests, corner cupboards, hutches—are the basis of today's flourishing antiques trade.

Early houses, framed, sided, and roofed with wood, showed this luxury of wood. Inside were good wood floors, even in rude backwoods cabins. Furniture was varied and abundant, except where it was eschewed, as by Pennsylvania German farmers. Wood fences surrounded the house and the field. Early industry was water-powered. Mills ground grain, made tools and furniture, sawed wood, and made paper. The waterwheel and all gears and pulleys were also made of wood.

In transportation wood was used for carriages, wagons, and even roadbeds—the corduroy roads to fight the mud. Railroads burned wood and also made ties or sleepers, tanks, stations, and cars from wood.

The colonies were a wood culture unlike any before, and also a comparatively warm and comfortable society, with good wood houses and lots of firewood. Later, industry and the railroads were at first fired by wood, and the early iron-making industry used wood to produce charcoal. The juxtaposition of iron deposits and plenty of wood for fuel and charcoal made iron-making a major Pennsylvania industry. This wood culture was taken for granted. To appreciate it, consider how different colonial life must have been in the deserts of the American Southwest or even today in sub-Saharan Africa.

Native plants played other important roles in the early culture of Pennsylvania. Leather was the plastic of the colonial period, used for so many pliable things we no longer need it for (saddles, harnesses, traces, and saddlebags) as well as for clothing, shoes, boots, and tools. Tannin from trees is essential for curing leather, to make it durable and workable. Pennsylvania provided tanbark to other states and to Europe and the Orient as well. Almost all dyes came from plants. European dyes were usually too expensive for rural settlers. Until the coal-tar dyes of the mid-nineteenth century, these women used hemlock, willow, red maple, walnut shells, goldthread, and many other native plants to dye the homespun cloth—wool, linen (from flax), and linsey-woolsey.

Eighteenth-century medicine relied heavily on botanical materials, whether they worked or not. Indians did likewise, and settlers soon learned of and adopted Indian medicines. A medicinal use was ascribed to almost every plant. Some worked.

Native plants were also used for commerce. Europeans wanted to have them in their gardens, and a few Pennsylvanians, such as John Bartram of Philadelphia, made a profitable business of collecting and selling plants and seeds to European gardeners.

Pennsylvania's Lumber Industry in the Late Nineteenth and Early Twentieth Centuries

For decades, local sawmills served the needs of Pennsylvania, but beginning about 1860 there was a growing market for lumber in distant markets. By that time the only large area of virgin timber remaining was in north-central Pennsylvania. The West Branch of the Susquehanna drained thousands of square miles of virgin forest, where the white pine and hemlock were supreme. Many of these superb trees stood 150 feet tall and yielded more than 5,000 board feet of lumber.

The development of these forests lying within the deeply dissected plateau of north-central Pennsylvania was totally dependent on the use of water transportation, for land transportation was essentially nonexistent. Although the streams during the high-water periods were sufficiently large to float the logs, a means to catch the logs as they entered the Susquehanna from the tributaries had to be devised. Initially, workers were stationed in the Susquehanna in boats to catch the logs as they floated by. The logs were then formed into a crude raft and anchored to the shore. This was not only a dangerous operation but also ineffective, for a large percentage of the logs were lost as they were swept downstream during the high waters in the spring.

More effective control over the movement of the logs was achieved with the development of the lumber boom stretching across the river. The boom was an arrangement of cribs with chains strung between to catch the logs. When the boom was closed, all logs were caught until the boom was opened. Workers, known as "boom rats," armed with large pikepoles and wearing boots with sharp calks were stationed to keep the logs moving and identify the logs belonging to different owners.

The logs were caught in booms built at points where the streams from the Plateau entered the Susquehanna River. These included the Lock Haven and Linden booms, but the greatest of all was the Williamsport boom, which was six miles long stretching from one shore to another and was the major collecting point for the entire West Branch. In 1862, the sawmills of Williamsport produced 87,863,000 board feet of lumber. At the peak of the lumbering era, about 1885, more than 225,000,000 board feet of lumber were produced. After that year, production declined steadily.

Williamsport was not only the supreme transit center but also a sawmill and wood-milling center. Nowhere else in the world produced as much lumber. A vast complex of woodworking shops lined the riverbank. Planing mills produced doors, sashes, shelving, and a multitude of other wood items. In the late nineteenth century, Williamsport was the lumber capital of the world, but not all wood could be processed at Williamsport, so lesser centers developed at Lock Haven, Montoursville, Muncy, Montgomery, Watsonville, Lewisburg, and elsewhere.

By the early 1900s, the timber had been exhausted

near the streams. The next stage in Pennsylvania's lumber industry came with the development of the lumber railroad. The first known lumber railroad began operating in Jefferson County in 1864, and the last one was abandoned in Elk County in 1948. The industrial logging railroad penetrated the most isolated areas. The railroads were short-lived, for after the timber of a local area had been harvested, they were abandoned. At the turn of the century, these railroads supplied the timber for more than 600 sawmills, tanneries, and wood chemical companies. Although many single lumber operations were quite small, the penetration of hundreds of tracks into the forests by the logging railroads made it possible to produce a vast quantity of lumber. The hemlock forests of Pennsylvania provided the bark for a large leather-tanning industry as well.

The exploitation of Pennsylvania's virgin forests was a dynamic period in the history of lumbering. Hundreds of small settlements grew up. Because the practice of the day was to remove the trees as quickly as possible and move on to a new tract, the settlements gave little appearance of being permanent, and because the timber resources of the vast forests seemed inexhaustible, no thought was given to conservation—there would always be another tract to exploit. The lumber industry was continuously migratory. Though hundreds of small settlements were built, there was no economic activity to sustain them once the forests were gone, and most settlements disappeared.

Modern Forest Resources

The initial lumbering era in Pennsylvania ended about eighty years ago. Much of the deforested land was not suitable for farming, and reforestation began, mostly by natural revegetation. Only a few small portions of the secondary and tertiary forests are the result of the planting of new trees. According to the most recent U.S. Department of Agriculture Forest Service Inventory, the state of Pennsylvania is now 58 percent forested. Nevertheless, there are great variations from one county to another (see Fig. 6.2). Thirty-five counties are above the state mean of 58 percent, and thirty-two counties below that mean. Within the state, Cameron County, with 97.4 percent of its land area in forests, is the most heavily wooded area, while Philadelphia County, with

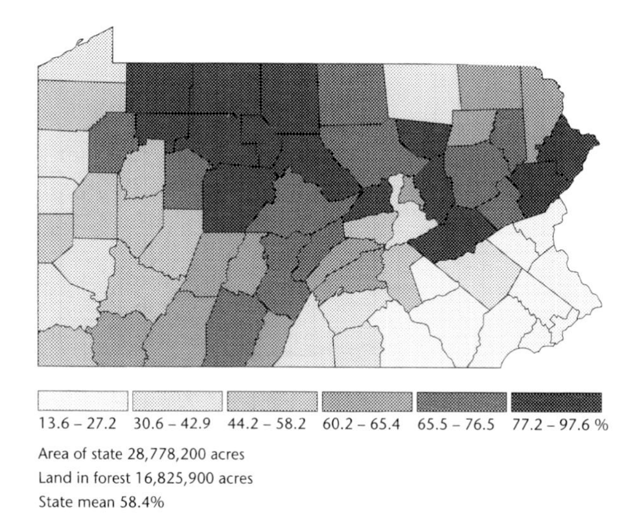

Fig. 6.2. Percentage of forestland in Pennsylvania.

forests covering 16 percent of its land, is the least forested. Most counties in north-central Pennsylvania have two-thirds of their land in forests. In southeastern Pennsylvania, most counties have less than one-third of their land in forests.

The quality of the forest varies greatly. Because of lack of management in most forest areas, about 61 percent of the forest is classified as having poor productivity on an annual basis, while only 5 percent is classified as having very good productivity. The highest productivity occurs in western Pennsylvania; the areas of lowest productivity are found in north-central Pennsylvania, where a combination of natural revegetation of undesirable species and a rigorous environment reduces the annual yield.

Ninety-five percent of the forest, or about 15.9 million acres, is classified as commercial. Nearly 22 percent of Pennsylvania's commercial forests (3.5 million acres) is privately owned, with the highest total percentage in the northeast. Nearly half of the commercial forestland is in sawtimber stands; less than 2 percent is scrub forest.

The available commercial stock timber is placed at 21.8 billion cubic feet, an average of 1,366 cubic feet per acre. In addition, the sawtimber volume is placed at 46.4 billion board feet, an average of 2,916 board feet per acre. Ninety-two percent of the stock volume is in hardwood species; red maple is the species with the most volume—3.3 billion cubic feet. Ninety-one percent of the sawtimber is also in hardwood species; northern red oak has the largest volume, with 7.7 billion board feet.

Stock growth is estimated to be about 2.2 times

timber removals. Select white oak is the only major species showing removal exceeding growth. Sawtimber growth is estimated to be 2.7 times removal. Northern red oak shows the largest annual growth of any species. Sawlogs are the major timber product and account for more than 40 percent of all growing-stock removals.

Forest Alteration

Pennsylvania's second- and third-growth forests are not the same as the original primeval covering. Only a few thousand acres, at best, of virgin woods still exist in the state, in such areas as Cook Forest in Clarion County. What appears as a lush, thick, fine forest is usually only a hint of its former grandeur. After logging, the light and moist environment of the forest is very different. Seedlings of would-be canopy trees, such as the hardwoods, are close together. Little room or light is left for the plants of the lower layers. In the intense competition, many of the tall, spindly trees die. This is the initial step that will lead back to the mature, stratified forest. In such conditions, eighty years is not enough time to reproduce what took thousands of years to form.

One tree species will be forever lacking. In the 1890s, a fungus capable of attacking and killing all American chestnuts was inadvertently introduced from China into New York City. By 1906, it was in Pennsylvania and causing great devastation. By 1940, it had extended itself through the entire eastern deciduous forest, killing every chestnut tree. The enormity of the loss is underscored by the knowledge that, in the Pennsylvania woods, chestnuts accounted for about one tree in five, and farther south, one tree in three. This huge loss must be ascribed to human activity and carelessness.

There are now more species of plants growing in Pennsylvania than before. How can this be? Land-clearing, logging, and farming all change the environment greatly, making habitats for other plants. Such species are called *adventive*. For example, land-clearing made habitats for herbs and grasses that could never have survived in the dark, primeval forest. Introduced plants, particularly European plants that came in for gardening and that escaped cultivation, or that arrived as seed contaminants in imported commodities and fell from railway cars, greatly enriched the state's flora.

The Modern Lumber Industry

The timber and wood products industry is once again expanding in the state. Employment has increased from 13,837 in 1969 to 31,300 in 1992. In 1992 the annual wages of production workers totaled $385 million, and the value of shipments was at $2.801 million. The industry is highly decentralized, employing at least some workers in 64 counties. The number of lumber workers in those counties ranged from a high of 2,550 in Lancaster County to zero. The county mean employment in 1990 was 352. Twenty-two counties were located above the mean and 45 counties below the mean. In 1990, the U.S. Department of Commerce reported 1,444 establishments producing timber wood products, of which 1,110 had fewer than 20 employees. There were 440 sawmills and planing mills, of which only 85 had more than 20 employees. The sawmills and planing mills produce about 22 percent of the value of timber and wood products.

These operations are migratory. The timber operators move into an area, cut the mature trees, and then move to another tract. Because of the high cost of transporting logs, it is usually more economical to move the sawmills to the raw material sources. As a consequence, the spatial pattern can change over a period of a few years. Because the industry is migratory, it does not provide a stable economic base for a community.

A wide variety of wood products are produced. Because the timber is hardwood, millwork emphasizes the production of cabinets, hardwood veneers, and containers. These products make up about 60 percent of the value of timber and wood products. Wood buildings and mobile homes are an additional 20 percent, and miscellaneous wood products are about 20 percent. The wood products industries are oriented to markets rather than to raw materials. The leading counties for these items are Lancaster, York, Berks, Lebanon, Franklin, Lycoming, and Philadelphia.

Forestland Ownership

A recent survey estimated there were 490,100 private owners holding 12,452,800 of the 16,825,900 acres of commercial forestland in Pennsylvania. Sixty-three percent of these owners hold less than 10 acres each and collectively control only 8 percent of the

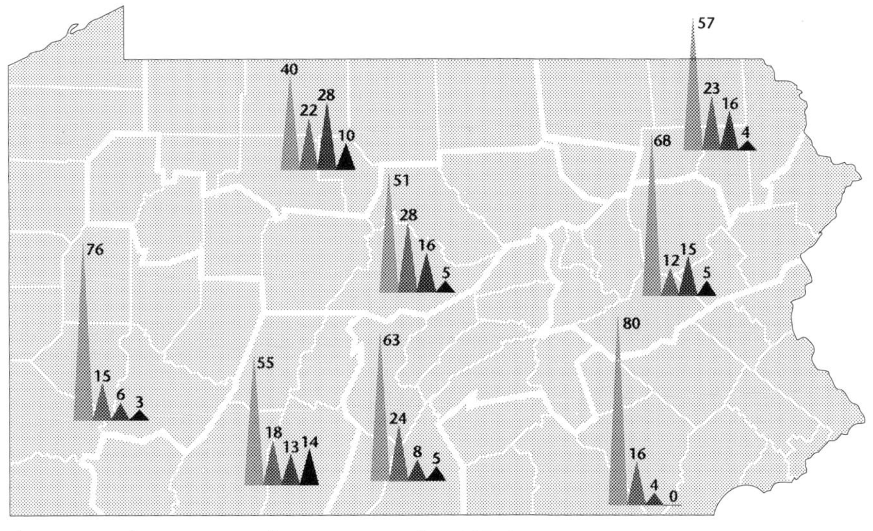

1 – 50 51 – 100 101 – 500 > 500 Acres
Spike heights indicate % commercial forest land.

Fig. 6.3. Average size of Pennsylvania forest tracts. For example, in western Pennsylvania 76 percent of the forest tracts are 1 to 50 acres in size, 15 percent are 51 to 100 acres, 6 percent are 101 to 500 acres and 3 percent are over 500 acres.

private forestland. Another 600 owners holding 1,000 or more acres each own 20 percent of the privately owned commercial forestland.

The average private forestland ownership is 25 acres. If ownership of less than 10 acres is excluded, the average rises to 63 acres. The average size varies considerably, ranging from 11 acres in the southeast to 55 acres in north-central Pennsylvania (Fig. 6.3).

Eighty-six percent of the private owners are individuals, collectively holding 69 percent of the privately owned commercial forestland. There are an estimated 9,000 corporations owning just over 2.2 million acres of forestland in Pennsylvania. The average corporate holding is 323 acres. A large portion of the corporate ownership is in the north-central area of the state, where corporations own 783,800 acres, or about 40 percent, of the corporate-owned forestland. The remaining commercial forestland is owned by partnerships, individual estates, clubs, and associations.

As is common in other states, retired people own a significant portion of Pennsylvania's forestland. There are an estimated 72,100 retired owners holding 21 percent of the state's individually owned forestland. Woodland acreage in farms is declining, particularly since 1959, but farmers still hold a significant portion of Pennsylvania's forestland: 1.5 million acres.

Fully one-fourth—about 4.2 million acres—of Pennsylvania's forestland is publicly owned—the greatest proportion and acreage in public holdings of any state in the Northeast. The Pennsylvania Bureau of Forestry manages 2 million acres of state forests for such diverse benefits as timber production, wildlife habitat, outdoor recreation, water, and minerals. The other large multiple manager is the U.S. Department of Agriculture's Forest Service, which administers the 489,000 acres of forestland in the Allegheny National Forest. The Pennsylvania Game Commission manages 1.1 million acres of forests in state game lands primarily to improve wildlife habitat. Small areas of forest are publicly held for recreation (state parks) and in freshwater municipal watersheds.

Nearly 83 percent of the public forestland is classified as commercial forest. The remaining 17 percent is noncommercial, due to its location, low productivity, or administrative designation. Nearly all the noncommercial forestland is publicly owned.

Forest Management

The forests of Pennsylvania today are generally the result not of forest management but of the natural regeneration of the land. Because the forests have made a recovery from the devastating exploitation of the nineteenth century, a basic question is: Why should there be forest management in Pennsylvania?

Part of the answer to this question is that the present situation may lead to wrong conclusions in the future. While nearly all forests are renewable, it takes from 70 to 120 years or more to renew a mature, commercial hardwood stand. Since timber is a crop that takes so long to mature, careful and thoughtful planning and management can help ensure relatively steady, reliable supplies. Within the next forty years, millions of areas of Pennsylvania's forests will be harvested. While a repeat of the destructive exploitation of the forest will not occur, most of the mature timber will be harvested. If the industry is to remain stable, a management program to continually renew the forests must be implemented.

In addition, many of the forests harvested by clear-cutting or killed by such insects as the gypsy moth are not regenerating satisfactorily. Natural regeneration of commercial hardwood trees means there are many undesirable tree species. Striped maple, American beech, black locust, and other undesirable species are growing on many sites. Further additional research is needed on how to control deer, acorn weevils, rodents, and other destructive agents to keep them from inhibiting the establishment and development of a desirable forest cover. These problems demonstrate a need for sound forest management.

Besides the biological factors in the regeneration of forests, there are also socioeconomic considerations. The USDA Forest Service has projected a 2.8 percent annual increase in demand for timber for the next thirty years, but this estimate may be low because the demand for Pennsylvania's wood is growing. Both national and international factors are important in shaping demand. Pennsylvania's hardwood timber is producing a size and quality that is well suited for fine furniture. Many foreign countries are now importing wood from Pennsylvania.

There is also an indication that Pennsylvania's hardwoods will experience a growing demand for commercial and residential fuels. The state's forests can certainly contribute to the national forest biomass energy program. Because the demand for softwood conifers will outdistance supply, there will be greater utilization of the hardwoods.

It must also be remembered that most of the timber volume will not be available to the lumber industry, given the objectives of today's private landowners, harvesting technology, and market conditions. Many private landowners never intended to harvest their timber. Furthermore, most of the timber privately owned is located in inaccessible areas—on steep slopes or along roads and streams where logging would detract from the scenery or damage water resources. In addition, not all the projected timber volume is of a desirable species. Many private forest owners believe that forests provide benefits other than economic benefits, such as a place for a diverse songbird population, scenic views, an unpolluted and productive trout stream, a well-used deer trail, or simply a retreat.

Forest management can be used to enhance both tangible and intangible benefits. For landowners who want to realize from their forests not only economic benefits, but such benefits as aesthetic enjoyment, plentiful wildlife, clean water, or a wilderness experience, forest management is needed.

Bibliography

Birch, T. W., and D. F. Dennis. 1980. *The Forest-Land Owners of Pennsylvania.* U.S. Department of Agriculture, Forest Service Resource Bulletin NE-66.

Bones, J. T., and J. K. Sherwood Jr. 1979. *Pennsylvania Timber: A Periodic Assessment of Timber Output.* U.S. Department of Agriculture, Forest Service Resource Bulletin NE-59.

Braun, E. L. 1950. *Deciduous Forests of Eastern North America.* New York: Hafner.

Considine, T. J., Jr., and D. S. Powell. 1980. *Forest Statistics for Pennsylvania, 1978.* U.S. Department of Agriculture, Forest Service Resource Bulletin NE-65.

Ferguson, R. H. 1958. *The Timber Resources of Pennsylvania.* U.S. Department of Agriculture, Forest Service, Northeast Forest Experiment Station.

———. 1968. *The Timber Resources of Pennsylvania.* U.S. Department of Agriculture, Forest Service Resources Bulletin NE-8.

Haines, D., W. A. Main, and E. K. McNamara. 1978. *Forest Fires in Pennsylvania.* U.S. Department of Agriculture, Forest Service Research Paper NC-158.

Illicik, J. S. 1923. *Pennsylvania Trees.* Pennsylvania Department of Forestry, Bulletin 11, fourth edition.

Marquis, D. A. 1974. *The Impact of Deer Browsing on Allegheny Hardwood Regeneration.* U.S. Department of Agriculture, Forest Service Paper NE-308.

Marquis, D. A., and P. L. Roach. 1976. *Acorn Weevils, Rodents, and Deer All Contribute to Oak-Regeneration Difficulties in Pennsylvania.* U.S. Department of Agriculture, Forest Service Research Paper NE-356.

Nichols, J. O. 1980. *The Gypsy Moth.* Harrisburg:

Pennsylvania Department of Environmental Resources, Bureau of Forestry.

Powell, D. S., and T. J. Considine Jr. 1982. *An Analysis of Pennsylvania's Forest Resources.* U.S. Department of Agriculture, Forest Service Resources Bulletin NE-69.

Severinghaus, C. W. 1978. "Problems in Penn's Woods." *Pennsylvania Forestry* 68:17–21.

Wharton, E. H., and J. T. Bones. 1980. *Trends in Timber Use and Product Recovery in Pennsylvania, 1966–77.* U.S. Department of Agriculture, Forest Service Research Notes NE-297.

Landis Valley Folk Fair

Courthouse in Somerset County

Orthodox church in Mount Carmel

Recreation on Presque Isle beaches, Lake Erie

PART TWO

THE PEOPLE

In moving from the opening chapters, which deal with Pennsylvania's natural habitat, to five chapters that treat the people of the Commonwealth and their geography, it is important not to forget the physical environment. Human beings have always had a considerable range of choices in coping with their surroundings, but the nature of the land—its climate, soils, vegetation, topography, drainage systems, and mineral resources—has obviously played a major role in shaping the spatial patterns of human activity in Pennsylvania. The even wider theme of the interrelatedness of all the things discussed in this volume should be kept in mind while reading about the demography, politics, and ethnic and general cultural patterns of the state, because all such human phenomena are intimately involved with the workings of the economy as well as the configuration of the habitat.

As we examine the human geography of Pennsylvania, a paradox becomes evident. In the course of three centuries, the state's demography has undergone great transformation in size, trends, urbanization, migration, fertility, and mortality. Although change has also been persistent enough, and quite rapid and profound in some other nondemographic phases of the human scene, Pennsylvania's society is also notable for its degree of conservatism, in a basic social sense rather than just in political behavior. After several generations, the enduring power of the early settlement fabric is still evident, while the lasting imprint of the ethnicity and religious preferences of the initial settlers is still observable in the style of the visible landscape, in social attitudes, and in the political impulses of voters.

As noted in Chapter 1, Pennsylvania is America in microcosm, arguably more so than any other state. The justification for this claim is that the state has contributed so much to the remainder of the nation through the outpouring of its people, attitudes, and inventions, especially during the first two centuries of its existence. But the state is also representative of the nation in another sense: in the fine-grained complexity of its human geography, in the localisms of Pennsylvania. In Pennsylvania, there are sharp contrasts among peoples, cultures, and economies that also exist throughout the United States but that in Pennsylvania are packed more closely together. What better vantage point for the America watcher?

The topics covered in Part Two may be the most basic for the understanding of the human landscape, but the five chapters do not provide a total, definitive treatment of the subject. For lack of sufficient space, but equally for lack of adequate data, some significant subjects are missing or have been accorded only minimal attention. The discussion of population topics is reasonably complete, again within the constraints of space, but there is much more to be learned and said about the political scene, ethnicity, language, religion, and other facets of culture than is available in the following pages. For example, omitted are serious discussions of the rich variety of Pennsylvania's vernacular architecture, much of which still remains unexplored; of the geography of personal consumption (i.e., patterns of food, drink, and choice of commercial commodities and services); of the spatial array and other characteristics of Pennsylvania's cemeteries; and of the geography of voluntary associations, folklore, and social customs in the state. Is it too much to hope that some of our readers will eventually make good these deficiencies?

7

GROWTH AND CHARACTERISTICS OF PENNSYLVANIA'S POPULATION

Paul D. Simkins

Total Population Growth

More than a half-century after the landing of the *Mayflower,* Pennsylvania still held only a few hundred people of European stock, perhaps no more than 700 in 1680 (Table 7.1). Delayed by conflicting territorial claims and by a perceived lack of easy penetration into the interior, effective settlement came later in Pennsylvania than in any of the colonies except Georgia. After the confirmation of Penn's charter by Charles II in the 1680s, however, the population of Pennsylvania grew rapidly, by natural increase but especially by immigration. During the eighteenth century Pennsylvania probably attracted more immigrants than any other American colony, and by the time of the American Revolution it was second in population size only to Virginia.

Although population growth in Pennsylvania was persistent throughout the period of colonial control, after 1680 there were pulses of exceptionally rapid growth. The first of these came immediately following the founding of the colony, when large numbers of English Quakers and German Pietists joined Penn's new colony. By 1700 the population of Pennsylvania numbered some 18,000 people and comprised about 7 percent of the total population of the American colonies. The next major pulse occurred after 1718, when the large immigration of Scots-Irish began. Between the date and the outbreak of the American Revolution, nearly a quarter-million Scots-Irish moved to the American colonies, largely into Pennsylvania. Between 1720 and 1740, the population of the Pennsylvania colony more than doubled, increasing from 31,000 to more than 85,000. By 1740, the leading edge of effective settlement in Pennsylvania had reached Blue Mountain, a diagonal range extending from Northampton to Franklin County (see Fig. 7.1).

Further westward settlement was slowed during the French and Indian War but once that conflict ended, rapid population growth resumed in Pennsyl-

Table 7.1. History of population growth, Pennsylvania–United States.

	Total Population (in 000s)		% Increase over Previous Period		No. Increase over Previous Period (Pa., in 000s)	% Urban	
	Pa.	U.S.	Pa.	U.S.		Pa.	U.S.
1680	0.7	152	—	35	—	—	—
1700	18	251	58	14	17	—	—
1720	31	466	27	41	13	—	—
1730	52	629	67	35	21	—	—
1750	120	1,171	40	29	68	—	—
1770	240	2,148	31	35	120	—	—
1790	434	3,929	33	41	194	10.2	5.1
1800	602	5,308	39	35	168	11.3	6.1
1810	810	7,240	35	36	208	12.8	7.3
1820	1,049	9,638	30	33	239	13.0	7.2
1830	1,348	12,861	29	33	299	15.3	8.8
1840	1,724	17,063	28	33	376	17.9	10.8
1850	2,312	23,192	34	36	588	23.6	15.3
1860	2,906	31,443	26	36	594	30.8	19.8
1870	3,522	38,558	21	23	616	37.3	25.7
1880	4,283	50,189	22	30	761	41.6	28.2
1890	5,258	62,980	23	26	975	48.6	35.1
1900	6,302	76,212	20	21	1,044	54.7	39.6
1910	7,665	92,228	22	21	1,363	60.4	45.6
1920	8,720	106,022	14	15	1,055	65.1	51.2
1930	9,631	123,203	11	16	911	67.8	56.1
1940	9,900	132,165	3	7	269	66.5	56.5
1950	10,498	151,326	6	15	598	70.5	64.0
1960	11,319	179,323	8	19	821	71.6	69.9
1970	11,801	203,302	4	13	482	71.5	73.6
1980	11,864	226,546	1	11	63	69.3	73.7
1990	11,882	248,710	0	10	17	68.9	75.2

SOURCES: For 1680–1770, U.S. Department of Commerce; for 1790–1980, U.S. Census, 1980; for 1990, U.S. Census, 1990.

vania. Despite a continuing large drain of Pennsylvanians southward into Virginia and the Carolinas, the population of the colony again doubled in size between 1750 and 1770, to reach 240,000 people. By the time of the Revolution, Pennsylvania contained more than 10 percent of the total population of the American colonies.

The first census of the United States was taken in 1790, when more than 434,000 people were enumerated in Pennsylvania. Nearly three-fourths of that total were confined in the area southeast of Blue Mountain. Philadelphia County alone held more than 12 percent of the state's population, and another 17 percent resided in Lancaster and York counties (from which Adams County had not yet been detached). Northumberland County, which in 1790 included all of north-central Pennsylvania, held only 4 percent of the state's population.

From 1790 to 1800 the population of Pennsylvania grew more rapidly than that of the nation as a whole, and settlement quickly filled in the northeastern and northwestern corners of the state, leaving an area in the northwest centered on Cameron County that would not become effectively settled until after 1820. After 1800, however, Pennsylvania's population growth rate fell below that of the nation as people moved westward and new states were added. Nevertheless, rates of growth remained consistently high in both Pennsylvania and the nation until 1850, when the effect of declining birth rates began to reduce the rate of intercensal population growth.

Following 1850, population growth rates for Pennsylvania and the United States as a whole moved downward in a roughly parallel fashion to bottom out during the period of low birth rates during the depression decade of the 1930s, when the

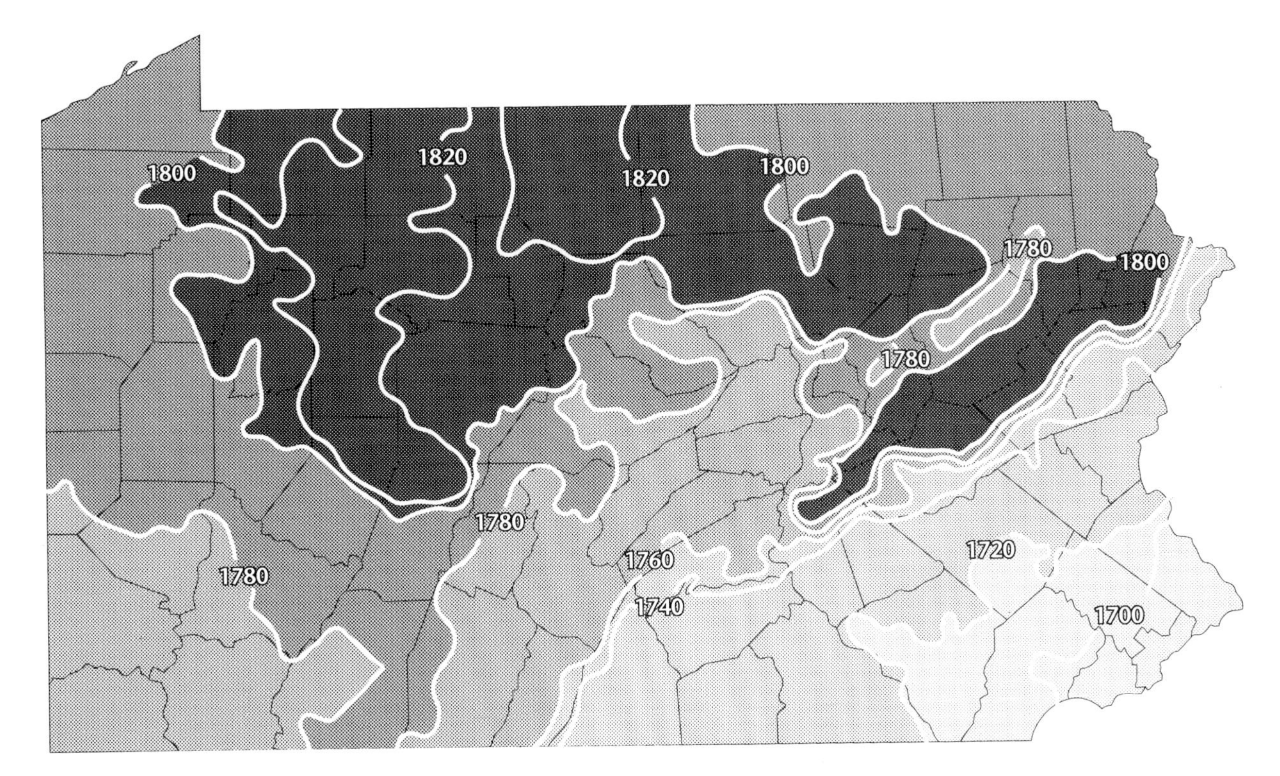

Fig. 7.1. Settlement dates.

nation's population increased by about 8 percent but Pennsylvania's increased by only 3 percent.

Yet for most of that period of declining growth rates, the absolute number of people added to Pennsylvania's population during intercensal decades increased, the largest absolute growth occurring during the thirty years from 1890 to 1920, when more than a million people were added to Pennsylvania's population each ten-year period (see Table 7.1). A substantial share of these new additions were immigrants, the interval between 1890 and 1920 being the time when immigration into the United States reached its all-time peak. For example, the increase in the number of foreign-born in Pennsylvania between 1900 and 1910 was equal to more than one-third of the population growth of the state over that decade. Further, children born to immigrants in Pennsylvania would have added considerably to the natural increase in the state.

After World War II, population growth rates in the United States climbed with the postwar "baby boom," which peaked in 1957. Thereafter, with declining fertility, population growth rates again slowed so that the nation's population increased by only 11 percent during the 1970s. Although Pennsylvania participated in the baby boom, its fertility levels were lower than the national average, and net out-migration removed much of that natural increase. The result was that population growth rates in Pennsylvania after 1950 fell further below those of the nation than at any previous time. During the 1970s Pennsylvania's population growth rate fell to less than 1 percent, and only 63,000 people were added to the total population, fewer than in any decade since 1770.

By 1980 Pennsylvania had fallen to fourth place in population size among the states and contained a bit more than 5 percent of the nation's population. Slow population growth continued in Pennsylvania during the 1980s, during which time the total population of the state increased by fewer than 17,000 persons, or only 0.14 percent.

Regional Population Growth

Changing boundaries caused by the creation of new counties make it difficult to follow the patterns of regional population change in Pennsylvania during the early nineteenth century. The present county boundaries of Pennsylvania did not become fixed until 1878, when Lackawanna County was detached from Luzerne County. However, from 1800 to 1860 the areas with the greatest relative population gains were clearly in the northern part of the state—at first in the northwest, later shifting to the Northern Tier and northeastern counties. These large, relative gains represent the continued filling in of the parts of the state settled only later, and were made possible by the small size of the initial populations. Although percentage gains in the southeast were relatively modest during the first half of the nineteenth century, absolute gains in population continued to be large. For example, Lancaster County, which had rates of population growth below the state average every decade from 1800 to 1860, nevertheless increased its population during those years from 43,000 to 116,000 people.

From 1860 to 1880 the areas of greatest relative population gains became more diffuse. Rapid gains were made in the northern anthracite region (Luzerne County), in the more rural counties of the central west, and in the Pittsburgh and Philadelphia areas. The 1880 decade that followed, however, marked the first time that a substantial number of counties of Pennsylvania lost population. During the 1880s the 12 counties of population loss were scattered throughout the state, but in the subsequent decades until 1930, there was an increase in the number of counties losing population each decade, and the areas of loss became more concentrated in the Northern Tier counties and in the more rural and agriculturally based counties in the central part of the state. As the lumber industry declined and agriculture became less labor-demanding, large numbers of people migrated from the rural areas of the state to the western states or to the more attractive economic opportunities of the cities. The Northern Tier counties from Warren to Susquehanna reached their maximum share of the state's population in 1880 with about 6 percent of the total. Thereafter, the proportion of the state's population included in these counties decreased, until in 1990 only 2 percent of the state's population lived in these counties (a smaller proportion than this area had in 1810). Among these counties, Potter reached its maximum population in 1900; Tioga in 1890. The combined population of these counties in 1990 was 7 percent smaller than it had been in 1900.

During the depression decade of the 1930s the closing of factories and loss of urban jobs led many to leave the cities and return to their rural origins. Most of the counties that had posted population losses earlier gained population between 1930 and 1940. Slower growth characterized counties that included larger cities. Indeed, Philadelphia County lost more than 19,000 people during the 1930s. The only concentrated area of population decline within the state during the depression years occurred in the anthracite counties of the Northeast. Of the 10 counties losing population during the 1930s, 5 were located in the anthracite region: Lackawanna, Luzerne, Carbon, Schuylkill, and Northumberland. These 5 counties had consistently gained in population from the early years of the nineteenth century and by 1910 contained nearly 13 percent of the total population of the state. Although population growth continued until 1930, it did so at a rate below the state average. After 1910 the share of the state's population contained in these counties dwindled, so that by 1990 only about 7 percent of the state's population lived in the anthracite counties, a smaller proportion than in 1860.

After 1940 the area of most rapid population growth in Pennsylvania became the southeast, especially the counties surrounding Philadelphia, where massive suburbanization occurred. Despite this suburbanization, the five counties of the Pennsylvania portion of the Philadelphia metropolitan area (Bucks, Chester, Delaware, Montgomery, and Philadelphia) have held a remarkably consistent share of the state's population during the period of record. In 1810 some 28 percent of the population was included in those five counties, a share that remained almost unchanged until 1940, after which the region increased its portion to 31 percent in 1990.

Within the Philadelphia region there has been a significant redistribution of people. From the early 1800s until about 1940 Philadelphia County gained in population more rapidly than the surrounding counties. Its share of the state's total population remained consistently about 20 percent, whereas the portion in the other four counties of the region generally declined. In 1900, Bucks, Chester, Delaware, and Montgomery counties held only 6 percent of the state's population. With the spread of the suburbs and the subsequent loss of population in Philadelphia, the population distribution in the area has changed greatly, so that by 1990 only 13 percent of the state's population was in Philadelphia County and 18 percent was in the surrounding counties.

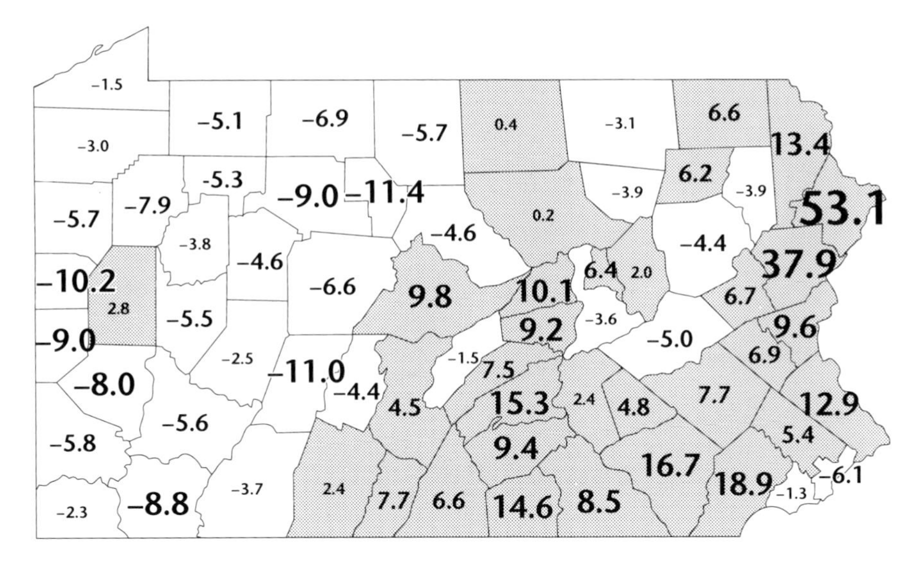

Fig. 7.2. Percentage change in Pennsylvania population, 1980–1990.

Three other trends have affected population distribution within Pennsylvania since 1940. First, there has been a decrease in population of the larger cities. In 1990 there were 21 Pennsylvania cities with populations of 25,000 or more since 1940. Of these, only one (State College) reached its maximum population in 1990. Indeed, of the 66 cities with 10,000 or more people, only 4 had their largest-ever population in 1990. The two largest cities in the state, Philadelphia and Pittsburgh both reached their peak populations in 1950; by 1990 their combined populations had decreased by 790,000.

A second major recent trend in regional population change has been a turnaround from population gain in the 1970s to a loss in the 1980s in most of the western half of the state. During the 1970s some 26 of the 33 westernmost counties gained population. Significant population losses were limited to Cambria, Allegheny, and Beaver counties and, in the north, to McKean and Cameron counties. During the 1980s, however, 27 of the 33 counties lost population. In the most western counties, only Butler County managed population gains in the 1970 and 1980 decades (Fig. 7.2).

The third trend in population redistribution within Pennsylvania in recent years has been the growth in recreation and amenity areas. This growth has been most marked in the Poconos, where Pike and Monroe counties have greatly increased their populations since 1940. Both counties had population increases exceeding 50 percent between 1970 and 1980. During the 1980s Pike County's population again grew by more than 50 percent. The growth rate in Monroe County slowed somewhat in the 1980s, but in absolute numbers more people were added to the county during the 1980s than during the previous decade. Moreover, population growth is spreading into parts of neighboring counties. Wayne County, for example, increased its population by more than 13 percent during the 1980–1990 interval.

The net result of regional population change in Pennsylvania during the twentieth century has been an increasing concentration of the state's population in the area southeast of Blue Mountain, the first part of the state to be settled. In 1790 this part of the state contained about 74 percent of Pennsylvania's total population. As settlement spread westward and northward, that share of the state's population progressively declined to about 40 percent in 1910. After 1910 the part of the state southeast of a diagonal drawn between Northampton and Franklin counties, and including all of Dauphin County, had gained in population faster than the state average, so that by 1990 the region included a bit over half the state's total population. Of the 11 counties of Pennsylvania having continuous population gains since 1790, or since their subsequent creation as counties, 9 are in this southeastern region.

The major regional changes in population in Pennsylvania during the twentieth century show up clearly on a map of the 1990 population by county as a percentage of the 1900 population (Fig. 7.3). Nine counties of Pennsylvania had a smaller population in

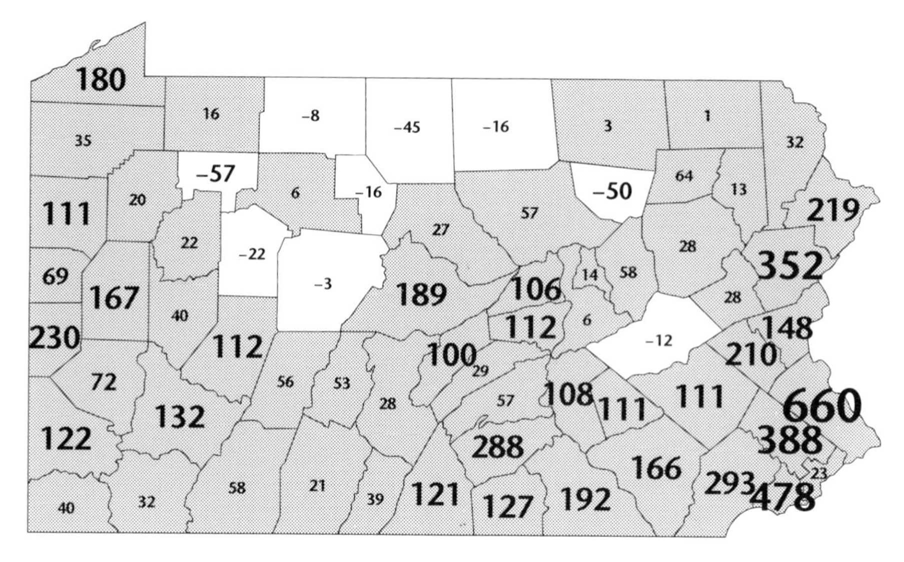

Fig. 7.3. Percentage change in Pennsylvania population, 1900–1990.

1990 than in 1900. These counties are located almost exclusively in the northern and west north-central parts of this state. Outside this area, only Schuylkill County had fewer people in 1990 than in 1900. The greatest relative gain in population within the state since 1900 has been in the suburban counties of Philadelphia. The four counties (Bucks, Montgomery, Delaware, and Chester) combined have had population increases of more than five-fold since 1900. Similarly, Cumberland County, absorbing much of the suburbanization from Harrisburg, has more than tripled in size since 1900.

Urban Population Growth

Few estimates have been made of population in towns before the first census in 1790. The only Pennsylvania town of considerable size during the colonial period was Philadelphia, which may have included 4,000 people by 1700, and a few more than 13,000 by 1750, when it had surpassed Boston in size to become the largest city in the American colonies. Lancaster had an enumerated population of 3,773 in 1790. Only two other Pennsylvania towns—Reading (population 2,225) and York (2,076)—were approaching urban status by the time of the first census.

The state as a whole had about 10 percent of its population classified as urban in 1790, or double the proportion for the United States (Table 7.1). Among the states, however, Pennsylvania was the fourth most urban state, behind Rhode Island, Massachusetts, and New York.

The urbanization of the United States in general and of Pennsylvania in particular was quite slow from 1790 to 1820. By 1820 some 7 percent of the U.S. population and 13 percent of Pennsylvania's population lived in urban areas. But after 1820 the degree of urbanization in both the nation and the state began to move sharply upward, with the state consistently being more urban than the nation. The period of most rapid rates of urban growth for both Pennsylvania and the nation occurred between 1820 and 1870. During the peak decade of growth, 1840–1850, the urban population of the nation increased by 90 percent, and that of Pennsylvania by 85 percent. Some 30 percent of Pennsylvania's population was urban by 1860.

After 1850, despite a progressive increase in the numbers of people added to the urban population each decade, the decennial rate of urban growth both in the United States and in Pennsylvania fell sharply to a low in the 1930s, with rates of growth in Pennsylvania generally slower than those of the nation. Nevertheless, the population of Pennsylvania became predominantly urban by 1900, a full two decades before that level was achieved by the United States as a whole.

The rate of urban growth in the United States rebounded after 1940 during the period of rapid population gain associated with the baby boom, only to decline again as birthrates dropped after 1960. Unlike the earlier trends, urban growth rates in

Pennsylvania after 1940 fell well below those of the nation. Indeed, since 1970 the urban population of the state has decreased in size. By 1990 some 69 percent of the population of the state was living in urban places, down from 71.6 percent in 1960. The unexpected result of these trends is that Pennsylvania became less urban than the United States in 1970 and by 1990 had fallen further below the national average.

Within Pennsylvania the evolution of urban places (incorporated places with populations of 2,500 or greater) proceeded slowly at first, their locations generally following the pattern of settlement expansion across the state. By 1800 York had been added to Philadelphia and Lancaster as an urban place, followed by Reading and Pittsburgh in 1810. In 1840 there were still only 3 urban places existing in the western half of the state (Pittsburgh, Allegheny City, and Erie), whereas 9 urban places were located in the southeastern quadrant. Some 43 urban places existed in Pennsylvania by 1860; only 9 were in the western half of the state. Most of the urban places remained small in size in 1860. Of the 43 urban boroughs, only 6 exceeded 10,000 in population.

The urban structure of Pennsylvania in 1860 was clearly dominated by Philadelphia, with its population of 500,000. In that year Philadelphia had nearly 20 percent of the entire state's population, more than 60 percent of the state's urban population, and was more than 10 times the size of the second-ranked city, Pittsburgh (49,217).

Between 1860 and 1880 the number of urban places in Pennsylvania more than doubled, reaching 96 places by the end of the period. Of the 53 new towns added, nearly half were located in the anthracite belt, which in 1880 contained about one-fourth of all urban places in Pennsylvania. The period from 1860 to 1880 was a time of rapid growth for Pittsburgh, which tripled in size from 49,000 to more than 156,000 people. As a result of Pittsburgh's rapid growth and the emergence of many new towns, Philadelphia's dominance of the state's urban structure was greatly reduced. Down from 63 percent in 1860, Philadelphia in 1880 contained about 41 percent of the state's urban-dwellers. Despite the development of new towns, Pennsylvania in 1880 still had 20 counties that could be classified as completely rural because they had no town with more than 2,500 people.

In the last two decades of the nineteenth century there was continued rapid increase in the number of urban places, the total growing from 96 to 185. By 1900 three major clusters of urban centers had emerged: the Pittsburgh area, the anthracite belt, and the counties adjoining Philadelphia. These three regions contained more than half the urban places of Pennsylvania in 1900. Most of the 185 urban places remained small; three-fourths contained fewer than 10,000 people. On the other hand, Philadelphia had exceeded one million people by 1890 and, despite the growth of other cities, still contained about 38 percent of the urban population of the state in 1900. Only 10 of the 67 counties of Pennsylvania remained completely rural in 1900.

During the twentieth century, the growth rates of Pennsylvania towns and cities slowed. Even by 1920 several of these urban places had reached their maximum population and had started to decline in size (Table 7.2). Most of these early declines were in

Table 7.2. Number of Pennsylvania urban places reaching maximum population, by decade.

Year of Maximum Population	No. Places by Size[a]					% of Places		
	2,500–10,000	10,000–25,000	25,000–50,000	50,000–100,000	+ 100,000	< 10,000	> 10,000	Total
1990	45	5	1	—	—	16.0	9.5	14.8
1980	19	1	—	—	—	6.8	1.6	5.8
1970	48	8	—	—	1	17.1	14.3	16.6
1960	33	7	1	1	1	11.7	15.9	12.5
1950	27	10	3	2	2	9.6	27.0	12.8
1940	28	6	1	—	—	10.0	11.1	10.2
1930	45	6	2	3	—	16.0	17.5	16.3
1920	26	1	1	—	—	9.3	3.2	8.1
Pre-1920	10	—	—	—	—	3.6	—	2.9

SOURCE: U.S. Census data, 1880–1990.
NOTE: Urban places incorporated after 1920 not included.
[a]Population size as of 1990.

Table 7.3. Population change in major cities of Pennsylvania, 1930–1990.

	1930	1990	Change	Per-cent	Maximum Population
Philadelphia	1,950,961	1,585,577	− 365,384	− 18.7	1950
Pittsburgh	669,817	369,879	− 299,938	− 44.8	1950
Scranton	143,433	81,805	− 61,628	− 43.0	1930
Erie	115,967	108,718	− 7,249	− 6.3	1960
Reading	111,171	78,380	− 32,791	− 29.5	1930
Allentown	92,563	105,090	+ 12,527	+ 13.5	1970
Wilkes-Barre	86,626	47,523	− 39,103	− 45.1	1930
Altoona	82,054	51,881	− 30,173	− 36.8	1930
Harrisburg	80,339	52,376	− 27,963	− 34.8	1950
Johnstown	66,993	28,134	− 38,859	− 58.0	1930
Lancaster	59,949	55,551	− 4,398	− 7.3	1950
Chester	59,164	41,856	− 17,308	− 29.3	1950
Bethlehem	57,892	71,428	+ 13,536	+ 23.4	1960
York	55,254	42,192	− 13,062	− 23.6	1950
McKeesport	54,632	26,016	− 28,616	− 52.4	1940
New Castle	48,674	28,334	− 20,340	− 41.8	1950
Williamsport	45,729	31,933	− 13,796	− 30.2	1930
Hazelton	36,765	24,730	− 12,035	− 32.7	1940
Norristown	35,853	30,749	− 5,104	− 14.2	1960
Easton	34,468	26,276	− 8,192	− 23.8	1950
Wilkinsburg	29,639	21,080	− 8,559	− 28.9	1950
Nanticoke	26,043	12,267	− 13,776	− 52.9	1930
Sharon	25,908	17,493	− 8,415	− 32.5	1950
Lebanon	25,561	24,800	− 761	− 3.0	1960
Washington	24,545	15,864	− 8,681	− 35.4	1950
Pottsville	24,300	16,603	− 7,697	− 31.7	1940
Butler	23,568	15,714	− 7,854	− 33.3	1940
Dunmore	22,627	15,403	− 7,224	− 31.9	1940
Oil City	22,075	11,949	− 10,126	− 45.9	1930
Shenandoah	21,782	6,221	− 15,561	− 71.4	1910
Kingston	21,600	14,507	− 7,093	− 32.8	1930
Duquesne	21,396	8,525	− 12,871	− 60.2	1930
Shamokin	20,274	9,184	− 11,090	− 54.7	1920
Monessen	20,268	9,901	− 10,367	− 51.1	1930
Ambridge	20,227	8,133	− 12,094	− 59.8	1930
Homestead	20,141	4,179	− 15,962	− 79.3	1920
Carbondale	20,061	10,664	− 9,397	− 46.8	1930

Source: U.S. Census, 1930 and 1990.

smaller towns scattered across the state, with no regional concentration.

After 1920, however, losses became increasingly common in the larger cities of the state. In 1930 there were 37 urban places in Pennsylvania with populations of 20,000 or more. Of these 37 places, only 2 had larger populations in 1990 than they had in 1930 (Table 7.3). Two regions were particularly affected by urban loss: the anthracite belt and the Pittsburgh metropolitan area.

By 1930 anthracite production had crested, and that region of production was increasingly racked by unemployment and economic distress. The total population of the five anthracite counties reached its peak in 1930 and began a long decline thereafter. Of the 63 places that were urban in 1920, 11 had peaked by 1920; 32 more reached their maximum population in 1930. Only 3 of these places reached their peak populations after 1940.

In the Pittsburgh metropolitan area, 76 incorporated places were large enough in 1920 to be classified as urban. Of these, 38 had reached their peak populations by 1940 and another 30 by 1960. Since 1970 only 8 of these 76 urban places have had any growth; only 3 gained in population during the 1980 decade.

Table 7.4. Population change by size class of minor civil division, 1970–1980, 1980–1990.

Boroughs/cities (size)	1970–1980[a]				1980–1990[b]			
	No.	No. Losing	No. Gaining	% Change	No.	No. Losing	No. Gaining	% Change
> 250,000	2	2	0	− 14.5	2	2	0	− 7.4
100,000–250,000	3	3	0	− 9.0	2	1	1	− 4.1
50,000–100,000	8	8	0	− 11.4	7	5	2	− 3.3
25,000–50,000	15	13	2	− 8.4	15	12	3	− 7.1
10,000–25,000	65	61	4	− 9.9	59	48	11	− 6.9
5,000–10,000	141	110	31	− 6.9	139	103	36	− 4.1
2,500–5,000	152	110	42	− 5.5	163	116	47	− 4.7
Total urban places	386	307	79	− 11.2	387	287	100	− 6.0
Nonurban boroughs	616	344	272	− 0.4	655	479	176	− 5.1
Townships	1,548	191	1,357	15.0	1,549	571	978	6.7

SOURCE: U.S. Census, 1970, 1980, 1990.
[a]City size as of 1970.
[b]City size as of 1980.

Whereas before 1950 the larger share of urban places experiencing declining populations were generally smaller, decreases in population size after 1950 became increasingly common among the larger cities as their basic industries decreased in profitabilty and as more families moved to the suburbs. Both Philadelphia and Pittsburgh reached their peak population size in 1950. Of the 21 Pennsylvania cities incorporated by 1940 and having populations exceeding 25,000 in 1990, 7 reached their maximum populations before 1950; an additional 8 reached their maximum in 1950. The only one of these largest cities to reach its peak population size in 1990 was State College. The combined population of these 21 cities in 1950 was a bit more than 4 million; by 1990 this population had decreased by more than one million.

During the 1970s losses in population occurred among all size classes of urban places (Table 7.4). Of the 386 places classified as urban in 1970, some 307 lost populations during the decade. And losses tended to be more severe in the larger cities. The smallest relative losses were in the nonurban incorporated places having populations smaller than 2,500 people. It should be noted that Table 7.4 does not take into account the urbanization areas, the built-up areas around major cities. No small part of the population gains in the townships of Pennsylvania during the 1970s took place in the suburban townships around the larger cities. Nevertheless, the total urban population of the state decreased by 2.6 percent during the 1970s.

The decrease in the urban population of Pennsylvania continued through the 1980s at a reduced rate. During that decade, the urban population of the state declined by 32,000, a loss of about 0.4 percent. Although all size classes of urban places continued to lose population, the rate of loss was reduced in every size class except for the smallest category: incorporated places having fewer than 2,500 inhabitants. Whereas in the 1970s no city with more than 50,000 people gained in population, during the 1980s three such cities—Allentown, Bethlehem, and Lancaster—gained. Townships as a class continued to gain in the 1980s, but at a much reduced rate relative to that of the 1970s. Many of the townships that lost population in the 1980s were in the western part of the state, where population growth rates turned negative. Moreover, a number of large townships bordering Philadelphia also lost population during the 1980 decade. The suburbs with the most rapid gains since 1970 have been the outer suburbs at some distance from the central city.

Because of the rapid growth of unincorporated suburbs during the twentieth century, the U.S. Census Bureau in 1950 introduced two new residential classes. "Urbanization areas" included the built-up areas around major cities that were unincorporated but had population densities in excess of 1,000 people per square mile. "Standard Metropolitan Areas" included entire counties that contain a city (or, since 1980, an urbanized area) of 50,000 or more people together with adjoining counties functionally tied to the central city. (For precise definitions of

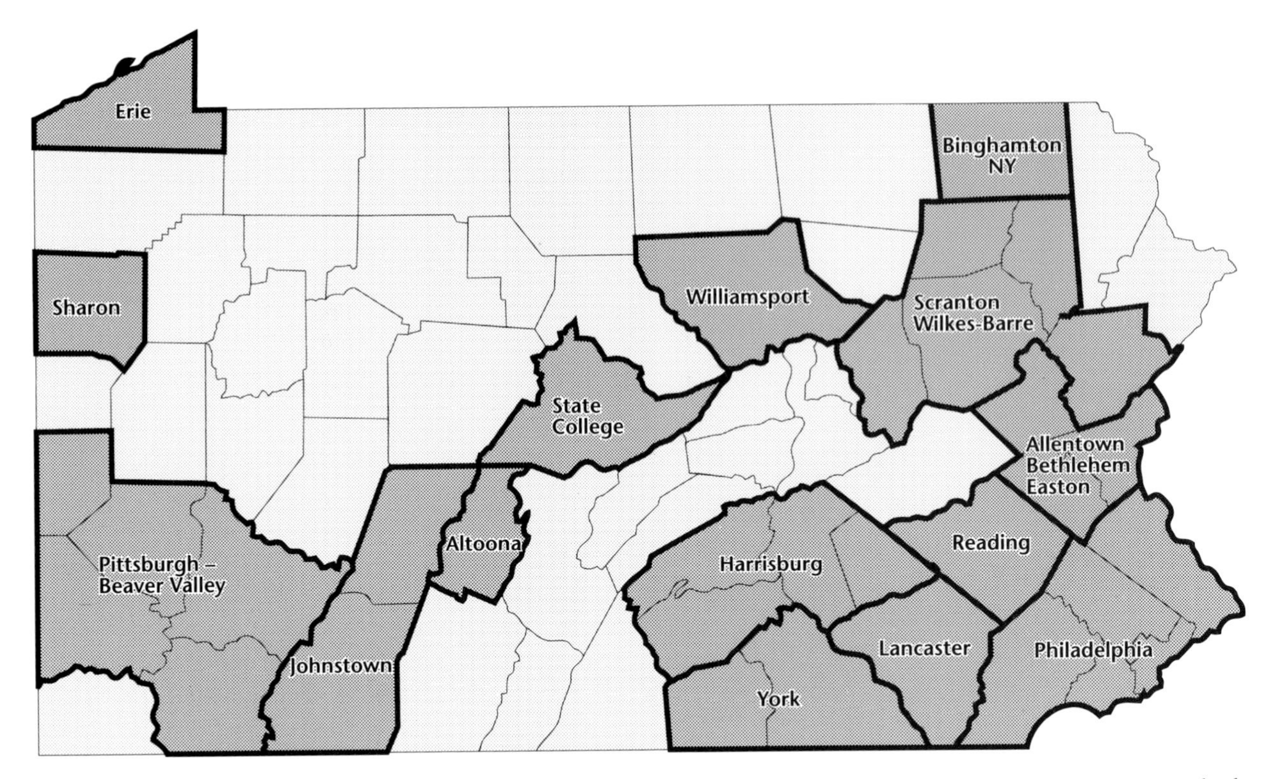

Fig. 7.4. Standard Metropolitan Areas of Pennsylvania, 1990. Darker shading indicates counties included in "Standard Metropolitan Areas." Names given to the metropolitan areas, i.e., Pittsburgh–Beaver Valley, are so designated by the U.S. Census Bureau.

urbanized areas and standard metropolitan areas, see the appropriate census volumes.)

In 1990 some 85 percent of Pennsylvania's people lived in the 14 standard metropolitan areas of the state, together with Susquehanna County, which is part of the Binghamton, New York, metropolitan area (see Fig. 7.4). That metropolitan portion of the state's population was only very slightly larger than in 1980. Of these metropolitan areas, Altoona, Erie, Johnstown, Pittsburgh–Beaver Valley, and Sharon had smaller populations in 1990 than in 1980. For the nation as a whole, about 77 percent of the population is classed as metropolitan. Pennsylvania, therefore, has the unusual distinction of being less urban, but more metropolitan, than the national average. The explanation is that more of the state's rural population is included in its metropolitan counties than is true for the nation.

Within these metropolitan counties the 20 central cities contained 24 percent of the state's population in 1990. Only 5 of these central cities gained population between 1980 and 1990. Around these central cities, the urbanized areas (i.e., suburbs) had about 37 percent of Pennsylvania's people in 1990 and were the residential class that held the largest share of the state's population (Table 7.5). That proportion of the state's population contained in suburbs is somewhat more than the average for the United States. Because the boundaries of the urbanized areas, based on population density, may shift with each census, it is difficult to determine the nature of the intercensal population shifts. Nevertheless, half of Pennsylvania's urbanized areas had a smaller population in 1990 than in 1980.

Rural Population Growth

In 1790 about 95 percent of the nation's population and nearly 90 percent of Pennsylvania's people were classified as rural. Thereafter, with decreasing birth rates and increasing rural-urban migration, the growth rate of the rural population of both the United States and Pennsylvania underwent irregular but progressive decline. The rates of growth gener-

Table 7.5. Percentage of population by residential class, 1990.

Class	Population	% of Total
Metropolitan	10,077,002	84.8
Central cities	2,917,873	24.6
Urbanized areas	4,345,805	36.6
Other urban	445,732	3.8
Rural	2,367,592	19.9
Nonmetropolitan	1,804,641	15.2
Urban	478,885	4.0
Rural	1,325,756	11.2
Urban	8,188,295	68.9
Rural	3,693,348	31.1
Total	11,881,643	100.0

SOURCE: U.S. Census, 1990.

ally paralleled each other, but the growth rate of the rural population of Pennsylvania was less than the national average. Despite the early decline in the rates of rural population growth, the numbers added to Pennsylvania's rural population continued to grow larger each decade until 1850. Indeed, the numbers added to the state's rural population each census interval from 1790 to 1850 were larger than those added to the urban population. After the Civil War and until 1920, the numbers added to the rural population each decade generally grew smaller. Nevertheless, from 1870 through 1920, Pennsylvania continued to have the largest rural population of any state.

During the decade 1910–1920, quite likely stimulated by wartime demand for factory production, the urban population of Pennsylvania grew by more than one million persons. The rural population grew by only 13,122, the smallest intercensal gain previously recorded for the state. By 1920 the rural population of the state was surpassed in size by that of Texas. There was a mild recovery of growth rates in the rural population of Pennsylvania during the 1920s, although an increasing share of this rural growth, as in the nation, was in unincorporated suburbs around the larger cities which were beginning to expand with the proliferation of transportation routes and automobile ownership.

By 1930 nearly one-third of Pennsylvania's population was classified as rural. Of that rural population, only a bit more than one-fourth was resident on farms. Indeed, only Fulton County had the majority of its population living on farms, although three other counties (Wayne, Bradford, and Susquehanna) had more farm residents than people living in either rural nonfarm or urban places. While counties with large proportions of rural nonfarm populations were concentrated in the central and western parts of the state, it is noteworthy that one-tenth of the total rural nonfarm population of the state lived in the four counties surrounding Philadelphia. It seems reasonable to assume that many of these people were living in unincorporated suburban communities.

During the depression years of the 1930s, there was a large increase in the size of the rural population of Pennsylvania as reduced urban opportunites caused fewer people to leave, and more to return to, the rural areas. Of the 60 Pennsylvania counties that were not completely urban or completely rural, 48 experienced an increase in the rural proportion of their populations between 1930 and 1940. Substantial growth of the rural population continued on into the 1940s, boosted by a continuing spread of suburbs. It was this expansion of suburbs—urban in character but rural by census definition—which led the U.S. Census Bureau to change its definition of "urban" in 1950 to include the built-up but unincorporated areas around larger cities which had previously been classed as rural. The effect of this change was to remove more than 400,000 people from the category of "rural" population of Pennsylvania and transfer them to the urban category. Fully half of the people transferred by this definitional change were in four counties (Allegheny, Delaware, Luzerne, and Dauphin) where past suburban growth had been considerable.

Had the definition of "rural" not changed, in 1960 both New York and California would have had larger rural populations than Pennsylvania. But because both those states had a larger share of their rural populations reclassified as urban, Pennsylvania reemerged in 1960 as the state having the largest rural population, a distinction it maintained in 1990.

With the new definition of "rural," Pennsylvania continued to gain in rural population from 1950 to 1970, contrary to the nation as a whole, in which the rural population was reduced in size. Since 1970 there has been a resurgence in the growth rate of the rural population both in the state and in the nation, although the large out-migration from Pennsylvania during the 1970s brought the rural growth rate for the state below that of the nation. Nevertheless, between 1970 and 1980 more than 280,000 people were added to the rural population, the largest increase in rural numbers since 1880. During the 1980s the growth of rural population in Pennsylvania greatly slowed as most of the counties in the western part of the state, which were predominantly rural, lost population. Nevertheless, the state added another 50,000 people to its rural population during the 1980 decade. It is interesting to note that Pennsylvania has experienced continuous growth in the size of its rural population since the beginning of record, a distinction shared only with New Jersey and North Carolina among the 50 states.

In 1990 nearly 3.7 million people in Pennsylvania (31 percent of the total population) were classified as rural. Even though more than two-thirds of Pennsylvania's population was urban, 42 of the state's 67 counties remained predominantly rural. Indeed, 7 counties did not have a single incorporated place with a population as large as 2,500 and thus were classed as completely rural. In addition, Perry County, which has no urban place as such, did have 6 percent of its population included in the urbanized area of Harrisburg.

Only a small part of Pennsylvania's rural population lives on farms; in 1980 only 1.3 percent of the state's total population was included in the rural-farm category, a somewhat smaller proportion than the average for the nation (1.67 percent). Only 14 counties of Pennsylvania had as much as 5 percent of their populations living on farms and nearly all these counties were in the eastern half of the state. Even in these 14 counties, more than one-third of the farm operators worked more than 100 days off-farm in 1978 in all but Lancaster and Mifflin counties. The largest rural farm population in the state in 1980 was in Lancaster County, where the 24,556 farm residents made up more than 15 percent of the total rural farm population of the state.

Natural Increase

Fertility

The primary reason for the declines in population growth rates across most of the nineteenth century was a progressive reduction in fertility levels. The birth rate of the United States early in that century is thought to have been quite high. Some authorities estimated that the crude birth rate (number of births per 1,000 total population) of the United States around 1800 was at least 50 per 1,000. Already by 1800, however, birth rates may have started to drop in parts of the Northeast under the impact of growing farmland shortage and increasing distance to cheap land on the frontier. Once under way, fertility levels continued to drop, hastened downward later in the century by increasing urbanization. So, for more than 100 years, beginning in the Northeast and moving westward, birth rates in the United States progressively fell until bottoming out in the 1930 decade of the depression. Precise measurement of changing fertility levels in the United States, or more specifically in Pennsylvania, is difficult because of inadequate records. By 1830, however, censuses record enough age data to allow calculation of child-woman ratios (the number of children less than five years of age per 1,000 women age 15–49 years). Child-woman ratios are not sufficiently sensitive to detect minor changes in fertility levels, for any improvement in infant mortality would result in increasing the number of children under five years of age surviving to be enumerated in the census and result in partially masking the effect of declining fertility. Nevertheless, child-woman ratios can be used to discern major trends.

By 1830 Pennsylvania and other areas of the Northeast had child-woman ratios well below the national average. Throughout the remainder of the nineteenth century, child-woman ratios continued to decline, with the ratio in Pennsylvania consistently falling below the national average. Toward the end of the nineteenth century, however, the decline in the child-woman ratio tapered off in Pennsylvania, and then rose during the 1910 decade while the corresponding ratios for the United States continued to decline. Even by 1930, following a substantial drop during the 1920s, the child-woman ratio of Pennsylvania remained slightly above that of the nation. The reason for this apparent higher fertility in Pennsylvania than in the United States generally is not clear. Certainly the answer does not lie in the influx of higher-fertility immigrants, for New York State, which received far more immigrants than did Pennsylvania, had a child-woman ratio well below that of Pennsylvania.

Toward the end of the 1930s, fertility levels in both the United States and Pennsylvania began to move upward, only to drop back a bit during the final years of World War II. Then, with the conclusion of the war came the tremendous vital revolution resulting in birth rates climbing rapidly to crest in 1957 at a level near that of 1920, only to decline again

Table 7.6. Total fertility rates for whites, 1940–1980.

	1940	1950	% Increase 1940–50	
U.S.	2,220	2,970	33.8	
Pennsylvania	1,960	2,577	31.5	
New York	1,656	2,504	51.2	
New Jersey	1,611	2,419	50.2	
	1960	**% Increase 1950–60**	**% Increase 1940–60**	
U.S.	3,527	18.8	50.4	
Pennsylvania	3,294	27.8	68.1	
New York	3,252	29.9	96.4	
New Jersey	3,359	38.9	108.5	
	1970	**% Decrease 1960–70**		
U.S.	2,380	28.7		
Pennsylvania	2,325	29.4		
New York	2,314	28.8		
New Jersey	2,307	31.3		
	1980	**% Decrease 1970–80**	**% Decrease 1960–80**	**% Decrease 1940–80**
U.S.	1,774	25.5	46.9	20.1
Pennsylvania	1,598	32.3	51.5	18.5
New York	1,603	30.7	50.7	3.2
New Jersey	1,554	32.6	53.7	3.5

SOURCE: *Vital Statistics of the United States.*
NOTE: Total fertility = Average number of children born per 1,000 women during their lifetimes given the age-specific fertility rates of specified years.

to a historic low in the mid-1970s and to fluctuate somewhat above that level thereafter.

Pennsylvania experienced the revolution in vital rates after World War II, but at a somewhat different tempo and intensity than its neighbors and the nation in general (Table 7.6). In 1940 Pennsylvania was among the states having below-average fertility, albeit at a level above the levels in the neighboring Middle Atlantic states. During the initial part of the baby boom, 1940–1950, Pennsylvania's white total fertility (defined in Table 7.6) increased substantially, although the relative increase was somewhat below that of the nation and was well below the increases in the neighboring states of New York and New Jersey. In the latter part of the baby boom, 1950–1960, the relative increase in the national white fertility rate slowed considerably. Although white total fertility rates also slowed during the latter part of the baby boom in Pennsylvania and its neighboring states of New York and New Jersey, the rates of increase remained well above the national average. Indeed, the state's rate of increase in white total fertility between 1950 and 1960 was nearly as large as in the previous decade. Thus, the Middle Atlantic states as well as most of the states of the Northeast where 1940 fertility rates were low, experienced a much greater increase in fertility during the baby boom than the states of the South and West, where 1940 fertility rates were already relatively high.

After 1960 fertility levels throughout the nation fell sharply. The decrease continued into the 1970s with somewhat greater intensity in the Middle Atlantic states than in the nation at large. Thus, by 1980 the white total fertility rate in Pennsylvania remained below the average for the United States and was intermediate between those of New York and New Jersey. Although birth rates in Pennsylvania remain below the national average, there has been a consistent increase in rates from 1984 to 1990. By 1990 the general fertility rate (number of births per 1,000 women of reproductive age) was 63.6, the highest of any year since 1972. The corresponding rate for the United States in 1990 was about 71.

Following World War II, Pennsylvania also underwent changes in its age-specific fertility patterns. In 1940 the age-specific fertility rates (the number of children born per 1,000 women in specific age-groups) climbed sharply with age to peak in the 20–24 age-group of mothers. However, the age-specific fertility rate for the women in the 25–29 age-group, contrary to that of the United States as a whole, was only marginally lower than the rate for the 20–24 age-group. Thus, by 1940, Pennsylvania women had adopted a pattern of later marriage and family formation than was characteristic nationally.

During the period of the baby boom, age-specific fertility rates increased in all age-groups except the 40–44 age-group in both Pennsylvania and the United States with the greatest relative increase occurring among women 20–24. In Pennsylvania the age-specific fertility rate for that group of white women almost doubled. Yet in 1940 the age-group 20–24 was responsible for 31.9 percent of all births to white women in the state and, despite the great increase in the fertility rate, only 31.4 percent in 1960. The explanation of this apparent puzzle is that the women 20–24 in 1960 were born during the depression, when birth rates were low and few females were born. Whereas in 1940 more than 435,000 white females in the state were between the ages of 20 and 24, in 1960 there were fewer than 300,000 in that same age-group. Thus, in 1960, a much smaller cohort of women produced almost the same proportion of total births in Pennsylvania as their more numerous counterparts in 1940. The baby boom was therefore most pronounced in the northeastern states and among younger women.

After 1960 age-specific fertility rates in Pennsylvania and the nation decreased in all age-groups, especially in the 20–24 age-group and among women 35 years and older. Thereafter, the pattern of delayed marriage and family formation that was characteristic earlier of the Middle Atlantic and southern New England states had diffused more widely through the nation, so that the peak fertility rates in the nation as well as in Pennsylvania had shifted from the 20–24 to the 25–29 age-group.

Table 7.7. Age-specific birth rates, Pennsylvania, 1970–1990.

Age-group	1970	1990
Under 15	0.8	1.2
15–19	53.2	45.1
20–24	156.1	94.5
25–29	149.6	118.5
30–34	75.7	80.8
35–39	32.3	29.7
40–44	7.5	4.5
All ages	80.6	63.6

SOURCE: Pa. Dept. of Health.

Between 1970 and 1990 the greatest change in age-specific fertility rates occurred within the 20–24 age-group, whose fertility level in 1990 had dropped by more than one-third below its 1970 level (Table 7.7). In 1970 mothers younger than 25 years old accounted for 52 percent of all live births, but by 1990 that share had dropped to 35 percent. Although between 1970 and 1990 there was a small increase in fertility rates for girls younger than 15, few births occurred in that age-group in either year. A much more significant increase occurred among women 30–34, among whom the age-specific birth rates in 1990 were higher than in 1970. Indeed, in 1990 there were almost as many births in Pennsylvania to women in that age-group as there were to women 20–24, although birth rates among the latter group still exceeded those of women 30–34. Further, birth rates to women 35–39 remained almost as high in 1990 as they had been in 1970, indicating that the pattern of delayed marriage and family formation has further strengthened since 1970.

Regionally within Pennsylvania there remains wide variation in fertility levels. Among the counties, the numbers of births per 1,000 women of reproductive age ranges from a high of 78.6 in Philadelphia to a low of 44.5 in Centre County. However, the rate for Centre County is depressed because of the large number of young, unmarried females attending the Pennsylvania State University. Although there are no discernible regional patterns in fertility levels across the state, rural counties have had a larger share of their births to women 24 years or younger compared with the more urban counties. Suburban counties in particular have had a large proportion of their births to women 25 years or older. Thus, women in rural areas tend to marry and have their first birth at younger ages than their counterparts in urban areas, who marry later, reach their peak fertility in the 25–29 age-group, and continue to bear children at older ages than the women in the rural areas.

Mortality

Few studies have examined mortality trends in the United States during the nineteenth century, and the estimates that have been made vary significantly. It is generally assumed that in our early history death rates were relatively high and subject to wide annual fluctuations, depending on the presence or absence of epidemics. Port cities commonly were more subject to epidemics than were inland areas, as disease often disembarked with ships' passengers. Not only were infections and parasitic disease common, but many of these diseases took an especially heavy toll among infants and children.

It is difficult to determine exact causes of death reported by the cities of Pennsylvania in the late nineteenth century because diagnosis was often imprecise and nomenclature has since greatly changed. For example, among the causes of death reported by the cities of Pennsylvania in the late nineteenth century were indigestion, exhaustion, teething, and menopause. Nevertheless, some idea of the nature of the more important killers can be gained from looking at the leading causes of death in cities of Pennsylvania around 1890 (Table 7.8). Noteworthy among the leading causes were tuberculosis and pneumonia, which were major killers not only in Pennsylvania but also generally across the United States; they commonly alternated as the principal cause of death. Indeed, the prevalence of tuberculosis may well have been understated. The 1906 report of Pennsylvania's Commissioner of Health, states that tuberculosis among children was often reported as marasmus because physicians were reluctant to put tuberculosis on death certificates since that would make it difficult for surviving family members to obtain insurance.

The prevalence of diseases that were especially lethal to children is also conspicuous on lists of principal causes of death in Pennsylvania in the late nineteenth century. Most of the data on reporting cities of Pennsylvania in the late 1880s and early 1890s suggest that 40 percent or more of total deaths in those cities occurred to children less than 5 years old. In 1898, the leading causes of childhood deaths, in order, were diarrhea, pneumonia, asthenia infantile (general debility), whooping cough, and convulsions.

Registration of deaths became progressively more comprehensive in Pennsylvania toward the end of the nineteenth century, and in 1906 the state was admitted to the Death Registration Area—states and cities deemed by the U.S. Census Bureau to register at least 90 percent of the total estimated deaths. The previous year, the Pennsylvania Department of Health was established and later issued its first annual report covering the year 1906. That year the leading causes of death in the state were, in order, diarrhea and enteritis, tuberculosis, and pneumonia, each being responsible for about 9 percent of the total reported deaths in the state. Diarrhea and enteritis were particularly concentrated among children. Of the 11,249 deaths in 1906 from that cause, 87 percent were children less than 2 years old. One out of every 40 children in the state in that age-group died from diarrhea and enteritis in 1906. Seventy percent of the deaths from this cause occurred in the summer months of July, August, and September.

From 1906 onward, with a few brief interruptions such as the influenza epidemic of 1918, infant mortality rates progressively declined (Fig. 7.5). Childhood mortality, however, continued to remain high for the first quarter of the twentieth century. In 1906 about 34 percent of all deaths in Pennsylvania occurred to children less than 5 years old, a

Table 7.8. Five leading causes of death, selected Pennsylvania cities, c. 1890.

Altoona 1890	Philadelphia 1890	Pittsburgh 1889	Reading 1889	Scranton 1887
Cholera infantum	Tuberculosis	Pneumonia	Tuberculosis	Pneumonia
Pneumonia	Inflamation of lungs	Cancer	Infant diarrhea	Cholera infantum
Tuberculosis	Cholera infantum	Typhoid fever	Pneumonia	Dysentery
Diphtheria	Old age	Diphtheria	Heart disease	Convulsions
Convulsions	Typhoid fever	Choleraic diarrhea	Apoplexy	Tuberculosis

SOURCE: Pennsylvania State Board of Health and Vital Statistics.

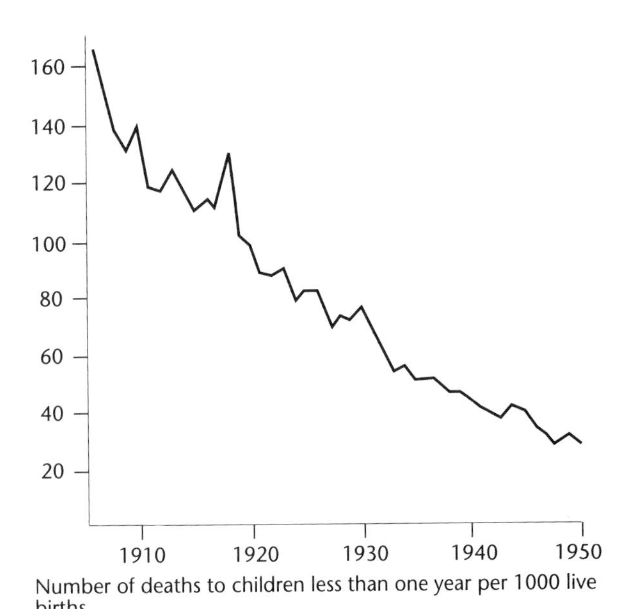

Fig. 7.5. Infant mortality in Pennsylvania, 1906–1950.

proportion that was maintained in 1910. For 1910 we have the first data available showing the deaths to children under 5 as a percentage of total deaths by county. Among the Pennsylvania counties, Crawford County had the lowest proportion of deaths to young children (about 16 percent of total deaths), and Cambria had the highest proportion (nearly 51 percent). Regionally, childhood deaths as a percentage of total deaths were highest in the southwest quadrant of the state and in the anthracite counties, and lowest in some of the Northern Tier counties. Two-thirds of the 53 cities of Pennsylvania for which 1910 data exist had a higher proportion of deaths among children than did the counties in which they were located. In ten of these cities, more than half the total deaths were children under 5 years old.

By 1920 pneumonia and tuberculosis were still included among the five leading causes of death in Pennsylvania, but diarrhea and enteritis had dropped to tenth place and most other childhood diseases took reduced tolls. The pattern for most contagious diseases was for an irregular, interrupted decline from early in the twentieth century until about 1920, then a steep decline to the low levels at present (Fig. 7.6). By 1984, infectious and parasitic diseases were responsible for only 1.6 percent of total deaths in the state, and many of these deaths were from septicemia; few of these latter deaths were to children. Tuberculosis, which in 1920 was responsible for nearly 8 percent of the total deaths in Pennsylvania, has almost disappeared as a cause of death. In 1990 only 74 of the total 121,000 deaths occurring in the state were charged to tuberculosis.

Largely as a result of improvements in diet, sanitation, and control of contagious disease, the infant mortality rate in Pennsylvania, which in 1920 was nearly 100 (nearly 100 infants out of every 1,000 born alive died before attaining their first birthday), had dropped to 9.5 per 1,000 in 1990. In the 1980s, infant mortality rates for Pennsylvania tended to be near the national average or slightly above. However, considerable regional differences in the level of infant mortality remain within the state. Relatively high infant mortality is characteristic of many of the larger cities. Indeed, nearly all the cities having populations in excess of 25,000 have higher infant mortality rates than the counties in which they are located.

While in the twentieth century the death rate from contagious diseases was declining in Pennsylvania, deaths from degenerative disease in the older population increased. Changing classifications of disease, and improvements in diagnoses, make it impossible to follow the rates of heart disease and malignant neoplasms (cancer), now the leading causes of death in Pennsylvania, precisely for the twentieth century. Nevertheless, by 1920 heart disease was ranked second only to pneumonia as the principal cause of death among reported deaths in the state, with malignant neoplasms ranking seventh. Since 1940

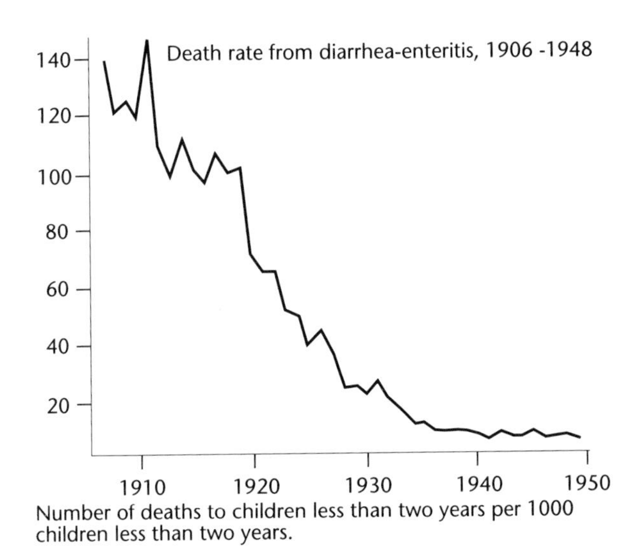

Fig. 7.6. Mortality under age 2 in Pennsylvania, 1906–1948.

Table 7.9. Age-adjusted death rates (per 100,000 population), Pennsylvania–United States, 1990.

	Pennsylvania		United States[a]	
	Rate	Rank	Rate	Rank
Heart disease	160.1	1	150.3	1
Malignant neoplasms	140.4	2	133.0	2
Accidents	27.5	3	32.7	3
Cerebrovascular disease	25.8	4	27.6	4
Chronic obstructive pulmonary disease	17.6	5	19.7	5
Diabetes mellitus	13.1	6	11.7	7
Pneumonia/influenza	12.0	7	13.5	6
Suicide	11.1	8	11.3	8
Liver disease and cirrhosis	7.3	9	8.3	10
Homicide	7.3	10	10.6	9
All causes	519.7		515.1	

SOURCE: Pa. Dept. of Health.
[a]Provisional.

those two conditions have consistently ranked first and second, respectively, among the causes of death in Pennsylvania, and in 1990 they were responsible for 60 percent of all the deaths in the state. Rates of death from heart disease in the United States generally and in Pennsylvania peaked in the 1950s and 1960s but have since dropped back, slowly but progressively. In Pennsylvania the heart disease death rate per 100,000 population in 1960 was 454.4; in 1990 the corresponding rate was 363.3. On the other hand, death rates from malignant neoplasms have climbed persistently during the century in both the state and the nation. Malignant neoplasms were responsible for nearly 15 percent of all deaths in Pennsylvania in 1950; by 1990 some 24 percent of the deaths in the state were from this cause. Death rates for heart disease and malignant neoplasms in the state have consistently been above average for the nation in recent decades. These elevated death rates in Pennsylvania can be misleading, however, because the state has a large share of older people, among whom the risk of the degenerative diseases is great. Still, even age-adjusted death rates, which remove the effect of differential age compositions among populations, indicate that death rates from heart disease and malignant neoplasms in Pennsylvania are above the national average (Table 7.9). The same remains true for all causes of death combined; the rate for Pennsylvania remains somewhat above that for the nation, although the difference has narrowed since 1950.

In Pennsylvania as in the nation, death rates for males are higher than for females, for nearly all causes of death and for all age-groups. Table 7.10 shows the differential between the death rates for men and women 65–84 from the 10 major causes of death for that age-group in 1990.

There are no distinct regional patterns in age-adjusted death rates among the counties of Pennsylvania, other than a concentration of low rates among the more prosperous counties of the southeast. Philadelphia County had the highest age-adjusted death rate for the period 1986–1990 among the counties of the state, whereas its neighbor, Montgomery County, had the lowest. Resident death rates for the larger cities of the state are almost universally higher than the rates for the counties in

Table 7.10. Male-female death rates (per 1,000), ages 65–84, selected causes, 1990.

	Male	Female
Heart disease	1,896.0	1,184.4
Malignant neoplasms	1,423.4	854.2
Cerebrovascular disease	283.3	240.2
Chronic obstructive pulmonary disease	272.7	137.2
Pneumonia-influenza	142.2	85.1
Diabetes mellitus	111.7	114.2
Unintentional injuries	82.4	47.5
Septicemia	68.5	48.2
Nephritis, nephrosis, etc.	60.4	41.9
Liver disease, cirrhosis	49.4	22.5
All causes	5,076.7	3,212.0

SOURCE: Pa. Dept. of Health.

which these cities are located. These latter rates are not age-adjusted; differences in the two areas may be due in part to differentials between them in age and sex composition. Average annual age-adjusted death rates for heart disease for the period 1986–1990 show a distinct concentration of low rates in the southeast counties, with the singular exception of Philadelphia County, which has rates well above the state average. The corresponding death rates for malignant neoplasms also show considerable variation among the counties, with Philadelphia and Allegheny counties showing rates well above the state average.

Net Migration

Over the entire period of U.S. history a number of major migratory flows have resulted in significant relocations of people: the westward movement of people to the frontier; the move from rural to urban areas, particularly from the South to the larger cities of the Northeast; the flight from the central cities to the suburbs and beyond; and, more recently, the flight from the Northeast and Midwest to the "sun belt" states of the South and West. Pennsylvania has been strongly influenced by all these major relocations of people, beginning quite early in the state's history.

Even before the American Revolution, as settlement was still spreading across southeastern Pennsylvania, a vanguard of Pennsylvania-born Germans, Scots-Irish, and others had already begun to move southward into Maryland and Virginia on either side of the Blue Ridge, a movement that intensified when the westward push of settlement reached the Ridge-and-Valley region, where terrain obstacles slowed expansion to the West. The Great Valley offered not only easy access to the South but also led to largely empty, good-quality land. Ultimately, the flow of Pennsylvania-born settlers southward would reach into the Carolinas and spread farther westward across the southern highlands.

The move westward into Kentucky and Ohio developed somewhat later, but after 1785, when the Ohio River route began to draw a sizable flow of migrants, large numbers of Pennsylvanians joined the flow. Dunaway (1948, 195) estimates that 15 percent of Kentucky's population in 1798 had been born in Pennsylvania. Movement of Pennsylvanians into Ohio began as early as 1778 and gathered force over time to contribute a larger share of Ohio's population in the early nineteenth century than any other state.

The number of migrants Pennsylvania contributed to the expanding frontiers in the late eighteenth and early nineteenth centuries can only be guessed at, given a lack of records. The first crude data regarding population relocation within the United States came with the 1850 census, the first time a question was asked concerning place of birth. By comparing the state of birth with the state of current residence, we can get some notion of population redistribution. The problem in interpretation, however, is that each enumeration would count the survivors of all previous migrants, so that a redirection of migration streams can be identified only some time after the fact.

In 1850 some 60 percent of Pennsylvania natives outside the state were living in adjoining states—that is, states that shared a common border with Pennsylvania. Indeed, 48 percent of these out-migrants from Pennsylvania were enumerated in Ohio, a state that was still being settled. A decade later, 1860, Ohio was still the leading destination of migrating Pennsylvanians, although its share of the total out-migrants from Pennsylvania had dropped considerably. Since many of the native-born Pennsylvanians in Ohio were survivors of migrations before 1860, it is apparent that out-migrants from Pennsylvania were redirecting their migration elsewhere. During the years leading up to 1860 the frontier had moved into and through the states of Illinois and Iowa; by 1860 the leading edge of the frontier, defined as the outer limit of settlement having a density of at least two people per square mile, had moved into the eastern part of the northern Great Plains. Table 7.11 indicates that large numbers of Pennsylvanians had joined the migration to those areas.

From 1860 to 1900, however, at a time when the frontier was rapidly moving westward, the percentage of people born in Pennsylvania but living in other states decreased. The reason for that decrease was that over time the distance between Pennsylvania and the frontier increased, as did the costs and hazards of reaching the frontier. Moreover, the interval between 1860 and 1900 was one of expanding opportunities in the factories and mines of the state, so that increasing numbers of Pennsylvania-born substituted in-state migration for the costly move to the frontier.

Table 7.11. Percentage born in Pennsylvania and living in another state, 1850–1970.

	1850	1860	1870	1880	1890	1900	1910
%Born in Pa. living							
In another state	18.6	20.3	19.9	19.1	17.4	16.3	16.6
In adjoining states	62.3	43.9	39.7	37.3	37.4	44.4	48.0
New York	6.2	5.2	5.4	7.0	8.6	11.8	14.7
New Jersey	3.6	4.2	4.9	5.9	7.9	10.1	11.9
Ohio	47.5	30.0	22.4	17.3	14.2	14.0	13.0
In New England	0.9	0.6	0.6	1.3	1.8	2.6	3.0
In Remaining South[a]	5.7	7.3	3.3	4.5	5.3	6.2	7.5
Florida	< 0.1	< 0.1	0.5	0.1	0.3	0.2	0.4
In Remaining North Central[a]	29.5	45.6	53.0	51.9	46.8	37.9	28.8
In West	1.3	2.7	2.9	5.0	8.7	9.1	12.7
California	1.1	1.9	1.7	1.9	2.5	2.7	4.6

	1920	1930	1940	1950	1960	1970
%Born in Pa. living						
In another state	17.0	18.8	18.4	21.7	24.3	26.3
In adjoining states	54.1	59.6	60.9	56.9	53.3	50.1
New York	16.1	18.8	21.2	18.0	15.1	12.7
New Jersey	13.4	16.7	16.3	16.2	16.8	16.5
Ohio	16.5	16.8	15.3	13.5	12.8	11.5
In New England	3.1	3.0	3.4	4.1	4.2	4.6
In Remaining South[a]	8.4	7.6	9.4	12.7	16.7	19.1
Florida	0.8	1.4	2.1	3.6	6.7	8.0
In Remaining North Central[a]	22.7	18.8	16.0	13.7	11.6	11.0
In West	11.7	11.0	10.5	12.7	14.1	15.1
California	5.4	6.7	7.0	8.8	9.9	10.5

SOURCES: U.S. Census data, 1850–1970.
[a]States other than those that border Pennsylvania.

Changes in the destinations chosen by Pennsylvania out-migrants through the latter part of the nineteenth century and into the twentieth illustrate the temporal variation in the opportunity surface of the United States. Until the mid-nineteenth century, the major migratory flow in the nation was westward to the frontier. Thereafter, the major destinations of Pennsylvanians progressively became the rapidly expanding cities of the Northeast. But by the mid-twentieth century the old, basic industries of that region were becoming increasingly depressed, so that a major part of the outflow of people from Pennsylvania was redirected to the new opportunities opening up in the South and West. These two regions held only 15 percent of the Pennsylvania-born out-migrants in 1900 but grew to contain more than one-third by 1970.

Throughout the period for which state-of-birth records are available, 1850–1970, Pennsylvania has lost consistently in the migration exchange with other states. For most of the decades involved, two or more Pennsylvania-born were enumerated in other states for every one person born in another state but living in Pennsylvania.

It is interesting to note that regardless of the apparent continuous net out-migration of Pennsylvanians from the beginning of record in 1850, the proportion of people born in Pennsylvania and living in other states has consistently been below the average of the United States of persons enumerated in states other than that of their birth. In 1990 nearly 85 percent of the people living in Pennsylvania had been born in the state. Indeed, most of the counties in the western interior of the state had 90 percent or more of their populations born in Pennsylvania. Counties of the west that bordered other states generally had below-state-average Pennsylvania-born, presumably because short-distance moves from adjoining states lowered the percentage of Pennsylvania-born among their people. Pike County, which has drawn large

Table 7.12. Net intercensal migration, Pennsylvania, 1870–1980 (in thousands).

	Total	Native White	Foreign-born White	Black
1870–1880	19.1	− 105.2	115.6	8.7
1880–1890	285.1	− 70.0	334.3	20.8
1890–1900	262.0	− 60.2	282.9	39.2
1900–1910	444.6	− 178.1	589.8	32.9
1910–1920	51.9	− 199.4	168.7	82.5
1920–1930	− 252.9	− 380.2	25.6	101.7
1930–1940	− 301.0	− 260.9	− 60.4	20.3
1940–1950	− 447.2	− 531.3	− 5.5	89.6
1950–1960	− 475.2	—	—	76.9
1960–1970	− 378.8	—	—	34.9
1970–1980	− 511.0	—	—	—

SOURCES: For 1870–1950, U.S. Department of Commerce; for 1950–1960, Bowles et al. 1965; for 1960–1970, Bowles et al. 1975; for 1970–1980, U.S. Department of Commerce.

numbers from neighboring New Jersey and New York, had the lowest proportion of Pennsylvania-born (25 percent) among its people. Only three counties, Pike, Monroe, and Centre, had fewer than half of their people living in the same house in 1990 as in 1985.

During the latter part of the nineteenth century and until 1920, the continuous net loss of the native-born population of Pennsylvania through migration to other states was more than offset by immigration from Europe, so that the state continued to grow in population by migration (Table 7.12). For example, during the 1900–1920 period, when immigration into the United States was at its peak, the net out-migration of more than 377,000 native whites from Pennsylvania was more than compensated for by the immigration of more than 758,000 foreign-born whites. During the period 1880–1920, assuming that our data on immigration into Pennsylvania are reasonably correct, about 30 percent of Pennsylvania's population growth resulted from the immigration of foreign-born.

After 1920, with the sharp reduction in the numbers of immigrants entering the state and the loss of people in the migration exchange with other states, Pennsylvania has been characterized by net out-migration. It is not unlikely that during the time since 1920 Pennsylvania has lost more than 2.5 million people in out-migration to other states. Net out-migration from Pennsylvania was especially large during the 1970s when the state lost more than 511,000 net out-migrants.

Data on net migration for Pennsylvania during the 1980s are not available as yet. During the 1980–1990 census interval, however, some 1.6 million children were born in the state and more than 1.2 million people died. Thus, had there been no migration into or out of Pennsylvania the state's population should have gained some 390,000 people, but in fact the state gained fewer than 18,000. During the 1980s, therefore, the indicated net out-migration was about 370,000 people. If this estimate of net migration is reasonably close, it would appear that the average annual net loss of population by out-migration to the state slowed during the 1980 decade, compared with the previous decade.

Since World War II remarkable shifts have taken place in the migration patterns of Pennsylvania and the nation. Continued growth of urban places, especially suburbs, and the exodus of rural people meant that during the 1950s and 1960s the common migration pattern for most states was net in-migration to combined metropolitan counties, and net out-migration from nonmetropolitan counties. Pennsylvania departed from this common pattern in having net out-migration from both residential classes. Many of the larger cities of the state began to experience net out-migration by the 1950s. The rapid growth of suburbs during the two decades from 1950 to 1970 offsets to a degree the migration loss of the larger cities, so that the net out-migration rate of the combined metropolitan counties was less than that of the nonmetropolitan counties as a class. Of the 45 counties classed as nonmetropolitan in 1960, only 8 had a positive net migration during the preceding decade. The areas with the highest rates of net out-migration were several counties of the Northern Tier, the counties of the anthracite belt, and, especially during the 1960s, most of the nonmetropolitan counties of the Appalachian Plateau. The greatest volume of net out-migration was from Philadelphia and Allegheny counties, where during the 1960s the combined net migration losses totaled more than 335,000 people. Net out-migration from the Pittsburgh metropolitan area was large enough that the population of the region declined. The Pittsburgh metropolitan area was the only one of the nation's 70 largest to lose population between 1960 and 1970.

Large net migration gains during the 1950s and 1960s were in the southeastern quadrant of the state, although Delaware County, especially the older, inner suburbs, was characterized by net out-migration during the 1960 decade. The other region of large in-migration rates during the two decades was the Poconos area centered on Pike County.

Between 1970 and 1980 there was a marked change in the national pattern of population redistribution, with nonmetropolitan counties as a class growing faster than the combined metropolitan counties. Pennsylvania followed this national trend, after a fashion. During the 1970s the metropolitan counties combined lost population even though only the Philadelphia and Pittsburgh metropolitan areas actually decreased in population. Nevertheless, the population of the nonmetropolitan counties, as a class, had a faster rate of growth than did the metropolitan areas that did gain in population during the 1970s.

With respect to net migration during the 1975–1980 census migration interval, the distinction between the two residential classes is less marked. Essentially half of both classes of counties had net out-migration during the 1975–1980 interval. Both classes were characterized by net out-migration during the period, although the rate of net out-migration was less for nonmetropolitan counties than for metropolitan counties.

The most interesting migrational change during the 1975–1980 migration interval occurred in the western nonmetropolitan counties. Of the 22 counties involved, 16 improved their migration status, experiencing smaller rates of loss, greater rates of in-migration, or, in the case of 6 counties, changing from migration loss to migration gains during the 1975–1980 period relative to the previous migration interval, 1965–1970.

The turnaround in growth rates of nonmetropolitan counties during the 1970s has been credited to a growth in opportunity in rural areas and the lengthening of commuter ranges made possible by the extension of limited-access highways.

Whatever the causes that led to the improvement in the migration status of nonmetropolitan counties during the 1970s, they did not persist into the 1980s. Estimates made by the Census Bureau suggest that all the counties of western Pennsylvania, save Bedford and Fulton, had net out-migration between 1980 and 1988. Indeed, of the 37 nonmetropolitan counties that existed in 1980, only 11 had estimated net in-migration between 1980 and 1988. All these, except two, were in the eastern part of the state.

Assuming that the Census Bureau estimates are reasonably correct, it appears that both the metropolitan and the nonmetropolitan residential classes experienced net out-migration between 1980 and 1990, although the rate of out-migration was less for the combined metropolitan counties than the nonmetropolitan counties. Indeed, the metropolitan counties outside Philadelphia and Allegheny appear to have had positive net migration between 1980 and 1988, suggesting again that the cities of Pennsylvania have slowed in their rate of decline since 1980.

Age and Sex Composition

Over time, changes in birth and death rates, combined with variations in the nature of migrations affecting Pennsylvania, have had profound effects on the age and sex composition of the population of the state. Most notably the people of Pennsylvania have become older and increasingly female.

In the 1830 census, the first to give detailed age data, nearly 45 percent of the total population of Pennsylvania was less than 15 years old. Even by that time the birth rate in Pennsylvania had apparently begun the decline that would continue almost without interruption for another century. As the birth rate declined, so did the youthful proportion of the state's population. By the end of the nineteenth century, those under age 15 represented a bit fewer than one-third of the state's total population. Still, by 1900, the age pyramid retained the broad bottom and regularly tapering sides characteristic of a high-fertility population (see Fig. 7.7).

The decline in the relative size of the young component of Pennsylvania's population continued in the twentieth century, although the downward trend was interrupted after the two world wars by a temporary increase in the birth rate. During the

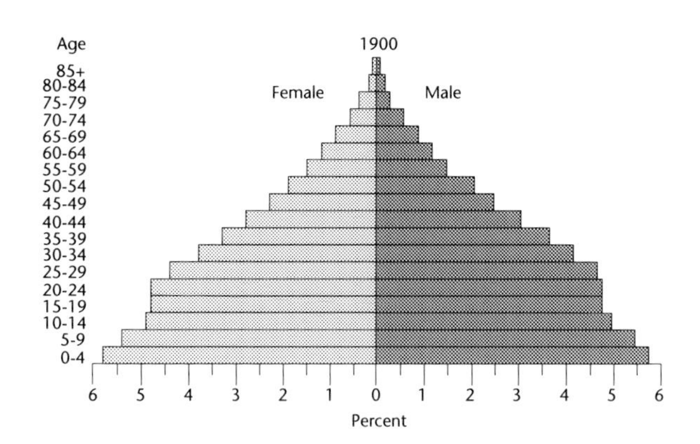

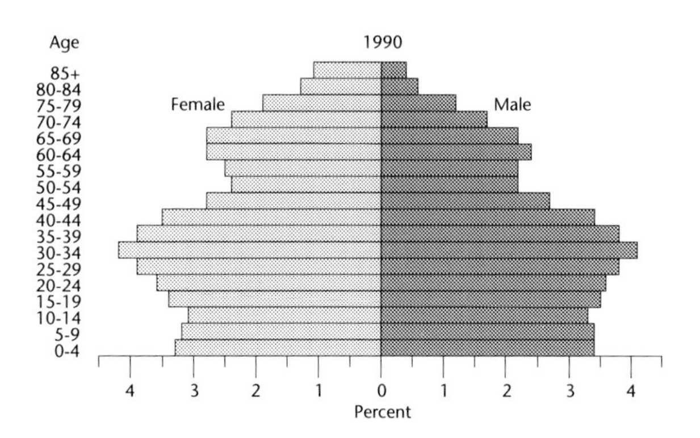

Fig. 7.7. Age structure in Pennsylvania, 1900–1990.

1910–1920 decade the increase in the birth rate was sufficient to bring the child-woman ratio of the state to slightly above the national average. As a result of the increased births, the proportion of Pennsylvanians under 15 years old rose from 30.9 percent in 1910 to 32.1 percent in 1920. Some part of that increase in fertility, and the consequent increase in the youthful proportion of Pennsylvania's population, was due to the large influx of immigrants from southern and eastern Europe (among whom birth rates were relatively high) during the early twentieth century.

After 1920 three things combined to further reduce the youthful component in Pennsylvania's population: a resumption of fertility decline, a sharp reduction in immigration into the state, and a shift from net in-migration into the state to a net out-migration, which by selectively removing young adults reduced the size of the potential parent population, thereby reducing further the numbers of children born in the state. By 1940 only 24 percent of the state's population was less than 15 years old.

The great increase in fertility after World War II profoundly altered the age composition of Pennsylvania's population. By 1960, just after the crest of the baby boom, the proportion of the state's population less than 15 years of age had risen to 29 percent, approximating its 1930 level. Following 1960 subsequent declines in the birth rate progressively reduced the relative size of the youthful component in the state's population. In 1990, despite an increase in the birth rate during the previous six years, only about 20 percent of the state's total population was younger than age 15, a level well less than half of what it had been at the beginning of detailed age records in 1830.

Changes in migration and natural increase over time have also affected the older component of Pennsylvania's population. Censuses in the early part of the nineteenth century provide scant detailed information on age. By 1830 less than 4 percent of Pennsylvania's people had reached the age of 60. That proportion remained remarkably consistent through the latter part of the nineteenth century and into the early twentieth. As late as 1920 only 4.5 percent of the state's population was 65 years or older. Thus, the most rapid rates of growth of the state's population during the late nineteenth and early twentieth centuries were among people of working ages, 15–64. No small part of that working population was contributed by immigrants from Europe. In 1920, for example, about 16 percent of Pennsylvania's population was foreign-born. Among these people, 90 percent were between the ages of 15 and 64.

After 1920, with a progressive decline in both birth rates and death rates, the abrupt slowing of immigration, and the loss of young people in the migration exchange with other states, the proportion of aged in Pennsylvania's population began to increase sharply. Not even the great increase in the youthful population following World War II was sufficient to slow the relative increase in the state's older population. In 1940 a bit less than 7 percent of Pennsylvanians were 65 or older. By 1960 this share had risen to 10 percent; by 1990 it was a little more than 15 percent, a proportion exceeded by only one state, Florida.

The distribution of ages within any large population, barring catastrophes such as war, is largely the result of past changes in fertility. The decrease in both rates during the 1930 depression decade, and the increase in both the young and the old sectors of Pennsylvania's population during the postwar baby boom, have altered the age structure of the state's population considerably. Fig. 7.7 compares the age structure of Pennsylvania's population in 1900 with that of 1990. The age pyramid is a series of bars, the length of each bar indicating the share of the total population included in each age-group, commonly divided into the male and female components. For example, in 1920 a bit less than 6 percent of the total population was male between the ages of 0 and 4, with a slightly smaller share made up of females in the same age-group.

The 1900 age pyramid of Pennsylvania, with its wide base and regularly tapering sides, is typical of a high-fertility population. In this instance the age structure is somewhat distorted by large numbers of foreign-born who were predominantly male, and among whom few were children. If the foreign-born would be excluded, the base of the pyramid would be wider and it would taper more rapidly upward among the ages.

The age pyramid for Pennsylvania in 1990 is quite irregular in shape. Among the older ages, females make up a larger share of the population than males do. Among people 65 years or older in 1990, some 60 percent were female, which attests to their greater resistance to the major degenerative diseases. The most conspicuous features of the 1990 pyramid are the smaller cohorts of people now in their fifties who were born in the depression years of reduced fertility, and the large bulge in the 30–39 age-group, the result of the postwar baby boom. The progressive reduction in the size of the youngest age-groups is the result of declining fertility levels after 1960, although the increase in birth rates during the last few years before 1990 shows up at the very bottom of the pyramid.

The postwar baby boom will continue to affect the age composition of Pennsylvania for years to come as that cohort ages. The leading edge of that age-group will turn 65 shortly after the turn of the twenty-first century. By the year 2030, according to population projections published in the *1985 Pennsylvania Abstract,* more than 18 percent of the state's population will be 65 or older, nearly equal to projections for the age cohort younger than 15 years.

Recent changes in the age composition of Pennsylvania's population has significantly altered the dependency ratio of the state (the number of persons younger than 15 and those 65 or older for every 100 people age 15 to 64). In 1940 the dependency ratio for the state was 45, the lowest ever attained in the state. By 1960, with the addition of large numbers of children during the postwar fertility increase, that ratio had risen to 64, the highest of any census date after 1880. Declining birth rates after 1960 have allowed the dependency ratio to drop again, reaching 54 in 1990.

Among the counties of the state, the nature of migration strongly shapes the age composition. For the state as a whole, the median age was 35 in 1990, compared with 30 for the United States. Within Pennsylvania, the median age varies considerably from one county to the next. Counties with in-migration tend to have median ages below the state average; counties with long-term out-migration generally have relatively high median ages (Fig. 7.8). Centre County shows the lowest median age (27 years), a result of the large number of college-age people in its population. The highest median age (40.7) occurs in Forest County, which has long experienced out-migration. An apparent exception to the rule, Pike County shows the highest positive net migration rate among the counties yet has a median age slightly above the state average. However, among the in-migrants to the county are sufficient numbers of retired persons to raise the median age.

One effect of the differences in median age among the counties of the state is illustrated in Fig. 7.9, which shows the average number of deaths for every 100 births during the 1986–1990 interval. Note that counties of the anthracite region and the long-term out-migration counties, such as Cambria and Forest, have had more deaths than births for the four-year period, thereby reflecting the large proportion of elderly people in those counties.

The early nineteenth-century population of the United States included more males than females; the sex ratio (number of males per 100 females) for the country was 103.3 in 1820. Over the next four decades, the sex ratio gradually increased, largely because of an increase in immigrants among whom males predominated. During the 1860 decade, because of casualties during the Civil War and a reduced volume of immigration, the sex ratio declined, only to rise again to a record high of 106 in 1910 as growing numbers of immigrants entered the United States. Since 1910, the sex ratio of the nation has declined consistently, reflecting in large measure the aging of the population and the lengthening of the life span of women to a greater degree than that of men.

The pattern of change in the sex ratio of Pennsylvania is similar to that of the nation, although the ratio

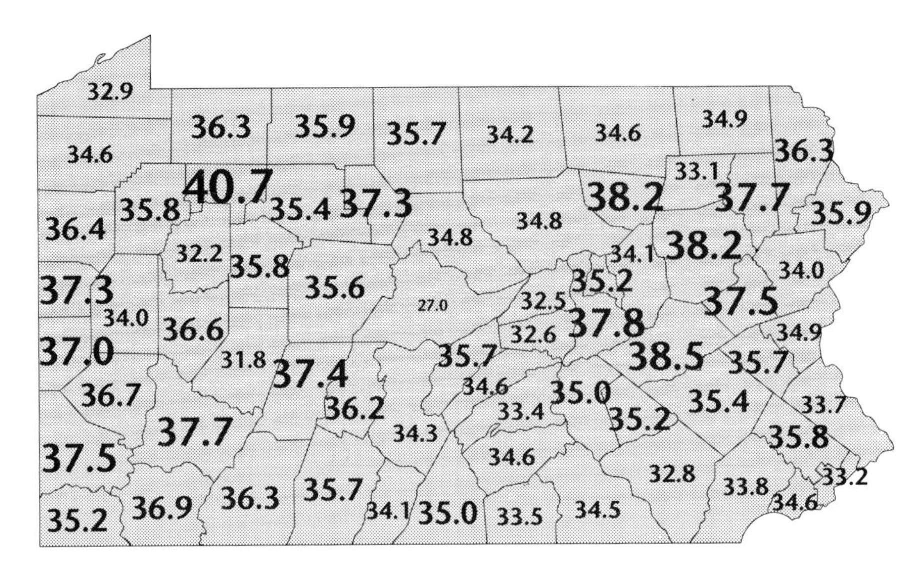

Fig. 7.8. Median age in Pennsylvania, 1990.

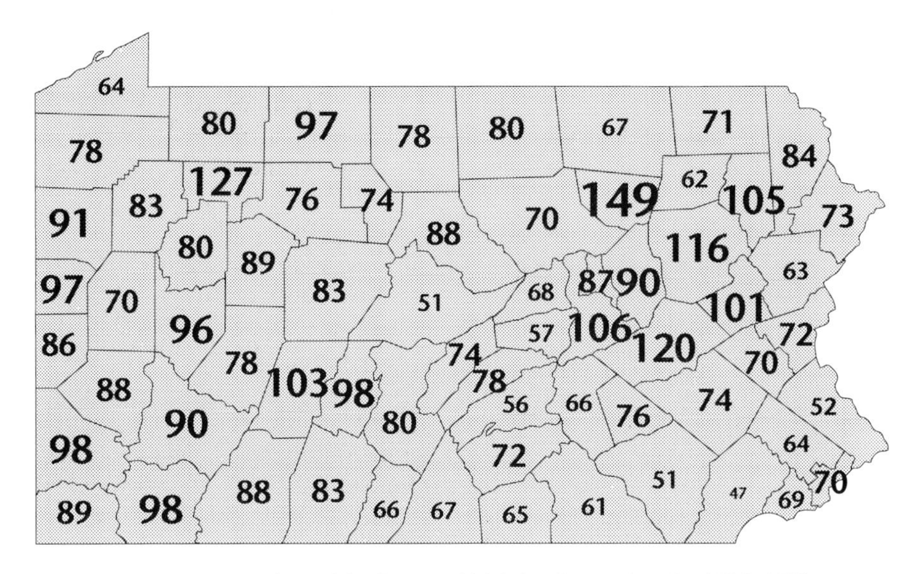

Fig. 7.9. Average number of deaths per 100 births, Pennsylvania, 1986–1990.

has consistently been lower than the national average due to the selective out-migration of males, earlier to the frontier and later to the regions where opportunities beckoned. Toward the latter part of the nineteenth century, the growing opportunities in the mines and factories of Pennsylvania drew increasing numbers of male immigrants, which brought the sex ratio of the state closer to the national average. The impact of the male-dominant immigration shows up strikingly in a comparison of the sex ratios by age in Pennsylvania in 1900 and in 1990 (Fig. 7.10).

Among the young, the sex ratio was higher in 1990 than in 1900 largely as a result of improvement in death rates late in pregnancy and in early infancy, when death rates among males exceed those of females. As more infant males survived as the century progressed, the sex ratios in the younger ages rose. The sharp decline in the sex ratio in the

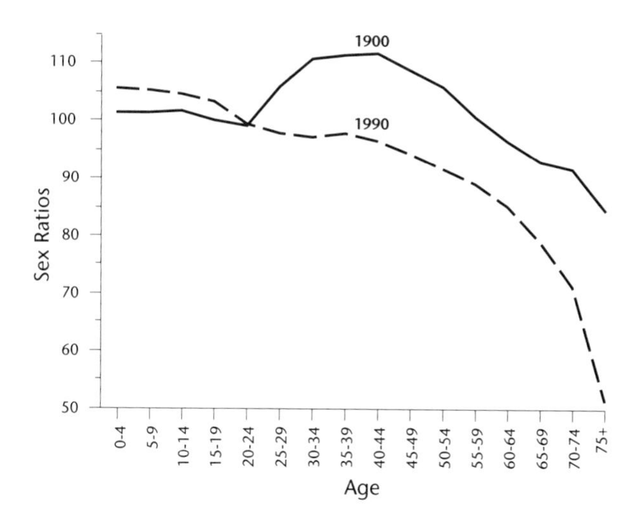

Fig. 7.10. Sex ratios by age-group, Pennsylvania, 1900 and 1960.

late teens and years of early adulthood in 1990 results from the net out-migration from the state selective for males.

However, in 1900 the sex ratio rose among young adults after age 20, to peak in the 40–44 age-group at 111.7. This increase was due entirely to an overwhelmingly male foreign-born component in Pennsylvania. Sex ratios in that segment of the population were 150 or more for ages 30 to 44 in 1900.

When one excludes the foreign-born from consideration, the sex ratio for the remaining population of Pennsylvania in 1900 remained at about 100 for ages 10 to 54. For as long as infectious and parasitic diseases were the major causes of death, males and females suffered more or less equally, so the population contained about as many men as women. In the older ages women did outlive the men, so that in 1900 the sex ratio among people 85 years and more had dropped to 84 in the total population and to 61 in the native-born population.

As infectious and parasitic diseases were brought under increasing control, degenerative diseases became the major killers, and the differential between male and female life spans widened. Among all ages and from nearly all causes of death, the male death rate exceeds that of females. Thus, in 1990 all age-groups in Pennsylvania above age 23 contained more females than males; ages above 82 had more than twice as many females as males.

The predominantly rural counties of Pennsylvania generally have sex ratios above the state average. Those that gained population between 1980 and 1990 combined that relatively high sex ratio with a median age that was below the state average. Both these characteristics are common in the major suburban counties and most of the metropolitan counties of the southeast, which also gained in population during the 1980s. Both areas functioned as regions of opportunity that drew substantial numbers of male in-migrants and/or retained a large share of their young male residents.

Rural counties losing population during the 1980s had sex ratios above the state average but had median ages above that average as well, suggesting a relatively greater out-migration of young females than of males. Metropolitan counties with long-term out-migration, such as Lackawanna, Cambria, and Allegheny, generally exhibited median ages above the state average, but relatively more females in their populations than was characteristic of the state as a whole.

Thus, over more than 300 years, first as a colony and later as a state, Pennsylvania has undergone a remarkable demographic evolution. The state has grown to include nearly 12 million persons, or a bit more than one of every twenty people in the United States. That growth in numbers has resulted from both an excess of births over deaths and, until recently, a net in-migration of people from other states or from abroad.

In the early years, Pennsylvania's birth rate was quite high, although by the beginning of the nineteenth century the rate had probably fallen below the nation's average and continued to fall faster and further than the national average, at least until 1920. After World War II, Pennsylvania participated strongly in both the baby boom and the following fertility decline, in its relative increase in fertility level from 1940 to 1960, as well as in its relative decrease from 1960 on, both being greater than the national averages. By 1985, the fertility rate of Pennsylvania was one of the lowest in the nation.

Perhaps the most remarkable change in Pennsylvania's demographic history has been the decline in mortality rates, especially among infants and young children. The early history of Pennsylvania was marked by relatively high death rates punctuated at intervals by epidemics of infectious diseases concentrated in the larger cities and especially lethal to children. Reports from individual cities late in the nineteenth century suggest that 40 to 50 percent of all deaths were to children less than 15 years of age.

After 1920, these infectious killers were brought under progressive control and virtually eliminated as a cause of death among children. However, as infectious diseases declined in incidence, degenerative diseases, particularly among the elderly, have increased. Heart disease and cancer are now responsible for three-fifths of the deaths to Pennsylvanians.

The United States has been described as a nation of immigrants, and Pennsylvania has been the chosen destination for a substantial share of those who came to America. During the colonial period, Pennsylvania probably attracted more immigrants than any other colony. Throughout the latter part of the nineteenth century and until 1920, about 10 percent of all the foreign-born enumerated in the United States were counted in Pennsylvania. Only since the major sources of immigrants have shifted away from Europe have foreign-born become a rather small part of the total population of the state.

But not long after settlement began, Pennsylvanians began to move in large numbers to the frontiers of the south and west, contributing significantly to the population and culture of those regions. As the frontier became more distant, and as the Northeast became more industrialized, rural-urban migration was increasingly substituted for frontier settlement, and migration of Pennsylvanians was redirected to the growing cities within the state and in neighboring states. By 1900, Pennsylvania became predominantly urban; in 1990 some 69 percent of the state's population lived in urban places while almost 85 percent lived in metropolitan areas.

After 1920, with the restructuring of the American economy and the declining fortunes of the mines and the mills that had fostered the earlier growth of Pennsylvania, the state became one of net out-migration as Pennsylvanians joined in the move to the sun belt. From 1920 to 1984 more than 2.5 million people have been lost to Pennsylvania's population by net out-migration. Data from the 1990 census suggest, however, that the migration drain from Pennsylvania may have slowed and that several of the metropolitan areas have reversed their population loss. The demographic evolution of the state continues.

Bibliography

Bogue, D. J. 1969. *Principles of Demography*. New York: Wiley.

Bogue, D. J., et al. 1957. *Subregional Migration in the United States, 1935–1940,* Vol. 1, *Streams of Migration Between Subregions*. Oxford, Ohio.

Bowles, Gladys K., et al. 1965. *Net Migration of the Population, 1950–1960, by Age, Sex, and Color,* Vol. 1, Part 1: *Northeastern States*. Washington, D.C: U.S. Department of Agriculture Economic Research Service.

———. 1975. *Net Migration of the Population, 1960–1970 by Age, Sex, and Color,* Part 1, *Northeastern States*. Athens: University of Georgia Institute for Behavioral Research.

Department of Health of the Commonwealth of Pennsylvania. (Various years.) *Biennial Report*. Harrisburg.

Dunaway, W. F. 1948. *A History of Pennsylvania*. Second edition. New York: Prentice Hall.

Florin, J. 1977. *The Advance of Frontier Settlement in Pennsylvania, 1638–1850: A Geographic Interpretation*. Papers in Geography 14. University Park: The Pennsylvania State University, Department of Geography.

Pennsylvania Department of Health. 1992. *Pennsylvania Vital Statistics Annual Report 1990*. Harrisburg: State Health Data Center.

Pennsylvania State Board of Health and Vital Statistics. (Various years.) *Annual Report*. Harrisburg.

Pennsylvania Statistical Abstract 1985. Harrisburg: Pennsylvania State Data Center.

Slater, C. M., and G. E. Hall, 1992. *1992 County and City Extra Annual Metro, City and County Data Book*. Lanham, Md.: Bernan Press.

Statistical Abstract of the United States 1987. 1986. Washington, D.C: U.S. Bureau of the Census.

U.S. Census Bureau. *Census of Population*. (Various years.)

U.S. Department of Commerce. *Historical Statistics of the United States Colonial Times to 1957*. 1960. Washington, D.C: U.S. Census Bureau.

8

ETHNIC GEOGRAPHY

Wilbur Zelinsky

From the very beginning of European settlement to the present time, Pennsylvania's ethnic composition and its geographic expression have been remarkably varied and complex. We are concerned here almost exclusively with the peoples who arrived in Pennsylvania from other parts of the world, either directly or indirectly. The indigenous Indian population—the Native Americans—have played only the most minor of roles in the evolution of Pennsylvania society.

Native Americans

The basic reasons for the relative unimportance of the Native Americans were their small numbers, the thinness of settlement, and the absence of effective political organization. We can only conjecture about the size and detailed identities of the populations who lived within the present boundaries of the state before the impact of the invading Europeans made itself felt, because the archaeological and documentary records are so scattered and incomplete. What is clear, however, is that European diseases, which arrived well before the newcomers did, tragically decimated the indigenous peoples. Moreover, most of the survivors in the Pennsylvania interior had been shifted about and demoralized, socially and culturally, by thrusts from other groups to the north and south, who in turn had been seriously disturbed by Europeans.

Fig. 8.1 depicts the general pattern of tribal territories at contact time—usually some point in the seventeenth century. The most important group, which spoke various Algonquian dialects and came to be known as the Delawares, inhabited the eastern section of the state after migrating westward from their original homeland and totaled perhaps 11,000 in 1600 (Trigger 1978, 213). "To the west [in the lower Susquehanna Valley] the Susquehannocks, who had dominated a broad region during the Dutch

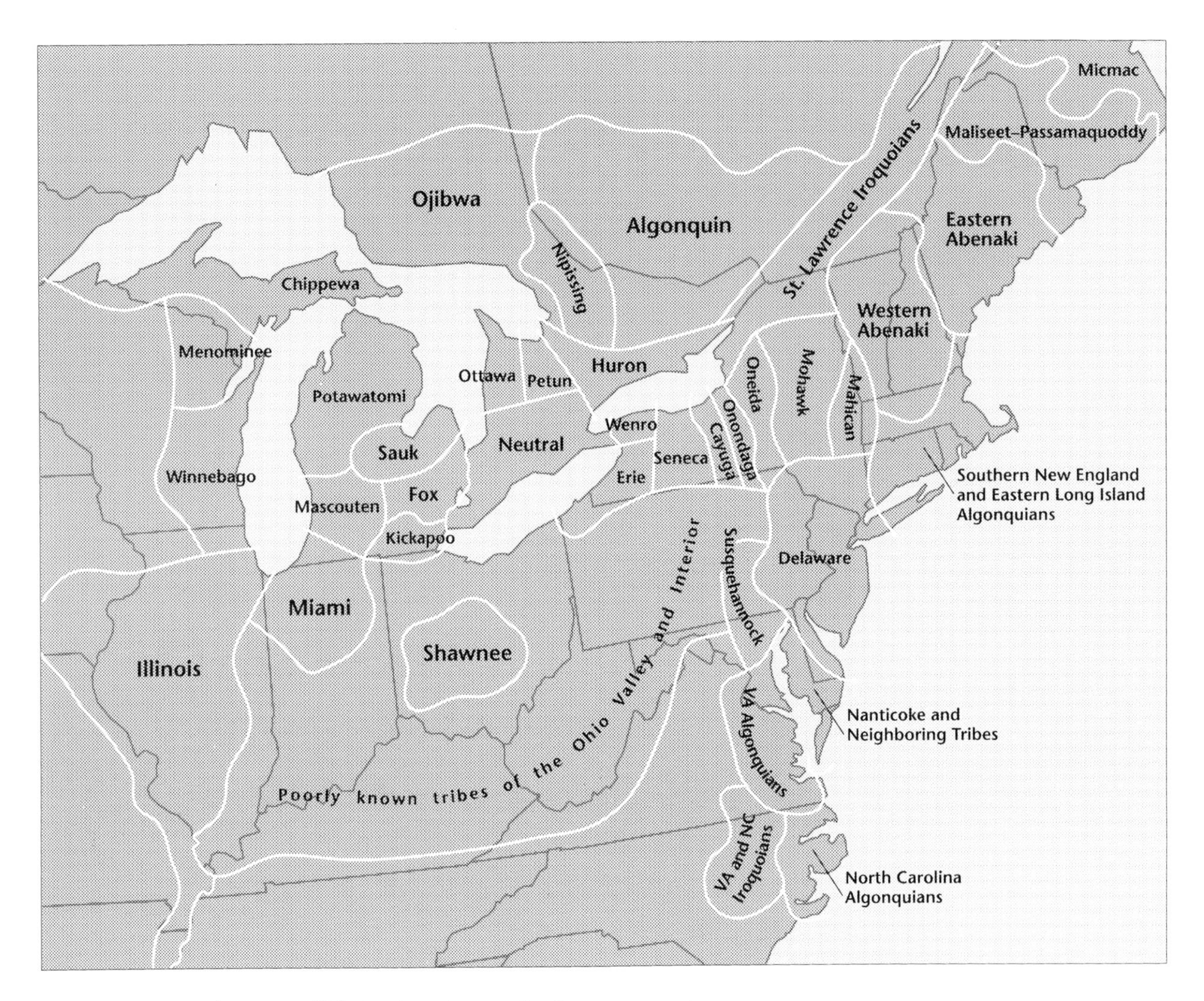

Fig. 8.1. General pattern of Native American territories at contact time.

and Swedish eras, had suffered heavily in conflicts with the Iroquois to the north and the English settlers of the Chesapeake to the south. . . . Most of them had shifted to the middle Susquehanna and become clients of their recent enemies, serving to guard the southern margins of the Iroquois Confederation" (Meinig 1987, 135). The situation is poorly known for the areas farther west, but it was evidently one of considerable flux and disarray.

The history of relations between Europeans and Native Americans in Pennsylvania was quite benign and largely free of the bloodshed that was so common elsewhere along the Atlantic Seaboard. In part, this was the result of conscientious diplomacy and fair dealing by William Penn and his heirs, but it was even more a product of the numerical and technological weakness of the aborigines. As the result of loss of land through treaty and purchase, and the constant pressure of the advancing European settlers, the Native Americans moved westward, no longer able to maintain their traditional economy of hunting, fishing, and crop cultivation or the short-lived fur trade. After a complicated series of moves involving western New York, Ontario, Wisconsin, Kansas, and Texas, most of the greatly reduced population of Delawares and other groups ended up in Oklahoma, where their descendants still reside.

With the departure of the last recognizably aboriginal groups by the middle of the nineteenth century and the lack of any reservations, Native Americans have become an inconspicuous part of Pennsylvania's social landscape, even though some racial intermixture undoubtedly occurred in the frontier zones. Only 1,639 individuals identified themselves as Indian in the 1900 census, and this value reached an all-time low of 443 in 1940. It is surprising that the number of Native Americans in Pennsylvania has rebounded sharply in recent years, with 10,684 being enumerated in 1990. Changes in definition or self-identification, along with a high birth rate, may account for some of this increase, but it is largely a matter of migration from other parts of the United States, and migration to urban centers at that. With totals of 5,891 and 1,452, respectively, the Philadelphia and Pittsburgh metropolitan areas now account for more than half of the state's Native Americans.

Serious European interest in what was to become Pennsylvania materialized long after colonies were firmly in place in New England, New York, Virginia, Maryland, and elsewhere. Such neglect was caused in part by the difficulties encountered in navigating Delaware Bay, which in any case did not offer an especially inviting route into the continental interior, but also by the absence of rich fur territories or large Indian populations. After the Dutch West India Company failed to capitalize on its tenuous claims to the Delaware Valley, a New Sweden began to take shape under Scandinavian patronage in 1638 along the lower Delaware, in what is now Delaware and the southeasternmost section of Pennsylvania. Shortly thereafter, a scattering of Swedish, Finnish, and Dutch farms, forts, and trading centers appeared, so that the area took on a polyglot character from the very outset. For the next 40-odd years, the settlements along the lower Delaware experienced a confused political and military history with control alternating among the Swedes, the Dutch, and the British. To add to the confusion, a group of New Englanders from the New Haven colony also ventured into the neighborhood in 1641 in an effort to extend the reach of Puritanism. When Great Britain finally attained firm control of the area, and William Penn initiated his colony in 1682, the number of Europeans residing along the west bank of the Delaware was still quite small, even by comparison with other North American areas—probably between 2,000 and 3,000.

Because of astute advertising and management by Penn and the fine agricultural and commercial possibilities of southeastern Pennsylvania, the city of Philadelphia and its immediate hinterland were instant successes. A major reason for such swift demographic and economic expansion was Penn's open-door policy for immigrants from all lands and of all ethnic and religious persuasions. But the founder's greatest concern was to provide a safe haven for his co-religionists, because the Friends (or Quakers) had been undergoing persecution in Great Britain and elsewhere in Europe as well as in most of the other North American colonies where they had tried to settle. Prominent among the first wave of Penn colonists were three distinct groups of Quakers: the English variety from the home country itself along with many from the West Indies, New York, and West Jersey; a Welsh contingent, which developed the "Welsh Tract" west of Philadelphia; and a community of Quakers from the lower Rhineland, who populated Germantown just a few miles north of the colonial capital. Although the influx of Quakers soon subsided and came to be outnumbered by British colonists of varied religious persuasions and other ethnic groups, they were to maintain strong political and economic influence in the entire colony, and especially in the Philadelphia area, throughout the colonial era and even beyond.

Within a few years of its founding, non-English immigrants responded as eagerly to the social and economic opportunities offered by Pennsylvania as did the English Quakers, Anglicans, Presbyterians, and others. The Welsh and Rhineland Quakers were the precursors of the much greater wave of German-speaking folk who arrived from various small political entities in western Germany and Switzerland, a movement that by 1720 was fully developed. Almost entirely Protestant in character, this major flow contained representatives of a wide array of denominations extending from the mainstream Lutherans and (Calvinist) Reformed to various pietistic groups, including the Mennonites and the Dunkards, among others. As was to happen repeatedly later throughout North America among any number of immigrant groups, the Pennsylvania Germans did not begin to recognize their ethnic commonalities until they began the process of cultural assimilation in the New World. Coming from a region of many small sovereign principalities, they had previously regarded themselves as Hessians, Alsatians, Bavarians, or residents of a particular Swiss canton or some other small locality with its own special set of traditions and self-awareness. They became Germans, or rather German-Americans, only after they were so envisioned by their neighbors.

Although the Welsh and the pre-Penn Swedes, Finns, and Dutch were to lose their identity and blend into the general population within a generation or two after the birth of Pennsylvania, the German presence has been much more enduring. The combination of a substantial population of German stock and a certain degree of clustering helps account for such persistence. Such large clumps of settlers who appeared so alien to British-American eyes were for a time considered by many to be too difficult to assimilate into the larger society. It was an unfounded fear, but it did generate some anti-German agitation, quite possibly the first of that long series of xenophobic attacks that have afflicted Americans for more than 200 years. Hard on the heels of the Germanic invasion came the next major ethnic addition to the Pennsylvania mosaic: the Scots-Irish. From about 1718 on great numbers of Presbyterians left Ulster (Northern Ireland) for the New World, only several decades after their ancestors had left Scotland to populate a war-ravaged land. Driven by economic hard times and religious intolerance in Ireland, they found Pennsylvania to be the most nearly ideal refuge of all the British colonies.

Ethnic Areas

By 1740, when all the desirable farmland southeast of the Ridge-and-Valley region had been claimed and occupied, there were three quite general ethnic zones, roughly parallel bands. The first, and oldest, included Philadelphia and most of Bucks, Montgomery, Delaware, and Chester counties, an area in which people of English and Welsh extraction were dominant. Farther inland, especially in Northampton, Lebanon, Berks, Lancaster, York, and western Chester counties, there were pluralities or majorities of German stock. Still farther northwest the Scots-Irish tended to outnumber other groups. Such a description is necessarily oversimplified, though, because there was—and still is—a good deal of spatial interdigitation of these communities in certain areas settled in colonial times and the early Republican era. Moreover, this tripartite scheme ignores a number of smaller but not insignificant ethnic groups found in early Pennsylvania, including Huguenots from France and elsewhere, Jews, the Dutch, Swedish, and aboriginal remnants, and African-Americans both slave and free.

The initial and frequently permanent location of the various ethnic groups was primarily the result of timing: where the nearest accessible and affordable lands open to settlement happened to be when members of the group disembarked at Philadelphia or New Castle. A secondary factor of some consequence was the location of kinfolk and neighbors from the homeland who had already settled somewhere in the state, for like tends to gravitate to like. In spite of the stubborn myth to the contrary, we have learned, from the study by James Lemon (1972), that there were no significant differences in the types of land the Germans selected, as opposed to the types the Scots-Irish or British chose. All groups were represented on the frontier and shared proportionately many kinds of soil and terrain. After initial settlement, however, there might well have been interethnic differentials in the propensity to migrate farther. Although all the ethnic groups of colonial Pennsylvania participated vigorously in the settlement of the remaining 80 percent or so of the state, and also streamed southwestward in considerable numbers via the Great Valley and the Piedmont to people portions of Virginia, the Carolinas, and the trans-Appalachian Upper South, there are suggestions that the Scots-Irish may have been more mobile than the others. Before that question can be finally settled much research remains to be done.

We may be unable to attach reliable numbers to the process, but it is safe to assume that the varied stream of immigrants who entered Pennsylvania to reside in the colony or move farther on was substantial and grew in annual volume from 1682 to the eve of the War of Independence. As we shall see, students of American ethnicity have always had difficulty obtaining usable statistics. For the colonial period, we have only local documents (church, property, tax, and militia records, gravestones, and the like) to furnish at best only indirect clues to place of origin and probable ethnicity. With the first of the decennial national censuses in 1790, there was finally comprehensive coverage of the population—but scant information on its composition: only sex and numbers of white and black persons, the latter by slave and free status.

In 1790 Pennsylvania's 10,274 African-Americans formed a small but significant portion (2.3 percent) of the state's total population. Despite the strong antislavery attitudes of most Quakers, legally held

Table 8.1. Distribution of identifiable white U.S. population by ethnic group, 1790 (percentile).

Ethnic Group	Me.	N.H.	Vt.	Mass.	R.I.	Conn.	N.Y.	Pa.	Md.	Va.	N.C.	S.C.
Welsh	4.3	3.9	5.0	5.0	6.2	6.2	4.8	5.9	7.1	9.5	11.6	8.8
Scots	15.3	16.4	14.5	10.9	13.9	8.7	15.5	26.7	19.1	22.8	27.6	32.9
Irish	5.5	5.4	4.1	3.1	3.0	2.6	5.5	10.2	12.5	9.4	13.3	11.7
Subtotal Celtic	25.1	25.7	23.6	19.0	23.1	17.5	25.8	42.8	38.7	41.7	52.5	53.4
German	1.3	0.4	0.2	0.3	0.5	0.3	8.2	33.3	11.7	6.3	4.7	5.0
Dutch	0.1	0.1	0.6	0.2	0.4	0.3	17.5	1.8	0.5	0.3	0.3	0.4
French	1.3	0.7	0.4	0.8	0.8	0.9	3.8	1.8	1.2	1.5	1.7	3.9
Swedish	—	—	—	—	0.1	—	0.5	0.8	0.5	0.6	0.2	0.6
Total Non-English	27.8	26.9	24.8	20.3	24.9	19.0	55.8	80.5	52.6	50.4	59.4	63.3
English	72.2	73.1	75.2	79.7	75.1	81.0	44.2	19.5	47.4	49.6	40.6	36.7

Source: Adapted from McDonald and McDonald 1980.

black slaves had been present throughout the state's first century in fair number. But as time went on an increasing percentage of blacks were in the category of "Free Negro." Indeed, southeastern Pennsylvania had become one of the two principal concentrations of Free Negroes in the Northeast, along with Rhode Island; and Philadelphia was second only to New York City among American urban places in terms of size of Free Negro population (Zelinsky 1950). Some of these individuals or their parents had been manumitted by their Pennsylvania masters, but we can assume that many others were migrants from Delaware, Maryland, and Virginia.

The surnames of heads of white households listed in the manuscript schedules of the 1790 census have been scrutinized by a number of scholars intent on working out the ethnic makeup of the early American population, and most recently by Forrest and Ellen Shapiro McDonald (1980). Aside from the unavoidable subjective biases of the analysts, and the fact that records for New Jersey, Delaware, and Georgia have not survived, there are certain problems with this approach. It does not enable us to detect the amount of intermarriage among ethnic groups, which was undoubtedly significant even by 1790, and there are many surnames we cannot confidently associate with a particular country of origin or ethnic group. Thus, for example, the "Millers," who are so numerous in Pennsylvania, may hail from England, Germany, the Netherlands, Scandinavia, Poland, or Russia. But making all due allowances for shortcomings of data and methodology, Table 8.1 leaves little doubt that Pennsylvania was the most ethnically heterogeneous of all the states during the early Republican era, and probably for some time thereafter. In contrast to New England and the South, the Middle Atlantic states have always been quite diversified ethnically. With 80 percent of the identifiable white heads of household categorized as non-English, and 37.7 percent non–British Isles in origin, Pennsylvania easily outclasses neighboring New York and Maryland and almost certainly New Jersey and Delaware as well. Fig. 8.2 indicates the sections of the state in which the various groups were most strongly represented as of 1790.

With the approach of the Revolution and then all the disturbances of the Napoleonic period and the consequent military, political, and economic turmoil, large-scale emigration from Europe to North America dwindled to a mere trickle until about 1820. From that point on, until the eve of World War I, the number of immigrants entering the United States increased steadily, with temporary downturns occurring only during wars and economic depression. Pennsylvania received a good share of these newcomers (Fig. 8.3). With the resumption of international migration at a brisk pace, there were also important changes in the source areas of the new Pennsylvanians and their selection of places in which to live and work. During the colonial period, the great majority of immigrants took up residence in rural areas; only a minority chose the cities. By the 1820s they and their descendants had occupied almost all the land fit for permanent cultivation. Increasingly, then, the next wave of immigrants found employment in the building of canals and railroads, in the emergent manufacturing, mining, and forestry indus-

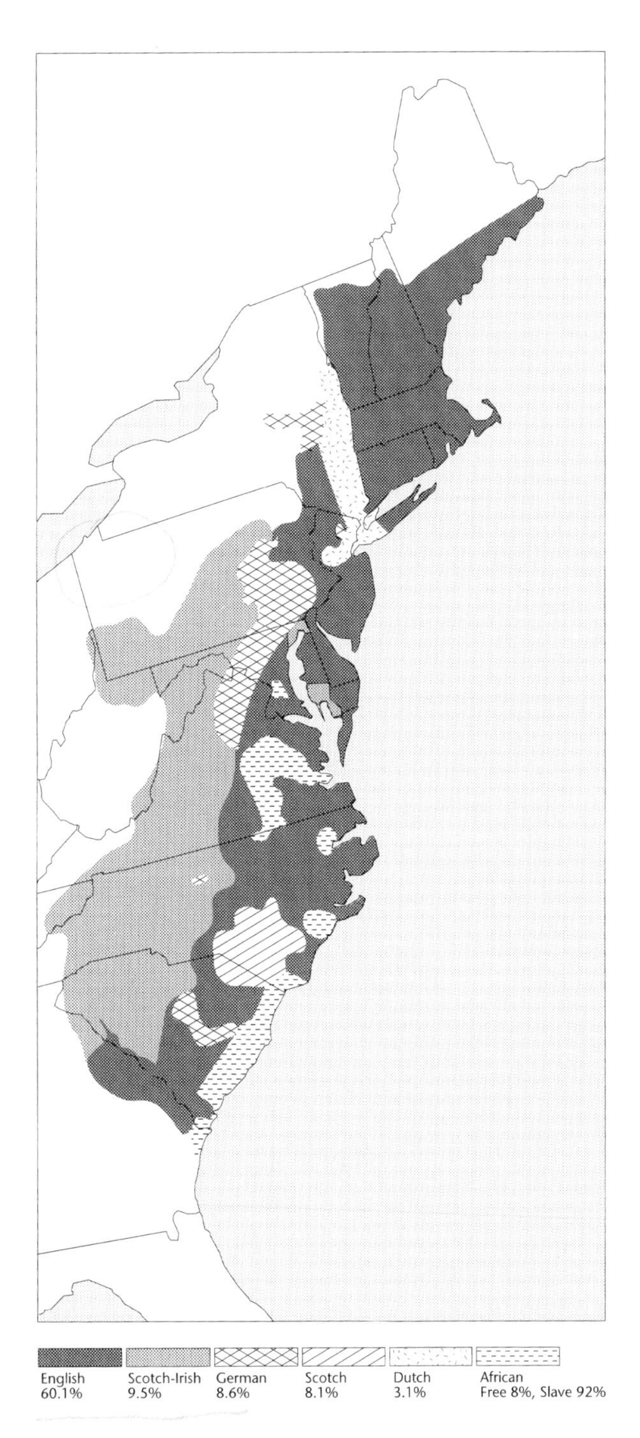

Fig. 8.2. Ethnic areas by national origin, 1790.

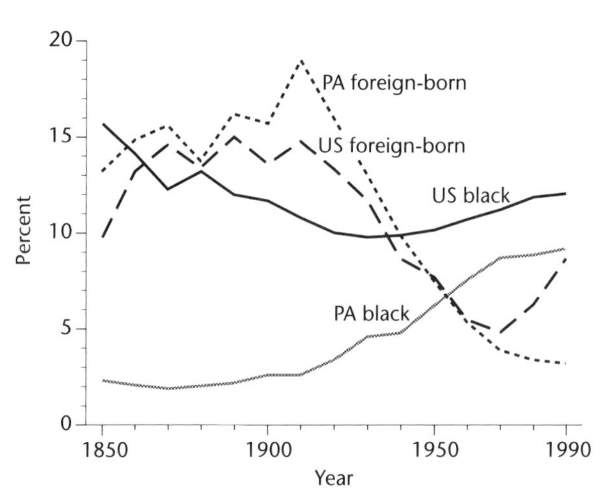

Fig. 8.3. Foreign-born and blacks as percentage of total population, United States and Pennsylvania, 1850–1980.

tries, and in a variety of economic niches in the state's rapidly growing cities.

Ethnic Groups

Our first firm data on nativity and country of origin comes from the 1850 population census, but as with comparable tabulations in the decennial enumerations to follow, certain caveats must be kept in mind when using these statistics. They refer to country of origin, not specifically to ethnic identity; and in a number of cases (e.g., Poland, Russia, or Austria-Hungary) we find several ethnic groups dwelling within a given country. Moreover, the international boundaries of some of the major homelands of immigrants to America have changed with disconcerting frequency.

The 1850 data indicate quite emphatically that northwestern Europe was still the dominant source of immigrants for Pennsylvania, as for the United States in general (Table 8.2). But there is a dramatic change from the eighteenth-century situation. Although Great Britain and the German principalities accounted for 17.9 percent and 25.9 percent, respectively, of the Pennsylvania foreign-born, more than half the total originated in Ireland. The vast majority of these Irish folk were Roman Catholics from the southern counties rather than Protestant Scots-Irish from Ulster. The mass exodus began in the 1830s as a result of severe population pressure

Table 8.2. Origin of Pennsylvania's foreign-born, selected census years, 1850–1980.

	1850	1880	1920	1950	1980
England	38,048 12.6%	80,102 13.6%	90,738 6.5%	48,752 6.3%	18,218 4.5%
Wales	8,920 2.9%	29,417 5.0%	21,167 1.5%		1,256 0.3%
Scotland	7,292 2.4%	20,735 3.5%	28,448 2.0%	20,231 2.6%	7,585 1.9%
Ireland	151,723 50.1%	236,505 40.2%	121,601 8.7%	46,385 6.0%	13,560 3.4%
Germany	78,592 25.9%	168,426 28.7%	120,194 8.6%	59,532 7.7%	38,071 9.5%
France	4,083 1.3%	7,949 1.4%	12,830 0.9%	6,550 0.8%	3,613 0.9%
Scandinavia	124 —	8,901 1.5%	28,176 2.0%	12,704 1.6%	1,382 0.3%
Other northwest Europe	1,297 0.4%	7,963 1.5%	13,194 0.9%	6,587 0.8%	2,077 0.5%
Austria	49 —	2,317 0.4%	122,755 8.8%	60,738 7.8%	11,597 2.9%
Hungary	—	1,168 0.2%	71,380 5.1%	32,134 4.1%	9,392 2.3%
Czechoslovakia	—	1,058 0.2%	68,869 4.9%	48,634 6.3%	10,433 2.6%
Poland	—	3,790 0.6%	177,770 12.8%	87,947 11.3%	21,214 5.3%
Italy	172 —	2,794 0.5%	222,764 16.0%	163,359 21.0%	67,829 16.9%
Greece	—	31	13,893 0.9%	10,474 1.3 %	10,277 2.6%
Yugoslavia	—	—	36,227 2.6%	21,412 2.8%	6,782 1.7%
Other eastern Europe	—	—	42,113 3.0%	27,252 3.5%	—
Canada	—	10,528 1.8%	15,100 1.1%	13,989 1.8%	13,626 3.4%
Latin America	—	1,277 0.2%	6,299 0.5%	6,093 0.8%	24,234 6.0%
Asia	—	404 —	3,674 0.3%	9,362 1.2%	60,360 15.1%
Total	303,105 100.0%	587,824 100.0%	1,392,557 100.0%	776,609 100.0%	401,016 100.0%

SOURCE: U.S. Census Bureau.

and economic distress in that agrarian island, then reached unprecedented intensity with the potato famines of the late 1840s. European lands, other than those in the northwestern part of the continent, were poorly represented in the 1850 figures, and exceedingly few people had arrived from other corners of the globe.

Within the next several decades, the provenience of Pennsylvania immigrants changed considerably. There was a gradual decline in the importance of the northwestern European source areas. Scandinavians tended to bypass Pennsylvania for localities in the north-central and northwestern states, but natives of eastern and southern Europe, notably Poland, Austria-Hungary, and Italy, began to flood into certain parts of the state. Eventually they were joined by Asians, Latin Americans, and Canadians. These trends paralleled shifts in immigration patterns for the nation as a whole.

There are two obvious reasons for Pennsylvania's failure to attract its previous large share of the national total of immigrants during the post–World

War II period. Given the recent rise of Latin America and Asia to dominance as source areas, relative remoteness from these homelands would tend to reduce the number of arrivals here even as relative proximity would stimulate a buildup in the western and southern states. But if such places as Los Angeles, New York City, and Miami claimed a disproportionately large share of the newcomers, the 1980s did witness a substantial increase in the number of Latino and Asian immigrants and refugees residing in Pennsylvania. It is also worth noting that not all the foreign-born or their children necessarily remain in their original American abode. Almost certainly a significant percentage of the Chinese, Japanese, or Mexicans, among others, living in Philadelphia and Pittsburgh today reached those metropolises after stopovers elsewhere. It is also likely that for most of these newcomers from abroad the economic opportunities in other parts of the country would seem to be more interesting.

The addition to the Pennsylvania population of hundreds of thousands of aliens from the 1830s to the 1910s, and in decreasing volume thereafter, has meant major revisions in the ethnic map, but without fully canceling out the larger patterns laid down during the colonial era. Indeed, both the older and newer areal pockets of ethnic identity have been remarkably persistent over the years, whether in the "coal patches" of southwestern Pennsylvania, the backwater valleys of the Ridge-and-Valley region, the mill towns lining the Allegheny, Monongahela, and Ohio rivers, the industrial and mining communities of the anthracite region, or the distinctive and often colorful neighborhoods of Philadelphia, Pittsburgh, and Erie. And this variegated ethnic geography continues to play a meaningful role in the political life of Pennsylvania.

The county-level distribution of white foreign-born residents in 1910 (the peak census year for that population) conveys some useful hints about the latterday ethnic geography of the state (see Fig. 8.4). In twelve counties dominated by mining and manufacturing, the foreign-born accounted for 20.0 to 29.2 percent of the total population, and if we add their offspring the figures would be even more impressive. At the other extreme are the south-central and central counties, where the foreign-born formed only a tiny fraction of the total, the minimum being 0.2 percent in Snyder County. For the most part, the ethnic situation in these Appalachian and inner Piedmont localities was determined during the years of initial settlement and has scarcely changed since then, much like the situation in so much of the American South.

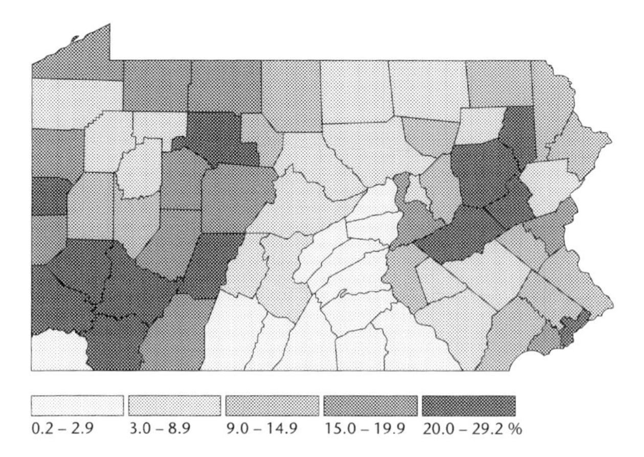

Fig. 8.4. Foreign-born whites as percentage of total population, Pennsylvania, 1910.

Throughout the history of Pennsylvania, there have been interesting differentials in the appeal of major urban areas to various ethnic or national groups (see Table 8.3). Once again, keep in mind that country of origin does not necessarily correspond to ethnic identity. For example, the "Austrians" may belong to any of a dozen groups, while the "Russian" category includes both Christian and Jewish individuals. Aside from the shifts in the relative representation of the British, Irish, Germans, and Italians in the Philadelphia, Pittsburgh, and anthracite areas, one is struck by the fact that Germans bulk so large among Lancaster's foreign-born even today, almost three centuries since the first large arrival of German stock in its vicinity. Evidently we have here another instance of chain-migration, of persistent lines of communication between origin and destination fostering a steady flow of migrants. The apparently anomalous composition of the foreign-born population in the State College Metropolitan Statistical Area (MSA) (as of 1990) is a matter of the occupational opportunities for professional Asians and Latin Americans in a community dominated by a large university. (Data on foreign-born by country of origin as of 1990 were not yet available at time of writing.)

A comparison of the map of the 1910 foreign-born (Fig. 8.4) with the analogous item for 1990 (Fig. 8.5) provides some useful insights into the changing social geography of the United States. Although urban areas continue to claim a large share of a diminished cohort of newcomers, some new pat-

Table 8.3. Area of origin of Pennsylvania foreign-born in selected Pennsylvania cities, 1880, 1910, 1980.

	1880		
	Phila.	Pittsburgh	Scranton
Total foreign-born	204,335	44,605	15,857
United Kingdom	11.1%	18.3%	34.5%
Ireland	49.8	38.4	42.7
Germany	27.3	35.8	19.9
Austria	0.3	0.4	0.1
Hungary	0.1	0.2	0.1
Poland	0.3	0.8	0.4
Russia	0.1	0.5	0.2
Italy	0.8	0.5	0.1
Greece	< 0.1	< 0.1	< 0.1
Canada	0.9	1.0	0.9
Latin America	0.5	< 0.1	0
Asia	0.1	0.1	< 0.1

	1910			
	Phila.	Pittsburgh	Scranton & Wilkes-Barre	Lancaster
Total foreign-born	384,707	140,924	51,219	3,214
United Kingdom	12.2%	10.6%	22.7%	5.7%
Ireland	21.6	13.4	13.5	4.5
Germany	16.0	20.9	13.1	62.0
Austria	5.1	15.2	11.3	1.4
Hungary	3.2	4.7	2.9	1.2
Russia	23.6	18.7	23.6	12.1
Italy	11.8	10.0	8.8	6.1
Greece	0.2	0.5	0.3	1.2
Canada	1.0	1.3	0.2	1.0
Latin America	0.4	0.1	—	—
Asia	0.5	0.2	—	—

	1980				
	Phila. MSA[a]	Pittsburgh MSA	N.E. Pa. MSA	Lancaster MSA	State College MSA
Total foreign-born	198,591	79,099	18,613	7,059	3,661
United Kingdom	7.4%	7.1%	11.2%	8.2%	8.9%
Ireland	4.2	1.9	1.5	1.2	1.4
Germany	8.8	7.8	10.0	13.5	7.5
Austria	1.7	4.2	4.6	0.8	1.4
Hungary	1.6	3.2	1.5	1.0	0.7
Poland	4.9	5.8	11.2	1.3	1.1
Yugoslavia	1.0	3.5	0.9	0.8	0.1
U.S.S.R.	10.1	4.0	3.8	2.0	7.6
Italy	15.3	224.5	19.9	5.3	3.3
Greece	2.5	2.6	17.5	5.7	1.3
Canada	2.9	2.8	2.9	10.2	6.1
Latin America	8.7	5.1	3.2	9.3	11.7
Asia	15.8	11.9	9.2	28.7	33.5

Source: U.S. Census Bureau.

[a]Pennsylvania portion only.

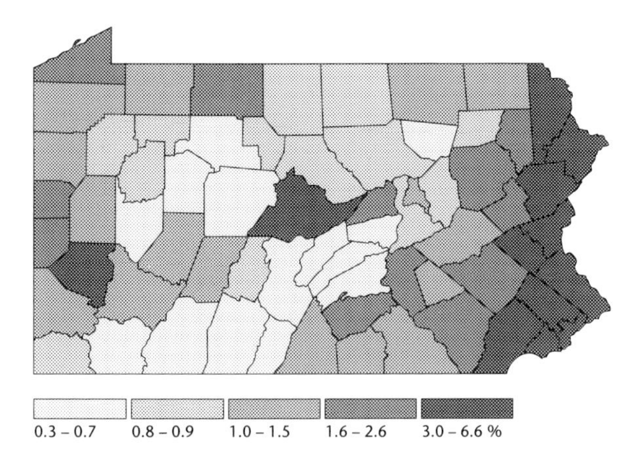

Fig. 8.5. Foreign-born as percentage of total population, Pennsylvania, 1990.

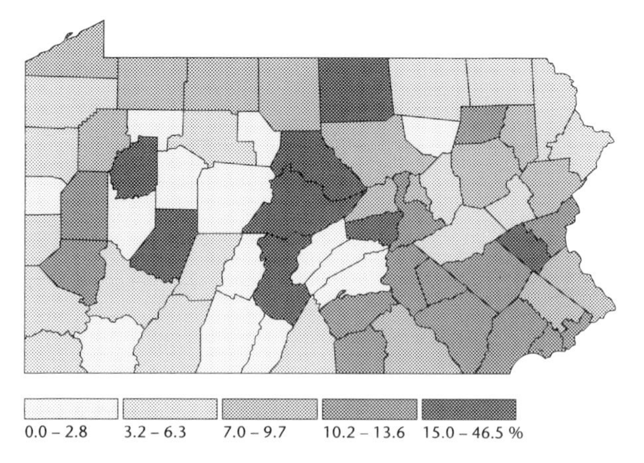

Fig. 8.6. Percentage of foreign-born Pennsylvanians immigrating between 1987 and 1990.

terns have emerged. Counties that are suburban to major cities (e.g., Bucks, Montgomery, Chester, Cumberland, and Westmoreland) have begun to receive a noticeable complement of first-generation immigration, in addition to second- and third-generation individuals. The high values plotted for the Poconos counties almost certainly reflect retirement migrations and long-distance commuting on the part, presumably, of immigrants formerly residing in the nearby northeastern megalopolis. A related map portraying the location of the 12.2 percent of the 1990 foreign-born who arrived in the United States during the period 1987–1990 (Fig. 8.6) tells us something about the vitality and socioeconomic attractiveness of the dynamic northwestern frontier of megalopolis.

The apparent suburbanization of the foreign-born as of 1990, and thus, inferentially, of "ethnic" communities in general, is simply the latest stage in a process that has been ongoing in major American cities for more than a century: ethnic succession and a centrifugal shift of one community after another from central to peripheral locations. The case of Philadelphia is a classic example, and parallel to the experience of other large cities in the Northeast and the Midwest (Wurman and Gallery 1972, 68–69). During the first 150 years of its existence there was little or no spatial segregation of Philadelphia's rich mixture of ethnic groups, or of religious communities or social classes for that matter. The Catholic Irish who came in substantial numbers from the 1830s on clustered together in some of the older, less desirable residential neighborhoods toward the metropolitan core. A generation or so later, after they had climbed a few rungs on the social and economic ladder, many of these Irish-Americans moved outward and upward into the more respectable sections of town. Ultimately, this group assimilated into the general population, both socially and spatially. The large wave of German immigration that materialized in Philadelphia by the 1850s fully repeated the process. Subsequently, much the same sequence of events took place as Italians, Jews, Poles, and other Slavic communities occupied in turn the inner-city tracts that had once been occupied by the Irish and Germans. Although there has been much dispersal of these later groups into the outer portions of the city and into the suburbs, their old, first-generation neighborhoods still retain much of their ethnic character and many descendants of the original immigrants. Whether they will vanish in another generation or two is impossible to predict.

To a certain degree, some of Philadelphia's suburbs and suburbs of other large Pennsylvania cities have attracted noticeable concentrations of specific ethnic populations, despite the substantial amount of intermarriage and spatial intermingling that has gone on among the various ancestral strains of white American society in recent decades. Even though the phenomenon has not yet been studied intensively, there is some evidence that the outward migration of ethnic groups from central city to suburb has been sectoral in nature. That is, a given ethnic group initially inhabiting a neighborhood, say, to the northwest of the central business district would tend to migrate northwestward toward whatever suburban territory might be available, and so on along other sectors.

The great troublesome exception to the foregoing pattern of outward metropolitan dispersal is that of the African-American population. Two factors have prevented, or greatly retarded, any replay on their part of the standard scenario. For a variety of reasons, most blacks have not been able to advance economically as rapidly as nearly all other sectors of the population, so that only a minority have succeeded in buying their way out of inner-city neighborhoods previously abandoned by other groups. Second, even for blacks who are relatively affluent, the racial antipathies of the larger society, sometimes institutionalized in not-too-subtle forms, have blocked mobility within metropolitan areas. The net result, as in so many other large American cities, has been a racial polarization of the metropolitan population along spatial as well as other lines. The recent so-called "white flight" to the suburbs has brought about a situation in which the black population dominates extensive portions of the central city. In fact, in the larger central cities they have become the largest single ethnic or racial group. Thus, in 1990, blacks constituted 39.7 percent of the population of Philadelphia proper, but only 6.5 percent in the four adjacent counties that comprise the Pennsylvania remainder of the MSA. The disparities are perhaps even more striking in the cases of Pittsburgh and Harrisburg, the figures being 25.8 and 6.8 percent and 50.6 and 3.6 percent, respectively.

The presence of Latin Americans in Pennsylvania cities is so recent and still so slight that it would be hazardous to predict their locational future, but notable concentrations are now present in Reading, Lancaster, and other eastern Pennsylvania localities. On the other hand, Asian groups, though limited in numbers, have given every indication of rapid economic and social progress accompanied by gravitation away from slum areas to the outer city.

The shortfall in suburbanization is not the only respect in which the historical geography of Pennsylvania blacks has failed to parallel the experience of other groups. Unlike immigrants from eastern and southern Europe, Canada, Asia, and Latin America, African-Americans have been present in significant numbers since the early colonial period. For the entire period 1790–1910 there was little consistent movement in the black share of total population, with percentile values hovering between 1.9 and 2.9. One must assume relatively slight interstate movement, even though hundreds of runaway slaves may well have found refuge in Pennsylvania in antebellum days. The great demand for labor engendered by World War I (and the interruption of the normal immigration of European workers) stimulated a flow of Southern blacks to the urban-industrial centers of the North, including those in Pennsylvania. Once begun, the northward stream, predominantly from the South Atlantic states, continued to operate throughout the 1920s and 1930s, stimulated no doubt by poor economic and social conditions in the states of origin. World War II greatly expanded the process (for Southern whites as well as blacks), and once again the momentum failed to slacken in the postwar era. Thus, by 1970, after more than half a century of heavy in-migration, blacks accounted for 8.6 percent of Pennsylvania's people, and a far greater percentage in the major cities. It is likely that no other identifiable social group has ever entered Pennsylvania in larger volume from elsewhere in the United States. Unlike the situation in New York or Florida, we can ignore as statistically insignificant the number of Latin American blacks moving to Pennsylvania.

From 1970 to 1990, however, the demographic trends of Pennsylvania blacks are less easily described or explained. The total number did grow, but only slightly, from 1,015,576 to 1,089,795, a rate of increase somewhat ahead of the rest of the population but well below whatever natural increase may have occurred. Among other things, this signified a declining share of the national total of blacks. Clearly, then, there was a net outflow of blacks from Pennsylvania during the 1970s, with fewer persons entering the state and a substantial number leaving. Presumably most returned to the South to work or retire in a region where economic and social conditions had materially improved. The dynamics of the largely urban black population within the state (see Fig. 8.7) were quite mixed during the 1980s. Thus the totals enumerated in the cities of Philadelphia and Pittsburgh declined by 1.8 and 6.5 percent, respectively, while Reading and Lancaster recorded increases of 21.3 and 34.7 percent (from small initial bases).

Ancestry

From 1980 on, the Census of Population has included a question on ancestry. Although we cannot accept

Fig. 8.7. Pennsylvania's black population, 1990.

the tabulated results uncritically because there were some obvious biases for or against particular ancestral groups, quite apart from the fact that many Americans have hazy notions as to the identity of their forebears further back than three or four generations, the results are instructive. In interpreting the data in Table 8.4, one must remember that the 36.1 percent of Pennsylvanians reporting two ethnic/national ancestries are not included.

In the discussion that follows, we refer to maps derived from 1980 census data. Unfortunately, the 1990 ancestry materials became available at such a late date that updated maps could not be included here. Luckily, only minor locational shifts occurred during the 1980–1990 period for the groups we have depicted.

Pennsylvanians claiming German ancestry far outnumber those identified with any other group, but they are not evenly distributed within the state (see Figs. 8.8 and 8.9). German stock accounts for a plurality or absolute majority in most counties and is well represented even in the Northern Tier (where people of British descent predominate), Philadelphia, and the anthracite region. Individuals of Irish and English origin are also found in strength throughout the state, but they fall into spatial patterns that differ from those of the Germans (see Figs. 8.10 and 8.11). As already suggested, persons of Italian, Polish, and Russian derivation are more unevenly located, with a perceptible degree of economic determinism appearing, because major concentrations still exist in areas of initial employment (see Figs. 8.12, 8.13, and 8.14).

It is interesting to note Pennsylvania's rank-order among the states with respect to specific ethnic groups. Ranking fifth among the 50 states in terms of total population in 1990, Pennsylvania was home to more persons of Slovak and Ukrainian origin

Table 8.4. Pennsylvania population by claimed single ancestry and racial classification, 1990.

German	3,485,430
Irish	1,270,330
Hispanic	1,161,853
African-American	1,089,795
Italian	1,047,893
English	749,786
Polish	632,518
"American"	309,814
Slovak	295,843
Scots-Irish	195,394
Dutch	172,084
Russian	156,394
French	136,174
Scottish	132,813
Welsh	109,613
Hungarian	92,006
Ukrainian	89,780
Swedish	73,648
Lithuanian	66,899
Greek	44,265
Austrian	43,549
Swiss	40,610
Yugoslav	32,181
Arab	30,798
Chinese	29,562
Asian Indian	28,396
Czech	28,356
Korean	26,787
French-Canadian	22,293
Norwegian	18,777
West Indian	17,550
Vietnamese	15,887
American Indian	14,733
Sub-Saharan African	13,088
Other Asians and Pacific Islanders	12,538
Filipino	12,160
Danish	11,941
Romanian	10,447
Portuguese	9,209
Canadian	6,956
Belgian	6,933
Japanese	6,613
Cambodian	5,495
Finnish	5,471
Other	450,010
Unclassified	911,105

Source: U.S. Census Bureau.

than any other state, and ranked second in quantity of Germans, Lithuanians, and Welsh. On the other hand, there was a notable deficiency of Scandinavians, Native Americans, Chinese, and French-Canadians.

Given the growing extent of intermarriage among various ethnic groups, and the powerful forces of social assimilation that are constantly at work in the United States, it is logical to assume that the checkered ethnic mosaic Pennsylvania has been for so long must be evolving into something quite different. Will the Pennsylvania of the twenty-first or twenty-second century be socially homogeneous, a nearly uniform mass of thoroughly Americanized citizens? We cannot dismiss such a prospect. On the other hand, there is always the possibility that some unforeseen new wave of immigration will confound our expectations and for a time negate the enormous power of mass communication and mass production of commodities and services in a mass society, which could well prevail in the long run.

In the short run, however, something surprising has been happening in recent times in Pennsylvania, in the United States as a whole, and in a number of other countries: a revival of ethnicity. Since the 1960s this "neo-ethnicity" has manifested itself in several ways. Inspired in part by the "Black Pride" movement, other groups of European, Latin American, and Asian origin have begun to rediscover, revive, and celebrate genuine (or possibly invented) traditions in addition to preserving surviving practices. The persons most involved seem to be third- or fourth-generation Americans in search of their roots and possibly reacting against loss of identity in a mass society—or, just as plausibly, expressing disillusionment with the stultifying effects of a centralized economy and government. The parallel revival of regionalism may be another expression of the same mood.

In any event, the results have been startling. Instead of disappearing, as one might have predicted, new periodicals aimed at ethnic markets are appearing. Foreign-language and ethnic radio broadcasts and television programs continue to flourish, and there has been a veritable explosion of restaurants featuring ethnic cuisines (Zelinsky 1985). No less impressive in its way has been the proliferation of ethnic festivals in Pennsylvania and other states, and the opening of museums with an ethnic commitment. At the moment, one can safely say at least that reports of the demise of ethnicity have been greatly exaggerated.

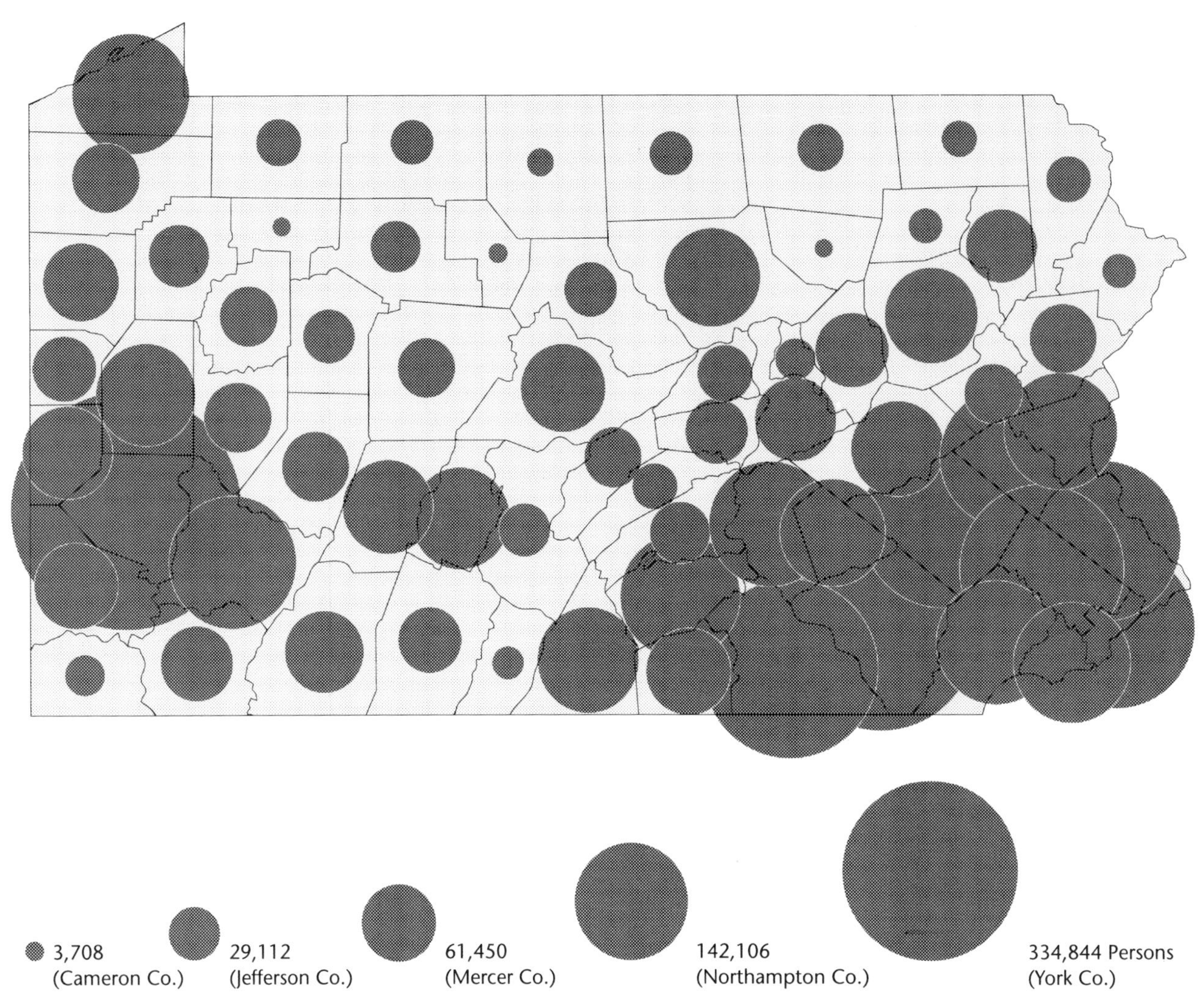

Fig. 8.8. Persons of German ancestry in Pennsylvania, 1990 (including both single and mixed ancestry).

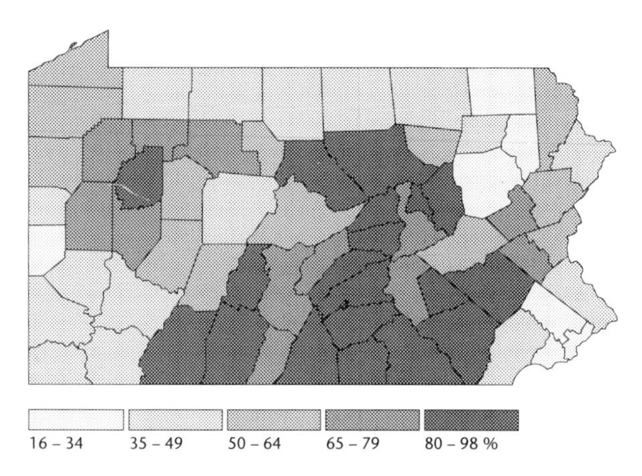

Fig. 8.9. Persons of German ancestry as percentage of Pennsylvania's total population, 1990 (including both single and mixed ancestry).

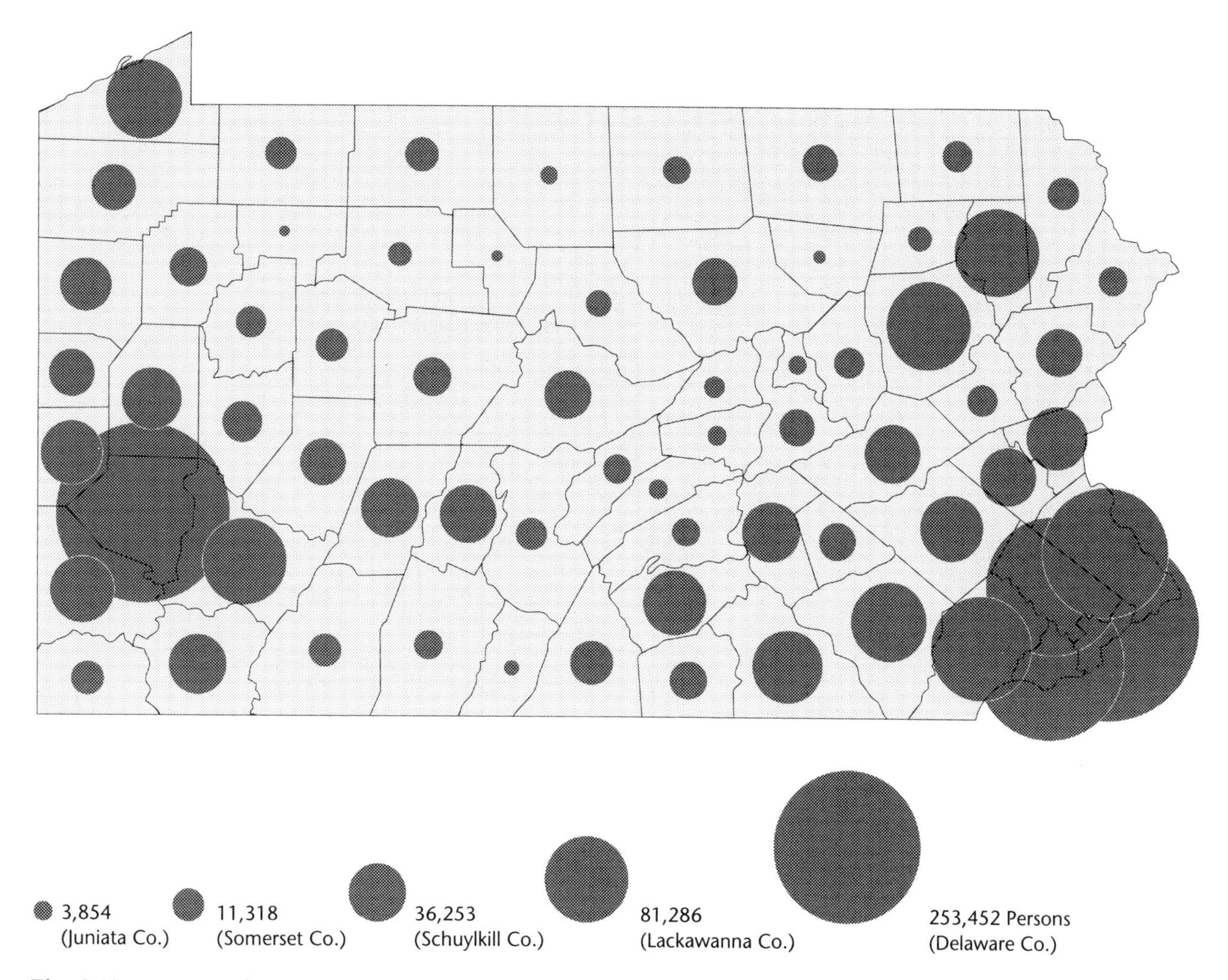

Fig. 8.10. Persons of Irish ancestry, Pennsylvania, 1990 (including both single and mixed ancestry).

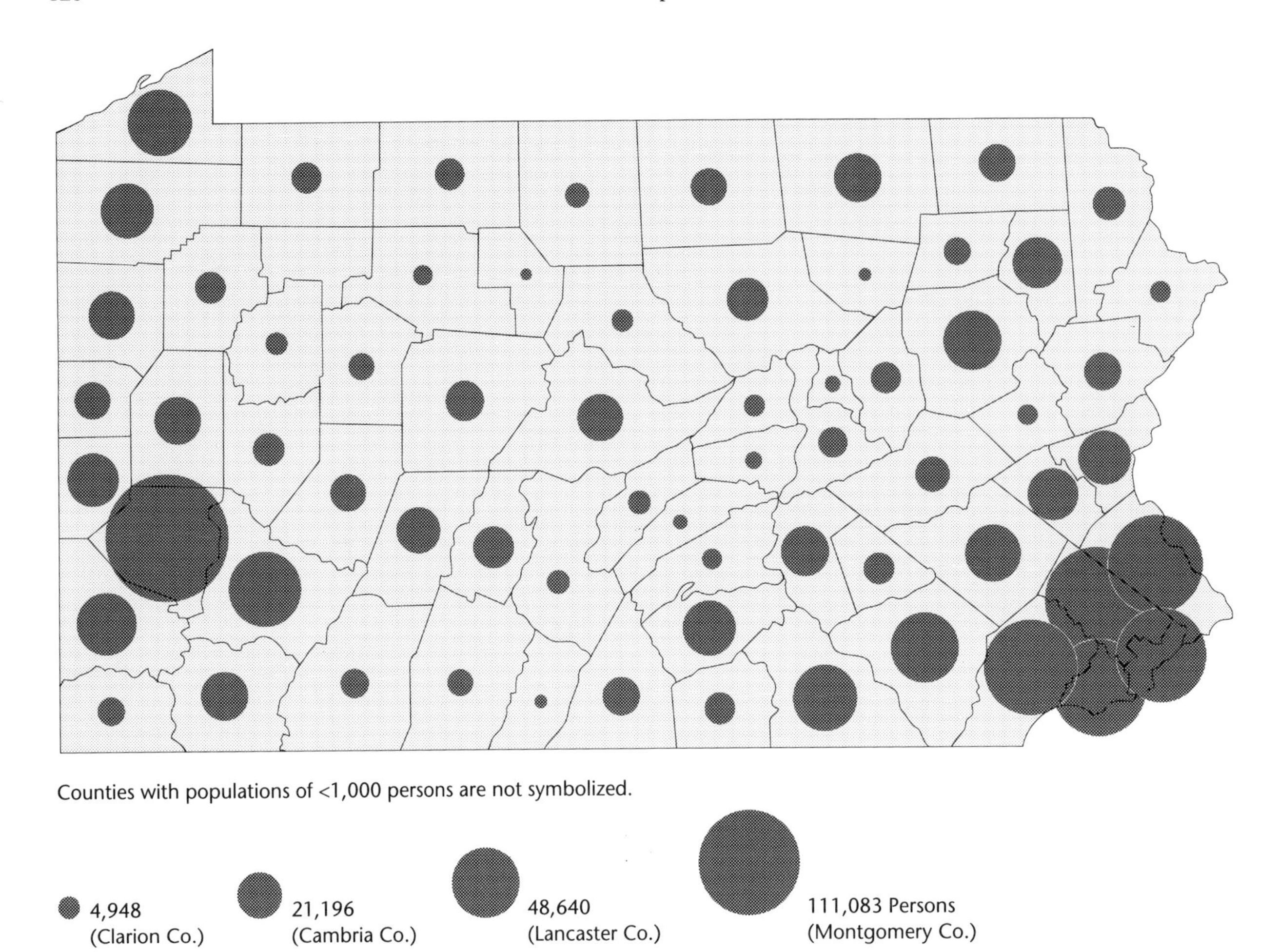

Fig. 8.11. Persons of English ancestry, Pennsylvania, 1990 (including both single and mixed ancestry).

Counties with populations of <1,000 persons are not symbolized.

Fig. 8.12. Persons of Italian ancestry, Pennsylvania, 1990 (including both single and mixed ancestry).

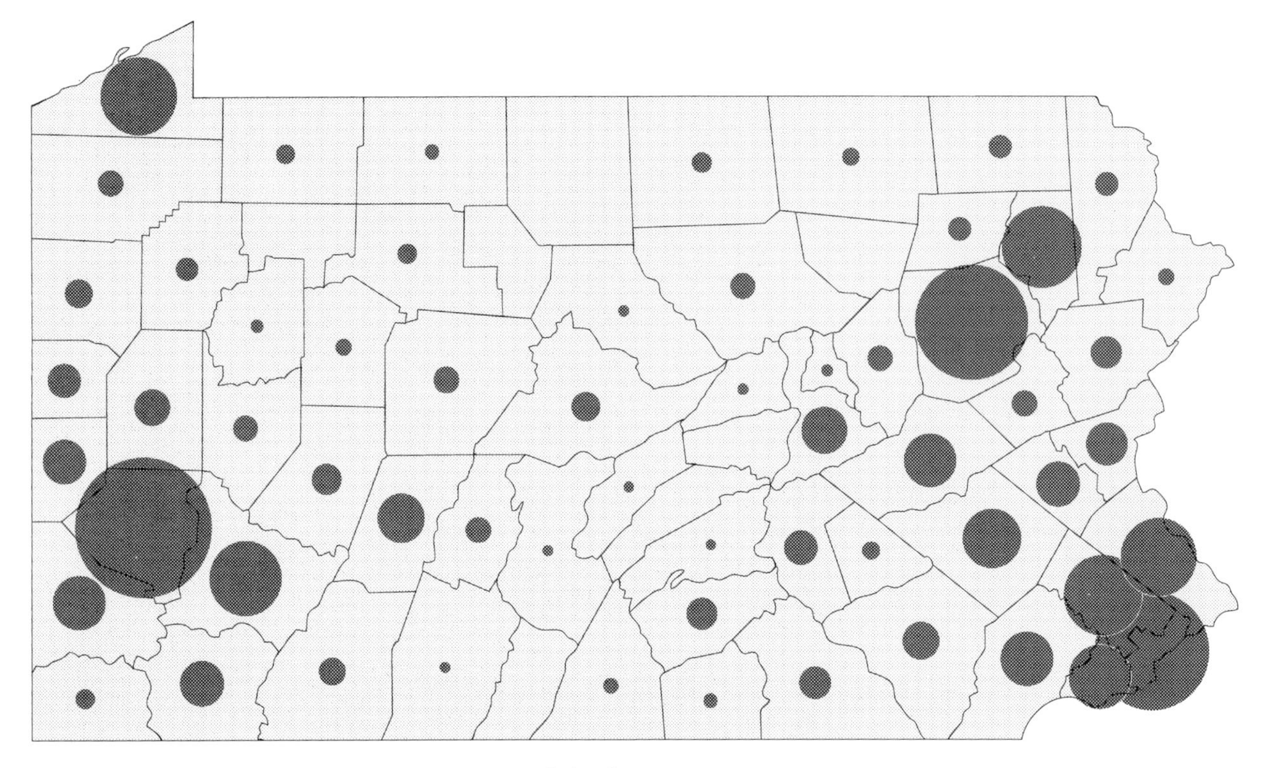

Counties with populations of <1,000 persons are not symbolized.

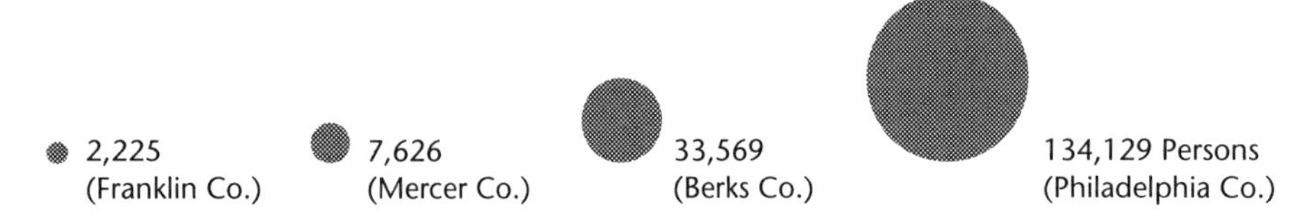

Fig. 8.13. Persons of Polish ancestry, Pennsylvania, 1990 (including both single and mixed ancestry).

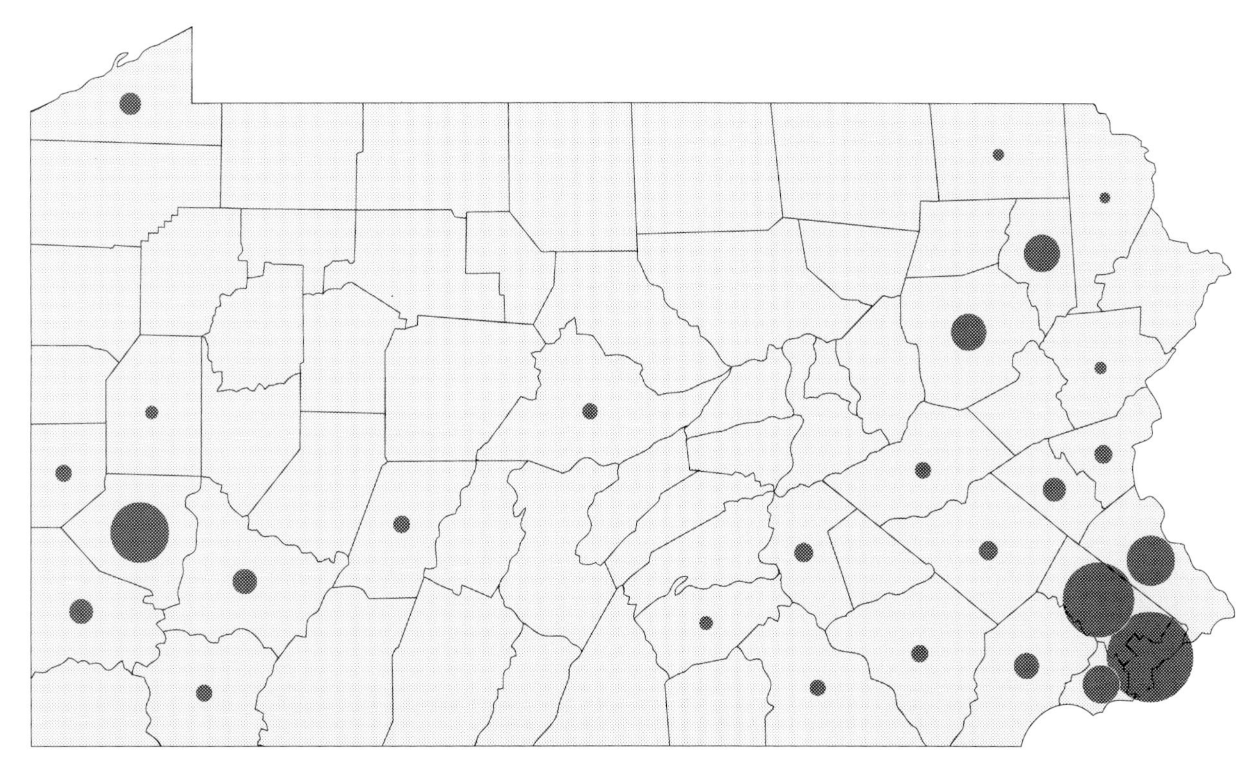

Counties with populations of <1,000 persons are not symbolized.

1,735 (Cumberland Co.) 5,594 (Washington Co.) 34,585 Persons (Allegheny Co.)

Fig. 8.14. Persons of Russian ancestry, Pennsylvania, 1990 (including both single and mixed ancestry).

Bibliography

Cappon, L. J., ed. 1976. *Atlas of Early American History: The Revolutionary Era, 1760–1790.* Princeton: Princeton University Press.

Lemon, J. T. 1972. *The Best Poor Man's Country: A Geographical Study of Early Southeastern Pennsylvania.* Baltimore: Johns Hopkins University Press.

McDonald, F., and E. S. McDonald. 1980. "Ethnic Origins of the American People, 1790." *William and Mary Quarterly,* 3d ser. 37(2):179–199.

Meinig, D. W. 1987. *The Shaping of America: A Geographical Perspective on 500 Years of History,* Vol. 1, *Atlantic America, 1492–1800.* New Haven: Yale University Press.

Paullin, C. O., and J. K. Wright. 1972. *Atlas of the Historical Geography of the United States.* Washington and New York: Carnegie Institution and American Geographical Society.

Raitz, K. B. 1979. "Themes in the Geography of European Ethnic Groups in the United States." *Geographical Review* 69:79–94.

Thernstrom, S. 1980. *Harvard Encyclopedia of American Ethnic Groups.* Cambridge, Mass.: Belknap Press.

Trigger, B. G., ed. 1978. *Handbook of North American Indians,* Vol. 15, *Northeast.* Washington, D.C.: Smithsonian Institution.

Wurman, R. S., and J. A. Gallery. 1972. *Man-Made Philadelphia.* Cambridge: MIT Press.

Zelinsky, W. 1950. "The Population Geography of the Free Negro in Ante-Bellum America." *Population Studies* 3:386–401.

———. 1985. "The Roving Palate: North America's Ethnic Restaurant Cuisines." *Geoforum* 16(1):51–72.

9

CULTURAL GEOGRAPHY

Wilbur Zelinsky

Is there something distinctive about the culture of Pennsylvania? Why and how does this "something" matter to the rest of the nation? And what useful and interesting things can we say about the cultural geography of the state, of its peoples and landscapes? The answers to these queries are of more than casual interest, for they transcend the parochial. In studying the cultural makeup of Pennsylvania, we take a giant step toward understanding the character of the entire United States.

A simple working definition of "culture" and "cultural systems" would certainly be helpful, but social scientists have wrestled with the problem for more than a hundred years and are not a great deal closer to a definitive solution today than when they started. Nevertheless, we can take certain central notions for granted. Culture is an immensely large, complicated set of ideas, symbols, values, and attitudes—conscious and subconscious—of practices, customs, memories, and traditions, including the material results thereof, that is the unique property of the human species. It is an entity acquired, modified, and transmitted socially, through learning and inventing, rather than by means of any strictly biological process. In addition, culture is a historical, time-dependent phenomenon that is constantly changing and evolving.

This huge conglomeration of mental attributes and their products also varies from individual to individual, changing substantially over one's lifetime, so that each of us is a unique cultural being always in the process of becoming someone different. Nevertheless, both simple and methodical observation reveal a fair amount of similarity (but not identity) among considerable numbers of persons at various levels of aggregation from small neighborhoods to entire continents. Such *relative* homogeneity, which is normally the result of generations of social interaction, is something we can call a cultural system without worrying about all the difficult details that make life exciting and contentious for anthropologists and cultural geographers.

Such systems tend to be rooted in particular territories. The genesis of a cultural system is an extremely complex phenomenon involving such processes as interaction between residents and the local, often changing habitat, social transactions with other groups near and far, political and military events, accidents of "cultural drift," and the passage of a generous amount of time. In any event, when we can connect a particular cultural system with a certain segment of the earth's surface, we are dealing with a "culture area" (or subregion, or some other such term, depending on the scale of observation). The great majority of the world's recognized culture areas consist of single contiguous chunks of territory, but there are exceptions, cultural systems occupying two or more widely separated areas. Luckily, we find only examples of the predominant pattern in Pennsylvania.

A question then logically follows as to whether Pennsylvania—that is, the territory encased within a particular set of interstate boundaries and the inhabitants thereof—can be regarded as a distinctive culture area. The answer for Pennsylvania is the same one that applies to virtually all 50 American states, with the possible exceptions of Hawaii and Alaska (which owe their social and cultural uniqueness to special physical circumstances). Within the United States, the factors involved in interstate boundary-making decisions seldom coincided with whatever processes happened to have shaped territories with a recognizable degree of cultural homogeneity. Indeed, several states, such as Florida, Maryland, and Illinois, are so wildly dissimilar from one end to another that it is impossible to apply uniform cultural labels to them.

On the other hand, the political entities we call states are not totally irrelevant to students of cultural geography. Many persons do identify with their specific political jurisdiction, be it state, city, or county and may be pleased to style themselves Kentuckians, Coloradans, or Michiganders, for example, even though such terms are little more than locational tags. In one extreme case, the term "Texan" masks dramatic cultural and social diversity even though it may convince the bearer of his or her singularity. But quite apart from whatever psychological impact the fact of residing in a state may have, the various states do exercise a certain measure of autonomy in handling some political and administrative affairs within a much more powerful federal system of governance. Thus there are noticeable differences among states in such items as the design and maintenance of highways, regulations concerning banking, liquor sales, hunting and fishing, strip-mining, and billboards, requirements for marriage and divorce, primary and secondary school curriculums, election practices, state holidays, and the operation of state hospitals and prisons. In parallel fashion, any number of business firms organize their operations in terms of state units. Finally, not the least of statewide phenomena are the various amateur athletic competitions (and state fairs) that engage the emotions of a goodly fraction of the population. Yet, paradoxically, the really passionate commitments to professional and college teams held by so many Americans display scant regard for state lines.

Culture Areas

Having said all this, it remains that the truly deep, important cultural divisions involving Pennsylvanians, those of particular interest to the human geographer, have little or no relationship to interstate boundary lines. What we find instead is that portions of at least three of the larger subnational American culture areas fall within Pennsylvania (see Fig. 9.1). None of them is conterminous with the state, and in only one instance—the entity known as the Pennsylvania Culture Area (PCA)—does the core of the area lie within the state. We shall have a great deal more to say about the PCA, especially since it accounts for much of the territory and population of the state, and even though it extends well beyond political Pennsylvania. But before dealing with this important region and the other, relatively marginal segments of neighboring cultural entities that chance to fall within Pennsylvania's borders, we should note a few of the other larger regional entities with which Pennsylvania has been associated.

Although the Census Bureau's designation of Pennsylvania as a component of its Middle Atlantic Division, along with New York and New Jersey, makes only a modest amount of sense in cultural terms, many historians and other scholars habitually refer to the "Middle States," a "Mid-Atlantic region," or some such term (sometimes including Maryland and Delaware) to distinguish the general zone from adjacent New England, the South, or the Midwest. This is really the process of regionalization

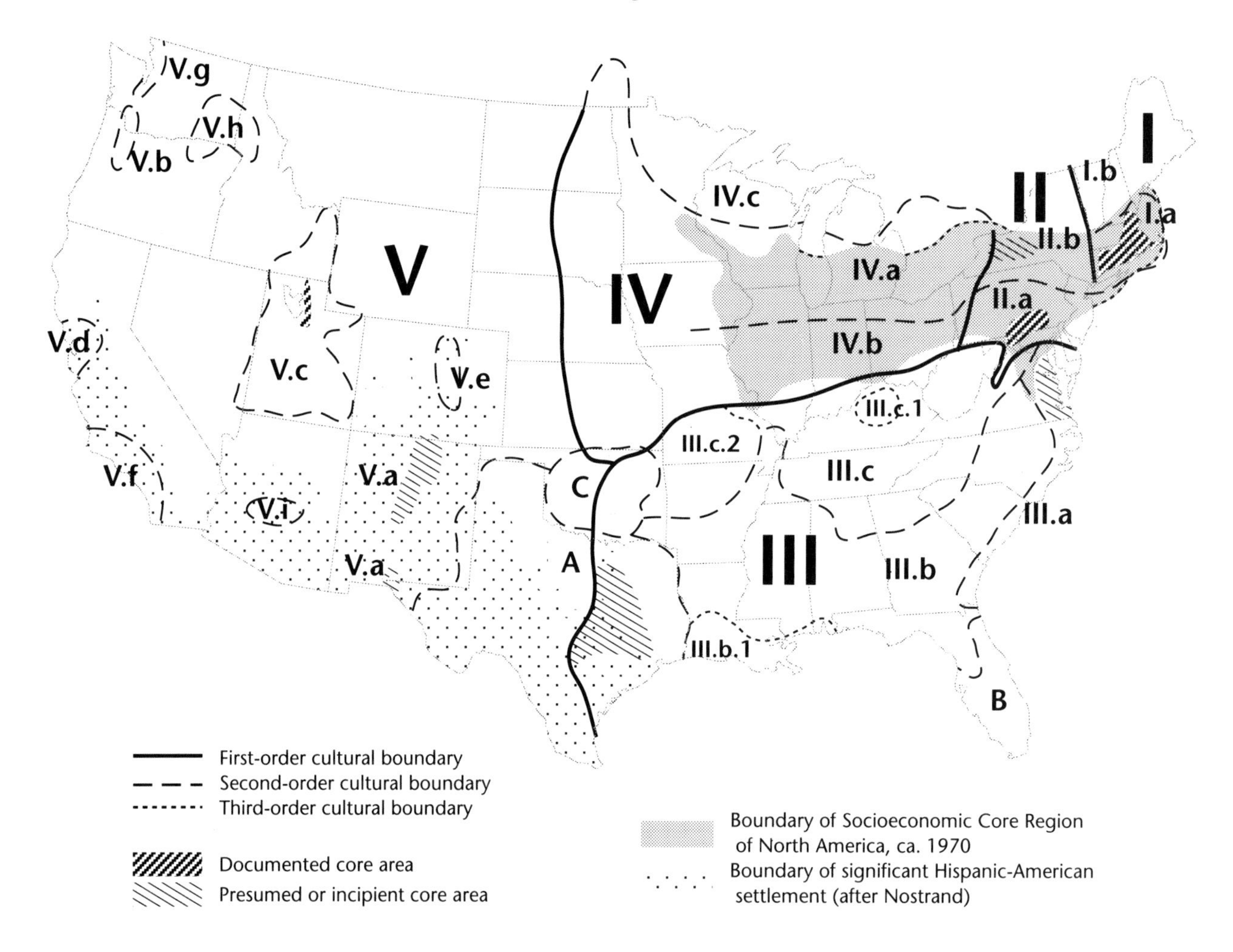

Region	Approximate dates of settlement and formation	Major sources of culture (listed in order of importance)
I. NEW ENGLAND	1620-1830	England
I.a. Nuclear New England	1620-1750	England
I.b. Northern New England	1750-1830	Nuclear New England, England
II. THE MIDLAND		
II.a. Pennsylvania Region	1682-1850	England & Wales, Rhineland, Ulster, 19th Century Europe
II.b. New York Region, or New England Extended	1624-1830	Great Britain, New England, 19th Century Europe, Netherlands
III. THE SOUTH		
III.a. Early British Colonial South	1607-1750	England, Africa, British West Indies
III.b. Lowland or Deep South	1700-1850	Great Britain, Africa, Midland, Early British Colonial South, aborigines
III.b.1. French Louisiana	1700-1760	France, Deep South, Africa, French West Indies
III.c. Upland South	1700-1850	Midland, Lowland South, Great Britain
III.c.1. The Bluegrass	1770-1800	Upland South, Lowland South
III.c.2. The Ozarks	1820-1860	Upland South, Lowland South, Lower Middle West
IV. THE MIDDLE WEST	1790-1880	
IV.a. Upper Middle West	1800-1880	New England Extended, New England, 19th Century Europe, British Canada
IV.b. Lower Middle West	1790-1870	Midland, Upland South, New England Extended, 19th Century Europe
IV.c. Cutover Area	1850-1900	Upper Middle west, 19th Century Europe
V. THE WEST		
V.a. Upper Rio Grande Valley	1590-	Mexico, Anglo-America, aborigines
V.b. Willamette Valley	1830-1900	Northeast U.S.
V.c. Mormon Region	1847-1890	Northeast U.S., 19th Century Europe
V.d. Central California	(1775-1848)	(Mexico)
	1840-	Eastern U.S., 19th Century Europe, Mexico, East Asia
V.e. Colorado Piedmont	1860-	Eastern U.S., Mexico
V.f. Southern California	(1760-1848)	(Mexico)
	1880-	Eastern U.S., 19th & 20th Century Europe, Mormon Region, Mexico, East Asia
V.g. Puget Sound	1870-	Eastern U.S., 19th & 20th Century Europe, East Asia
V.h. Inland Empire	1880-	Eastern U.S., 19th & 20th Century Europe
V.i. Central Arizona	1900-	Eastern U.S., Southern California, Mexico
REGIONS OF UNCERTAIN STATUS OR AFFILIATION		
A. Texas	(1690-1836)	(Mexico)
	1821-	Lowland South, Upland South, Mexico, 19th Century Central Europe
B. Peninsular Florida	1880-	Northeast U.S., the South, 20th Century Europe, Antilles
C. Oklahoma	1890-	Upland South, Lowland South, aborigines, Middle West

Fig. 9.1. Culture areas of the United States.

through exclusion—lumping together states that are not incontrovertibly New England, Southern, or Midwestern—rather than through positive identification. While there is some sharing of minor characteristics, there is scarcely any strong commonality of cultural personalities, taking the three states (Pennsylvania, New York, and New Jersey) as a group. The cultural interplay, past and present, among this trio of states has been complex and far from trivial, but it has not resulted in any of the grand blending of identities we observe elsewhere in North America.

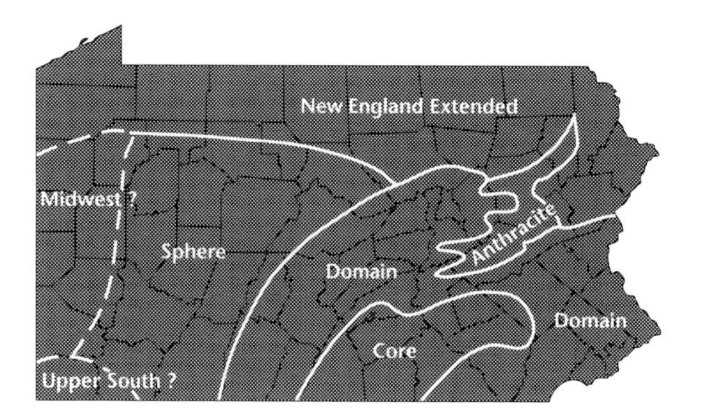

Fig. 9.2. Culture areas of Pennsylvania.

"Appalachia" is a regional term with a powerful appeal to the American imagination. If we define this great swath of rugged terrain stretching from central New York to central Alabama in physiographic, economic, or political terms, as has been done officially in the shape of the Appalachian Regional Commission, it is obvious that half or more of Pennsylvania's territory falls within Appalachia. But how genuine is it as a cultural concept? Despite the large literature on most other aspects of Appalachia, we still lack a clear, unambiguous answer to the question. It may be feasible to classify Appalachia as a subsection of the Upper South, a widely accepted regional concept, but such a decision assigns the Pennsylvania and New York portions of Appalachia to a cultural limbo because the two states are in no sense part of the South. Clearly more research on the identity of Appalachia is needed before we can decide whether it constitutes a valid culture area or simply the impoverished, relatively inaccessible fringes of a surrounding set of well-established regions.

The existence of Megalopolis has been universally acknowledged ever since Jean Gottmann's (1964) classic volume on the subject. Southeastern Pennsylvania occupies the central segment of this densely matted complex of urban, suburban, and exurban localities, the world's greatest conurbation, reaching as it does along the Atlantic Seaboard from southern Maine to somewhere in Virginia. Moreover, the outer frontier of the region has been marching steadily inland recently, so that it now embraces all of Piedmont Pennsylvania and has begun to encroach upon the Poconos, the anthracite region, and even the more seaward of the Appalachian valleys. If there is no question as to the reality of Megalopolis as a demographic and economic phenomenon, one must ask whether it has any meaning as a cultural entity. In actuality, it intersects (or rather overlies) a series of older, traditional culture areas, modifying but not obliterating them; yet it lacks any pervasive cultural personality of its own. Megalopolis is an outstanding representative of a family of latterday human regions that have added another layer of complexity to the North American scene.

Returning to the internal array of cultural areas within the state of Pennsylvania, we can detect two regions that straddle state boundaries and account for nearly all the territory not claimed by the Pennsylvania Culture Area. One of these *may* be the Midwest. The question marks appearing in Fig. 9.2 in the zone where the PCA shades into the tract farther west reflect the uncertainty about just where the eastern margin of the Midwest is located. In fact, some scholars are doubtful about including even Ohio within that important culture area (Shortridge 1985). The other major culture area that claims a sizable portion of Pennsylvania is "New England Extended"—the northeastern corner of the state, the Northern Tier, and much of its northwest. But in order to grasp its significance, we must survey the broader pattern of the evolution of American society over time and space.

Introducing Landscape I

We can trace both the peopling and the initial cultural imprinting of at least the eastern half of the United States to three principal "culture hearths" along the Atlantic Seaboard—Southern New England, Southeastern Pennsylvania, and the Chesapeake Bay area (Fig. 9.3)—a process expertly treated by Meinig (1987, 91–160) and Mitchell (1978). These are not the only significant beachheads for the Europeanization

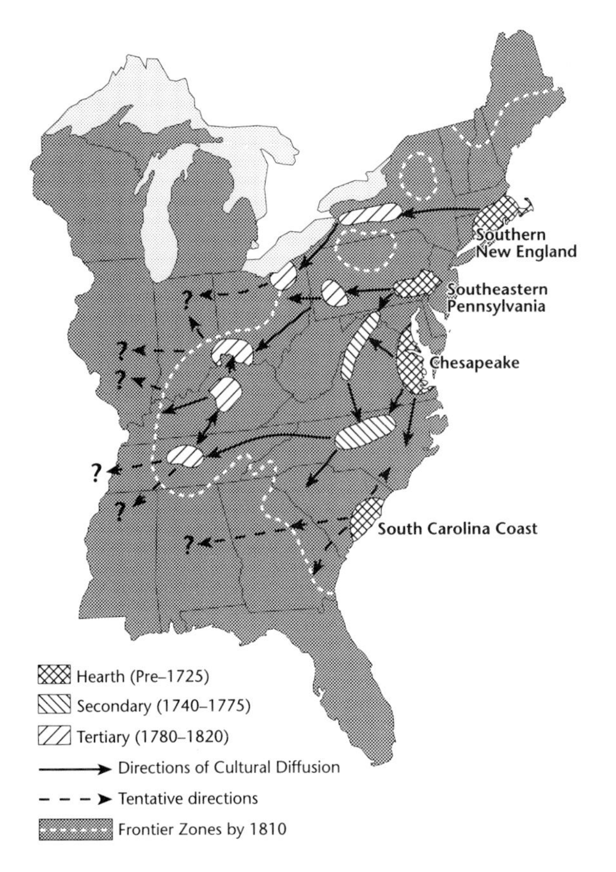

Fig. 9.3. Cultural diffusion circa 1810.

of North America. The South Carolina coast, the St. Lawrence Valley settlements, the French and Spanish foothold in southern Louisiana, the Mexican source area, and the somewhat problematic New York City / Hudson Valley colonial area (not shown in Fig. 9.3) all had their impact, at least within limited ranges. But all the evidence indicates that the trio of early European nodes of implantation and dispersion previously noted had the widest and most critical effect on the nation as a whole.

In the case of the New England thrust, migrants and ideas emanating from portions of Massachusetts, Connecticut, and Rhode Island, and somewhat later from the three northern New England states, penetrated New York State, northern New Jersey, and northern Pennsylvania, then spread far and wide within the Great Lakes region and, following another vector, invaded the Maritime Provinces of Canada as well (Hudson 1988). They were to have a decisive impact on the human geography of such places as upstate New York, northeastern Ohio, and lower Michigan, but in the case of Pennsylvania the affected area, mostly glaciated uplands with little economic appeal at the time, attracted little settlement, and much of that relatively late compared with the more southern tracts of the state. Furthermore, because of the difficulties and expense of moving goods or passengers between north and south against the topographic grain, there has been relatively little interaction between New England Extended, in or out of Pennsylvania, and the PCA. This is indeed a textbook example of physical geography playing a crucial role in channeling the development of a region's human geography.

From its initial base in the Delaware and Susquehanna valleys, the Pennsylvania Culture Area expanded outward during the eighteenth and early nineteenth centuries, most vigorously toward the west and southwest. Although the elongated ridges of the Folded Appalachians did seriously retard westward advances, many migrants bearing cultural cargo from the southeastern half of Pennsylvania did eventually manage to wend their way into the upper Ohio Valley and beyond, fanning out into what was to become the Midwest and contributing in a major way to its formation after merging with streams of settlers from other colonial hearth areas.

Many of the early Pennsylvanians who were discouraged by the hazards of trans-Appalachian journeys were deflected southwestward to frontier settlement zones either via the relatively traversable Piedmont or along the Great Valley, that convenient natural highway running all the way from New Jersey to Alabama. The result was that virtually all of Piedmont Maryland, West Virginia's eastern panhandle, and most of the length of Virginia's Shenandoah Valley are unmistakably part of the Greater Pennsylvania Culture Area (Fig. 9.4). Indeed, we can detect traces of Pennsylvanianness in sections of eastern Tennessee and central North Carolina. And the peoples streaming southwestward from Pennsylvania proper and its overflow tract in Maryland contributed decisively to the genesis of cultural patterns throughout the Upper South, a regional entity that eventually extended from Virginia and the Carolinas to portions of Oklahoma and Texas.

It is worth noting in passing that, although the Mason and Dixon Line, Pennsylvania and Maryland's common boundary, may have earned its historic and symbolic renown by virtue of having separated free from slave territory in antebellum days, it is nearly invisible in terms of the present-day cultural landscape.

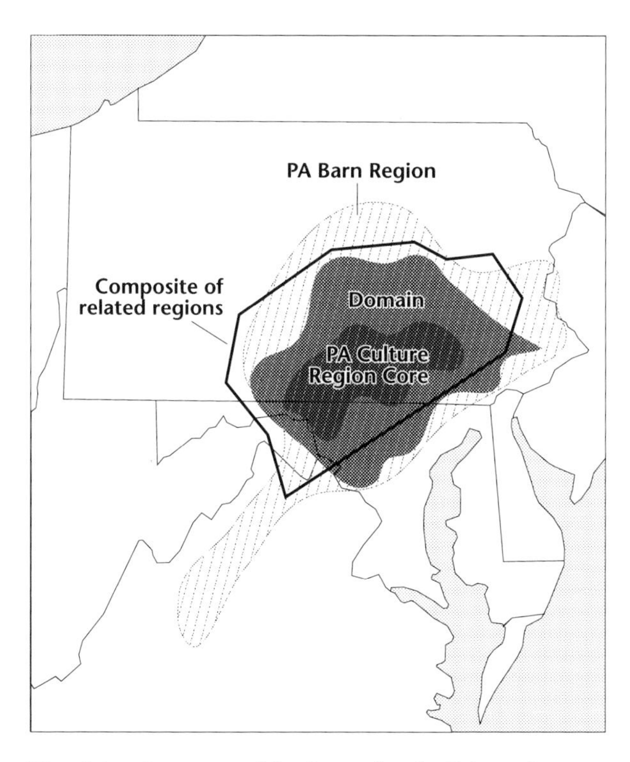

Fig. 9.4. Structure of the Pennsylvania Culture Area.

Like certain other culture areas in North America and elsewhere, the Pennsylvania Culture Area has a distinct, mappable internal structure. In certain respects, this structure duplicates the model postulated for the Mormon Culture Area by Donald Meinig (1965): three concentric zones—core, domain, and sphere—in which the purity and intensity of the culture lessens as one proceeds outward from the center. Such was the pattern discovered by Joseph Glass (1986) in his study of farmsteads and barns, in which he plotted the incidence of diagnostic PCA characteristics (see Fig. 9.4). And, as it happens, the limits of the PCA so defined accord remarkably well with the boundaries established by other scholars in research on other key attributes of the regional culture.

There is one interesting difficulty in applying the Meinig model to the Pennsylvania case. In the type example, we find that the concentric rings of the Mormon Culture Area are not just static phenomena, but that there is much interaction between Salt Lake City in the core and the peripheral areas. And much the same situation prevails within the New England Culture Area, with Boston acting as the point of central control. The historically logical site for the epicenter of the PCA's sociocultural life would be Philadelphia. But however crucial that city's contribution to the origin of the culture area may have been, it lies well outside the present-day boundaries of the region. Moreover, Philadelphia has been separated from the Pennsylvania Culture Area by a strip of territory in northwestern Bucks, Montgomery, and Chester counties, now being thoroughly suburbanized and exurbanized, where the relict cultural landscape bears little resemblance to that of the PCA. Instead, that tract is reminiscent of premodern rural England or Wales. In any event, unlike the Salt Lake City or Boston situation, there is no unambiguous command post somewhere in the middle Susquehanna Valley, nor any perceptible interplay nowadays between core and periphery. And we do not yet understand precisely how various raw materials from England, Ulster, and the Rhineland coalesced here to form a novel American culture, or what role Philadelphia may have played in the process.

Beyond the immediate effects on the Upper South and the Midwest through its role as launching pad for many of their earliest European settlers, the Pennsylvania Culture Area continued for some time to help mold the culture of the young republic by disseminating a variety of innovations and cultural axioms (see Chapter 1). Historical geographers have thus far documented only a handful of these items, among which we note the Conestoga wagon, several important advances in farming and mining technology, log buildings (Fig. 9.5), an important barn type, a distinctive courthouse plan (Price 1968), some truly basic concepts concerning town and city planning, and the idea of nationalistic place-naming (Zelinsky 1983).

The last of the three primal colonial bases for the demographic and cultural penetration of the continental interior, the Chesapeake Tidewater, has been the least consequential for Pennsylvania. The only section of the state where some signs of early Southern influence can be discerned is the southwestern corner. That territory fell within the huge expanse of western lands initially claimed by Virginia under the terms of its loosely worded royal charter. During much of the eighteenth century, Virginians in modest numbers attempted to ratify that claim by moving into the area, but after it had been ceded to Pennsylvania the evidence of early Southern occupancy has been largely erased by settlers entering from other directions.

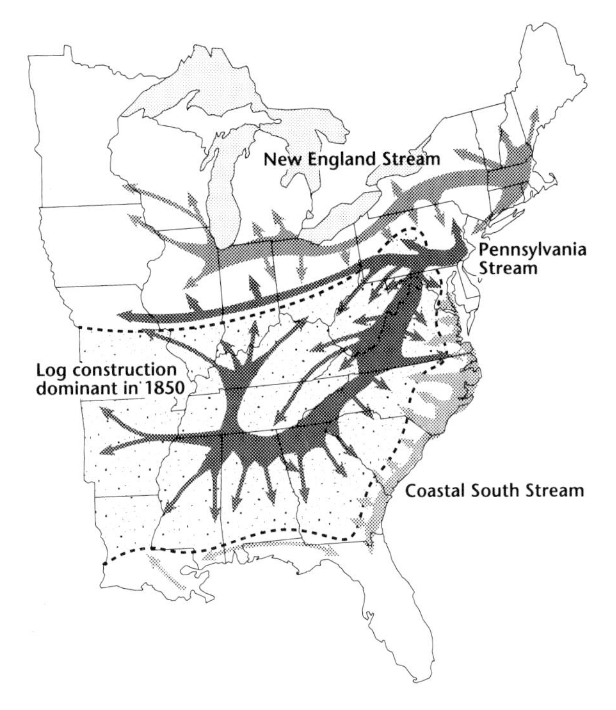

Fig. 9.5. Diffusion of different methods of wood construction.

Dominant Religious Groups

The special attributes of the Pennsylvania Culture Area deserve closer attention. This seminal zone in southeastern Pennsylvania not only lagged behind the other major colonial American culture hearths by a generation or two as far as date of founding is concerned, it was also different in kind, in the sorts of human material attracted to it. Whereas early New England and Tidewater Virginia were thoroughly dominated by settlers of English origin, the great majority of whom professed either the Anglican or Congregational faith, the newcomers to Pennsylvania were highly diversified in terms of both ethnicity and religion. In fact, the only possible competitors in any contest for social heterogeneity among the original thirteen colonies would have been two other Middle Colonies: New York and New Jersey. Polyglot Pennsylvania's early ethnic situation is described in Chapter 8, but we dare not overlook the equally important and persistent variety of religious groups that found a congenial haven in the state.

The notion of providing refuge and material opportunity for the beleaguered Quakers of the British Isles and Continental Europe was certainly one of the prime motives for initiating the Penn colony. Significant numbers of Quakers (or Friends) did arrive during the first decades of settlement, and their immediate and long-range impact has been considerable. Even though the Quakers were soon outnumbered by immigrants of other persuasions, and their ranks have tended to stagnate quantitatively ever since in lieu of vigorous proselytizing, they remained in firm political and economic command of affairs throughout the colonial regime, and their cultural impact was pervasive and lasting not only in Philadelphia and environs, the area of greatest concentration of co-religionists, but in the hinterland as well. Put another way, the Quakers gave the emerging personality of the PCA a decisive early twist. This contention has been documented in elegant fashion in historian E. Digby Baltzell's classic account of Quaker Philadelphia and Puritan Boston. Baltzell (1979) shows how the pervasive individualism, devotion to economic free enterprise, and distrust of central authority so typical of both early and later Pennsylvania—qualities that have permeated the psyche of much of the rest of the nation—a general, tolerant privatism so different from the mind-set of early New England, originated in the religious tenets of Pennsylvania's colonial elite. This atmosphere of economic liberalism, a piety allied with materialism, proved to be congenial to immigrants from the British Isles and Europe other than the Quakers, as James Lemon (1972) showed in his study of early Chester and Lancaster counties.

If William Penn dreamed of a tolerant colony that would serve as a new homeland for settlers of all shades of religious conviction and cultural background, as well as a sanctuary for Quakers, his vision soon came to pass. Indeed, it would be impossible to find any other part of the eighteenth-century world where adherents of so many different denominations resided side by side in such amity or a reasonable facsimile thereof. A count of the various congregations (no head count of members is available) extant in Pennsylvania as of 1775 provides vivid testimony to bolster this contention (Cappon 1976, 38):

Lutheran	142
German Reformed	126
Presbyterian	112
Friends	64
Mennonite	64
Anglican	24

Baptist	24
German Baptist Brethren	22
Moravian	13
Roman Catholic	11
Methodist	7
Dutch Reformed	3
Schwenkfelder	3
Jewish	2
Congregational	2
Independent	1

With the passage of time and the advent of many boatloads of people from ever more distant corners of the world, the complexity of the PCA's (and Pennsylvania's) religious composition has intensified, but not without retaining the regional peculiarities that set it apart from the rest of North America.

Although Anglicans (Episcopalians) and Presbyterians were, and are, well represented within the PCA, and Methodists and Baptists have become more numerous over the years, it is the strength of Protestant denominations of Teutonic origin that makes the area distinctive on any map of American religion, so much so that the term "Pennsylvania-German religious region" is not inappropriate (Zelinsky 1961, 193). Among the more conspicuous groups have been the Lutherans, the Evangelical and Reformed Church (merged with the Congregationalists in 1961 to form the United Church of Christ), the Evangelical United Brethren (united with the Methodist Church in 1968), the Church of the Brethren, the Mennonites (of whom the Amish are only one component), and the Moravians. Within the Pennsylvania portion of New England Extended, the representation of Methodists, Congregationalists, and other non-German groups has been relatively strong, while Presbyterians have been locally dominant in much of western Pennsylvania, thanks in large part to the importance of early Scots-Irish settlement.

From the mid-nineteenth century onward, a massive influx of immigrants from areas other than Great Britain and the German-speaking section of Europe has greatly modified the aggregate religious makeup of Pennsylvania's population and other aspects of Pennsylvania's culture as well. Arriving from Ireland, Italy, Poland, Czechoslovakia, Yugoslavia, Hungary, and Latin America, among other places, including the Catholic portions of Germany, the number of Roman Catholics in their multiethnic totality now far exceeds the largest of the Protestant groups. In fact, the 3,538,025 members of the Catholic faith listed in a 1990 survey accounted for no less than 48.5 percent of the total church membership reported for the state (Bradley et al. 1992). Furthermore, Roman Catholics constitute either the first or second largest group of church members in all but twelve counties. With the exception of Forest County, all of these predominantly Protestant counties are located within the south-central section of the state, the area with the smallest numbers of the foreign-born (see Fig. 9.6).

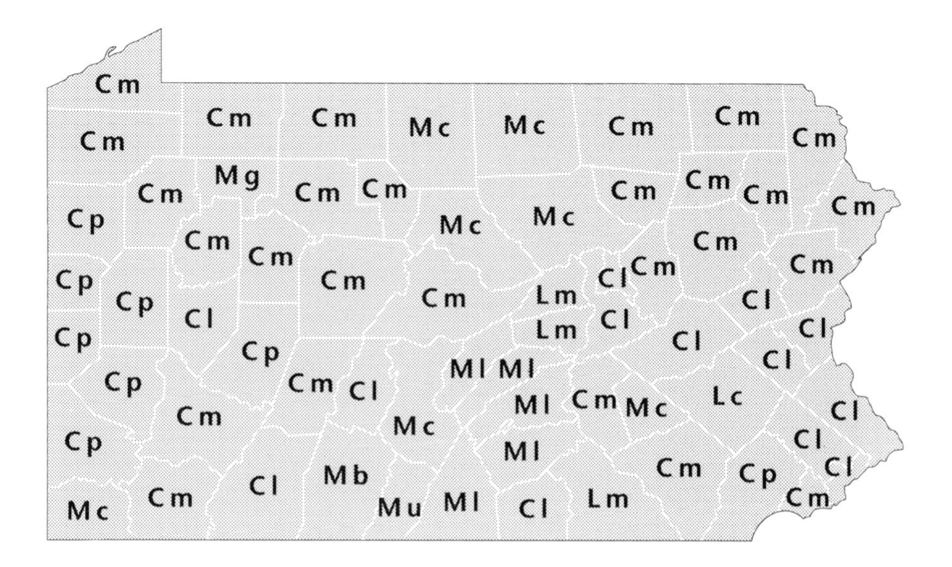

Fig. 9.6. Dominant church group by county, Pennsylvania, 1980.

Table 9.1. Reported membership in leading denominations: Pennsylvania, 1990.

Roman Catholics	3,538,025
Methodists	747,292
Lutherans	706,955
Presbyterians	405,672
Baptists	358,691
Jews	329,651
United Church of Christ	284,275
Episcopalians	139,448
Byzantine Ruthenian Rite	137,225
Mennonites	91,182
Assembly of God	74,616
Church of the Brethren	57,368
Christian & Missionary Alliance	37,006
Congregationalists	32,451
African Methodist Episcopal Zion Ch.	28,041
Church of the Nazarene	26,626
Mormons	24,488
Seventh Day Adventists	18,579
Disciples of Christ	17,619
Brethren in Christ	12,982
Church of God (Anderson, Ind.)	12,694
Friends	12,081
Church of God (Cleveland, Tenn.)	11,937
Moravians	11,724
Other denominations	174,071
Total reported	7,290,699

SOURCE: Bradley et al. 1992.

The ranks of the newcomers included many non-Catholics who had been poorly or totally unrepresented in pre-1840 Pennsylvania: Central and Eastern European Jews; adherents of various Eastern Orthodox churches from Russia, Ukraine, Serbia, Greece, the Middle East, and elsewhere; and black and white Southerners professing unfamiliar forms of the Protestant faith. The vast majority of these latterday Pennsylvanians settled in urban and mining localities, so that they did not greatly revise the religious complexion of agrarian and small-town Pennsylvania. In any event, the array of reported church members as of 1990 is a far cry from the situation in 1775, as indicated in Table 9.1. However, the reader must keep in mind the fact that many church groups were not included in this unofficial survey—notably all but a few of the Eastern Orthodox churches, most of the black and many of the white fundamentalist denominations—and that criteria for membership differ considerably among the organizations that did respond to the request for information. Consequently, these statistics give only an approximate impression of the actuality.

Nineteenth- and Twentieth-Century Developments

We have focused here mainly on the pioneering phase of the shaping of Pennsylvania's cultural map. By the 1840s, or 1850s at the latest, the settlement frontier had stopped advancing in the state, as was the case throughout the eastern United States. More precisely, all the lands deemed exploitable by means of contemporary technology had been claimed and occupied. The result was a set of what might be called "traditional culture areas," rough replications in a virgin setting of the much more venerable culture areas of the Old World, but a good deal less deeply layered and complex, and much more dynamic and broader in areal extent, than their European models. These newly hatched culture areas, however, were not destined to marinate forever in their own environmental and traditional cultural juices: too many technological, economic, and social changes were afoot. As we shall see, the PCA and other portions of the state have experienced immense change in every department of human activity, but the configuration of culture areas that had crystallized by 1850, which we can label Landscape I, is still readily recognizable. In fact, in many a backwater area, in the more secluded nooks and valleys of the countryside—indeed, in many stagnant villages—it is easy to believe that one is gazing upon a nineteenth-century scene.

The decade of the 1850s was a watershed period in the evolution of the human geography of Pennsylvania, and for the entire nation. The key development was the transition from the primacy of local or regional cultures and economies to the supremacy of national (and ultimately transnational) forces acting in the social, economic, and cultural arenas so as to downplay, or even obliterate, local peculiarities.

The closing of the agricultural frontier was followed quickly by the maturation of the railroad network (after a short-lived but consequential episode of canal-building), which brought about a revolution in the economy. A brief lumbering boom in the central and northern sections of the state created new wealth and new towns, but of much greater import was the widespread expansion of manufacturing enterprises and a greatly accelerated pace of urbanization. Closely associated with the rise of heavy industry and the demand for fuel for all the new factories and a growing population was the exploitation of Pennsylvania's vast reserves of coal. As already suggested, the labor needed to build all the railroads, operate the many mines and mills, construct the new urban infrastructure, and handle all the many chores of a bustling urban economy arrived mostly from culturally alien sources. The result of all these changes was the appearance of a new stratum in the cultural geography of the state, a novel set of landscapes, and in many ways a distinct and separate world: Landscape II.

Landscape II

In place of the virtual continuity of cultural characteristics between town and countryside that was common in eighteenth-century Pennsylvania, the great metropolises of Philadelphia and Pittsburgh acquired their own special identities that had little to do with older rural hinterlands. Much the same can be said about such one-industry railroad towns as Altoona or Renovo. The banks of the Ohio, the Allegheny, and the Monongahela and their major tributaries were lined with manufacturing communities that, in terms of ethnicity, religion, or physical texture, bore little resemblance to the traditional Pennsylvania scene. Similarly, throughout much of the Allegheny Plateau, previously quite thinly occupied, there materialized a liberal sprinkling of "coal patches"—small, concentrated mining settlements exploiting the rich bituminous resources of the region. But the most striking example of the creation of a totally novel cultural landscape was the emergence of the anthracite region in the final third of the past century as a district of closely spaced, densely settled small cities and towns entirely dependent on the mining and processing of the world's richest deposits of hard coal. Within a few years, what had been a near-wilderness, a cluster of ridges and valleys almost totally devoid of any settlement, became thickly packed with the houses, shops, and churches of a highly diversified population drawn largely from central and eastern Europe but under the financial and managerial control of "old stock" Americans. In every respect—vernacular architecture, ethnicity, religion, diet, dress, and dialect, among others—the contrast with the surrounding, conventionally Pennsylvanian territory was absolute. Crossing a single mountain ridge meant entering a different world. This second generation of cultural landscapes in Pennsylvania, essentially urban-industrial in character, may have had little in common with the older human patterns of the state, but it did resemble parallel developments in other sections of the nation. Thus one can say that, beginning in the 1850s, the nationwide forces of homogenization, in the cultural and all other human realms, began to play a decisive role in the life of Pennsylvanians.

Landscape III

The twentieth century has witnessed further striking alterations of a Pennsylvania cultural landscape that still retains the distinct personality imparted to it by two successive waves of settlers: the initial, largely agrarian folk of British, Scots-Irish, and Germanic origin, and the influx from the 1840s onward of a polyglot mixture of immigrants from all manner of places. If the first set of Europeans was associated with the primary industries and the second set with the secondary sector of the economy—manufacturing and construction, but with significant involvement in mining and transportation—the current phase of Pennsylvania's cultural evolution is being governed by the tertiary and quaternary sectors of the economy and their sociocultural fallout.

In this era of advanced transportation and communications and general affluence, perhaps the most visible and widespread landscape evidence of the latest cultural transformation is the growth and spread of suburbia. These peripheral developments are not limited to the environs of the larger metropolises but are to be seen girdling even the smallest urban centers with any sign of economic vitality. The larger examples are no longer merely aggregations of single-family homes but contain apartment complexes, shopping malls, office parks, factories, hotels, and recreational attractions as well as schools and churches. Indeed, the more elaborate suburbs have become veritable freestanding cities in their own right, with decreasingly intimate ties to the old central city. There may be no better example in Pennsylvania, or anywhere else in the nation for that matter, than the King of Prussia complex some 17 miles northwest of downtown Philadelphia. It has little or no connection with the traditional nineteenth-century countryside of Montgomery and Chester counties, but is instead closely akin to its burgeoning counterparts strewn the length and breadth of metropolitan America.

The outward surge of standardized new settlement does not stop at the outer edge of the built-up suburbs. As is happening in so much of the nation, extensive tracts of bucolic Pennsylvania are being "exurbanized." Relatively isolated, dispersed new houses—ranging in quality from the meanest of trailers and prefab structures to the most pretentious of estates—are occupied by people who do not earn their livelihood from the soil but who may be retired or are willing and able to commute long distances to their places of employment or have occupations they can carry on at home. They are living out the American dream of combining the joys of country life with the amenities of urban civilization. Needless to say, there is little that is regional about the houses or life-styles of the great majority of these exurbanites. A phenomenon closely linked to the exurban trend is the proliferation of seasonally occupied second homes, often informally rustic in design, which belong both to Pennsylvanians and to out-of-staters. The favored sites for these weekend and vacation retreats are alongside streams and lakes and in forested tracts.

If suburban and exurban settlement are generic features of Landscape III, other developments are much more localized because of their association with specific physical and social resources. Such is clearly the case for the many centers of tourist activity. Although organized tourism had begun by the mid-nineteenth century, when the relatively well-to-do frequented such places as Mauch Chunk and Bedford Springs, and the economy of Philadelphia was much enlivened by the millions of one-time visitors from almost everywhere attending the 1876 Centennial Exhibition, long-distance tourism for the masses as a year-in, year-out industry was hardly feasible before automobile ownership and paid vacations became universal. (During the nineteenth century there were the annual but localized excitements of the county fair, an institution that still flourishes throughout the less urbanized sections of the state.)

Because of its many historic buildings and sites, all carefully remodeled or restored to satisfy modern expectations, Philadelphia has certainly been one of the prime tourist targets among larger American cities. The techniques and wiles of historic preservation have also greatly boosted traffic to nearby Valley Forge and to Gettysburg, along with a variety of other towns such as Lancaster, Hollidaysburg, and Bethlehem. But no other section of Pennsylvania can rival Lancaster County in terms of touristic notoriety. The heart of the Pennsylvania-German countryside not only attracts hordes of the curious because of its seemingly "untouched" exotic inhabitants and their remarkable barns, farms, and crafts, but also obliges the visitors by transforming and commercializing itself—to the profit of some locals and the resentment of many others—into something of a caricature of the authentic early landscape. This is particularly obvious along a considerable stretch of U.S. 30.

Although it is not tourism in a strict sense, the craze for shopping at factory outlets (or reasonable facsimiles thereof) is emphatically a late twentieth-century phenomenon. Nowhere is it more fully expressed in landscape and other terms than in Reading and vicinity, but it has spread to other cities in the Northeast and is on the verge of becoming national.

Outdoor recreation has enlivened and radically altered large sections of the state in recent decades. Greatest in its areal impact is the annual deer hunt that begins in late November and draws many hundreds of thousands from nearby states as well as from within Pennsylvania into the virtually uninhabited forests of the Allegheny uplands and other wooded tracts. Smaller but still significant numbers of people frequent the wilderness and the many seasonal lodges and cabins, as well as the associated taverns and other enterprises, to hunt for small game or bear or to fish for trout and other species.

There is much of this robustious activity in the Poconos, but that northeastern corner of the state, a near-wilderness not too many years ago, has developed what may be a unique set of recreational opportunities that are being vigorously exploited because of a strategic location. Skiers and other winter sports enthusiasts from megalopolitan centers have swarmed into the region; second homes, rental cabins, resorts, and hotels all abound; and growing numbers of the elderly have elected to spend their golden years in the Poconos. But not the least of its claims to fame is the multitude of honeymooners attracted to the area, and the Poconos may have already overtaken Niagara Falls in that department.

Still another form of recreation has greatly modified at least one other corner of the state, but it is not for the masses: the gentleman farms that dominate much of rural Chester County along with the hunt culture associated with such establishments. The distinctive equine landscapes of this area, somewhat reminiscent of Kentucky's Bluegrass, are closely akin to those found in certain sections of Piedmont Maryland and northern Virginia. One can also see parallel trends in portions of Westmoreland County catering to Pittsburgh's elite, but as yet there has not been the same intensity of development.

The enormous expansion of higher education in Pennsylvania, as in the rest of the United States, has meant the growth of older institutions and the creation of many new ones—and some notable social and cultural impacts on the immediate localities. Although such effects are apparent enough in Lewisburg, home of Bucknell University, or in the communities adjacent to Dickinson, Franklin and Marshall, Villanova, Haverford, Lehigh, and Juniata colleges among others, there is no more nearly ideal example than The Pennsylvania State University and its satellite town of State College.

First there is the enormous campus itself, which in its architecture and landscape decidedly belongs to the twentieth century. The Penn State campus simply carries to a logical extreme the tendencies inherent in other colleges throughout the nation. Next to it one encounters a student shopping district and off-campus residential area, other late twentieth-century landscape features characteristic of present-day colleges, even those like the University of Pittsburgh and the University of Pennsylvania encysted within crowded central cities. In addition to the student body, faculty, staff, and business community dependent on the school, college towns like State College have attracted a growing population of the relatively well-to-do retired people who savor the mix of amenities available in such localities. Furthermore, Penn State, like most other major multiversities, has begun to generate clusters of research and development enterprises in its outskirts. Such industrial "campuses," with their offices, laboratories, and manufacturing facilities, are distinctive in appearance. The same phenomenon can be seen along a suburban corridor running northeast-southwest through suburban Montgomery and Chester counties. Such high-tech landscapes (with their relatively rootless, mobile professional work forces) are not yet as fully fleshed out in Pennsylvania as in California's Silicon Valley, North Carolina's Research Triangle, or Boston's Route 128 corridor, but one can anticipate further growth in that direction.

The cultural stratigraphy sketched above is not unique to Pennsylvania. Elsewhere in the older portions of North America we can observe much the same superimposition of cultural patterns brought about by more or less simultaneous shifts in demographic, technological, and economic factors. What makes the Pennsylvania situation, and specifically that of the Pennsylvania Culture Area, especially interesting, however, is the way Landscape I, the pre-1850 set of cultural items, has strongly persisted in many localities, though modified in many ways by latterday developments. Indeed, one of the more intriguing aspects of the PCA is how the three temporal layers of culture have affected one another without losing their essential personalities. There is still much to be learned about Landscapes II and III, but the resulting lessons are national in scope and only incidentally enrich our understanding of the specific Pennsylvania story. We return, then, to the earlier array of cultural patterns, which were firmly in place by the middle of the nineteenth century, and to their subsequent careers—to discover what it is that makes Pennsylvania and the Pennsylvania Culture Area so fascinating for the cultural geographer.

Settlement Pattern

Even the most casual tourist cannot help noticing how regionally distinctive the traditional settlement landscape of the Pennsylvania Culture Area happens to be (Noble 1984). While the countryside with its picturesque farmhouses and barns has received the lion's share of attention from both scholars and laypeople, it is in the towns and cities that we find the most convincing case for the particularity of the regional culture. Several physical characteristics set the Pennsylvania town apart from agglomerated settlements in all other sections of the United States (Zelinsky 1977). First, there is the sheer density of both structures and population per unit area in Pennsylvania towns large and small, values some 50 to 100 percent greater than for non-Pennsylvania towns of comparable size. Such compactness results from the practice of leaving little or no open space between adjoining buildings and from also omitting lawns between pavement and structure. Indeed row-housing is encountered more often in the PCA (and the kindred metropolises of Philadelphia and Baltimore) than in any other part of the United States. The townscapes so generated resemble those that appeared in Great Britain and other North Sea countries in the late seventeenth and early eighteenth centuries, and in this and other respects the Pennsylvania town, which took shape in the nine decades preceding the Revolution, is a product of contemporary Northwest European fashion, much more so than the colonial town types of any other part of the future United States.

The streets along which buildings are so thickly clustered are also well shaded, for residents of the PCA are accustomed to planting many maples and other hardy species by sidewalks or curbs, even in the most central blocks of the town. Certain characteristic street plans are also indigenous to the region. Although the larger places, such as Lancaster, Reading, and Allentown, adopted the Philadelphia model, a simple rectilinear grid that directly or otherwise so greatly influenced the shaping of the American urban scene nationwide, many of the smaller settlements have a street layout found only in the PCA (Pillsbury 1970). It is a rectangular lattice, relatively long, narrow, and linear in form (with a main street often straggling far into the coutryside) whose blocks are bisected lengthwise by alleys (Fig. 9.7). The alleys, which are so common in Pennsylvania towns of whatever magnitude, are consequential. They are usually named, well-maintained, lined with shops, dwellings, and other structures, and accommodate much pedestrian and vehicular traffic. Another noteworthy feature in many, but not all, towns (and much less frequently in other American regions) is the central diamond, an open space consisting of the right-angle intersection of two streets at or near the most functionally central point of a town, along with rectangular corners cut out from the four adjoining blocks. In its ideal form, the diamond is square in shape, but often it is elongated, so that its length can be several times greater than the width.

If rectangular grid plans are shared with the rest of the nation by the PCA, the same is true for the pattern of street names, which also originated in Philadelphia and flourishes in this region. Thus, there is a high incidence of a series of numbered streets intersected by another series bearing arboreal terms (e.g., Chestnut, Walnut, Maple, Cherry). The Pennsylvania town may be truly unique in the spatial scrambling of functions within central districts. We find something close to total randomness in the placement of retail, residential, professional, and governmental premises in all but the largest towns. Dwellings, shops, and offices consort cheek by jowl in adjacent buildings or under the same roof, at street level or above; but, with rare exceptions, churches, cemeteries, schools, parks, and playgrounds, if any, and manufacturing and wholesale enterprises are consigned to peripheral locations.

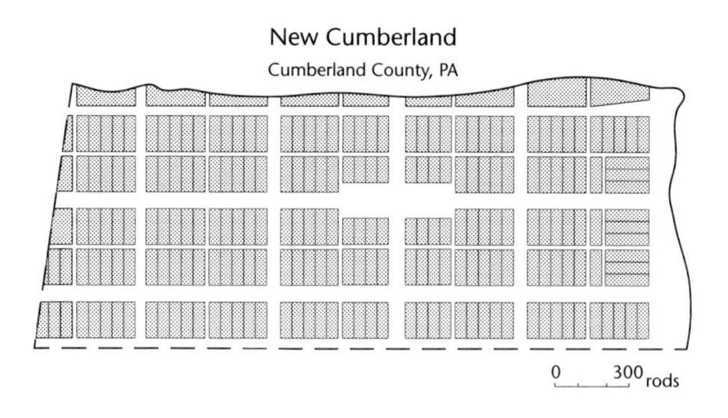

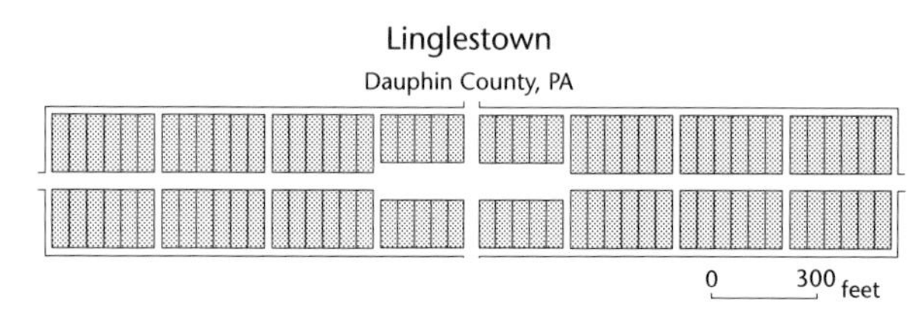

Fig. 9.7. Rectilinear (top) and standard linear-R (bottom) Pennsylvania towns.

Along with all its other special attributes, the Pennsylvania town merits attention for its architecture. In essence, we have here an elaborate series of variations on Georgian themes in both spirit and substance. Conceptually if not numerically dominant are two- or two-and-a-half-story structures, with roof ridge parallel to the street, with central doorway and hallway and symmetrical fenestration, and restrained ornamentation of classical inspiration. Although only the residences of the urban and rural well-to-do approximate the fashionable dwellings of the eighteenth-century British gentry and well-heeled merchant, it is easy enough to discern yearnings in that direction on the part of the lesser classes in Pennsylvania towns and their builders. In fact, this particular complex of architectural ideas persisted far into the past century and has not yet totally died out. It is another sensitive material index to the social conservatism of the population. The unusual popularity of nineteenth- and twentieth-century duplexes (symmetrical two-family structures as distinguished from the two-door Pennsylvania farmhouse) may be another sign of the durability of tradition. Later building vogues, such as the Classical Revival and the Gothic, never really caught on in the Pennsylvania

Culture Area, and even the eventually triumphant Victorian styles are often represented here by designs that somehow hybridize Georgian with later notions. Many of the older structures have undergone changes over time, of course, most conspicuously the addition of front or side porches, especially in the countryside where space for expansion is not a problem.

There has been, and still remains, a powerful preference for building in brick, another lasting heritage of eighteenth-century taste, and many PCA householders periodically paint the exterior walls, usually in red, but occasionally in white or yellow. Nowhere else in North America does brick prevail so strongly for buildings of all sorts, and for sidewalks as well. Although stone buildings, with or without stucco, are far less common, they are certainly not rare, occurring much more frequently in the Pennsylvania town and countryside than anywhere else in Anglo-America. Another indication of the grip of tradition is the persistence of older building fashions in suburban settings within such metropolitan areas as Reading or Lancaster. There, amid the predictable welter of split-levels and ranch houses, one can also find recent row housing and duplexes and various architectural nuances that obviously pay homage to Pennsylvania builders of another generation.

The fact that Pennsylvania towns can be easily identified does not imply a high degree of standardization. In fact, there is remarkable diversity of physical details within and among the several hundred places in question, so that no two of them are genuine twins. What they all share is the same basic personality, the Pennsylvanianness whose general traits have been described above.

If there is a close stylistic kinship between many of the PCA's urban dwellings and the farmhouses of the region, the renowned Pennsylvania barn is an entirely different matter. We know little about the barns and other outbuildings of the eighteenth-century farm, but one can speculate about the use of logs for some. What is known is that in the early 1700s southeastern Pennsylvania served as the zone for entry and early standardization for techniques of log construction originating in Central Europe and/or Scandinavia. The precise sources are still a matter of controversy, but recent research by Robert Ensminger (1992) has convincingly located the ancestral home of the Pennsylvania barn in eastern Switzerland. From an initial foothold in southeastern Pennsylvania, the practice of building log houses and barns spread rapidly with the advancing frontier throughout the wooded portions of the United States. Although a few venerable log houses still remain in Pennsylvania, the style was soon superseded by more socially acceptable modes of construction. In any event, the formidable Pennsylvania barn as we know it had materialized in its present-day form no later than 1800 as a wood, brick, or stone construction, or some combination thereof, superimposed on a stone understory. In terms of size, shape, and internal layout, these barns are admirably adapted to a productive agrarian economy in which livestock and field crops have been the main sources of income.

The essential geometry of the Pennsylvania barn is that of the symmetrical Georgian dwelling, but greatly magnified (Fig. 9.8). An entrance in the center of the longer side, frequently associated with an earthen ramp, leads into the middle level of a structure often built against a slope and oriented to maximize exposure to the sun on the opposite side. This far side, which usually overlooks a fenced or walled barnyard, is what makes the Pennsylvania barn truly distinctive, for it projects an appreciable distance past the understory to produce a "forebay." The shelter thus produced is presumably a convenient space in which livestock can huddle during inclement weather. Unlike the situation in most other parts of rural America, wood exteriors here are almost always painted, usually in red but occasionally in other colors. Popular myth to the contrary notwithstanding, the so-called hex signs and other artistic motifs that frequently appear on barns, sometimes with the construction date and proprietor's name, are purely decorative, another expression of the barn pride so pervasive in this region. As already suggested, these buildings are remarkable for their size and are probably the bulkiest vernacular structures to be found in North America. Among barns, the only competitors in the size department may be found in certain districts of Germany, Switzerland, and other Central European countries.

Other features of the PCA farmstead contribute to the distinctiveness of the region, as Joseph Glass (1986) has demonstrated in his definitive account of its farm landscapes. Not only do Pennsylvania house-types predominate throughout, there are also certain repeated patterns in the total ensemble of house, barn, and other artifacts and their orientation relative to one another and to the road and terrain. Glass has also shown how the zone of highest

Fig. 9.8. The Pennsylvania barn.

incidence of Pennsylvania barns tends to coincide with other measures for delimiting the Pennsylvania Culture Area. Although full-fledged PCA barns and farmsteads may appear outside the core and domain of the region, even if only sporadically, there can be little doubt that, in modified form, the barns and house-types of the PCA have been spread far and wide throughout the Midwest and the Upper South.

There is more to the settlement fabric of the Pennsylvania Culture Area than the items discussed above, and much room for further fieldwork and documentary research. Thus we may have a reasonably good grasp of the general character of the towns and farmsteads, but only the sketchiest knowledge about a number of other important landscape elements. For example, the study of the PCA's thousands of cemeteries is still to come. Are they regionally distinctive? Or are the funerary practices in this region indistinguishable from those in other sections of the country? The same questions could be posed about the fence types of early and later Pennsylvania or about its bridges, gristmills, monuments, land survey patterns, the geometry of fields, and the architecture of vernacular churches in both town and country, or of schools and shops.

What of the settlement traits of the residual portions of Pennsylvania, the fractions of the state beyond the Pennsylvania Culture Area proper? Less information is available for these areas, but it is clear enough that, in terms of the built landscape as well as other criteria, the crest of the lightly populated highlands running east-west most of the length of the state is indeed a sharp cultural divide. To the north, within the margins of New England Extended, the look of the humanized scene differs emphatically from that of the PCA. House types, barns, and townscapes are reminiscent of upstate New York and, less distinctly, New England itself, though much less richly developed. Toward the west, within the upper reaches of the Ohio Valley, we enter a zone of transition where the attenuated settlement patterns of the Pennsylvania Culture Area gradually merge into the urban and rural landscapes dominant throughout the Midwest.

Language

Although it may be far less tangible than such things as houses or fences, cultural geographers and anthropologists have come to realize that language, in all its many nuances, may be the most sensitive indicator of the identity and territorial as well as social structure of cultural communities. That would certainly appear to be true for Pennsylvania, especially when we examine the dialects of American English. Fortunately we have a reasonable abundance of evidence.

Despite the highly varied ethnic and linguistic origins of Pennsylvania's peoples, the English language, in its several varieties, is overwhelmingly dominant. According to the 1980 census, 93.1 percent of all inhabitants five years of age or older spoke English at home, and the next most popular tongue (Italian) accounted for only 1.3 percent of the population in question. But Pennsylvania does have the distinction, shared only with portions of Louisiana and the northern fringes of New England, of having a surviving enclave of non-English speakers of colonial origin: the Pennsylvania-Germans. The German-speakers enumerated in Pennsylvania in 1980 totaled only 83,392, a value far exceeded by speakers of both Italian and Spanish. This total undoubtedly includes some recent immigrants, but the great majority are descendants of prerevolutionary settlers from the Rhineland; and today virtually all are bilingual in English and the ancestral tongue. The present-day Pennsylvania-German language is an amalgam of some of the eighteenth-century dialects of mostly rural localities in western Germany and Switzerland, and in its spoken form it is about as different from the standard High German of the homelands as Standard English is from the Scottish tongue. Despite the efforts of local enthusiasts to preserve and encourage the survival of Pennsylvania-German, the size of this linguistic community seems to be dwindling. Within the area of strongest representation, Lancaster, Berks, and Lehigh counties, individuals five years and older speaking German at home in 1980 numbered only 5,900, 4,084, and 4,996, respectively. There are, of course, other pockets of non-English speech, especially among groups of foreign origin residing in the Philadelphia and Pittsburgh metropolitan areas, but they are almost entirely first- and second-generation immigrant stock undergoing rapid linguistic assimilation.

Fig. 9.9. Dialects of American English.

Most of our knowledge of the linguistic geography of English in the United States derives from two immense, ongoing research projects that are being published piecemeal: *The Linguistic Atlas of the United States* begun in the 1920s (O'Cain 1979), and the *Dictionary of American Regional English* (*DARE*) dating from the 1960s (Cassidy 1985, 1991). The former has been concerned primarily with vocabulary and secondarily with questions of pronunciation, while the latter has concentrated on vocabulary to an even greater extent. Consequently, our grasp of word choice and forms is much firmer than it is for pronunciation or grammar, but there are good reasons to believe that the territorial aspects of all these dimensions of human speech tend to coincide.

In his pioneering *Word Geography of the Eastern United States* (1949), based on field data gathered in the 1930s by the *Linguistic Atlas* investigators, Hans Kurath (1949) postulated three principal dialect regions: the North, the Midland, and the South. On the basis of the *DARE* evidence, Craig Carver (1987) suggested a revision of the Kurath typology (Fig. 9.9), maintaining that the most meaningful dialect divide within the eastern United States is that separating a region of Northern speech (consisting of the Upper North and Lower North, or Midland, subregions) from the territory of Southern speech (which in turn contains two principal subregions—the Upper and Lower South). Whether or not the area of Midland speech should be relegated to secondary status within the hierarchy of American dialect regions, there is no question about its actuality, or the fact that this particular way of speaking English has diffused widely into the central portions of the country, eventually contributing substantially to the more-or-less national version of our language that is especially prevalent in the western half of the United States.

The Midland, or Lower North, dialect region, as defined by Carver, stretches from the New Jersey coast to the Mississippi. (As might be expected, the Pennsylvania segment of New England Extended falls within another dialect region, the Upper North.) In his earlier study, Kurath recognized two subdivisions of the Midland: the Delaware Valley and the Susquehanna Valley areas. Many casual observers have also commented on patterns of pronunciation and vocabulary that are peculiar to the cities of Pittsburgh and Philadelphia, but we lack formal studies.

Carver's methodology for delimiting dialect regions involves mapping significant overlaps within groups of "dialect layers," which he defines as "the composite of a unique set of areal isoglosses." An isogloss, in turn, is the territory within which a particular word or usage is common, or the boundary thereof. Among the several dialect layers making up the Midland, one, the Southeastern Pennsylvanian, is of special interest because it coincides so neatly with the Pennsylvania Culture Area. Its isoglosses include such terms as:

all = completely gone, finished
baby coach = baby carriage
chicken corn soup
dressing = gravy
got awake = woke up, awakened
viewing = a funeral wake

The distinctiveness of Midland speech clearly emerges in one special vocabulary of paramount interest to geographers: place names (Zelinsky 1955). Thus we find "run" used extensively as the generic term for smaller streams, particularly in the outer reaches of the PCA and throughout the Midland, as opposed to "brook" in New England or "creek," "fork," or "branch" in the South for the same feature. In naming their settlements, Midlanders have displayed a striking partiality for the suffixes "-burg" and "-town," while within the PCA proper there is a remarkably high incidence of the terms "city" and "square" in the designations for small towns and villages. It is also quite likely that field investigation would reveal other regionalisms in the terms applied to landscape features in informal conversation but that fail to appear on maps.

We have seen that in at least three principal categories of culture—religion, settlement characteristics, and language—the Pennsylvania Culture Area can lay claim to regional individuality. The same statement applies, of course, to the Pennsylvania slice of New England Extended, but with the reservation that we are dealing only with the outer, rather dilute fringes of a cultural complex. Other truly vital aspects of Pennsylvania's cultural patterns merit discussion, but systematically collected and analyzed data are meager or nonexistent. One of these topics is political behavior as a reflection of underlying value systems. Data are abundant and accessible but seldom exploited. It would be useful to know, for example, how Pennsylvanians have voted for national candidates, their place-to-place variations in casting ballots for statewide officeholders and in various referenda, or the directions in which their state and congressional delegations have leaned in legislating certain diagnostic state and national questions. But there is more to politics than simply voting.

"Dry" Areas

The ways in which political jurisdictions are delimited, local regulations governing land-use and zoning, the licensing of various activities, decisions as to taxation and the allocation of public funds, levels of voter registration and citizen participation in public affairs—all these and many other issues fall within a realistic definition of politics and can enrich our understanding of Pennsylvania's cultural geography. One illustration of such possibilities is the mapping of the state's "dry" areas (Fig. 9.10). The legal sale of alcoholic beverages is controlled by the locality—by cities, boroughs, and townships. The areal pattern of partial and complete prohibition seems to be related,

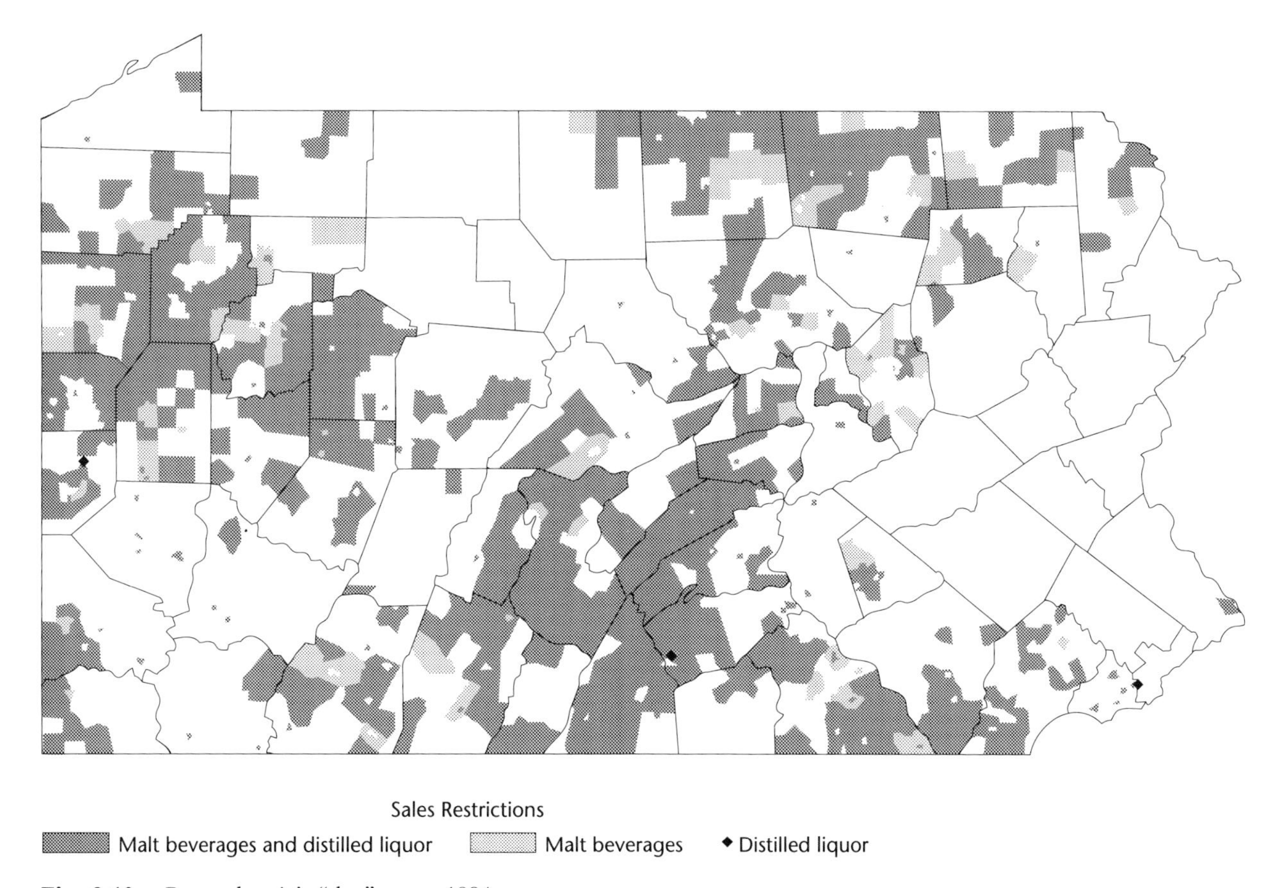

Fig. 9.10. Pennsylvania's "dry" areas, 1984.

to a significant degree, to ethnic and religious factors and to even more fundamental social attitudes, which do vary considerably among various Pennsylvania localities.

Political Cultures

Daniel Elazar (1984) came up with a provocative map showing the distribution of three distinct "political cultures" (see Fig. 1.3), although he is vague about data sources and mode of analysis. The three types are: Moralistic (most strongly represented in New England and the areas under the influence of New England's out-migrants and ideas), Traditionalistic (particularly strong in the South), and Individualistic. In keeping with Baltzell's thesis concerning Quaker Philadelphia, and inferentially much of the rest of the state, it is not surprising that the Individualistic theme manifests itself strongly in Pennsylvania. Among the seven Pennsylvania localities plotted on Elazar's map, the Individualist culture is symbolized as being dominant in five and secondary in the remaining two.

Foodways and Other Cultural Components

A strong argument can be made for the proposition that diet—or, in more general terms, foodways—is one of the most fundamental components of any cultural system. There is abundant anecdotal evidence for the claim that there is something special about the dietary patterns of the PCA. Many of us are aware of scrapple, funnel cakes, and chicken corn soup, and a few restaurants even feature Pennsylvania-German cuisine, but thus far there has been no methodical study of food and drink preferences within the region, of any of the many forms of behavior having to do with food and drink

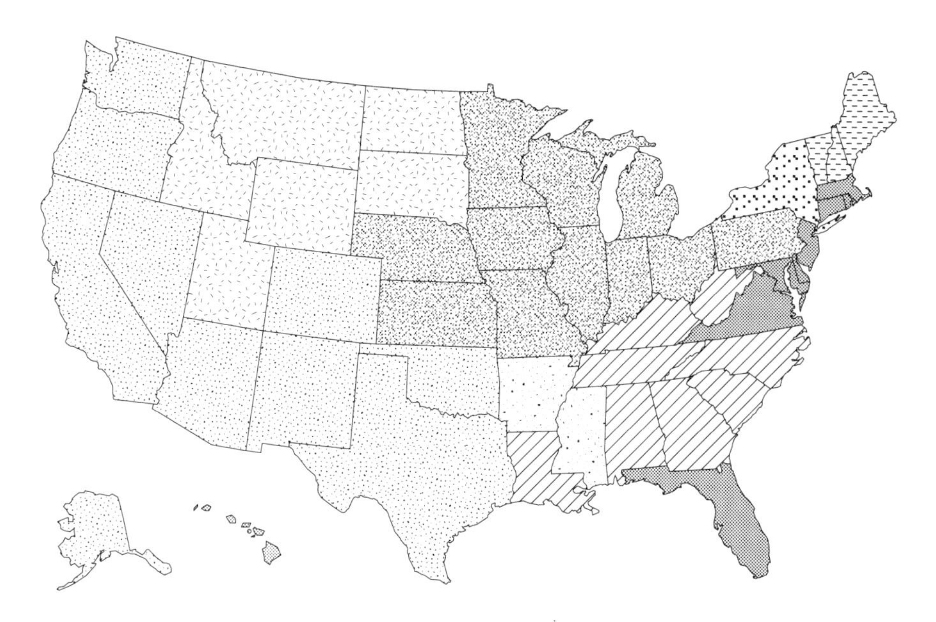

Fig. 9.11. Regions as defined by cluster analysis of special-interest magazines and voluntary organizations, United States, 1970–1971.

preparation and intake, or their geographical expression. Some exciting scholarly rewards await the enterprising investigator.

What has been said about foodways applies all too well to the geographical study of other important realms of the PCA's cultural system: interesting expectations but almost nothing to report. Thus there may be some measure of regional distinctiveness in various social customs and practices (e.g., the extraordinary importance of volunteer fire companies), in folklore, in games, outdoor recreation, and other leisure activities, or in costume (apart from the well-publicized quaintness of Amish dress), but for all these items the data cupboard is bare.

There is, however, one quite relevant state-level body of evidence offering insight into the general sociocultural configuration of adult Pennsylvanians, many of whom reside in the Pennsylvania Culture Area. Statistics were gathered for the period 1970–1971 on membership and circulation in each of the 50 states and the District of Columbia for 163 selected voluntary organizations and special-interest periodicals (Zelinsky 1974). When a cluster analysis was performed using seven sets of unweighted factor scores from an R-mode factor analysis of these 163 entities, Pennsylvania clearly entered the company of ten decidedly Midwestern states to form one of the three largest regional aggregates in the nation (Zelinsky 1974, 172) (see Fig. 9.11). Plainly, then, insofar as the proclivity of a certain fraction of the population to join or read certain things of their own free will is a clue to the psychic identity of all Pennsylvanians, this is a society whose members share essentially the same basic value system, much the same likes, dislikes, and interests, as do the people who live in Iowa, Indiana, or Kansas. Such a conclusion is congruent with the known facts of the historical geography of the settlement of the north-central states.

Vernacular Regions

A final line of inquiry into the cultural persona of Pennsylvania remains: identifying the vernacular, or popular, regions of this part of the world. In this respect, much of the state and nearby localities are in a peculiar and possibly unique situation. As defined by Terry Jordan (1978, 293), "Perceptual or vernacular regions are those perceived to exist by their inhabitants and other members of the population at large. Rather than being the intellectual creation of the professional geographer, the vernacular region is the product of the spatial perception of average people." Although such perceptions may coincide with culture areas as delimited by scholars, the vernacular region may simply be locational or an artifact of the print and broadcast media. In any event, if insiders and outsiders find it easy enough to

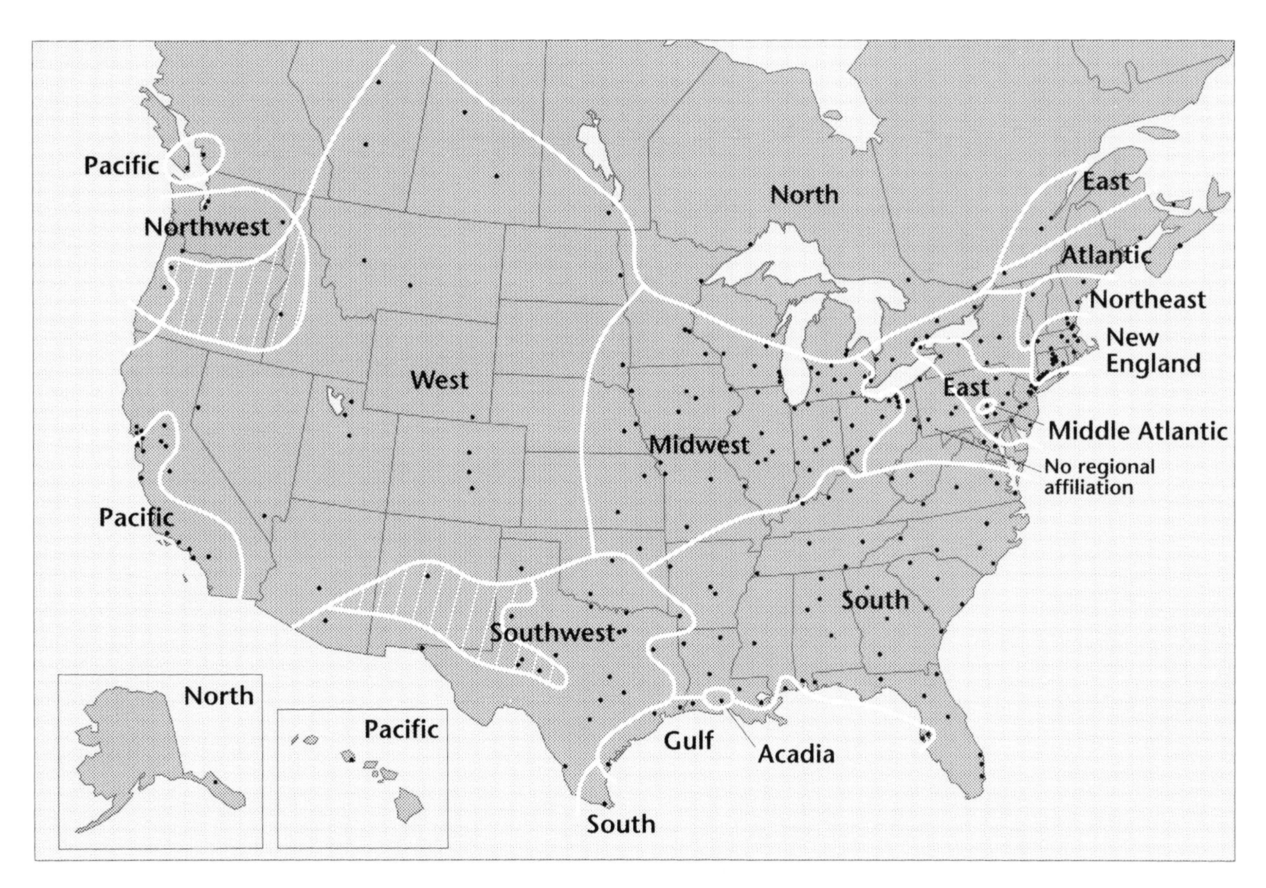

Fig. 9.12. Popular regions of North America, as indicated by names of metropolitan enterprises.

recognize such regions as the Southwest, the Northwest, the South, or New England unhesitatingly, the results are different in the Middle Atlantic area.

In an analysis of terms of locational and cultural significance appearing in the names of various enterprises listed in telephone directories for 276 metropolitan areas in the United States and Canada, I discovered that nowhere within metropolitan North America is there a weaker sense of regional affinity than in the cities of western Pennsylvania and adjoining portions of neighboring states (Zelinsky 1980, 14) (see Fig. 9.12). Furthermore, in the eastern half of the state the situation is not much better. Only in the Harrisburg metropolitan area do we find an appreciable number of enterprises labeled "Middle Atlantic." The most popular regional term for the other eastern Pennsylvania cities is the culturally anemic "East," and even that is not strongly represented.

How do we account for this apparent lack of self-knowledge or interest, so different from the insistent awareness of Southernness or New Englandness elsewhere? The failure to appreciate the regional personality of the PCA (aside from its barns and Mennonites) stems largely, I suspect, from its sheer middleness. In many ways the region is intermediate between the New England–New York cultural landscapes and those of the South, and thus less aberrant from eastern seaboard or national norms (Shryock 1964). It has been a major source, or at least channel, for the flow of migrants and ideas into the vast central and western reaches of the nation. Thus, much that was to become national and "mainstream" later is found in the PCA, too prosaic and normal to stir up comment.

A Regional Revival?

There are signs in the PCA, as well as in the rest of the state and in the nation as a whole, that some form of regional revival has been brewing. It is a phenomenon intimately connected with the ethnic revival discussed in Chapter 8. Here as elsewhere we can observe the recent proliferation of local festivals, the creation of periodicals catering to state and local interest and pride, university presses nurturing regional motifs, and the founding of historical museums focused on the glories of the locality. (A seemingly trivial but truly revelatory symptom of the trend is the adoption of promotional slogans and other such devices on vehicle license plates of virtually every state.)

To what degree such developments are spurious and driven by simple commercial greed or are manifestation of genuine, long-term sociocultural change, it is too early to say. But what can be said about Pennsylvania and every other section of the land is that the unfinished agenda for the student of cultural phenomena is rich and overflowing. In this particular instance, we have the challenge of not only documenting the cultural facts of Landscapes I, II, and III but also of charting their changes over time and of learning how these three epochs in the cultural chronicles of the state have interacted and coexisted. The answers are important not only for Pennsylvanians but also for all Americans. To a degree probably unmatched by any other state, Pennsylvania is America in microcosm, the seedbed for much that was to blossom later into our national way of life.

Bibliography

Baltzell, E. D. 1979. *Puritan Boston and Quaker Philadelphia*. New York: The Free Press.

Bradley, M. B., et al. 1992. *Churches and Church Membership in the United States*. Atlanta: Glenmary Research Center.

Cappon, L. J., ed. 1976. *Atlas of Early American History: The Revolutionary Era, 1760–1790*. Princeton: Princeton University Press.

Carver, C. M. 1987. *American Regional Dialects: A Word Geography*. Ann Arbor: University of Michigan Press.

Cassidy, F. G., ed. 1985, 1991. *Dictionary of American Regional English*, Vol. 1, *Introduction and A–C;* Vol. 2, *D–H*. Cambridge: Harvard University Press.

Cuff, D. J., et al., eds. 1989. *The Atlas of Pennsylvania*. Philadelphia: Temple University Press.

Elazar, D. 1984. *American Federalism: A View from the States*. Third edition. New York: Harper & Row.

Ensminger, R. F. 1992. *The Pennsylvania Barn: Its Origin, Evolution, and Distribution in North America*. Baltimore: Johns Hopkins University Press.

Gibson, J. R., ed. 1978. *European Settlement and Development in North America*. Toronto: University of Toronto Press.

Glass, J. W. 1986. *The Pennsylvania Culture Region: A View from the Barn*. Ann Arbor: UMI Research Press.

Gottmann, J. 1964. *Megalopolis: The Urbanized Northeastern Seaboard of the United States*. Cambridge: MIT Press.

Hudson, J. C. 1988. "North American Origins of Middlewestern Frontier Populations." *Annals of the Association of American Geographers* 78:395–413.

Jordan, T. G. 1978. "Perceptual Regions of Texas." *Geographical Review* 68:293–307.

Kniffen, F., and H. Glassie. 1966. "Building in Wood in the Eastern United States." *Geographical Review* 56: 40–60.

Kurath, H. 1949. *A World Geography of the Eastern United States*. Ann Arbor: University of Michigan Press.

Lemon, J. T. 1972. *The Best Poor Man's Country: A Geographical Study of Early Southeastern Pennsylvania*. Baltimore: Johns Hopkins University Press.

Meinig, D. W. 1965. "The Mormon Culture Region: Strategies and Patterns in the Geography of the American West, 1847–1964." *Annals of the Association of American Geographers* 55:191–220.

———. 1986. *The Shaping of America: A Geographical Perspective on 500 Years of History,* Vol. 1, *Atlantic America, 1492–1800*. New Haven: Yale University Press.

Mitchell, R. D. 1978. "The Formation of Early American Cultural Regions: An Interpretation." In *European Settlement and Development in North America,* ed. J. R. Gibson. Toronto: University of Toronto Press.

Noble, A. G. 1984. *Wood, Brick, and Stone: The North American Settlement Landscape*. 2 vols. Amherst: University of Massachusetts Press.

Nostrand, R. L. 1970. "The Hispanic-American Borderland: Delimitation of an American Culture Region." *Annals of the Association of American Geographers* 60:638–661.

O'Cain, R. K. 1979. "Linguistic Atlas of New England." *American Speech* 54:243–278.

Pillsbury, R. 1970. "The Urban Street Pattern as a Culture Indicator: Pennsylvania, 1682–1815." *Annals of the Association of American Geographers* 60:428–446.

Price, E. T. 1968. "The Central Courthouse Square in the American County Seat." *Geographical Review* 58:29–60.

Shortridge, J. R. 1985. "The Vernacular Middle West." *Annals of the Association of American Geographers* 75:48–57.

Shryock, R. H. 1964. "The Middle Atlantic Area in American History." *Proceedings of the American Philosophical Society* 108:147–155.

Zelinsky, W. 1955. "Some Problems in the Distribution of Generic Terms in the Place-Names of the Northeastern United States." *Annals of the Association of American Geographers* 45:319–349.

———. 1961. "An Approach to the Religious Geography of the United States: Patterns of Church Membership in 1952." *Annals of the Association of American Geographers* 51:139–193.

———. 1974. "Selfward Bound? Personal Preference Patterns and the Changing Map of American Society." *Economic Geography* 50:144–179.
———. 1977. "The Pennsylvania Town: An Overdue Geographical Account." *Geographical Review* 67: 127–147.
———. 1980. "North America's Vernacular Regions." *Annals of the Association of American Geographers* 70: 1–16.
———. 1983. "Nationalism in the American Place-Name Cover." *Names* 30:1–28.
———. 1993. *The Cultural Geography of the United States.* Revised edition. Englewood Cliffs, N.J.: Prentice Hall.

10

POLITICAL GEOGRAPHY

Anthony V. Williams

One major question about Pennsylvania politics is and has been why this state, so important historically and industrially, and among the top five in population (even by 1990), is so politically invisible on the national scene. There have been powerful people like David Lawrence and others mentioned later, but no nationally recognized political figures since Benjamin Franklin. Former President Dwight D. Eisenhower was a transplant, so he does not count. Why haven't we had a Ronald Reagan, a John Kennedy, or even a Robert Dole or a Gary Hart? Our only Pennsylvania president, James Buchanan, joins Taylor and Polk in being worthy challenges in presidential "Trivial Pursuit." Are there any reasons, geographical or other, that might help explain the state's lack of national political prominence?

Paul Beers, in his *Pennsylvania Politics Today and Yesterday* (must reading for those interested in the politics of the state), suggested thirteen reasons. Several are geographic, directly or indirectly, and include the tradition of strong local parties, surely related to the transportation network and topography. Then there is the influence of the declining economy as the nation shifted from reliance on the Pennsylvania standbys of steel, coal, and railroads. Perhaps most interesting is what Beers feels is a long-standing anti-Philadelphia sentiment, which is reciprocated and has derailed the ambitions of several who in other states might have attained national prominence. It has been difficult for the whole state to get together behind one of its own. Arlen Specter, while running for reelection in 1992, said that campaigning in Pennsylvania is like campaigning in six states.

The lack of identification most Pennsylvanians have with the state has been a critical factor in its national political, and perhaps also economic, fortunes. Ask an Ohioan, a Texan, or a Washingtonian where they come from, and they would reply with the name of their state. But in Pennsylvania, people are more likely to mention their home town or even their county. The more than 2,500 fiercely independent boroughs and townships, and their reluctance to cooperate, not to mention amalgamate, is a cause and consequence of this localism. The state's county structure, stable since the late 1800s, is another. And the existence of two large competing metropolitan areas has certainly been a major contributing factor in preventing Pennsylvania from achieving its potential political importance. Many might demur on the last point, noting the rising fortunes of Californians in national politics despite the rivalry of the Bay area and Los Angeles. But California and Pennsylvania have a critical difference. California represents dynamism and the future, and its people know it (even though some deplore the consequences). In fact, one can argue that California has come to political prominence despite the presence of its two competitive metropolises, and only since it has become recognized as the nation's leading state. This issue is discussed at the end of this chapter, but first we need to understand the state's basic political geography.

A Capsule Political History

Pennsylvania's early political role was propitious. Pennsylvania was the lynchpin of the English colonies, a bridge between the New Yorkers and New Englanders and the Southern group. It was preeminent in manufacturing, and such well-known tools of nation-building as the Kentucky rifle were from the state. Until 1840 it was the leading food-producing state. Its association with William Penn, and its tradition of religious and ethnic tolerance, were widely known and admired. The state was the site of the first two Continental Congresses and was the place where George Washington was chosen Commander-in-Chief and where the Declaration of Independence was pronounced. It was the state where the first American flag was unfurled, it hosted the Constitutional Convention, and its capital, Philadelphia, was the nation's first capital. In 1800, according to the second census of the United States, it was the third largest state in population, and Philadelphia was the country's largest city.

The state boomed in the nineteenth century. Pittsburgh and steel were the metaphors for Pennsylvania industry, but the coal, iron, forestry, oil, machinery, and the railroad industry all contributed to create an economic giant. Somewhere along the way, however, dynamism faded. Census after census has revealed that more people migrate from the state than to it. Major opportunities were missed in the automobile and chemical industries, to take two examples, and the state reached its economic peak in national status in World War I. Since then its relative status has continually declined. The artificial stimulus of the "Pittsburgh plus" steel-pricing policy kept that industry preeminent until World War II, but the rest of the state suffered and many of its best and brightest left for more promising areas. Other chapters tell the details of this story; our concern here is what happened politically.

Buchanan's presidency represented the peak of Pennsylvania's national visibility, but the state's real political influence was represented by what Paul Beers called narrow-minded and parochial bosses looking out for the industrial interests of steel, coal, and the railroads. These bosses were deal-makers rather than national leaders, and they did their job well but quietly. The classics of the nineteenth century were Matthew Quay of Beaver County and Boies Penrose of Philadelphia. Their most notable successors in this century are probably Joseph Grundy of Bucks County and David Lawrence from Pittsburgh. It is no accident that three of these four were Republicans. From the Civil War to Franklin Roosevelt's election, the state was held in what has been called the "Republican hammerlock." From 1860 until 1932, the only break in a string of Republican presidential victories in the state was the vote for Teddy Roosevelt in 1912. The two Democratic governors elected from Philadelphia in 1883 and 1891 were accidents. In the twentieth century, it was 1934 before the Democrats elected a governor and a senator.

Of course, there was a democratic presence in the state, but the financial muscle of the Republicans, the aftermath of the Civil War, and the Democrats' association with immigrants and Catholics kept it to a tolerable nuisance. Indeed, at times elements of the party were subsidized by the Republican bosses to prevent potential winning candidates from getting nominated. There may still be some of these conservative Democrats around.

The depression and the New Deal snapped the Republican stranglehold on the state. Previously faithful voting blocs, such as blacks, broke the traces to vote for Republican mavericks like Gifford Pinchot and for Democrats. Pittsburgh went Democratic for good in the 1930s, and Philadelphia did so in the 1950s. Still, even today it has been difficult for the Democrats to elect candidates to the governorship or the U.S. Senate, despite a registration edge and many close elections. They have done much better in the state legislature and the U.S. House of Representatives. In the 1992–1994 session, for instance, the Democrats had a majority in the 203-member House, with 103 seats, 11 of the 21 congressional seats, and managed to tie in the State Senate with the help of a turncoat Republican.

Let us now consider some selected electoral data, focusing on the presidency and the governorship, with one U.S. Senate race shown to illustrate those contests. One aspect to consider is whether there are areas of visible Republican or Democratic strength. Political analysts over the years have identified several. For the Republicans, the classic strongholds have been suburban Philadelphia, particularly Montgomery County, Dauphin, Lancaster, and York counties, the Northern Tier, and the Allentown and Williamsport areas. In addition, the rural parts of the state have traditionally voted Republican. Democrats have counted on Pittsburgh, Philadelphia, Lackawanna, Luzerne, and Northampton counties and the counties on the western border, especially the southwest.

Some Recent Presidential Elections

The election of Presidents every four years represents the most visible aspect of our voting tradition and attracts more voters than any other elections, although a disgracefully small proportion compared to other democracies. These elections are more big-issue oriented than the congressional or state legislative contests, and therefore probably better reflect underlying political attitudes. Five recent presidential elections were selected to explore state voting patterns. Two were runaways: the 1964 Johnson versus Goldwater campaign, where the Democratic candidate got almost two-thirds of the state vote, and the 1972 Nixon and McGovern contest, in which Nixon took almost 60 percent. The third election was the very close Carter and Ford election, which was barely won by the Democrat. The fourth was the also-close (in Pennsylvania) Bush-versus-Dukakis race won by George Bush with just 51 percent of the vote. And the fifth was the unusual three-way race of 1992 among President Bush, Bill Clinton, and Ross Perot, won by Clinton with 45 percent of the vote. Table 10.1 gives an idea of how Pennsylvania as a whole votes in national presidential elections.

Except for 1968, Pennsylvania voted with the winner. But these tabular results are of no value in understanding the spatial pattern of voting. The maps in Figs. 10.1 through 10.7 show how useful maps are in understanding politics. They show the distribution of the votes in Table 10.1 for 1964, 1972, 1976, 1988, and 1992 and how each county voted relative to the overall state vote. This makes it easy to see which counties are Democratic or Republican strongholds regardless of whether the election was a sweep or a squeaker.

What do the maps reveal, remembering the

Table 10.1. Pennsylvania votes in presidential elections, 1952–1992.

Election Year	Republican		Democrat	
1952	Eisenhower	2,415,789	Stevenson	2,146,269
1956	Eisenhower	2,585,252	Stevenson	1,981,769
1960	Nixon	2,439,956	Kennedy	2,555,282
1964	Goldwater	1,673,657	Johnson	3,130,954
1968	Nixon	2,090,017	Humphrey	2,259,405
1972	Nixon	2,714,521	McGovern	1,796,951
1976	Ford	2,205,604	Carter	2,382,677
1980	Reagan	2,261,872	Carter	1,937,540
1984	Reagan	2,584,323	Mondale	2,228,131
1988	Bush	2,290,904	Dukakis	2,182,915
1992	Bush	1,791,841	Clinton	2,239,164

SOURCES: Pearce, *Pennsylvania Manual,* 1988; Internet mosaic, 1992.
NOTE: In 1992 Perot (Ind.) had 902,667 votes.

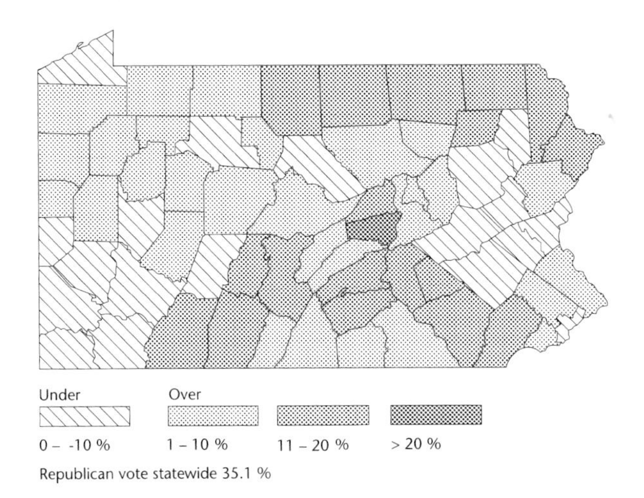

Fig. 10.1. Republican presidential vote, 1964: Percent under or over state average.

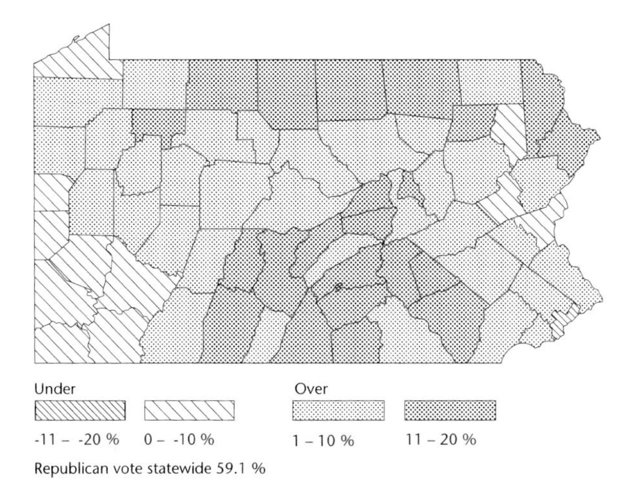

Fig. 10.2. Republican presidential vote, 1972: Percent under or over state average.

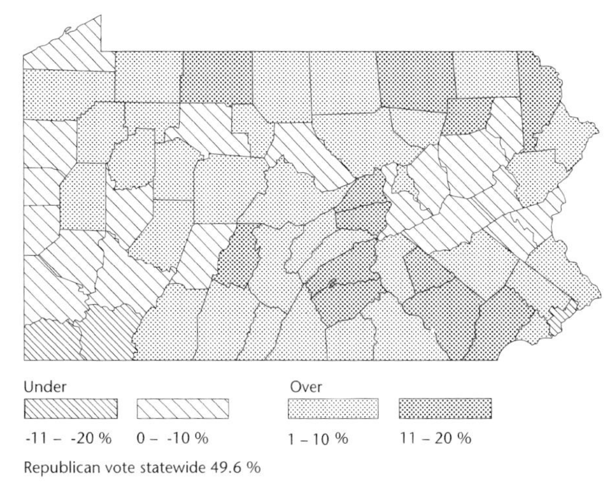

Fig. 10.3. Republican presidential vote, 1976: Percent under or over state average.

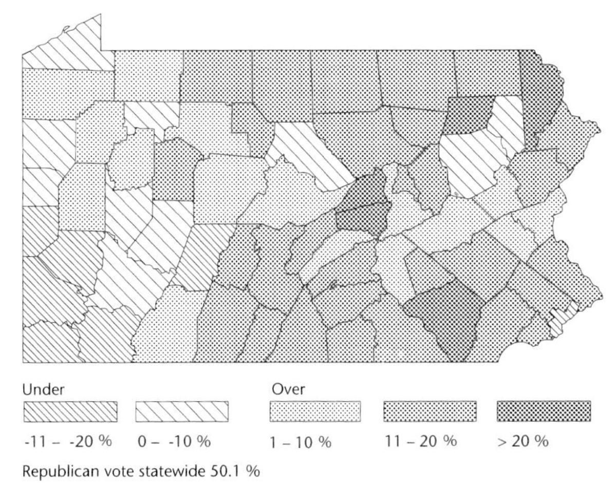

Fig. 10.4. Republican presidential vote, 1988: Percent under or over state average.

conventional political wisdom about Democratic and Republican strongholds? In reading each map, the thing to remember is that the darker the pattern the better the Republicans did. Putative strongholds do indeed exist, but with some interesting twists. The most consistent Republican county in these elections is Snyder, which gave the party at least 10 and sometimes more than 20 percent more of its votes than the state at large. The Northern Tier, with the exception of Erie County, is consistently more Republican than the state, with Wayne and Bradford staying very faithful. Other rural areas also lean Republican in these four elections, relative to the state, but many are hardly monolithic. Of the larger counties, the position of York and Montgomery as bastions of Republican strength seems a bit of an overstatement, although their large populations make them important. While both are more Republican than the state as a whole, they fall into the lowest of the plus categories for these elections, giving the Republican candidate 0–10 percent more of their votes than his or her statewide average. Indeed, the inner suburban ring of Bucks, Montgomery, and Delaware counties falls into this category.

The traditionally strong Democratic areas are consistent on these voting maps. There is indeed a

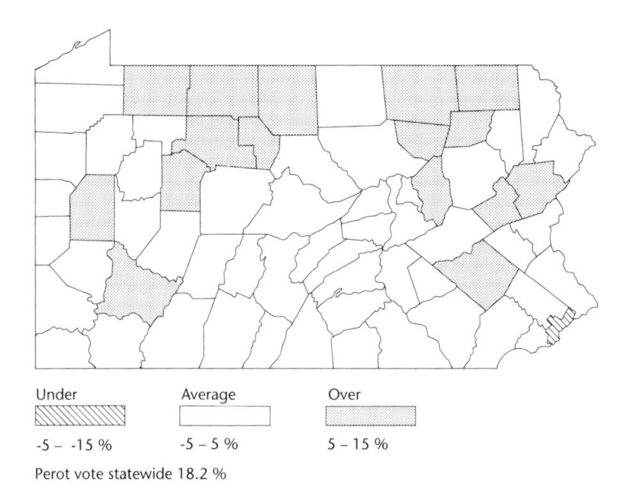

Fig. 10.5. Republican presidential vote, 1992: Percent under or over the state average.

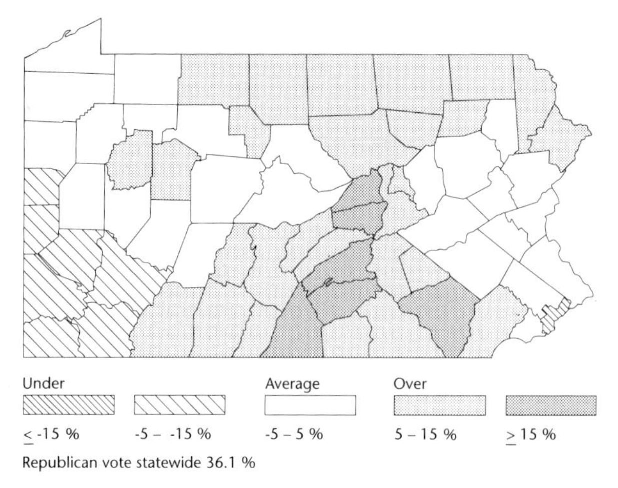

Fig. 10.7. Perot presidential vote, Pennsylvania, 1992: Percent under or over the state average.

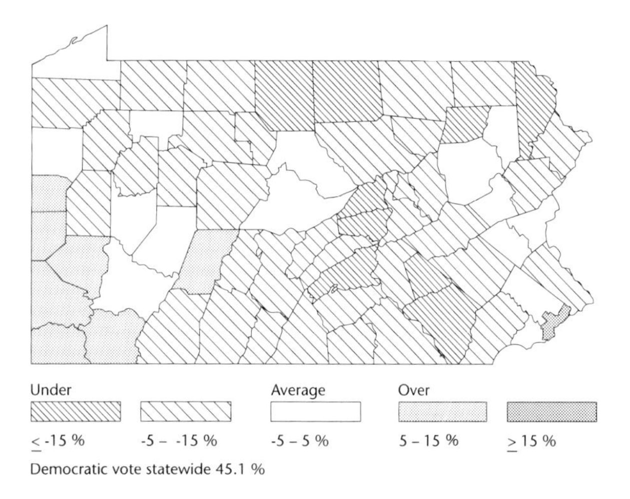

Fig. 10.6. Democratic presidential vote, 1992: Percent under or over the state average.

heartland in the southwestern counties of Greene and Fayette, and the counties to the north are consistent in delivering a smaller percentage of their vote to the Republican candidate than the state as a whole. Lackawanna is consistently more Democratic than the state, but the putative stronghold of Luzerne County is less stably so, although its defection during the Nixon-McGovern election can certainly be explained. In fact, the whole industrial and old mining area of the northeast was Democratically inclined, with the exception of that 1972 landslide, and Carbon and Northampton counties joined Lackawanna as consistently more Democratic than the state.

The only close parallel to the 1992 three-way campaign was in 1912, where the Bull Moose candidate, Former President Theodore Roosevelt, won against William Howard Taft and Woodrow Wilson. So one might have expected it to break patterns, but that hardly occurred. The Democrats polled between 5 and 15 percent higher than their statewide average of 45 in their southwestern heartland and Cambria County and attracted a satisfactory 68 percent of the vote in Philadelphia. The Republicans kept their strongholds in the northern and southern tiers. But what happened to them is that, in this state, Perot pulled his best voting percentages in normally Republican areas. Thus, rather than affecting both parties about equally as the conventional wisdom went, in Pennsylvania, he hurt the Republicans. Clinton actually got more votes than Dukakis did in 1988 in a two-man race. Bush—unfortunately for him—ended up a half-million short of his 1988 tally. Without Perot, he would have won.

But the key area of Democratic strength is Philadelphia. Its importance is reflected in the oft-repeated saying that if the Democrats don't carry the city by at least 200,000 votes they will lose any statewide election. Yet it is visually easy to overlook the fact that Philadelphia generates large Democratic majorities in each election. That, it turns out, is an artifact of standard maps like those in Figs. 10.1 through 10.7, where an area's physical size governs

Fig. 10.8. Population of Pennsylvania counties, 1983.

how prominent it is. As a final aid in interpreting such maps, the population cartogram (Fig. 10.8) may be of help in that it not only shows each county in its proper location but also provides, via the shaded outlines, information on its population size.

Where the Governors Come From

The gubernatorial election process can be examined the same way as the presidential election process. The 1978, 1982, and 1986 elections show the same general patterns that appear in the presidential vote, with one interesting exception: the effect of what has been called "friends-and-neighbor voting." This term is used to reflect the often-noted fact that candidates tend to do well in areas close to their home base. Thus, while Pennsylvania's Governor Thornburgh, who came from Allegheny County, did not manage to overturn the traditionally strong Democratic voting pattern in the western and southwestern parts of the state, he did better there than Republicans who do not come from that region. The votes for the Democratic gubernatorial candidate in 1986, Robert Casey, show the same effect. That election, though, also demonstrated another factor in statewide races: the importance of the local party leaders. The loser, Bill Scranton, would ordinarily have been favored, especially with a popular Republican U.S. Senate candidate running in the same election and with the benefit of his incumbency as lieutenant governor under Dick Thornburgh. What undid him was the effective use of television spots focusing on his admitted use of marijuana as a young man, and possibly an anti-abortion effect. Finally, there was the enmity of many Republican county chairmen, who felt he was too aloof and too arrogant and would be unlikely to be partisan enough. Their lack of enthusiasm in drumming up the vote for him probably brought him the narrow defeat.

It is possible to obtain some interesting insights into the highest office in the state using very simple information. Consider, for instance, the home county of each governor since 1861, the beginning of the modern era in the state's history. What the raw data communicate most clearly is the dominance of the Republicans over the period. But one should also remark on the concentration of governors' home counties. Twelve of the 67 counties provided all the Republican governors. Rural Centre County alone provided 4 of the 24 (counting Andrew Curtin twice), with Allegheny, Luzerne, and Philadelphia counties tying for runner-up spots. The Democrats, with only 7 governors, display an equally impressive concentration, with 4 governors from the neighboring counties of Montgomery and Philadelphia.

It seems hardly likely that this distribution arose by chance, but there is not enough room in this chapter to tell the whole story in the detail it deserves, and the interested reader should refer to such authors as Beers, Cooke, and Klein. But consider some of the reasons for selecting candidates for this statewide office. Most obvious is the ability of the candidate and his backers to pull together a coalition out of the separate interests represented in the party. Equally important is the ability to generate the millions of dollars needed for a statewide campaign. If the candidate is from a moderately to highly populated area, that helps because of the "friends and neighbors" factor. Today, of course, one would add a telegenic personality!

But are these practical grounds really the bases? The evidence is mixed. In recent gubernatorial contests, we *have* had candidates selected by the party regulars on the basis of electability. A classic case is the selection of the first-term congressman from Scranton, William Scranton, to run in 1962 after the Republicans had lost two gubernatorial elections, a senatorial election, and a presidential election in the previous eight years. His relative youth and urbanity were assets to a party desperate to counteract the Democrats' Kennedy image, as was his home base, a largish county in an area that tended Democratic and that was not involved with the Pittsburgh and Philadelphia rivalry. Family wealth and contacts were not hindrances either. But there have been less pragmatic decisions too. Again, staying with the Republicans, the candidacy of Ray Broderick in 1970 represents the occasional triumph of ideology over electability. After eight years of moderate Republicanism represented by Bill Scranton and Ray Shafer's governorship, a return to the old verities must have seemed overdue. Broderick's defeat seems to have had an effect in the selection of subsequent Republican gubernatorial candidates.

The Democrats, as the lagging party in this competition, would seem to have a real stake in careful calculation of candidates' electability in the November sweepstakes. But such has not proved uniformly the case. The maverick candidacies of Milton Shapp from Montgomery County in 1966 and 1970, and of Pete Flaherty of Pittsburgh in the 1978 election, indicate that for them, as well as Republicans, the practical route is not always taken. And even more than is true of their rival party, the Democrats are bedeviled by the rivalry between Philadelphia and the western part of the state. Shapp's money, Broderick's selection by the Republicans, and the recognition built in Shapp's losing 1966 campaign gave him victory in 1970, and incumbency and Watergate proved decisive in 1974. (Shapp was the first governor elected under the provisions of the 1968 constitution, which allowed for two terms.) But Flaherty, in addition to conflict with the party in his primary campaign, came from the west (which lessened Philadelphia enthusiasm), was running against a moderate Republican from his own area (which reduced his vote in the west), and ran in the middle of Jimmy Carter's presidency.

This means that within Pennsylvania at least, the gubernatorial contests are even more complicated than the presidential elections. To win, it is necessary to have the party united and sufficient money to mount a reasonable campaign in a television era. But it is also helpful to run in the right election, to avoid the prospect of endangering votes in the Pittsburgh or Philadelphia areas while getting very strong support in at least one, and not to be seen as too ideological. And as the governors elected since the 1968 constitution have shown, incumbency is a major asset, as it is for other elections.

A U.S. Senate Election

The 1986 U.S. Senate race is a good contest to examine to show the influence of national trends in a non-presidential-election year, the importance of incumbency, and (as in gubernatorial races) the effects of friends-and-neighbors voting. Senator Arlen Specter was running in an election where the Democrats overturned a six-year-long Republican majority in the senate, yet he won reelection handily. He did this because of a demonstrated independence that helped with independents and swing voters, the Republican attempt to keep control of the Senate, which helped with regular Republican voters, and his long association with Philadelphia as a prosecu-

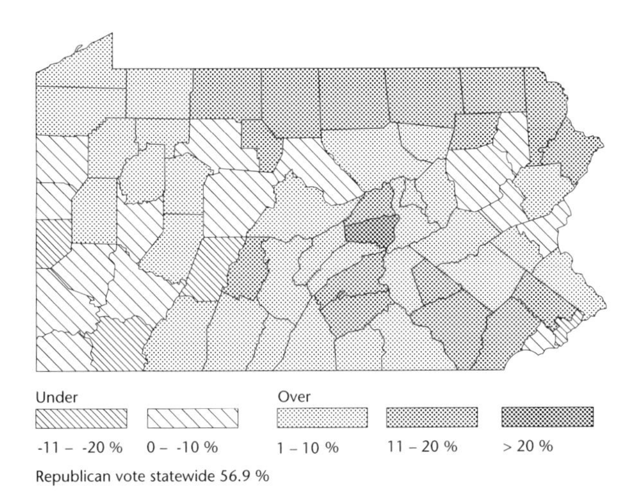

Fig. 10.9. Republican senatorial vote, Pennsylvania, 1986.

tor. While he failed to carry Philadelphia, as would be expected for a Republican, he received 44.5 percent of the vote there. This can be contrasted with the performance of the Democratic candidate, who got 55.5 percent but ran well behind the gubernatorial Democratic candidate, Robert Casey, who got 63.7 percent. Given the advantages of incumbency, I had predicted years before the 1992 election that it would take a strong Democratic candidate from the Philadelphia area to defeat Specter. His 1966 pattern of statewide strength (Fig. 10.9) was in strong support of this prediction. In the event, a maverick Democrat, Lynne Yeakel from the Philadelphia suburbs, came close to winning. But even in the year of the woman and with Specter's criticized comments in Supreme Court Justice Clarence Thomas's hearing against his accuser, Anita Hill, his moderate image cultivated over the years and his home base protected him.

Leadership in the State Legislature

Members have to vote to pass bills, but the bills that are voted on, and the results, depend largely on the legislative leadership of both parties. Surely the governor has an important role, but the principle of separation of powers is as embodied in the state system as it is in the federal. As many governors have discovered to their dismay, or chagrin, they are often regarded at best as first among equals, and those who do not play the game with the leadership of their party in the state House and Senate (and sometimes with the other party as well) do not succeed in getting their will.

Given that the leadership in the House and Senate are so important, let's look at some geographic aspects of it. As with the earlier discussion of elections, we can hope to do no more here than ask some interesting questions, provide some data, and try to reveal that geographic factors play a role in defining the action.

According to informants in Harrisburg, both parties have unwritten but well-known and generally followed rules in selecting the leadership. Seniority is important, but not as much as other factors, among which is a "good" geographical distribution of leadership posts. In questioning politicians, a geographer is pleased to find that they have the equivalent of "mental maps" of their territory. They feel, justifiably so, that to get elected and reelected they must know the territory and how it is changing. So such basic geographic knowledge as where ethnic and religious blocs are, how many people they represent, what the economic underpinnings of the area are, and what things are affecting that economy are a sine qua non for a successful politican. The state parties and their leaders know that, to be successful, a proper regard for such factors at the local as well as the state level is critical. And to help avoid blunders, the leadership should be distributed not only by taking into account important interest groups in the party but also by careful geographic selection of who will hold important positions of power.

Whether these rules really operate as stated can be discovered in a number of ways. One could interview the political leaders and ask direct questions. But that runs the risk of eliciting rejection or false information, because the topic is obviously sensitive. Another way would be to get information from official documents identifying individuals in the formal leadership positions, to label them by personal characteristics, power base, and home area, and to analyze a number of legislatures to tease out regularities that might represent the bases for selection decisions. Among other problems, this runs the risk of seeing regularities where none exist, or of picking an unrepresentative sample of legislatures.

For present purposes, a combination of these two

Table 10.2. Leadership in the Pennsylvania Senate, 1985–1987, with county identified.

Position	Republican	Democratic
President Pro Tem	Jubelirer Blair	
Floor Leader	Stauffer Chester Montgomery	Zemprelli Allegheny Washington
Whip	Leeper Delaware	Lincoln Fayette Somerset
Caucus Chair	Moore Adams	Mellow Lackawanna
Caucus Administrator	Hess York	Lynch Philadelphia
Caucus Secretary	Brightbill Lebanon York	Ross Beaver
Appropriations Committee Chair	Tilghman Philadelphia	Fumo Philadelphia
Policy Committee Chair	Wilt Crawford Mercer	Stapleton Armstrong

SOURCE: Pearce, *Pennsylvania Manual,* 1985–1986.

approaches is used because the aim here is only to get a basic insight into how the state legislative leadership selection operates. In 1987, Democratic and Republican informants, former legislators who know the system but are not currently involved, were asked to provide a verbal description of the selection rules. Then the current leadership of both parties in the state House and Senate were mapped to provide a geographic picture of the formal power structure in the legislature.

The Democratic informant and his colleague agreed that both parties attempt to achieve some kind of geographic spread in the leadership. For the Democrats, the following rules were seen as operating in the most recent decade at least. First, because black voters are so critical to Democratic successes in statewide races and in seeking control of the two houses, a black leader must be in a top post. The Speaker of the House, Leroy Irvis from Allegheny County, filled that role and at the same time represents the important Allegheny County constituency. The two financial committee posts go to Philadelphia, as must one additional leadership position. One position must go to a western Pennsylvanian, and there should be at least one rural Democrat from a generally Republican area. The Republicans have a somewhat simpler set of desiderata, but that too reveals a geographic agenda. As with the Democrats, the two financial committee posts go to the southeast, but to the Philadelphia suburbs rather than the city proper. One leader must be from the Pittsburgh area and one from Philadelphia or the suburbs. The rural areas must have at least one prominent post, and the southeastern stronghold outside the immediate Philadelphia area should have one too.

Are these rules actually followed? Not having the space to analyze several legislative sessions, we will use the most recent session available at the time of writing. Only the half-dozen or so formal top positions will be looked at. Tables 10.2 and 10.3 list these posts in the state Senate and House and the 1985–1987 incumbents, and the county (or counties) they represent.

The absence of major committee chairpersons blurs the picture in that they would help confirm the professionals' understanding of the rules as stated earlier. But the picture that does emerge is an interesting one. For these highly visible leadership positions, the Democratic legislators from western Pennsylvania dominate. The Republicans in both House and Senate have spread their net wider but focus on the southeast and south-central parts of the state. This distribution will hardly be permanent, but the patterns revealed here are not atypical of recent sessions. Those interested in the relative

Table 10.3. Leadership in the Pennsylvania House, 1985–1987, with county identified.

Position	Republican	Democratic
Speaker		Irvis Allegheny
Floor Leader	Ryan Delaware	Manderino Westmoreland
Whip	Hayes Blair Huntington	O'Donnel Philadelphia
Caucus Chair	Noye Cumberland Perry	Itkin Allegheny
Caucus Administrator	Bowser Erie	Dombrowski Erie
Caucus Secretary	Cessar Allegheny	Fee Lawrence
Policy Committee Chair	Brandt Lancaster	Wright Armstrong Clarion

SOURCE: Pearce, *Pennsylvania Manual*, 1985–1986.

power positions of various parts of the state would find it useful to repeat this exercise for future legislatures.

A Glimpse of Possible Futures

Only the surface of the fascinating world of Pennsylvania's political geography has been revealed, but it would be remiss not to offer some speculation as to future possibilities, hopefully well founded. Of the many topics available, only four are discussed: the changing state legislature, voting trends, the influence of the west on the Democratic party, and the possible impacts of the 1990 census.

The Pennsylvania legislature has not traditionally had a good reputation—that dates back at least to the days of the bosses, such as Matt Quay and Boies Penrose, and extends to the present, in the minds of many people. It was in 1969 that *Philadelphia* magazine first called the Philadelphia legislature "The House of Ill Repute," a characterization that was repeated in 1974. It is generally felt that while not yet up to the standards of California or perhaps even New York, the legislature is better and more professional. Higher salaries and expense money have made it possible for a broader class of people to serve. Staff help is better, and the use of microcomputers has enabled the average legislator to control his or her activities and respond to constituents much more efficiently and effectively. One outcome is an increase in the advantage of incumbency; a more debatable outcome may be a higher regard for the welfare of the state as reelection becomes less of a gamble.

Predicting election outcomes is a chancy business, but some interesting things are happening. Party registration means less now than it once did—the Democrats' lack of success in the statewide races, despite their official registration edge, is one evidence of that. It is now easy to change one's party registration, and there is good evidence that many voters do so in order to be able to participate in primaries that interest them. Young voters have recently been registering Republican in larger proportions, although given their miserably low participation rate the impact is uncertain. A Democratic informant noted that Jews, Catholics, and blacks were becoming more linked to his party—the result, he believes, of a fear of the new, fundamentalist right. If true, that would have important effects. The same person noted that the Republican party might not be able to tailor its policies to hold together the broader base of voters accumulated during the two Reagan presidential wins, and felt that his own party would be better at targeting potential supporters. Geographically, an examination of recent elections at several levels from the presidential to the local indicates that the southeastern part of the state is in considerable flux. The western and, especially, southwestern regions, on the contrary, seem rather stable and should continue to elect Democrats at a high rate.

The reason for this last observation reduces to one man—David Lawrence—and his successors. Seeing the dreadful state of Pittsburgh and the surrounding areas after the second world war, Lawrence and such business leaders as the Mellons put together a coalition to revitalize the city and, to a much lesser extent, the immediately surrounding area. While the economic success of this effort has not lasted, the political alliances formed back then have. Both business and labor have cooperated on a range of agendas all focusing on development. The political spinoff has been a sense of regional cohesion missing in other parts of the state, with the partial exception of the Northeast industrial belt. In effect, Lawrence installed parochialism in the west. He is reported to have said that if he could not control the state he would control the rump. And electoral results of that effort are still visible. In effect, the west sticks together, enforces solidarity, and does not split its vote unnecessarily. Whether the wrenching economic setbacks of basic industry in the area intensifies that kind of behavior, or causes it to break down, will have a major impact on future Pennsylvania politics.

Equally important will be the results of the 1990 census. As other sections of this book indicate, the state's demographic picture has not been a favorable one for many years. While the population has grown slowly over the years, the rate has lagged well behind that of the nation, and even behind that of neighboring states. The demographic picture has important impacts on such things as industrial dynamism as the population grows older at a faster rate than rival states. But its impact on Pennsylvania politics is even more direct in the wrenching reapportionments that take place after each decennial population census. The most obvious impact is on representation in the U.S. House of Representatives, where, other things being equal, the voice of Pennsylvania must inevitably decline. The reapportionment affects the political geography within the state too. The mechanism chosen for reapportionment is a reasonably fair one—equal Democratic and Republican representation with a mutually agreeable selection of a neutral chairman. For students of reapportionment, the process is of interest. In the 1970s the Democrats used traditional "hand" methods and precinct voting maps, while the Republicans depended more on computer algorithms. Both sides seemed reasonably satisfied with the results, and it will be interesting to see whether the same choices will be made next time.

Of more significance in the context of this discussion is the impact of regional population growth or decline on regional political power. Based on the information available between censuses, it is safe to say that while the west may remain solidly Democratic, its influence on statewide elections will continue to decline. That is also true of Philadelphia. Meanwhile, the southeast and some developing rural areas will gain. For Pennsylvania itself, the predictable slow growth or perhaps even a loss of population will mean a continued low profile at the national level unless some miracle creates a spirit of dynamism and a regard in the minds of citizens and politicians for the welfare and influence of the whole state, rather than a focus on its constituent parts.

Bibliography

Archer, J. C., and F. M. Shelley. 1986. *American Electoral Mosaics*. Washington, D.C.: Association of American Geographers.

Beers, P. 1980. *Pennsylvania Politics Today and Yesterday*. University Park: The Pennsylvania State University Press.

Cooke, E. F., and G. E. Janasik. 1957. *Guide to Pennsylvania Politics*. New York: Holt, Rinehart & Winston.

Klein, P. S., and A. Hoogenboom. 1973. *A History of Pennsylvania*. New York: McGraw-Hill.

Pearce, N. R. 1972. *The Megastates of America*. New York: W. W. Norton.

———. *The Pennsylvania Manual*. Annual. Harrisburg, Pa.: State Department of General Services.

11

RECREATION AND TOURISM

Wilbur Zelinsky

Among all the developments that have transformed our lives in recent times, few can rival the boom in leisure-related activities in their immediate, visible impact on the humanized landscapes and economies of so many parts of the world. The evidence for such a statement is rich and persuasive for Pennsylvania, even though the state is not as heavily dependent on recreation and tourism as certain other sections of the United States and foreign lands.

How has such a veritable revolution come about in our allocation of time, money, and natural and human resources, in the reshaping of places, through the pursuit of happiness? The answer lies in the convergence of several factors. An unprecedented rise in the personal income of Americans and residents of other advanced countries in recent generations has accompanied the arrival of new technologies in travel, communication, and business organization that have rendered indulgence in recreation by great masses of people both convenient and relatively inexpensive. At the same time, the work week has been shortened to 40 hours or less for the great majority of the gainfully employed. Paid vacations and weekend jaunts have become routine for many of us. Furthermore, early retirement and the growing absolute and relative numbers of the retired have swelled the ranks of pleasure-seekers. Less obvious but also quite consequential is the delayed entry into the labor force of many young adults who, as college students, are also frequent participants in recreational pursuits on and off campus.

The geographic consequences—in social, cultural, economic, and landscape terms—of a modern lifestyle in which spending has come to be regarded as more rewarding than getting are extremely varied and complex, and obviously vary markedly from locality to locality. But we are hindered in efforts to map and interpret such areal phenomena by problems of definition and by the supply and quality of the needed data. As a concept, recreation is more difficult to capture than might appear at first. In the broadest terms, it embraces the entire, vast range of activities we pursue for pleasure (another broad, elusive concept) during the waking hours not spent at work or routine chores. But when we examine specifics, we discover that it is not at all easy nowadays to draw sharp boundaries between recreation and other departments of human endeavor. Thus, for example, how do we classify the time devoted to church picnics? As churchgoing or as recreation? Are we engaging in horticulture or recreation when puttering about in the backyard vegetable garden? Are the professionals who participate in conventions for members of their disciplines doing so for pleasure or for business reasons, or some combination thereof? How do we classify the hours spent in restaurants? And aren't some types of shopping more fun than drudgery?

To complicate matters further, recreation exists at every level of complexity from the most formal and official, frequently involving great masses of people, down to the entirely private, individual, and extremely casual. Nevertheless, we can attempt a useful distinction between the relatively public, relatively regimented forms of recreation and those that are relatively spontaneous and/or practiced by single individuals or small socially bonded groups. If we seek statistics, even if only for the former category, we find that by the very nature of the pursuits in question there is no central agency collecting and dispensing data at the federal or state level, but rather a multiplicity of sources of differing levels of coverage and reliability. Turning to the less formal kinds of recreation, information may be totally lacking for many items or hopelessly general (e.g., national surveys based on small samples). Consequently, we may realize that such pastimes as card and board games, crossword and other puzzles, crafts and hobbies of all sorts, reading, party-giving, radio and television, workouts at health clubs, backyard sports and sunbathing, mushrooming, gossiping, courtship, saloon-hopping, or jogging around the block escape the statistical net. Moreover, some activities, such as gambling and cockfighting, are illegal, so that it would be foolish even to imagine searching for hard data.

As a matter of necessity, then, this approach to the geography of recreation (and tourism) in Pennsylvania is preliminary and incomplete, offering for the most part only a description of facilities, patronage, and impact as reported by those governmental and commercial organizations willing and able to divulge information on resources and turnover in money and participants. Thus this survey is simply a first, rather blurred approximation of a great, multifaceted phenomenon that is quite unevenly developed within the state. It will put forward perhaps as many unanswered questions as solid facts or generalizations.

What is abundantly clear is that recreation and tourism are big business in Pennsylvania and claim a healthy fraction of the state's territory. Thus it is reported that, during 1986, American travelers, including tourists and others, spent nearly $10.2 billion in the state. The employment of 209,400 people directly generated by these travelers accounted for 4.4 percent of total nonagricultural employment. But we must also realize that a multiplier effect operates here, so that recreation is actually responsible for a significantly larger share of the labor force and economy, and all the more so if we include the substantial contribution made by Pennsylvanians indulging in recreational activity in or near their homes. Furthermore, all the available statistical indexes show a strong, continuing upward trend in all enterprises associated with recreation and tourism.

In the discussion that follows, two themes dominate: the highly varied character of Pennsylvania's recreational resources, and their strategic location with respect to potential consumers. Let us begin with the recreational facilities created and managed by governmental agencies and private businesses. Although such developments originated almost entirely in the twentieth century, in response to deep-seated structural changes in our society, recreation—mostly of the private type—was not totally absent from the earlier scene. But what tourism may have existed in centuries past was limited almost entirely to the affluent, as was the case throughout the eastern seaboard. Thus, Philadelphia was an obligatory stop for the visiting foreigner during the first half of its existence and usually earned glowing reports. From the late eighteenth century onward, Bedford Springs (Bedford County) attracted a fashionable crowd of Americans whose excuse to socialize in pleasant surroundings was the reputed therapeutic effect of the mineral waters. Several decades later the scenic and technological wonders of Mauch Chunk (now Jim Thorpe, in Carbon County) in the anthracite region drew a steady stream of tourists from a wide area.

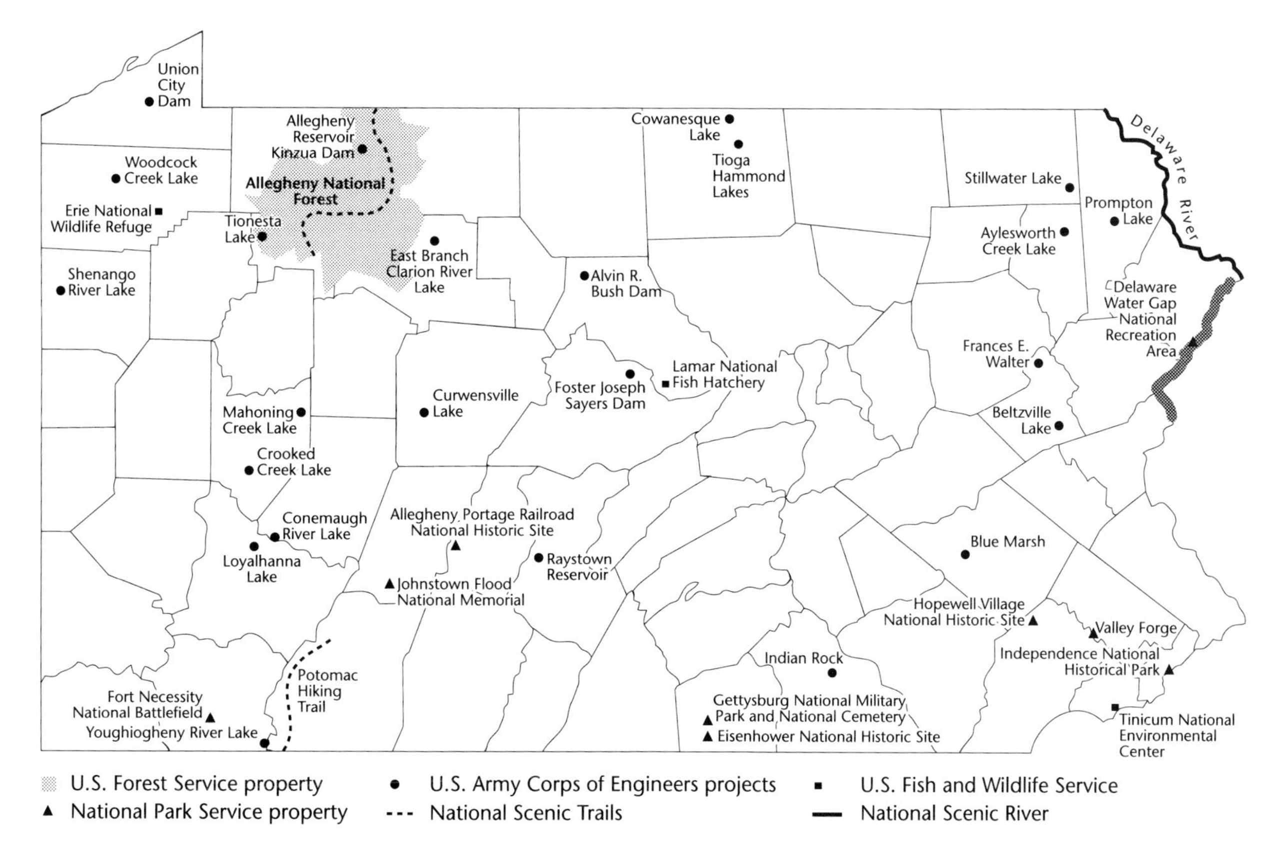

Fig. 11.1. Pennsylvania federal recreation areas and facilities.

Outdoor Recreation

A remarkable amount of Pennsylvania's territory is devoted entirely or predominantly to recreation: no fewer than 9.7 million acres (35 percent) of a total expanse of 28 million acres. Somewhat more than half of such land (54.5 percent) is in private hands. State recreational lands total some 3.7 million acres (37.5 percent); the federal holdings amount to more than 600,000 acres (6.2 percent), and some 180,000 acres (1.8 percent) are held by local (county, township, and municipal) authorities. With some notable exceptions, these tracts tend to be located in the emptier and more thinly settled portions of the state. Thus, for the most part, the larger parcels of recreational land are topographically rugged or wooded areas that may have been tried by farmers and found wanting or have been logged over at least once, then left to regenerate through natural processes or occasionally with some human help. Indeed, much of this real estate has reverted to local governments for nonpayment of taxes.

Although most of these recreational sites are relatively offside (see Figs. 11.1, 11.2, 11.3, 11.4, 11.5), at least as far as the bulk of Pennsylvania's population is concerned, the state does enjoy a considerable advantage over most southern and western competitors in the recreational marketplace. Its parks, forests, rivers, and other attractions are situated within short distances, in terms of both time and miles, not only from major population concentrations inside the state but also from those in several other states in the Northeast, the Midwest, and the Upper South. In fact, nearly half the American population lives within a half-day's drive of Pennsylvania. The result is a substantial influx of tourists and vacationers from a broad sweep of territory, in addition to local patrons. It is not possible to say whether the out-of-staters outnumber the Pennsylvanian pleasure-seekers who patronize other states and countries.

The public lands available for recreation are administered by a variety of jurisdictions. At the national level, the U.S. Forest Service manages the 508,593-acre Allegheny National Forest, the largest single parcel of recreational land within the state, and

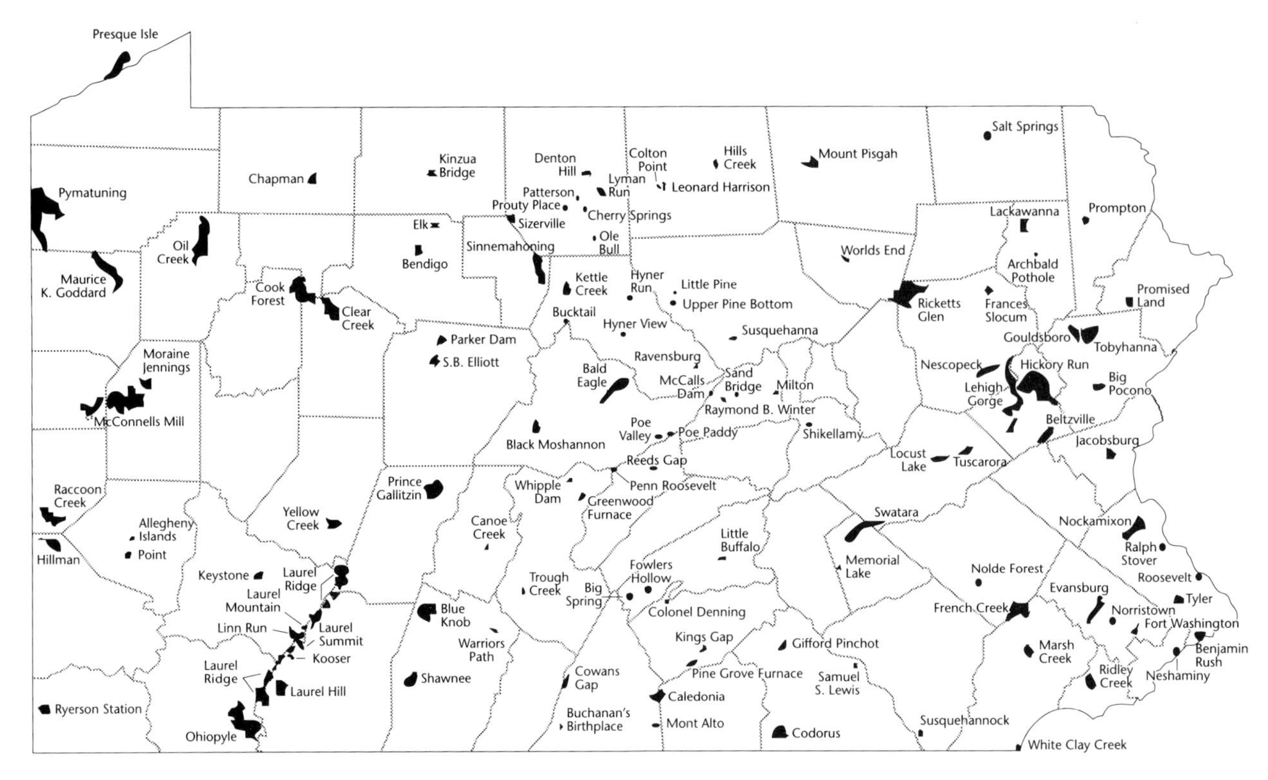

Fig. 11.2. Pennsylvania state parks.

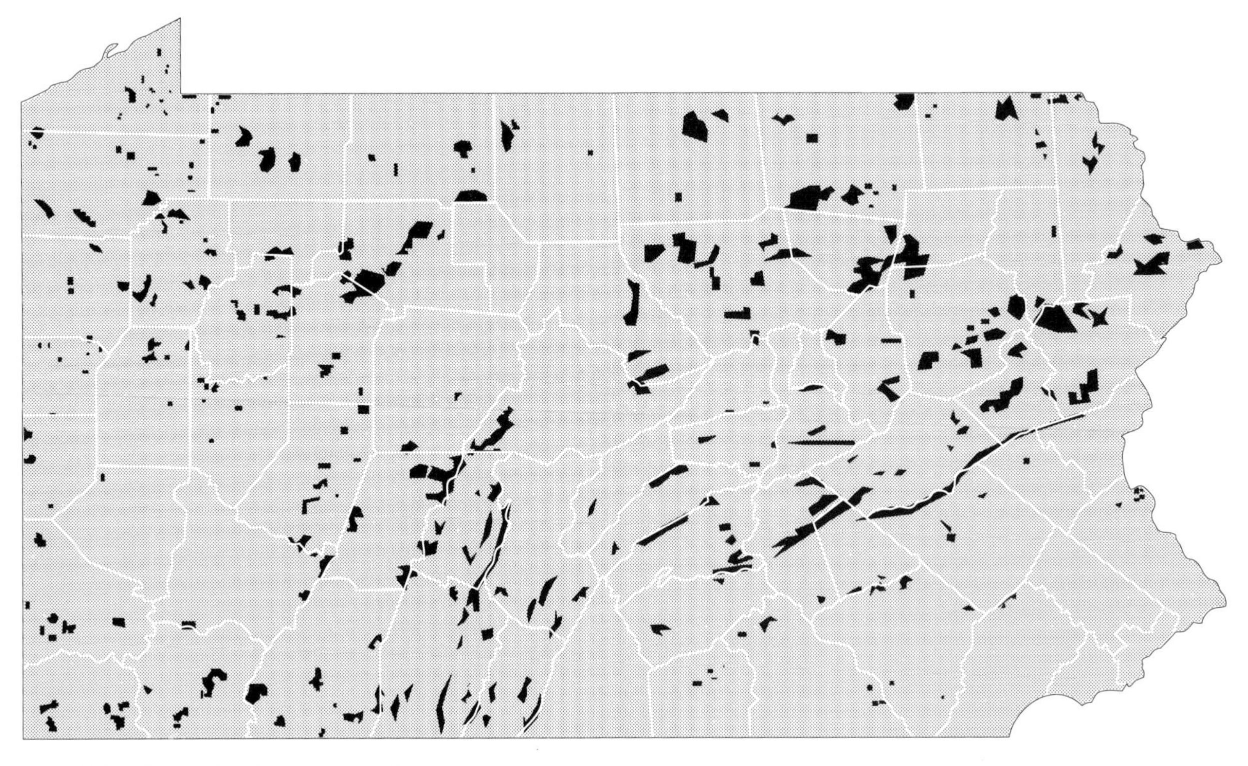

Fig. 11.3. Pennsylvania state game lands.

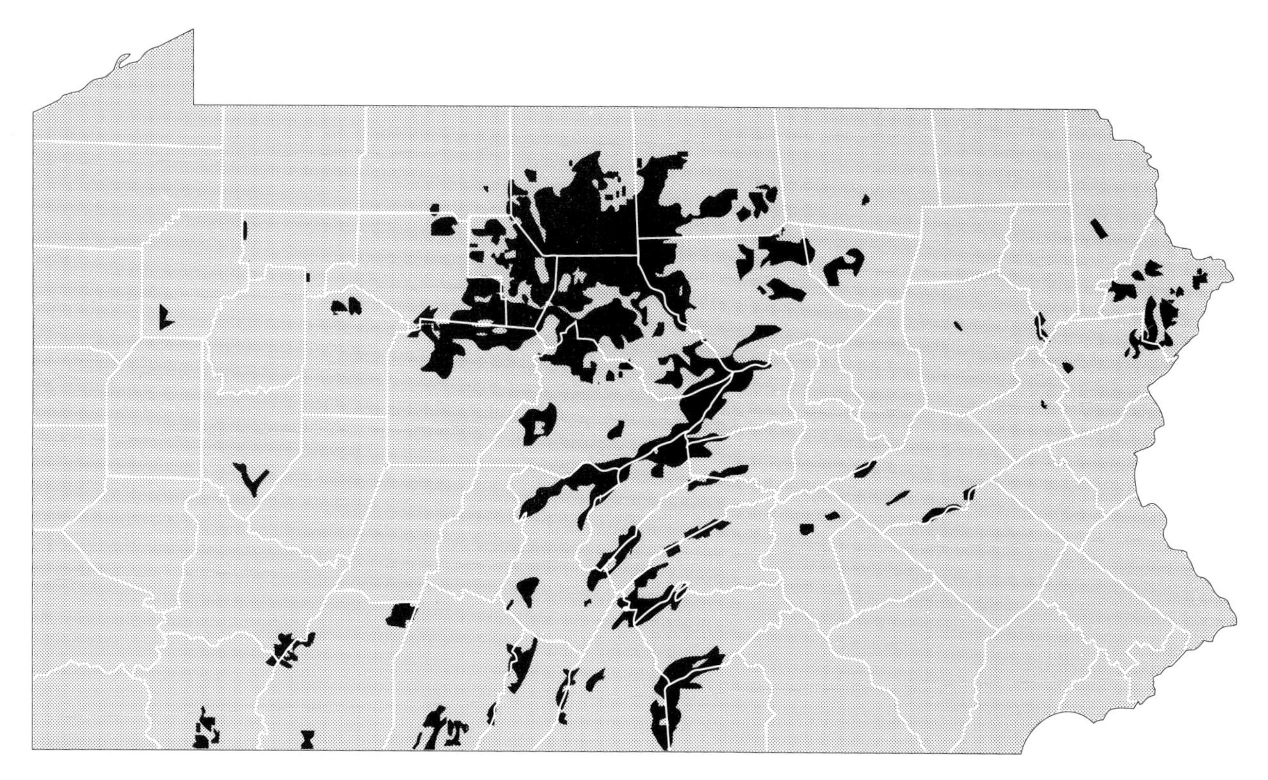

Fig. 11.4. Pennsylvania state forest lands.

one that accounts for much of the territory of four northwestern counties (see Fig. 11.1). The Corps of Engineers has completed twenty-four multipurpose projects in Pennsylvania, and the Soil Conservation Service has completed another twenty. The reservoirs impounded behind the dams in question provide a welcome set of opportunities for boating, fishing, and other aquatic diversions in the parts of the state that have no natural lakes. Other relevant federal agencies include the National Park Service and the U.S. Fish and Wildlife Service.

But the state government operates the greatest number and range of facilities, responding, as does the federal establishment, to a growing actual and anticipated demand for free or nominally priced recreational attractions. And it has done so more comprehensively, with greater sophistication and careful planning for the needs of the entire Pennsylvania citizenry, than is feasible at either the federal or the local level (Davis and Miller 1988). The Bureau of State Parks manages 113 tracts, some of considerable size, totaling 280,201 acres in all (Fig. 11.2) and recording 37,835,000 visits in 1985, while the state Game Commission owns several hundred parcels of relatively wild gamelands scattered liberally through the less-urbanized portions of the state, forests that are used for public hunting and other compatible forms of recreation (Fig. 11.3). Their aggregate acreage is an impressive 1.2 million. This is in addition to the more than 2 million acres of multipurpose State Forest Lands handled by the Bureau of Forestry (Fig. 11.4).

Although they are dwarfed by state and federal lands, public parks are operated by all the larger municipalities and by many counties. Especially noteworthy is Philadelphia's 8,000-acre Fairmount Park, the largest landscaped urban park in the world. Also of more than passing interest are the many college campuses and secondary schools with both outdoor and indoor facilities accessible to the general public at least on a part-time basis (Table 11.1). It is also worth noting that full-time directors and park and recreation agencies in 175 municipalities and counties provide year-round services to the local populations.

The recreational possibilities within all these public lands are varied indeed but usually involve some form of physical exertion. Prominent among them is an elaborate network of trails accommodating several types of locomotion (Pa. Department of

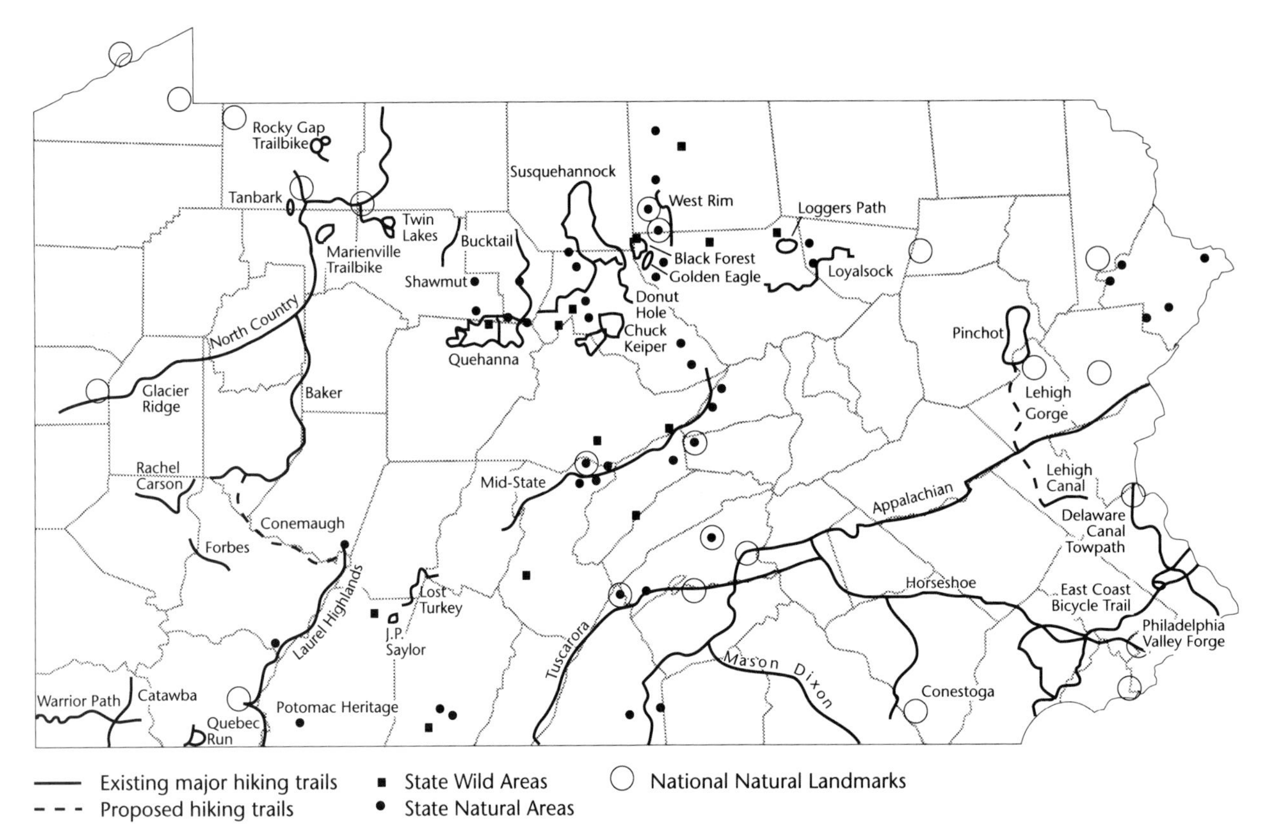

Fig. 11.5. Pennsylvania hiking trails and natural areas.

General Services 1980). If we count designated routes within private lands along with governmental lands, the total extent of these trails is 12,709 miles. In order of aggregate length, they include hiking trails, 3,773 (Fig. 11.5); snowmobile, 3,312; slow water, 2,273; bicycle, 1,286; fast water, 969; ski, 535; bridle paths, 481; trail bikes, 56; all terrain vehicles, 28. The Appalachian Trail, initiated in the 1920s, which challenges hikers all the way from Maine to Georgia, is the most famous as well as the longest within the state. In addition, many hikers, boaters, skiers, and riders avail themselves of usable terrain and water in places off the recognized routes, and wintertime sledding is feasible wherever children of all ages can find the proper snow-covered slopes.

Because only a relatively small fraction of Pennsylvania was glaciated, the inventory of natural lakes is limited. The only extended stretch of beach is that along the Lake Erie shore and, in particular, Presque Isle State Park. (Left unmapped here are the countless municipal, school, and private swimming pools.) On the other hand, the state does have a respectable number of man-made lakes, notably those (such as Raystown Reservoir) resulting from various governmental multipurpose projects, and they have earned much popularity. Furthermore, the combination of a suitably wrinkled and crevassed set of hills and ridges, and an abundance of precipitation, has endowed the state with more than its share of interesting streams, as well as ample opportunity for skiing and other winter diversions. Thus, the remarkable wealth of streams offers every grade of excitement for the canoeist and rowboater (and a plentiful supply of access areas) as well as the merely scenic. Indeed, especially picturesque stretches of the Delaware, the Lehigh, the Schuylkill, the Allegheny, the Youghiogheny, and other streams have been officially designated as wild or scenic (or are about to be so labeled) by state or federal agencies.

The incidence of boat registrations (Fig 11.6) reflects environmental opportunities, the adequacy of aquatic facilities, and their success in competing with alternative forms of recreation nearby. Also of note are the long, linear ridges of the Folded Appalachians—and Bald Eagle Mountain in particular—which create some of the most nearly ideal

Table 11.1. Pennsylvania public school acreage and recreation facilities.

Total acres	62,599
Water acres	45
Stream miles	102
Picnic tables	869
Hiking trails (miles)	199
Biking trails (miles)	26
Nature trails (miles)	338
Outdoor pools	12
Indoor pools	227
Football stadiums	535
Lighted stadiums	271
Seating capacity of stadiums	1,660,799
Gymnasiums	3,301
Multipurpose rooms	161
Tennis courts	2,001
Lighted tennis courts	327
Basketball courts	2,914
Baseball fields	3,395
Lighted baseball fields	55
Apparatus areas	3,255
Parking areas	293,827

SOURCE: *Statewide Recreation Areas and Facilities Inventory*, n.d.

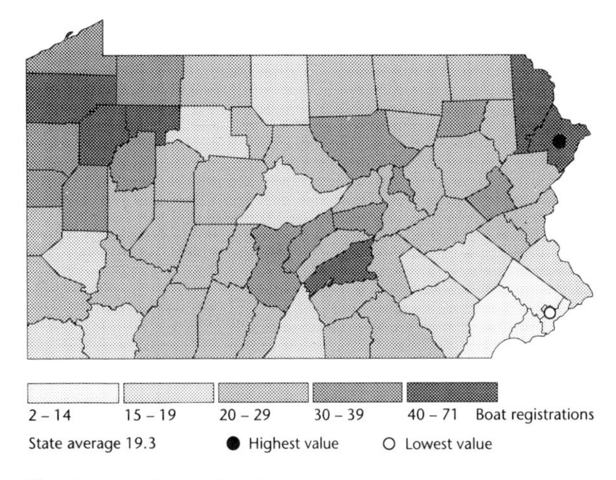

Fig. 11.6. Pennsylvania boat registrations per 1,000 persons, 1985.

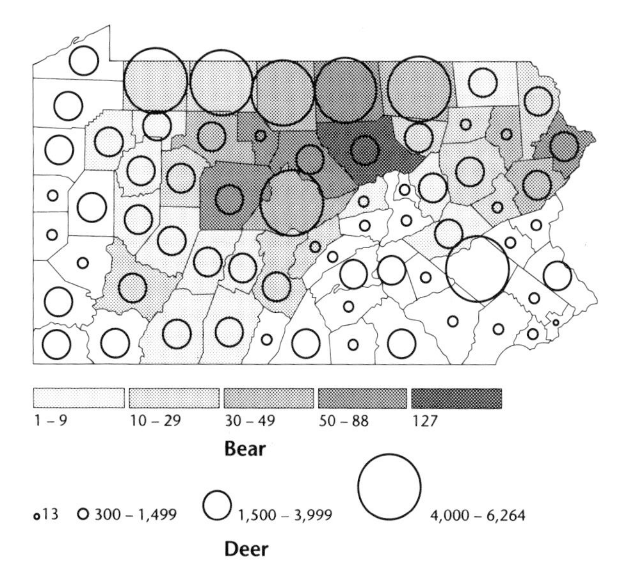

Fig. 11.7. Deer and bear harvest, Pennsylvania, 1985.

conditions to be found in North America for the pastime of gliding.

The wilder, wooded sections of Pennsylvania are home to an abundance of fish, fowl, and mammals that bring a gleam to the eye of those equipped with gun, rod, or camera. The annual deer "harvest," which may be the largest in the nation, is of unusual economic and social as well as recreational significance. At dawn on the Monday following Thanksgiving Day, hundreds of thousands of hunting enthusiasts (overwhelmingly male) who have gathered from many points in the Northeast and Midwest emerge from their campers, lodges, and cabins to stalk the elusive quarry on both public and private tracts. Although deer are killed in every single county, there is an essentially inverse relationship between the human population map and the number of deer carcasses (Fig. 11.7). Seasonal though it be, the deer season is a crucial source of income for many an economically marginal business. And "Deer Day" may be unofficial, but it is still a genuine holiday for which many schools and shops close their doors. Bears are fewer in number but also much sought after by the avid hunter, as are a wide variety of small game.

Pennsylvania's streams and lakes attract fishers from near and far, again during designated seasons. Especially prized are the trout, the majority of which originate in the twelve hatcheries operated by the state Fish Commission and are located strategically across the state. Each year the Fish Commission releases more than 5 million trout into more than 900 streams and nearly 100 lakes, in addition to an annual stocking of approximately 50 million warmwater species of fish. The incidence of fishing licenses by county (Fig. 11.8) offers insights into the life-styles of the northern third and south-central portions of the state.

Because of the presence of soluble limestone at or near the surface of so many localities, the state

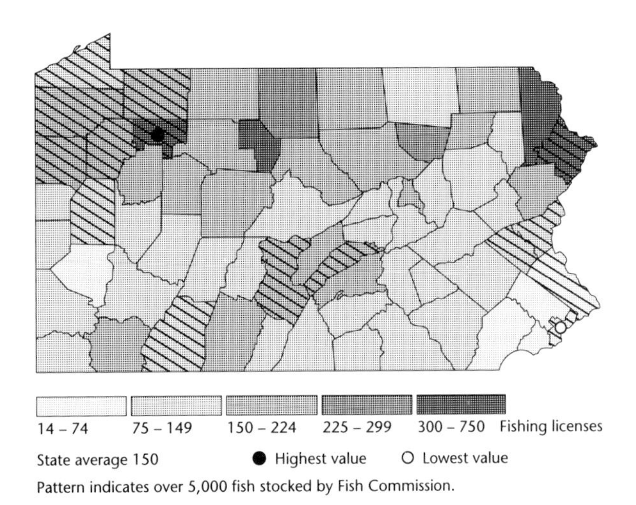

Fig. 11.8. Pennsylvania fishing licenses, 1985.

contains hundreds of caverns—and also countless sinkholes. A few of these are being exploited commercially, but many more await the venturesome spelunker.

Pennsylvania is among the states that have special appeal to horse-oriented folks. Equine activities, whether private or commercial, tend to cluster near the larger cities, but there is an especially elaborate development of thoroughbred farms, fox hunting, and associated phenomena bespeaking the well-to-do in suburban and exurban Philadelphia, and most emphatically in Chester County. This horsey landscape is actually the midsection of a larger belt reaching southwest from central New Jersey past Washington and culminating in northern Virginia.

Users of the relatively untouched reaches of the state include many people who neither hunt nor fish. In addition to the hikers, boaters, skiers, spelunkers, and other devotees of the strenuous life, there are those who delight in the simple joys of camping and picnicking, or in bird-watching, mushrooming, photography, and the contemplation of wild flowers. Of particular value to nature lovers are the protected "natural areas" set aside and administered by state and federal agencies (Fig. 11.5), places of unusual biological or scenic significance where one can see some semblance of the original plant cover. Furthermore, Pennsylvania has some of the finest hilltops from which to observe hawks, and these favored spots draw aficionados of this pastime from many points throughout eastern North America.

For the motorist as well as the hiker, there are the multiple delights of sightseeing in Pennsylvania. The spectacular colors of October bring many a traveler to the state's tree-lined highways, and throughout the year there is a bountiful supply of picturesque agricultural and urban landscapes on which to feast the eye, in addition to the scenic streams and some outstanding geological and topographic features, such as the Delaware Water Gap or Tioga County's Grand Canyon of Pennsylvania.

The existence of a good deal of privately owned land in or near areas with multiple opportunities for outdoor pleasures, real estate lying within commuting range of major population centers and available at reasonable prices, has meant a remarkable boom in seasonal housing and commercial resort developments catering to day-trippers, tourists, and vacationers. The most intensive development is to be found in the Poconos, where in a manner reminiscent of New York's Catskills, ski lodges, golf courses, summer camps for youngsters, motels, roadside enterprises of all sorts, and larger complexes of attractions centered on hotels have been springing up in great quantity, in addition to the swarm of isolated cabins and speculative clusters of cottages in woodsy or lakeside settings. The Poconos may be distinctive by virtue of the large traffic in honeymooning couples (a rather special form of recreation, it may be argued) and its successful competition in the newlywed trade with traditional Niagara Falls. But parallel changes have been going on at a rather less frantic pace elsewhere in many of what were once the emptier portions of Pennsylvania.

According to the decennial housing census, the number of seasonal housing units in the state more than doubled between 1950 and 1980, from 43,071 to 87,099. Such vigorous growth has also occurred throughout the sections of Appalachia that are southwest of Pennsylvania, but the increase has been much more modest in New York, New Jersey, and Maryland, where the stock of available, desirable, and affordable sites has dwindled considerably. Trends in the creation of second homes—year-round units held for occasional use, especially in the summer and on weekends, have been quite similar to those for the seasonal type. The intrastate geographical patterns for both forms of essentially recreational housing resemble each other (Fig. 11.9). Relative numbers are highest in the Poconos, the Northern Tier and adjacent parts of the Allegheny Plateau, and the Ridge-and-Valley region.

The proliferation of such part-time residences in

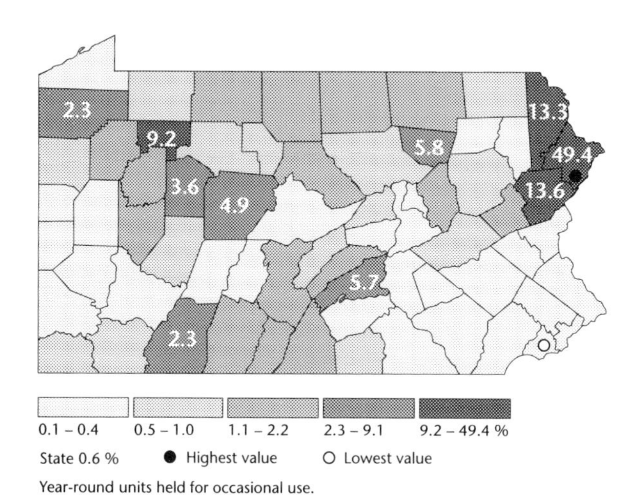

Fig. 11.9. Second homes as percentage of all dwellings, Pennsylvania, 1980.

recent decades seems to have followed a predictable sequence. Weekend visits and vacation sojourns are frequently followed by the building or acquisition of a structure for use as a second home. Eventually the second home may become the regular year-round residence for the retired person or couple, so it is not surprising to find some kinship between the spatial patterning of the elderly (see Chapter 7) and that for recreational housing.

Given the lack or weakness of land-use planning and control in most of rural Pennsylvania, the settlement patterns based on recreation (and retirement) display little spatial coherence or regularity. Instead, they are responsive mainly to market forces and to personal whim. Thus, we encounter many a haphazard subdivision in sylvan locales where highway strips contain businesses and homes in random sequence interspersed among vacant lots and are often lacking adequate arrangements for water or sewage. But of course there are some isolated cottages on secluded sites that are far from being eyesores. In general, however, there is a striking contrast between public and private lands in terms of general appearance.

Sport

Organized sport is a major fact in the life of Pennsylvania, as it is nowadays in so much of the world—and it is becoming more so. Moreover, it is a subject with exceptional possibilities for the geographer. The amount of time, money, and emotion expended on professional and college athletics has become truly extraordinary, but more to the point the popularity of specific sports varies regionally, and sense of place in contemporary America has come to be closely associated with loyalty to one's high school, the local college, and their athletic programs, or to a favorite professional team.

Although the topic has not received the scholarly attention it merits, organized sport has begun to transform the landscape in many places within Pennsylvania. In addition to the myriad of playing fields, the state has an abundance of outdoor and indoor stadiums and arenas of varying size and quality, the largest of which may be the most imposing edifice in a locality. They may even attain symbolic status, as does Pittsburgh's Three River Stadium (or, to look elsewhere, the New Orleans Superdome). One is seldom more than a few miles from one of the nearly 1,000 carefully landscaped public and private golf courses in Pennsylvania, whose grand total of 11,382 holes (as of 1980) so distinctively punctuate the scene. Fewer in number but also quite striking visually are the various county-fair harness tracks, commercial thoroughbred racing facilities, and the automotive drag strips and oval speedways, not to mention the adroitly engineered downhill ski runs.

The list of other sport-related artifacts that are altering the humanized landscape could be extended with such additional items as tennis courts, backyard basketball hoops, and urban jogging paths, but the less visible impact of big-time athletics on the collective psyche is a much more pervasive geographic phenomenon. It is the major-league professional baseball and football teams that generate the maximum fanaticism amid their regional fansheds, as well as commanding national attention. In the case of Pennsylvania, this means principally the Pittsburgh Pirates, the Philadelphia Phillies, the Pittsburgh Steelers, and the Philadelphia Eagles. By mapping the listening areas of radio stations regularly broadcasting accounts of the games in question, we gain some appreciation of the territorial reach of these powerful emotional bonds and discover that interstate boundaries may be irrelevant, because such outside teams as the Baltimore Orioles, the New York Yankees, the New York Mets, and the Washington Redskins have their loyal followers within Pennsylvania (Fig 11.10). Less intense but still significant are the devotees of

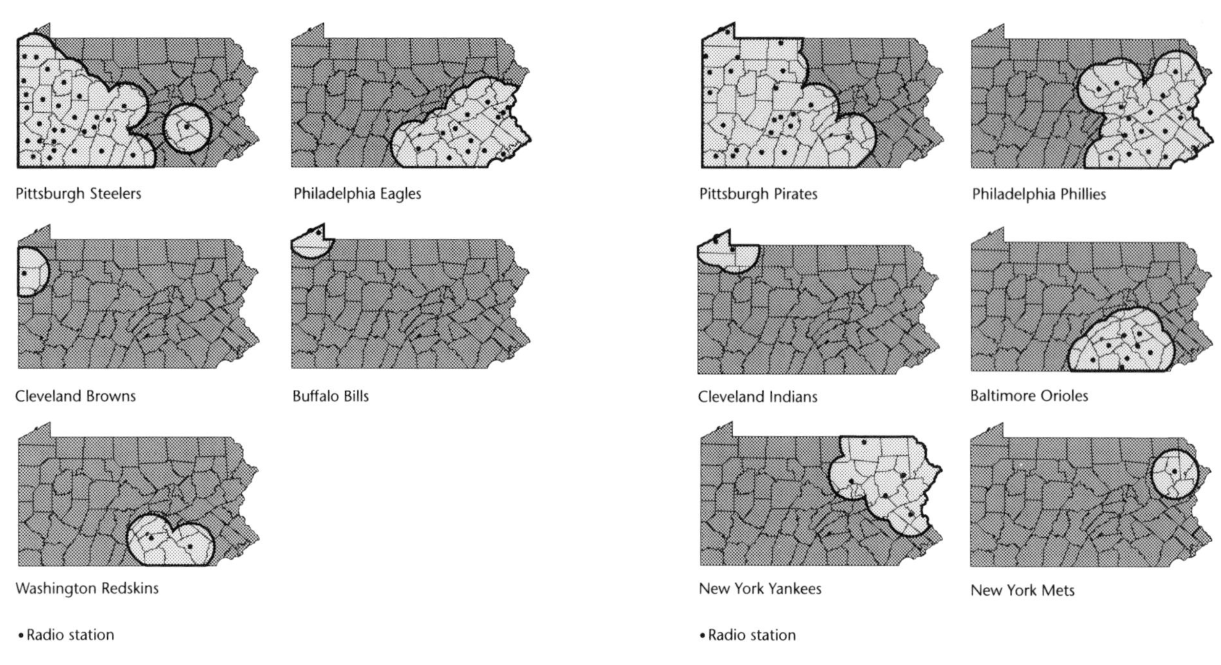

Fig. 11.10. Professional football and baseball fansheds, Pennsylvania, 1985.

professional basketball or ice hockey. As a matter of incidental interest, western Pennsylvania can lay claim, along with neighboring portions of the upper Ohio Valley, to being a crucial, formative region in the early history of professional football, and one that has remained important as a breeding ground for players (Rooney 1974).

State boundaries may be more meaningful for fans who are energized by intercollegiate sports. During their more triumphant football seasons, statewide pride in Penn State's Nittany Lions or the Pitt Panthers may transcend mere parochial enthusiasms. In any event, the volume of automobile traffic and commercial activity generated by the 49 greater and lesser Pennsylvania college football teams every fall—and to a lesser extent by the intercollegiate basketball programs—is truly formidable. Thus, for example, it is estimated that, on the average, recent football seasons have added around $20.5 million in direct expenditures to the income of the State College metropolitan area and that the total contribution, including indirect spending, amounts to some $40 million.

Although collegiate football and basketball produce the greatest public fuss, a variety of other sports engage the time and energies of students and townspeople in Pennsylvania's college communities: soccer, baseball, softball, volleyball, field hockey, lacrosse, tennis, track, cross-country, gymnastics, wrestling, swimming, and golf. We are just beginning to understand the geography of school sports in the United States, but it seems that lacrosse is a game still restricted largely to southeastern Pennsylvania as it gradually diffuses throughout the northeastern United States from an original hearth in Canada. But a passion for wrestling at both secondary school and college levels seems to characterize all sections of the state.

It seems safe to claim that high school athletics are as important in social and cultural terms in the urban and rural communities of the state as they are anywhere else in the nation, even though we have yet to study the geography of the phenomenon. The local competitions mean a great deal to local residents, and even more when the champions of the twelve PIAA (Pennsylvania Interscholastic Athletic Association) districts engage in statewide contests each year. In recent years, even elementary school youngsters have become involved in organized sports, such as soccer and baseball. The spatial array of Little League baseball within Pennsylvania (Fig. 11.11), an enterprise that happens to have its national headquarters in Williamsport, is thought-provoking. It appears that this predominantly middle-class activity is especially popular in most urban areas, and Scranton–Wilkes-Barre particularly, but is somehow unrepresented in the

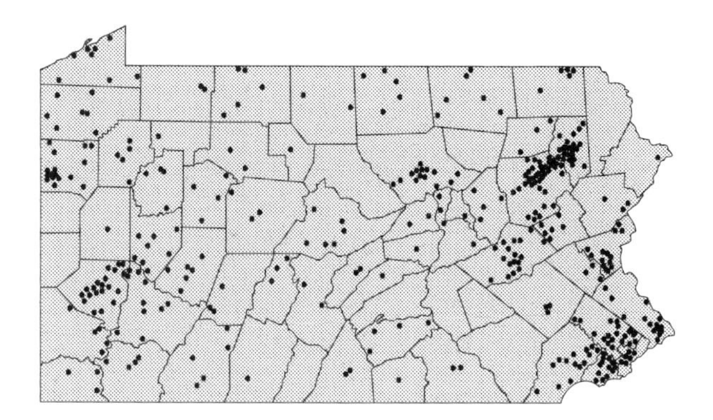

Fig. 11.11. Pennsylvania Little Leagues, 1987.

Lancaster and Harrisburg localities. It is also totally missing in Fulton, Clinton, Wayne, Sullivan, Lebanon, Beaver, and Lawrence counties, for reasons that are far from self-evident.

Thus far we have dealt only with spectator sports, formally organized activities where relatively passive onlookers usually far outnumber the actual participants. But the phenomenon of sport and games embraces a much broader range of things, including all sorts of casual, impromptu pastimes, sometimes involving only one or two people and no spectators. In the cases of golf, bowling, ice skating, swimming, and diving, we have sports that straddle the boundaries between amateur and professional or between formal and informal. Regrettably, little has been learned to date about their geography in Pennsylvania or anywhere else for that matter, but it is reasonably certain that geographical analysis would be rewarding. The same must be said about such casual adult diversions as horseshoes, running, croquet, ballooning, survival games, and water-skiing.

Research conducted by folklorists shows geographers could also benefit from exploring the rich world of children's outdoor games, which somehow survive despite the inroads of television, electronic toys, and adult supervision. It is likely that there are interesting regional and local diversities in such activities as marbles, skateboarding, pickup versions of softball and hockey, jump rope, hopscotch, and other street diversions, but we can only guess their nature.

Other Diversions

Enjoyment of the higher forms of "culture" certainly qualifies as recreational, and Pennsylvania can hold its own against any other state in terms of both the supply and the consumption of aesthetic experience. Within the performing arts, Philadelphia and Pittsburgh are home base for two world-class symphony orchestras, while there are scores of professional and amateur orchestras as well as a fair quantity of choral groups, ballet and opera companies, and an encouraging number of community and college concert, or artist, series. Similarly, the theater is alive and well throughout the state in terms of the professional, dinner theater, and college, community, children's, and summer enterprises. There is clearly an interesting relationship between the number and quality of such activities, on the one hand, and the size and character of the city, the presence of colleges, and basic sociocultural attitudes, on the other hand, which would bear looking into by geographers and other scholars.

The number and variety of museums in Pennsylvania are great enough to test the stamina of even the most resolute museum-goer. The admirer of fine paintings, sculpture, and other artful artifacts has a choice of sixteen museums to visit in the city of Philadelphia and six in Pittsburgh, some of international repute, and many more further afield in other localities. There is also an appealing assortment of museums dedicated to natural history, science, industry, and technology, and some rare and special topics.

Given its pivotal role in the history of the nation, it is not surprising to encounter an exceptional number of historical museums, large and small, general and specialized, within Pennsylvania. Indeed, the celebration of the past is another facet of recreation that figures prominently in the state's economy. The quantity of officially designated historical parks, monuments, museums, landmarks, and the like is remarkable. There are 1,867 Pennsylvania sites on the National Register of Historic Places; the Pennsylvania Historical and Museum Commission owns and manages 51 properties; and the U.S. Department of the Interior has designated no fewer than 107 National Historic Landmarks (45 of them in Philadelphia). In addition, there are many county and municipal museums and privately operated facilities of a historical character. But, beyond question, the most renowned and heavily patronized of all such places are three administered by the National Park Service: Independence National Historical Park (Philadelphia); Gettysburg National Military Park and National Cemetery (Adams

County); and Valley Forge (Montgomery County). The count of visitors to these revered sites in 1985 was 4.9 million, 1.4 million, and 12.9 million respectively. Commercial enterprises have proliferated along the periphery of some major historical attractions, most notably in the case of Gettysburg.

Another form of amusement and edification—the local festival—has flourished throughout the United States in recent times and thrives with special vigor in Pennsylvania. These are ordinarily annual affairs but may become more elaborate than usual when celebrating the centenary or some other major anniversary of a notable event. The most venerable of these institutions are the county fairs, which are held in most counties in the late summer or early fall. Originating in the mid-nineteenth century as agricultural exhibitions, they now combine agrarian concerns with pure entertainment and sporting events. Harrisburg's well-attended annual Farm Show, which serves the entire state, preserves the old format.

The newer breed of local festivals that have multiplied so lustily in the past few decades are extraordinarily varied in character. There is a decided seasonality in their scheduling. With the exception of those devoted to winter sports and certain holidays, and some that are essentially indoor, they tend to materialize from March through October, peaking in the month of August. Of special interest is the growing number of ethnic festivals, usually held in localities with noticeable concentrations of the groups in question, but such celebrations are open to all comers. Local foods, plant and animal life, and historical traditions are also fine excuses for these events. Every popular sport has its festival, with appropriate competitions; and almost every conceivable type of craft, musical, theatrical, or artistic endeavor is the occasion for an annual gala, some of which attract visitors from well beyond state borders. Apart from its role in nurturing local pride, the festival phenomenon is clearly generating much income for the places in question.

Pennsylvania offers the visitor and resident many more ways in which to wile away one's time pleasantly in addition to outdoor recreation, sport, the exploration of history, or the cultivation of the finer things of life. As already noted, the diverse urban and rural landscapes of the state (and the brilliant hues of October) appeal to great numbers of tourists. Among other attractions, one might list the various farmers' markets, such as that held every Wednesday morning in Belleville (Mifflin County), the resuscitated short-line steam railroads, such as the Orbisonia and East Broad Top (Huntingdon County), tours of mines and factories—for example, the aromatic establishments in Hershey (Dauphin County) that produce so much of the country's chocolate—the amusement parks, such as Blair County's Boyertown and Bland's Park, the roadside zoos and other tourist traps beckoning along well-traveled highways, and the many shops purveying antiques and craft items.

The sorts of attractions just noted are what might be expected in almost any of our older, populous American states, but other touristic magnets are unique to the state, or nearly so. They include the remarkable forebay barns of the southeastern and central portions of Pennsylvania (Glass 1986), the heavily exploited charms of the Pennsylvania-German country, especially in Lancaster, Lebanon, and Berks counties, the restored villages of yesteryear, such as Hopewell Village (Berks County), many old iron furnaces in various stages of decrepitude, the picturesque covered bridges (more numerous in Pennsylvania than in any other state), or uniquely important artifacts, such as the State Capitol in Harrisburg, Frank Lloyd Wright's Falling Water (Fayette County), the Pennsylvania Railroad's Horseshoe Curve near Altoona, Philadelphia's extraordinary Italian Market, and, at least temporarily, Three Mile Island. Awaiting future development, in the manner of Washington's Chesapeake and Ohio, are the many abandoned nineteenth-century canals crisscrossing the state.

Especially lively is the traffic to the factory outlets that are so numerous in the southeastern section of the state. This form of shopping, which is also a legitimate form of recreation, draws customers from all the major cities lying within a few hours' bus ride. It materialized earliest and most vigorously in Reading but has been spreading since to other towns in Pennsylvania and other parts of the Northeast.

Although one does not usually think of it in recreational terms, conventioneering must also be included here. The periodic gathering of members of business, professional, and other groups normally entails a good deal of sightseeing and other pleasurable activities. Philadelphia, Pittsburgh, Harrisburg, Hershey, and some of the major colleges are equipped with the necessary meeting facilities along with the ancillary hotels and motels, and all such places bid eagerly for a convention trade that can contribute much to the local economy via its recreational and other enterprises.

Despite the inadequacy of the available statistics, it is clear that recreation and tourism, in all their diversity, have been increasingly influential both economically and geographically in the life of Pennsylvania. This has come about because of millions of decisions by individuals and households about life-style and expenditures of time and money, but it has also been aided and abetted by the efforts of many entrepreneurs, trade associations, county and municipal promotional agencies, and chambers of commerce. Above all, the state government through its relevant agencies has been exceptionally vigorous in brightening the image of the state as an irresistible place to visit and experience and in providing many of the needed facilities. Among others, the Departments of Commerce, Transportation, and Environmental Resources have exercised much ingenuity in promoting recreation and tourism by means of, for example, the eight Information Centers next to the interstate borders, conducting surveys, and generating road maps, roadside signs, posters, pamphlets, advertisements in the mass media, and slogans on automobile license plates. But it is only fair to add that Pennsylvania is only one of many players, along with all the other states and many foreign countries in what is essentially a zero-sum game, seeking to gain an appropriate slice of an expanding pie.

A substantial number of Pennsylvanians and out-of-state visitors avail themselves of the state's recreational resources, even though it is difficult to arrive at any firm totals. A 1985 survey of visitors to some of the principal attractions indicated that 39 percent of these travelers were Pennsylvanians. As might be expected, the majority of visitors came from nearby states—namely New York, New Jersey, Maryland, Virginia, and Ohio—but there were appreciable flows from California, Florida, Michigan, and other relatively remote states. The distance-decay effect does not seem to operate in the cases of visitors from West Virginia and Kentucky, states whose outdoor recreation facilities largely parallel Pennsylvania's. The survey also elicited the fact that 87 percent of the trips to the state fell into the pleasure category, while conventions accounted for another 2 percent, leaving 11 percent for business purposes solely, far below the national value of 24 percent.

The impact of recreational travel and related activities varies markedly within the state, whether measured in absolute or relative terms. Data on dollar receipts for hotels, motels, and other lodging places collected by the 1982 Census of Retail Trade provide us with a rough surrogate measure of county-to-county differentials in this impact (Fig. 11.12). (Statistics are withheld for six counties because of the disclosure rule.) It is obvious how basic the income from transient guests is to the economy of the three Pocono counties, where the per capita values range from eight to sixteen times the state mean of $92. The high figure for Montour County is misleading because it reflects a brisk turnover in motel registrations by families of patients being treated at the large Geisinger Medical Center in Danville. The relatively low values in the northwest quadrant of the state obscure the actual significance of recreational dollars in a region where outdoor activities are a mainstay of the economy—and increasingly so. Thus, according to the figures published annually in the U.S. Department of Commerce's *County Business Patterns,* the number of motels and hotels in McKean County grew from 16 in 1959 to 196 in 1978, and the number of hotel and motel employees grew from 12 to 258 (suggesting both their small size and ubiquity).

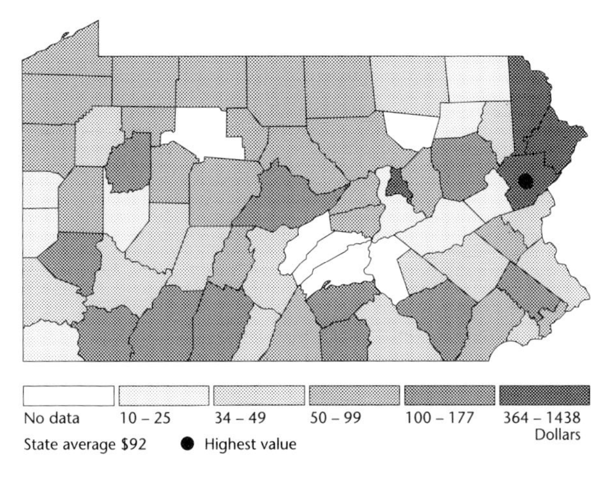

Fig. 11.12. Receipts per capita for Pennsylvania hotels, motels, and other lodging places, 1982.

The "travel impact factor" (per capita traveler expenditures as a percentage of per capita income) presents us with a more sensitive measure of the relative importance of recreation and tourism (Fig. 11.13). In 1985 this index varied enormously from a high of 70.0 in Pike County to a low of 0.8 in Northumberland. Once again, the Poconos are far

Fig. 11.13. Small-boat harbor, Presque Isle.

Fig. 11.14. Drake Well, Titusville.

Fig. 11.15. Mountain Laurel, Central Pennsylvania.

Fig. 11.16. Gettysburg battlefield.

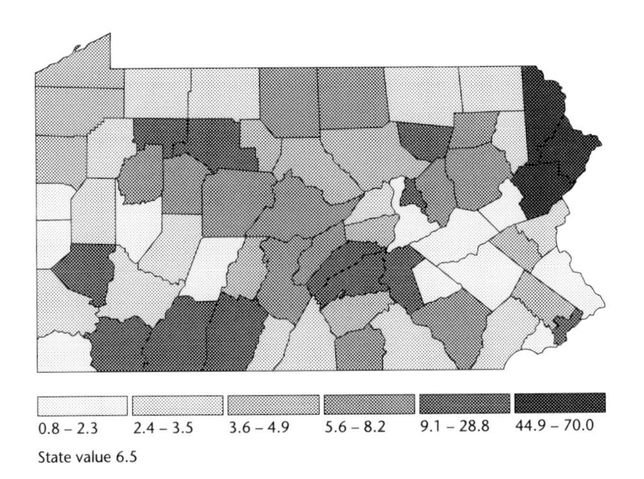

Fig. 11.17. Travel impact factor, Pennsylvania, 1985.

ahead of all remaining counties, but values are also well above the state figure of 6.5 in Forest, Elk, and Sullivan counties, where hunting, fishing, and other outdoor pastimes are so important, and in Philadelphia, Pittsburgh, and the Harrisburg area, metropolitan magnets that attract more than their quota of visitors from near and far. It is also not too startling to discover that the impact of travel on the local economy is minimal for counties, such as Delaware, Carbon, Cambria, and Beaver, that are or have been heavily dependent on mining or heavy manufacturing for their livelihood.

The study of recreation and its rapidly evolving geography is still in its infancy, however, and much remains to be done in terms of data-collecting, methodology, analysis, and interpretation. Because of the variety, complexity, and dynamism of its human and physical geography and of its pleasure-seeking activity, few sections of the world promise richer results for the student of the subject than those we can find in Pennsylvania.

Bibliography

Cuff, D. J., et al., eds. 1989. *The Atlas of Pennsylvania.* Philadelphia: Temple University Press.

Davis, A. A., and K. A. Miller. 1988. *Pennsylvania's Recreation Plan, 1986–1990.* Harrisburg: Department of Environmental Resources and Department of Community Affairs.

Glass, J. W. 1986. *The Pennsylvania Culture Region: A View from the Barn.* Ann Arbor: UMI Research Press.

Pennsylvania Department of General Services. 1980. *Pennsylvania Trail Guide.* Harrisburg.

Rooney, J. F., Jr. 1974. *A Geography of American Sport.* Reading, Mass.: Addison-Wesley.

Farm near Bird-in-Hand in Lancaster County

Mushroom houses near Kennett Square

Pennsylvania Turnpike
near Fort Littleton

Bradford Coal Company
mine near Philipsburg

PART THREE

THE ECONOMY

The economy of Pennsylvania has been evolving for more than 300 years. In the process of change, some industries and regions have declined while other industries and regions have grown. During the pioneer era, from the early settlements of the 1680s to about 1820, primary activities, especially agriculture, dominated the economy. Directly allied with agriculture was exploitation of the forests and the early endeavors to extract the mineral wealth of the state.

Between 1820 and 1920 the economy of Pennsylvania experienced an industrial revolution. A traditional way of life that had existed for centuries was completely altered as a predominately rural agricultural economy was replaced by an urban industrial society. The catalyst that initiated these changes was the substitution of mechanical power for human and animal power, made possible by the exploitation of the vast energy resources of Pennsylvania.

A complex industrial society evolved in which the component elements—mining, agriculture, manufacturing, transportation, business—reinforced one another. When one component changed, there was change in each of the other components. The nineteenth and early twentieth centuries were a dynamic period that permanently altered the character of life in Pennsylvania.

As the economy evolved in the nineteenth and early twentieth centuries, it advanced from a primary stage, based on the development of original resources, to the secondary stage, in which manufacturing was the major catalyst of growth. Since about 1920 a postindustrial society has been evolving in which tertiary activities have become the dynamic force in the economy. The growth industries of the nineteenth century, such as iron and steel and textiles, have declined significantly, and manufacturing has been sustained by the substitution of high-tech industries. Agriculture has evolved from subsistence farming to general farming and finally to specialized farming. There has been a change not only in employment in the traditional industries but also in location in the new economic activities.

Of the tertiary activities, growth has been greatest in the service industries, such as health, education, and government, and in trade. The sign of prosperity in a town is now the new services building, not a new factory.

The economy of Pennsylvania will continue to change. If the state is to remain prosperous, there must be continued adjustment to new economic conditions. Pennsylvania, with its skilled labor force of more than 5 million and a well-developed industrial structure, will continue to be of major importance in the economy of the United States.

12

AGRICULTURE

E. Willard Miller

Agriculture provided the first major occupation in Pennsylvania. Although its relative position has been in decline for more than 200 years, agriculture remains a significant part of the state's economy. As the economy of the state and the nation has evolved, the agriculture of Pennsylvania has adjusted to changing conditions. Gradually over the past century the crops and livestock that have the least competitive position have declined, and those that provide the greatest economic return to the farmer have grown to dominate the agricultural economy.

Evolution of Pennsylvania's Agriculture

Pennsylvania's agriculture has been in a continuous state of change since the colony was established in 1682. The changes have been gradual and evolutionary, but three distinct periods are recognizable. The Pioneer Era, in which a pioneer agricultural system was established, extended from 1682 to the 1830s. The second period, dominated by general agriculture, evolved with the establishment of an industrial-urban economy in Pennsylvania between 1830 and about 1920. The modern period has witnessed the rise of specialized agriculture that offers the greatest economic potential to the state's farmers.

Pioneer Agriculture

Agriculture dominated Pennsylvania's economy well into the nineteenth century. About 80 percent of the employed population were still engaged in agricultural pursuits in 1820. The growth of the agricultural economy was favored not only by the natural endowment of large areas of Pennsylvania but also by the immigrants who brought from Europe a tradition of farming. Pennsylvania was soon recognized as the "breadbasket" of the colonies.

During the pioneer period of agriculture, the first objective was to clear the land of its forests so that agriculture could begin. The clearing process began in the southeast and extended westward with the pioneer fringe, disappearing about 1840 in north-central Pennsylvania. The densely wooded areas of Pennsylvania required vast amounts of labor before crops could be planted and farms became productive. Because the settlers of Pennsylvania came from different countries of Europe, the method of clearing the land varied greatly. Generally, the English and Scots-Irish in southeastern Pennsylvania did not clear the land initially but used the Native American practice of girdling rather than felling the trees. The crops were planted among the trees, and the dead trees were removed over a period of time. In contrast, the German settlers cleared the land by felling the trees, grubbing out the underbrush and most stumps, and burning the wood, which added fertilizer to the soil.

The European settlers brought the agricultural system of Europe to America. Many of these practices were inappropriate, however, and it was not until about 1800 that an American system of crops and livestock husbandry was established. The principal crop in eastern Pennsylvania was wheat, for it was well adapted to the physical environment and a large market existed. In western Pennsylvania the major crop was corn. Corn culture had a number of advantages. It could be grown on newly cleared land. It was resistant to disease and required less labor for harvesting than many grains. Furthermore, the market for most grains was small, and shipping costs prevented their movement to the eastern market. Corn could be reduced in bulk by making it into whiskey, for which there was always a demand. Besides corn and wheat, the early Pennsylvania farm produced barley, rye, oats, flax, hemp, tobacco, and potatoes.

There was little land in hay, and pastures for these grass crops required too much labor for the economic return. The early farmer relied on native grasses to feed livestock that were free to roam in the woods most of the year. By the end of the eighteenth century, however, so much land had been cleared that not all of it was needed to produce grains. The result was a transformation of animal husbandry after 1790. The central feature of this was the introduction of red clover and grasses into the previously all-grain crop-rotation scheme. The growth of livestock began in southeastern Pennsylvania, and shortly after 1800 the area became the beef-producing center of the nation.

Swine were more important in the pioneer economy than livestock. They could survive the cold winters, needed less care than cattle, and could forage in the forests for roots, acorns, nuts, and other wild feed. They could also protect themselves from predators. They provided both meat to the family and a source of income. Pennsylvania led the colonies in the export of saltpork.

A major characteristic of agriculture in the pioneer period was the simplicity of the farm implements, many of which had not changed for centuries. In the early colonial period the plow and the harrow were the only implements drawn by animal power. Until the late eighteenth century the plow was made of wood. In 1797 a major advance occurred when the first plow was patented with the moldboard, share, and landside made of cast iron. Until the 1780s wheat, oats, and other grains were cut with a sickle that had changed little since the time of the Babylonians. After 1780, the cradle began to replace the sickle. Grass was cut with a scythe until after 1840. During the pioneer period, there was a gradual evolution from the predominance of hand labor on the farm to the increased use of animal power. At the beginning of the eighteenth century horses or oxen were used to pull a plow, harrow, or wagon. By the end of the pioneer agricultural era, animal-powered machines were beginning to appear. The threshold of the farm implement industry was in sight, signaling the decline in production of farm equipment in local blacksmith shops.

Agriculture in the Industrial Age

The development of an industrial economy brought vast changes to the state's agricultural economy. Although agriculture remained a significant endeavor, its relative importance in the state's economy declined sharply. Between 1820 and 1920 agricultural employment declined from about 80 percent of the total to about 18 percent of the state's total work force. Pennsylvania was evolving from a rural-dominated society to one in which urban activities were the catalysts of growth. At the same time, dynamic changes—technological, economic, and cultural—were taking place within the agricultural sector. The agricultural economy grew rapidly to supply the food demands of the expanding urban population.

The nineteenth century witnessed the development of animal-powered machinery. The golden age in the development of farm machinery occurred between 1840 and 1860. By 1860 patents had been granted on the fundamental principles of most of the modern farm machinery of today. Developments since then have been mainly refinements to improve the basic design. The availability of animal-powered machinery made it possible to cultivate more land. As a result, large areas with poor, acidic soil and steep slopes were cleared of forests and placed in cultivation after 1840.

The Pennsylvania farms of the nineteenth century produced a wide variety of crops and animals. The typical crop-rotation pattern on the farms consisted of corn, oats, wheat, and grass. By the 1880s corn, wheat, and oats reached their maximum acreage at 1,400,000, 1,500,000, and 1,330,000 respectively. Minor grains of the period included rye, barley, and buckwheat.

In the early part of the nineteenth century hogs continued to be an important animal on Pennsylvania farms. By 1840 it was estimated that 1,500,000 hogs were distributed throughout the state, with the greatest concentration in the southeast. After 1850, however, the number of swine declined sharply. It was estimated that it cost six cents to produce a pound of pork in Pennsylvania, compared with only two cents in the Midwest. Consequently, the center of production shifted westward.

Beef cattle were also found on essentially all farms of Pennsylvania. In the 1830s Chester County, with its rich pasture lands, was called the "fat-cattle capital" of the nation. After 1870, with the decline of grass for fattening cattle, the center of cattle production shifted to Lancaster and adjoining counties, where grain feeds were available. In the first half of the nineteenth century, small meatpacking establishments were widely distributed over the state. Without refrigeration, fresh meat could not be transported great distances. The development of refrigeration beginning in the 1870s led to the gradual disappearance of small plants. By 1910 the meatpacking industry was concentrated in Philadelphia, Pittsburgh, and Lancaster, three centers that processed 85 percent of the state's dressed livestock. Cattle and pigs continued to be raised in rural areas and small towns for home consumption.

In the nineteenth-century evolution of agriculture, a number of significant changes occurred in fruit and vegetable production and in the dairy industry. As the diet of Pennsylvanians improved, there was a growing demand for a greater variety of food products. Most significant was the rise of the potato from a feed for livestock to a principal human food. Potato culture was universal in Pennsylvania, and by the twentieth century potatoes were a leading cash crop in the state.

The dairy industry came into existence in the 1830s. Over the next 70 years it was transformed from a home industry to a highly organized commercial enterprise. It was not until the eradication of bovine tuberculosis in 1895, and the introduction of pasteurization after 1900, that milk was recognized as a safe food for human consumption. With education, the health value of milk was recognized and the modern dairy industry came into existence.

Until the 1850s vegetable and fruit production was mostly for home consumption. Only after 1850 did market gardens develop in the vicinity of the larger cities. In the same period most farms had an area devoted to orchards producing apples, cherries, pears, plums, and other fruits—first for home consumption, but with the decline of grain crops farmers turned to fruits as well as vegetables as a major source of income. Rapid transportation plus refrigeration expanded the market for Pennsylvania's vegetables and fruits.

Modern Specialized Agriculture

As Pennsylvania's agriculture adapted to changing demands within the state in the nineteenth and early twentieth century, it also had to adapt to evolving national trends of agriculture. While Pennsylvania farms could produce a wide variety of crops and livestock, it was soon recognized that many of these products could be produced only at costs exceeding those of other regions. As a response to competition each region in the nation began to produce agricultural products that brought the greatest economic return. The result was that general farming in Pennsylvania in the twentieth century declined. It has been replaced by a system of specialized agriculture in which the dairy and poultry industries dominate. Fruits and vegetables have thrived in localized areas, such as South Mountain and the Lake Erie Plain.

Pennsylvania's Status Today

In 1990 Pennsylvania's farm income of more than $4 billion put the state in nineteenth place in the nation. This was approximately the same farm income as New York, Michigan, North Dakota, South Dakota, Georgia, Kentucky, Mississippi, and Washington. There is, however, great variation in rank for the different livestock products and crops. Pennsylvania is a leader in number of dairy cattle and in volume of whole milk sold. Standing in fifth place, it is exceeded only by Wisconsin, California, New York, and Minnesota. In poultry products, the state is the third-largest producer of eggs and fifth in all chickens except broilers. Pennsylvania, with 52 percent of the nation's total value of production in 1990, is the nation's leader in mushroom production. In many other agricultural products, Pennsylvania plays a lesser role. In crop production, Pennsylvania ranks twenty-sixth in the nation. Even in corn output, the major grain in the state, Pennsylvania stands in fifteenth place nationally. Wheat production has declined significantly, and Pennsylvania is in twenty-sixth place. In total numbers of cattle, Pennsylvania ranks seventeenth, and in numbers of swine the state is fourteenth.

As Pennsylvania's agriculture has evolved, the structure of the industry has changed, reflecting the economic conditions of the period. In 1990 about 72 percent of the farm income was from livestock products, and 28 percent was from field and forage crops, mushrooms, fruits, and vegetables. Of the livestock products, dairy production was most important, with 60 percent of the total, followed by 21 percent from the sale of livestock (cattle, calves, pigs, and sheep) and 19 percent from the sale of eggs and poultry. Of the crops produced in the state, field and forage crops make up about 39 percent of the total value, followed by horticulture and mushrooms at 36 percent, fruits at 12 percent, vegetables and potatoes at 11 percent, and miscellaneous crops (including maple sugar) at 2 percent.

Land-Use Trends

In the development of specialized agriculture in Pennsylvania, it soon became evident that vast areas had been cleared that were highly unproductive. The peak in farm acreage was reached around 1880, when 19,791,800 acres were in farms—about 69 percent of the total land area of Pennsylvania. Since then there has been a continuous decline of acreage. The farm acreage dropped to 15,309,000 acres in 1930, and then declined slowly to 14,112,000 in 1950. Since 1950 the decline has continued to a low of 7,800,000 acres, or about 29 percent of the state's total acreage in 1990 (Figs. 12.1 and 12.2).

The decline in farm acreage has not been uniform, varying from 100 percent in Philadelphia County to 27 percent in Lancaster County (Fig. 12.3). The least decline in farm acreage is in southeastern Pennsylvania in the area from Lehigh and Chester counties, on the east, to Franklin and Cumberland counties on the west. In this area the farmland in each county varies from 43 to 69 percent of the total acreage. In contrast, in central and northern Pennsylvania 17 contiguous counties—from Warren and Venango,

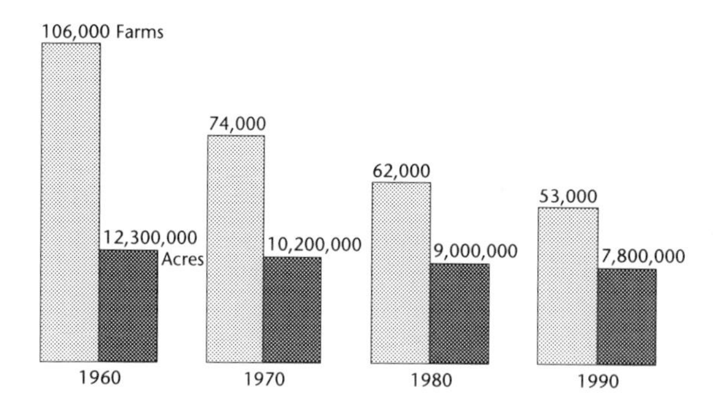

Fig. 12.1. Pennsylvania farms and farm acreage, 1960–1990.

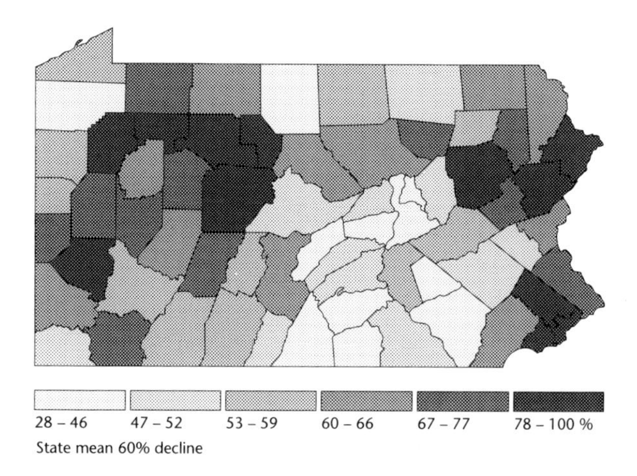

Fig. 12.2. Decline of agricultural acreage, Pennsylvania, 1880–1989.

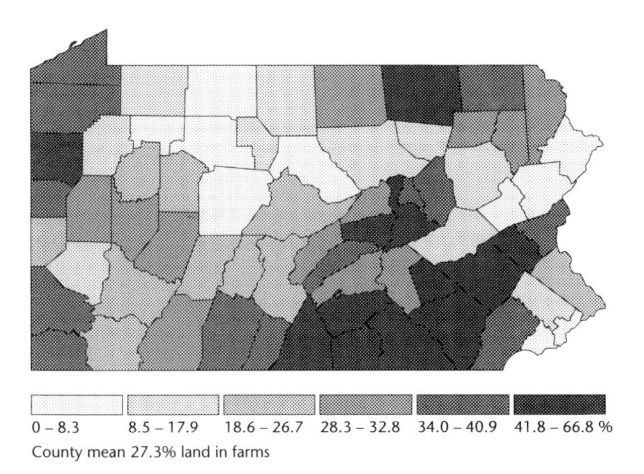

Fig. 12.3. Percentage of land in farms, Pennsylvania, 1987.

on the west, and from Blair, Cambria, and Huntingdon, on the south, to the New York border—have only between 2 and 28 percent of their land in farms.

The decline of farmland is attributed to two factors. First, the traditional abandonment of the poor farmland continues as field crops decline in importance and farms are abandoned. Second, the encroachment on farmlands for residential and economic activities, particularly in the richest farming areas of southeastern Pennsylvania, has been of major importance.

The decline in farm acreage has been a primary factor in the number and size of farms in Pennsylvania. The largest number of farms existed about 1900, when abut 224,250 were in existence, with an average size of 86.4 acres. This number has steadily declined to 146,887 in 1950, with the average increasing to 96.1 acres. By 1990 there were only 53,000 farms with an average acreage of about 146. Southeastern Pennsylvania has a concentration of about 40 percent of the state's farms. Lancaster County, with 4,940 farms, has the largest number. There are also concentrations of farms in the dairy regions of northeastern, northwestern, and southwestern Pennsylvania.

Lancaster County

Lancaster County, occupying the limestone lowlands of the Piedmont, is the prominent farming county in the state. This single county produces about 40 percent of the state's eggs, 42 percent of the broilers, 16 percent of the milk, 36 percent of the hogs, 11 percent of the field and forage crops, and 12 percent of the grain corn. The value of all farm products annually totals about $633 million or about 17.6 percent of the state's output. The next most important county, Chester County, has a total farm production of $250 million, of which two-thirds is for the production of mushrooms.

Lancaster is endowed with a remarkably favorable natural, cultural, and economic environment for farming. Of the elements of natural environment, topography presents few handicaps to farming. The gently rolling plain throughout the region is 350 to 400 feet in elevation. Local relief is rarely as much as 100 feet. The area is slightly lower than the surrounding Trenton Prong on the southeast and Triassic Lowlands to the northwest.

The region lies between the humid subtropics to the south and the humid continental to the north. The growing season ranges from 160 to 205 days. The January temperatures average 31.1 degrees Fahrenheit, and July averages 75.4. With mild winters, hot summers, a long growing season, and a precipitation of 40 inches with a summer maximum, the region has an excellent climate for middle-latitude agriculture.

With a limestone bedrock, the soils of Lancaster County are among the richest in the state. The Hagerstown series is most widespread and consists of clay loams and silt loams. These soils are well drained and have an excellent structure.

Besides the superb natural environment, the type of people who settled in Lancaster County is of paramount importance. The early settlers were German farmers of two distinct types. The "plain people" consisted primarily of Amish and Mennonites, while the other German settlers were mainly Protestants belonging to the Lutheran and Reformed churches. While both groups were thrifty and diligent farmers, the "plain people" have had a special relationship to the land. The way land is used has a spiritual significance to the Amish and the Mennonites: the land provides sustenance but also more, including aspects of pleasantness and orderliness of life. The care of the soil is based on biblical interpretation. The Amish believe that they are caretakers of the land for an absentee landlord. This stewardship is continuous and ends only at the Day of Reckoning, when one must account to God. An Amish farmer believes he must improve the quality of land when he farms it.

The economic location of Lancaster County,

close to the highly urbanized eastern United States market, is superb. The population is not only large but also quite prosperous. The high purchasing power of the region has encouraged high agricultural productivity.

Farmsteads are a conspicuous part of the agricultural landscape of Lancaster County. The barns are large, showing the need for storing hay and protecting the animals during the winter months. The silos are a striking feature of all dairy farms, and the huge chicken "coops" with tens of thousands of chickens are impressive sights. Houses are large and well maintained, providing an impression of well-being. The region has a dense network of roads, needed to move the perishable agricultural products.

Preservation of Farmland

To stem the loss of the good agricultural land in the state, the Pennsylvania state legislature in 1981 passed the Agricultural Area Security Law. This law stated:

> It is the declared policy of the Commonwealth to conserve and protect and to encourage the development and improvement of its agricultural lands for the production of food and other agricultural products. It is also the declared policy of the Commonwealth to conserve and protect agricultural lands as valued natural and ecological resources which provide needed open space for clean air, as well as aesthetic purposes.

The law authorized local governments to organize Agricultural Security Areas (ASAs). An ASA is an area created when township supervisors approve a petition presented by farmers who collectively own at least 500 acres of viable farmland. The ASAs are reviewed, and may be reestablished, every seven years. A participant in an ASA receives special consideration regarding local ordinances affecting farming activities and nuisance complaints, state agency rules and regulations, and state-funded development projects. Condemnation of farmland is discouraged in an ASA because four levels of reviewers must approve the condemnation: the township Agricultural Area Advising Committee, the township supervisors, the county commissioners, and the Pennsylvania Agricultural Land Condemnation Approval Board. Hazardous waste and low-level radioactive waste disposal areas cannot be sited in an Agricultural Security Area. As of 1992 there were 388 ASAs controlling 1,501,101 acres of agricultural land. This farmland is located on 15,210 farms in 57 counties.

An ASA offers farmers some protection against developers by reimbursing farmers for surrendering their developmental rights. In areas where pressure is strong to convert agricultural land to commercial development, a farmer is often offered money many times greater than the income he can expect from farming. The legislature in 1988 amended the ASA law creating the means for development of Agriculture Conservation Easements. The purpose of these easements, which are funded by a bond issue of $100 million passed in 1989, is to purchase farmland located in Agricultural Security Areas. An Agricultural Conservation Easement is a legal document that landowners sign when they voluntarily decide to protect their land from commercial development, thus ensuring that it will remain agricultural. The value of a perpetual easement is the difference between the development value and the farm value of the land. Easement land can therefore be used only for production of agricultural products.

The term of an easement can be perpetual or, in a limited number of cases, can have a duration of only 25 years. If the easement is for perpetuity, the purchase price cannot exceed the differences between the appraised agricultural value and the appraised commercial value. For 25-year purchases of easement rights, the price cannot exceed one-tenth of the calculated difference in appraisals.

To sell a conservation easement, the farmer must have approval at both county and state levels. It is the county board's responsibility to adopt rules and regulations for administrating a county-wide program. As of July 1991 some 29 Pennsylvania counties had established boards. In addition to the county board, a 17-member State Agriculture Land Preservation Board oversees the entire program. This board is responsible for the distribution of funds, approval and monitoring of the county programs, and specific easement purchases.

The Pennsylvania Agriculture Department's Bureau of Farmland Protection has prepared guidelines for counties to follow in establishing the farmland protection program. To apply to the county program to secure an easement, each farmer must:

1. Be located in an Agriculture Security Area
2. Have harvested either cropland, pasture, or grazing lands, and 50 percent of its sales must be of high quality
3. Be capable of producing sustained yields equal to the county average yield for crops produced

Farms selected for easement purchase are those that receive the highest scores in the county numerical ranking system. This system ensures that the excellent farms are located in areas targeted for preservation. It is important to recognize that the farmland preservation program is not an open-space program. The Conservation Easement Program has been widely accepted by farmers. By 1992 some 105 easements in nine counties consisting of 12,315 acres had been made at a cost of $27,181,719.

The Clean and Green Program is another program designed to help preserve Pennsylvania farmland. It was established in 1974, when Pennsylvanians passed a constitutional amendment permitting preferential assessment of farmland and forestland. The program's purpose is to preserve farmland, forestland, and open space by evaluating land according to its use value rather than the prevailing market value. The program is voluntary and generally requires a minimum of 10 acres in designated areas. Land taken out of the permitted acre area becomes eligible for a roll-back tax. Currently enrollment in the Clean and Green Program totals 35,000 landowners, with more than 3 million acres in 41 counties.

The Dairy Industry

The dairy industry provides the largest agricultural revenue in most Pennsylvania counties. Because Pennsylvania produces about 20 percent more milk than is consumed in the state, it is an important supplier of milk to the Northeast, particularly New Jersey and metropolitan New York.

Trends and Development

The trends of the Pennsylvania dairy industry are closely related to changes at the national level. The number of dairy farms has decreased from 32,500 in 1958 to about 13,000 in 1990, an annual decrease of 2 to 4 percent. The small dairy farms have become uneconomical and have been combined with larger farms or simply abandoned.

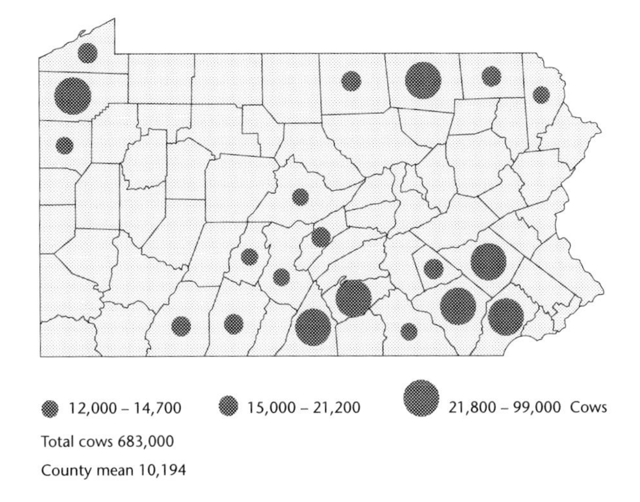

Fig. 12.4. Pennsylvania milk cows, 1990.

The number of milk cows has also declined, from 900,000 in 1958 to 683,000 in 1990 (Fig. 12.4). The decrease was not uniform; the major decline occurred in the 1960s, and the number has been declining slowly since 1970. The number of milk cows on commercial farms has increased to more than 95 percent of the total, in contrast to the once large number of milk cows on farms producing fluid milk for home consumption.

Although the number of milk cows has decreased, milk production has risen from 6,500 million pounds in 1958 to more than 9,900 million pounds in 1990—an annual increase of about 3 percent.

In 1958 the average annual milk production per cow was 7,240 pounds, but by 1992 this figure had risen to more than 14,500 pounds. In 1958 no herd in Pennsylvania averaged more than 650 pounds of butterfat per cow annually, but by 1992 a few herds were producing more than 900 pounds of butterfat per cow. A highly efficient breeding program, combined with a more intense feeding regime, has increased the milk and butterfat production.

There have also been significant changes in the processing and marketing of dairy products. The number of processing plants in Pennsylvania has declined from 511 in 1957 to fewer than 200 in 1992. With the development of refrigeration and improved transportation, milk is collected from a larger market area. The larger plants have the advantages of economies of scale in processing larger quantities of milk. In marketing milk, home delivery has essentially disappeared. The quart glass bottle has nearly disappeared from stores, replaced by the paper or plastic containers in quart, half-gallon, and gallon sizes. This innovation of disposable containers reduces the cost of collecting and cleaning glass containers. Pennsylvania is a leader in the production of many dairy products. The state ranks second in output of ice cream, exceeded only by California; third in ice milk; fourth in Italian cheeses, butter, and milk sherbet; and sixth in all cheeses and creamed cottage cheese.

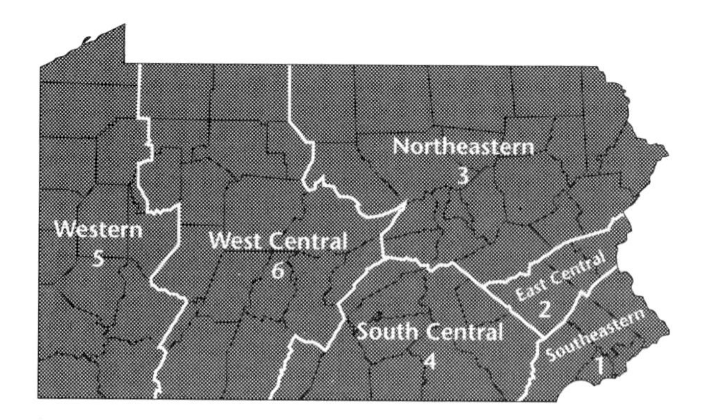

Fig. 12.5. Milk marketing areas, Pennsylvania.

The Pennsylvania Milk Marketing Board

The Pennsylvania Milk Marketing Law of 1937 stated that public control of milk pricing was being established because:

1. While milk is a vital human food, it is a most fertile field for the growth of bacteria and therefore its production and distribution have been surrounded by more costly sanitary requirements than those of any other commodity.
2. Milk consumers are not assured of a constant sufficient supply of pure, wholesome milk unless the high cost of monitoring sanitary conditions of production and standards of purity is returned to the producers of milk.
3. Milk dealers must handle constant surpluses to meet the requirements of normal variations in fluid consumption and seasonal variations in production.
4. Milk producers must make delivery of this highly perishable commodity immediately after it is produced, and must generally accept any market at any price.

In order to stabilize production and guarantee quality, public control of the price of milk was considered essential. To implement the program, the Pennsylvania Milk Marketing Board was established as an independent regulatory agency to set the raw milk price to farmers and minimum resale prices of milk in the state's wholesale and retail markets.

Pennsylvania is divided into six marketing areas (Fig. 12.5). The price of milk varies between regions and from season to season. The Milk Marketing Board periodically conducts public hearings in the six areas to determine the price structure for each region. The Enforcement Division plays a key role in the board, receiving and investigating complaints of violations of the law. It ensures that farmers are paid, issues licenses to and requires bonds from milk dealers, issues licenses to milk handlers and testers, and conducts a certification and testing program for milk-truck drivers.

The Pennsylvania Dairy Farm

The Pennsylvania dairy farm is typically a family-operated business. The average dairy farm has a labor force equivalent to about two persons and a herd of 45 to 60 cows. About 75 percent of the dairy labor force is provided by the farm family. Even on the larger farms with 90 to 100 cows, 50 percent or more of the labor is provided by the family. Furthermore, the farm operator manages the business as well as provides labor for milking and crop production.

The dairy farm has become a specialized endeavor. The cropping system is geared to the feeding requirements of the dairy animals. Corn and grass will occupy the greatest acreage on the farm. Pastureland is also necessary for supplying part of the forage of the dairy herd, but it normally occupies the poorest land on the farm. A typical farm will produce from two-thirds to three-fourths of the total value of feed consumed. An average of about three acres is required to feed a milk cow annually.

The modern dairy farm is highly mechanized. Specialized equipment includes pipeline milkers, milking parlors, bulk milk tanks, silo unloaders, and manure gutter cleaners. Mechanization makes it possible to increase the number of cows without increasing the labor force. To illustrate, manure was traditionally hauled from the barn daily to the fields. This resulted in lost nutrients, damaging soil structure and creating hauling difficulties during winter and rainy periods. Now an increasing number of dairy farmers are building storage tanks to hold manure four to six months. The manure is then spread on the fields at the time of its maximum value in fertilizing crops.

The capital investment on dairy farms has increased greatly due to higher land values, more expensive machinery and equipment, improved barns and sheds, construction of larger silos, larger herd size, and increased livestock values. Between the late 1960s and 1992 the average investment in land, livestock, machinery, feed inventories, fertilizers, and suppliers has risen from about $2,000 to about $7,500 per cow. This amounts to an investment of $375,000 for a 50-cow dairy herd. Capital to finance a dairy farm must be secured primarily by the owner-operator from local banks.

Dairy Regions

Regional variations in the Pennsylvania dairy industry can be attributed to differences in the physical environment and economic conditions. While the dairy industry is found in all counties, it has its greatest concentration in 18 southeastern counties of the Milk Marketing Board. These counties have 70 percent of the 683,000 milk cows and 68 percent of the value of milk produced in the state. Three other dairy regions are the Northeastern, Northwestern, and Central and Southwestern dairy regions.

The Southeastern Dairy Region

The Southeastern Dairy Region is a major dairy region that includes the lowlands of the Piedmont and the Ridge-and-Valley province. The core of the area centers on Lancaster County with its 112,300 milk cows, or about 16.3 percent of the state's total. Other major dairying counties in this region include Berks, Chester, Lebanon, York, Cumberland, Franklin, Centre, Mifflin, and Huntingdon.

Because dairying dominates in the region, the structure of agriculture is determined by this specialty. The basic goal of dairy farmers is to produce as much cattlefeed on the farm as possible. As a result, although corn is not the only grain produced, it normally occupies between 40 to 60 percent of the cultivated land in each county. Corn produced for grain occupies two to four times more land than corn grown for silage. Wheat and oats are minor grains in the region. The total grain acreage is 60 to 75 percent of the cultivated land. Hay crops occupy the second-largest acreage in this region. Because alfalfa is superior animal feed, it occupies three to four times more land than any other hay crop. Most of the land of this region is too valuable to be used as pasture for the milk cows. The milk cows are kept in small areas so the land can produce feed.

The Southeastern Dairy Region is central to the largest market areas for dairy products in eastern United States. Because of the large population, fluid milk is the major product. Milk is collected from the farms in huge tanker trucks, taken to processing points, and then marketed in the urban areas. Because fluid milk provides the highest return to the farmer, there is little production of butter and cheese.

The poultry industry is the second agricultural specialty of the region. This region has about 90 percent of the chickens and 90 percent of the egg production of the state. In addition, it has about 98 percent of the broilers produced in the state. The poultry industry began on the dairy farms of the region. In recent years, however, they have become separate establishments with tens of thousands of chickens. The small flocks have essentially disappeared.

The region also has about 52 percent of the beef cattle of the state. As with milk, the beef is marketed within the region. Although Pennsylvania is not noted for its production of hogs, the area has about 80 percent of the state's 950,000 hogs.

The Northeastern Dairy Region

Northeastern Pennsylvania is a land of rolling, glaciated hills. As the topography increases in ruggedness to the south and west, agriculture is limited to a few level upland areas and narrow floodplains. Large patches of woodland exist throughout the entire region and dominate the landscape in the more rugged areas and areas of poorest soils.

The Northeastern Dairy Region, consisting of the 20 counties of the Northeast milk marketing area, has about 140,000 (20 percent) of the milk cows of Pennsylvania. Of this number, about 60 percent are concentrated in Bradford, Tioga, Susquehanna, and Wayne counties. This region is characterized by farms of more than 200 acres with large barns and silos. From 70 to 75 percent of the land is in a hay crop. Of the hay crop, 65 to 75 percent of the acreage is in hay such as timothy and clover, and 25 to 35 percent is in alfalfa. The thin acidic topsoils are not well suited for alfalfa, but other hays thrive. Of the field crops, corn for grain and silage occupies from 75 to 80 percent of the cultivated land, with silage occupying about twice the amount of land as corn for grain. Oats is the next most important grain after corn, occupying from 15 to 17 percent of the cultivated land. Other field crops are of minor importance. Because grain production is limited, feed concentrates must be purchased from outside the region. The northeastern farmer must also sell more heifers than the dairy farmers of southeastern Pennsylvania, for the land will not support as high a milk-cow density. The milk production per cow is lower than in most counties in southeastern and central Pennsylvania. In Bradford County the pounds of milk produced per cow averaged 13,700 in 1990. This contrasts with 15,300 pounds in Mifflin Couny and 15,800 in Lancaster County.

The rural population of the area is low. Within the region the anthracite cities provide the largest market, but most of the milk is shipped to New Jersey and New York, all transported by truck. A portion of the milk is processed into butter and cheese, and a small portion is used to produce condensed and evaporated milk.

Closely allied to the dairy industry are all the poultry and livestock industries. The Northeastern Dairy Region has about 5.5 percent of the state's poultry. It is concentrated in Bradford and Wayne counties, which have about 60 percent of the region's total. The broilers and eggs supplement the dairy farmer's income. Because the market is the same as the market for milk, a single marketing system is effective for both products.

The Northeastern Dairy Region has about 16 percent of the state's beef cows, heifers, steers, and bulls. Many are produced on dairy farms, but a number of farms are devoted to beef animal production. As with the dairy products, the beef animals are marketed outside the region in the major urban centers of the Northeast.

The Northwestern Dairy Region

The dairy industry of northwestern Pennsylvania is rolling, glaciated country similar to that of northeastern Pennsylvania. The counties of the northwest have about 10.5 percent of the dairy cattle, with the major concentration in Erie and Crawford counties. Because the land is rolling, about 45 to 55 percent is devoted to crop production and an equal amount to hay crops. Corn is the most important crop, occupying about one-third of the crop area. Of the corn produced, 70 to 80 percent is grown for grain and 20 to 30 percent for silage. Because the density of animals on the land is not high, the demand for silage is modest. More corn is produced as grain to provide a cash crop for the farmers. Oats occupies between 10 and 12 percent of the area. Of the hay crops, other hay (timothy and clover) occupies from 60 to 80 percent of the area, and alfalfa has 20 to 40 percent.

The principal regional market for the dairy products is Erie, but the milk is also marketed in the Pittsburgh area. Because of stiff competition from the dairy farmers of New York and Ohio, little of the milk goes to Buffalo or Cleveland. The excess milk of the region is converted into butter, cheese, and condensed milk.

The poultry and beef livestock industries are poorly developed. The area, lying far from a major market for poultry products, has about 2.5 percent of the state's output, primarily eggs. The beef livestock industry is also small, with about 15 percent of the state's total.

The Central and Southwestern Dairy Region

The counties of the Central and Southwestern Dairy Region have about 22 percent of the milk cows of the state. As in the Northeast and Northwest regions, there is a concentration of the industry in the five counties of Bedford, Somerset, Blair, Huntingdon, and Centre which have nearly two-thirds of the dairy herds and also the greatest acreage of field crops. However, from county to county there is considerable variation in the acreage in hay and field crops. For example, 71 percent of the harvested cropland in Washington County is in hay, while in Westmoreland it is only 48 percent. Of the cultivated crops, corn occupies the largest acreage, varying from 20 to 30 percent of the total. In contrast to northeastern Pennsylvania, corn acreage for grain varies from 25 to 40 percent of the acreage. Of the hay crops, alfalfa occupies from 25 to 75 percent of the land. Corn and alfalfa compete for the better land, and the farmers decide what crop to produce. Oats are grown on a large number of farms using 5 to 15 percent of the cropped area. The region buys feed from outside the area because it does not produce sufficient feed for its poultry and livestock industries.

The milk of the area is marketed primarily in Pittsburgh and in the heavily industrial areas of western Pennsylvania. Some milk is processed into butter and cheese. The region also has a small poultry and livestock industry. The 18 counties have about 5 percent of the state's laying hens. The major poultry-producing counties do not coincide with the leading dairy counties—Indiana, Lawrence, Butler, and Westmoreland—which are located close to the major markets in urbanized southwestern Pennsylvania. Many dairy farms also produce beef cattle. The region has about 30 percent of the beef cows, heifers, steers, and bulls in the state. As with the poultry industry, the beef is marketed in the urbanized areas.

While a few hundred sheep are produced in every county in Pennsylvania, the only significant concentration is found in Greene and Washington counties, which have about 17 percent of the 140,000 sheep in the state. Merino sheep were imported into the region in the early 1800s. Because general farming could not compete with other areas in the state and in the Midwest, sheep provided the hill farmers with a source of income. Sheep thrived on the grass-covered hills, and the quality of the wool made it possible to absorb the high transportation costs from the isolated farms. Wool from the area commanded a high price, for the flocks were maintained by high-grade breeding stock. Sheep-raising required little labor, so the sparsely populated area could support the industry.

In spite of many factors favorable to the sheep industry, it has long been in decline in the region due to other factors. The growth of the mining industry, with its higher paying jobs, has attracted many farmers away from sheep-raising. And in recent years the number of sheep-killing dogs has increased, lowering the profits of the farmer. The demand for wool has also decreased as synthetic fibers have grown in importance. Competition of lower-priced wool from other areas of the world makes wool production in southwestern Pennsylvania, as well as in other areas of the state, less and less competitive. Unless there is a reversal of long-standing trends, the specialized sheep farms will disappear from Pennsylvania, and sheep production could disappear on all farms.

Fig. 12.6. Pennsylvania layers and egg production, 1925–1990.

The Poultry Industry

Pennsylvania is one of the leading poultry-producing states in the nation. The total number of chickens has risen from 17,443,000 in 1969 to 22,600,000 in 1990 (Fig. 12.6). With 18,710,000 layers producing 4,976,000,000 eggs annually, the state ranks fourth in the United States, exceeded only by Georgia, California, and Arkansas. The annual value of eggs was $251,947,000. Annual commercial broiler production in the state totals between $95,000,000 and $131,000,000, placing the state eleventh in the nation.

Although every county in the state produces laying chickens, 86 percent of the layers are concentrated in 18 counties in the Southeast Dairy Region from Chester on the east to Fulton on the west and Northampton to the northeast (Fig. 12.7). Within this region, Lancaster, with 8,172,000 laying chickens, has about 44 percent of the state's total.

Chickens are produced in two types of farming systems. One system is to produce chickens on dairy farms, where chickens provide a major supplementary income. The flocks will vary in size from about 100 to many thousands. Chickens fit well into the dairy-farming system, because corn is the major feed for both dairy cattle and chickens and because chickens do not take up a lot of space, leaving the land for crops and cattle. In addition, the market for eggs is the same as for fluid milk.

The specialized commercial chicken farm that

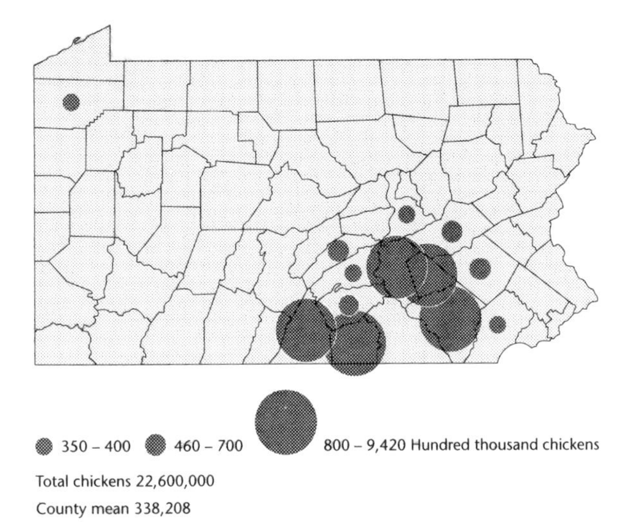

Fig. 12.7. Pennsylvania poultry, all chickens, 1990.

produces thousands of broilers and/or eggs has become increasingly important. The chickens may occupy small coops, but in recent years a single building can hold thousands of chickens. These highly mechanized operations, which in many ways resemble factory production, have been developed near the great urban markets of eastern Pennsylvania and adjacent states. They do not occupy much space, and the chicken feeds are purchased from the lowest-cost sources, not produced on the farm. Large trucks deliver the eggs and broilers to the nearby markets each day.

The poultry industry of Pennsylvania and adjacent states experienced a major problem when avian flu was confirmed in 46 hen (layer) and broiler flocks in Lancaster County in 1984. The disease, caused by a virus that originates from respiratory exudes and feces, is characterized by high mortality in broilers and great reduction in egg production in layers. The disease is spread when equipment, coops, vehicles, and people are contaminated with exudes and feces. Medication and vaccination of flocks was found to be of no value—quarantining the areas of the disease was the only way to prevent its spread. Within these areas extreme precautions were instituted to prevent its spread—for instance, cleanout crews were not allowed to leave poultry farms before removing clothes, and trucks could not visit poultry farms if they had recently been in the avian flu area. The only way to control the avian virus once it infected a flock was to kill the chickens. In spite of all precautions in the eleven months the disease existed, about 9,000,000 chickens were destroyed in Pennsylvania, and more than 17,000,000 in 452 flocks in Pennsylvania, Virginia, and adjacent states. In the eradication process federal funds to farmers to pay for loss of their flocks totaled $41.9 million, with an additional $22.6 million in federal support costs, for a total cost of $64.5 million. On May 29, 1985, the avian influenza epidemic was officially declared eradicated from the eastern United States by U.S. Department of Agriculture animal health officials. However, during the spring months of 1986 there was another, small outbreak of avian flu, which was quickly brought under control in southeastern Pennsylvania. To eradicate the disease completely, strict vigilance will be required for years.

Field Crops

The trends in field crop production in Pennsylvania reflect the evolution from general farming to a specialization in dairying and poultry production. One consequence has been a major change in the mix of crops. Patterns of production have changed, with animal-feed crops such as corn dominating, and cash-crop production such as wheat and rye experiencing notable declines (Fig. 12.8). Corn is a most important crop, with an annual output in 1990 of more than $480 million, followed by hay with an annual value of about $415 million. On most farms from 60 to 80 percent of the feed grains, 85 percent of the hay and alfalfa, and more than 90 percent of

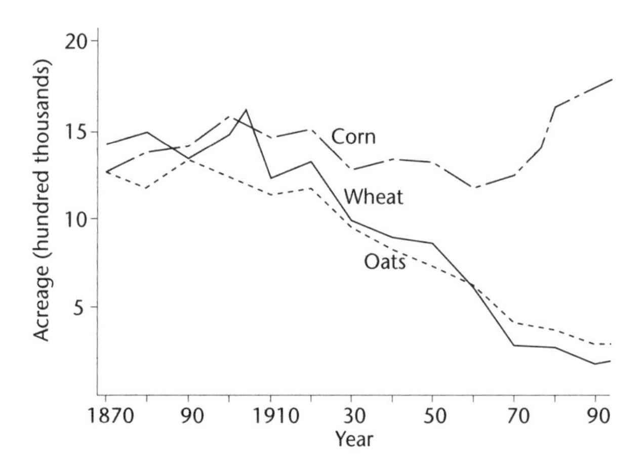

Fig. 12.8. Trends in grain acreage, Pennsylvania, 1870–1990.

the corn silage are consumed by animals on the farm where they are produced.

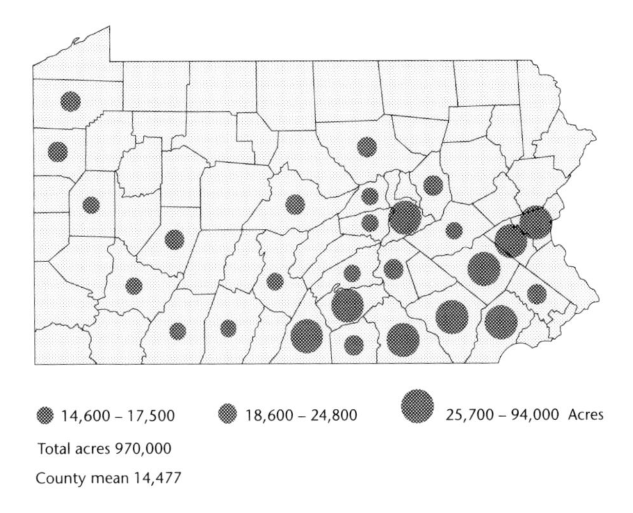

Fig. 12.9. Corn for grain acreage, Pennsylvania, 1990.

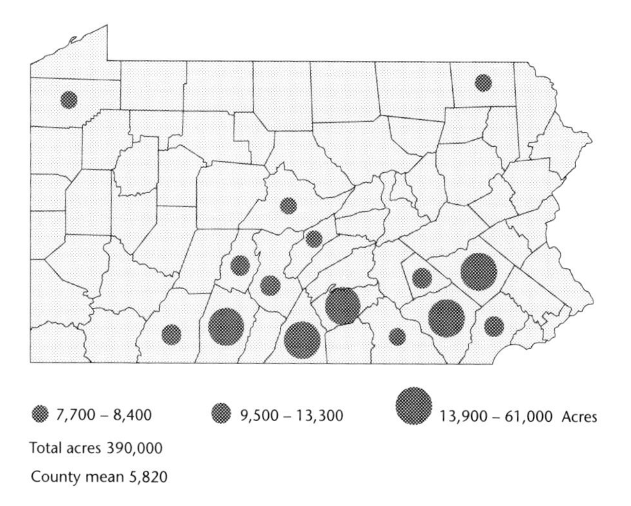

Fig. 12.10. Corn for silage acreage, Pennsylvania, 1990.

Corn for Grain and Silage

Since the early nineteenth century, corn has been the principal animal feed grain in Pennsylvania. Corn silage provides more annual nourishment per acre than any other crop, with the yield increasing on the more productive soils. Corn harvested as grain supplies the next most important animal feed and makes up most of the concentrate feed for livestock.

The total acreage planted in corn has varied in recent decades, depending on economic considerations and government influences. Between 1955 and 1960, when federal programs were implemented to limit grain production, total corn acreage declined from 1,313,000 to 1,083,000. Since then corn production gradually increased to more than 1,770,000 acres in 1990. Recently the economic and political environment has not only been more favorable, but the adoption of no-till and minimum-tillage production has lowered labor costs and made production more profitable. The practice of mulching steep slopes, which allows successful corn production on sloping and drier soils, has also been used.

While the total corn acreage changes gradually, the acreage of corn harvested for grain and that harvested for silage may change significantly from year to year. As the dairy industry has grown, the acreage in corn silage has increased gradually from 170,000 acres in 1920 to about 390,000 acres in 1990. Because of the importance of silage in the feeding of dairy animals, production remains stable regardless of silage yield. At harvest time, silos are filled and the remaining corn is harvested as grain. Because the demand for silage remains stable regardless of silage yield, more acres of corn planted to produce grain are harvested as silage in years when silage yields are low. Because corn silage is bulky, relatively low cost, has high transportation costs, and deteriorates rapidly outside the silo, it must be grown and consumed on the farm. In contrast, corn can be stored and easily transported, so it commands a national market.

Corn production is concentrated in southeastern and central Pennsylvania. Lancaster County, with 156,000 acres in corn, exceeds all other counties. Following Lancaster in acreage are York, Berks, Franklin, Cumberland, Chester, Northampton, Adams, Northumberland, and Lehigh counties. These 10 counties have about 45 percent of the corn for grain acreage in the state (Fig. 12.9).

The corn acreage for silage is concentrated in the southeast, south central, northeast, and northwest dairy regions of the state (Fig. 12.10). Lancaster County leads in acreage with 60,000, followed by Franklin, Berks, Cumberland, Bradford, and Lebanon counties. These 6 counties have about 40 percent of the corn for silage acreage in the state. There is a close correlation between the number of milk cows and the acreage devoted to corn for grain and silage.

Corn will remain a major field crop in Pennsylvania. The amount of acreage will depend on a number of factors, crucial factors being the demand in the dairy and poultry industries. A high percentage of the best land in the state is now utilized for corn production. Any increase in acreage must come from more rugged regions and less fertile soils. A number of Soil Conservation Service programs have stressed the need to reduce corn acreage due to excessive erosion. New technologies, such as strip-cropping and mulching, are now being utilized to maintain corn acreage. If a competing crop, such as soy beans, increases production, the importance of corn as an animal feed could be adversely affected. In spite of some problems, the future for corn as a silage and grain in the feeding of livestock appears bright in the farm economy of the state.

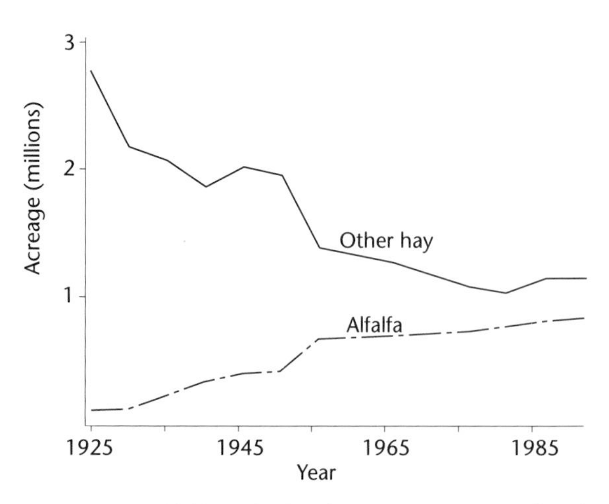

Fig. 12.11. Alfalfa and other hay acreage, Pennsylvania, 1925–1990.

Hay Crops

Alfalfa and other hay crops occupy more cropland than any other single crop. In the classification of the hay crops two major types are recognized: alfalfa and other hay. Alfalfa refers to pure stands as well as alfalfa grass mixtures. Other hay is made up of a mixture of grasses, such as timothy, and legumes other than alfalfa, such as clover.

The total acreage of all hay has declined from an all-time high of 3,250,000 acres in 1890 to about 1,990,000 acres in 1990, or about 45 percent of land devoted to crops. The average yield for all hay crops has increased from 1.10 tons per acre in 1900 to about 200 tons per acre in recent years. As a result, total production rose from 3,355,000 tons in 1900 to about 4,730,000 tons in the 1990. Production varies from year to year depending on climatic conditions. Even a modest drought can lower hay production by more than 500,000 tons for a particular year.

Although total hay acreage has declined, alfalfa acreage has increased while the decline is entirely in other hays (Fig. 12.11). Alfalfa acreage has grown steadily from 77,000 acres in 1925 to about 750,000 in 1960. Since then alfalfa acreage has grown slowly to about 810,000 acres in 1990. The growth in alfalfa acreage is due not only to higher yields per acre but also to a higher protein and energy content than other hays. The yield per acre has increased from 2.00 tons per acre in 1950 to 3.00 tons in 1990, while for other hays the yield went from 1.40 to 2.10 tons per acre.

With alfalfa's higher yield and quality feed advantages, a basic question is why there has been such a small acreage increase since 1960. A number of factors have limited growth. Production costs for alfalfa are much higher than for other hays because alfalfa requires more expert management, more fertilizer and pesticides, more labor because of a larger number of cuttings, and more expensive machinery. In many parts of the state, however, farmers consider the higher yield and quality feed of greater importance than the additional cost. The critical factor in limiting the amount of land planted to alfalfa is the availability of suitable land. Alfalfa requires more fertile and better-drained soils than other hay crops, but it must compete for this land with corn, a highly profitable alternative. For these reasons the land available for alfalfa production is limited.

Because most alfalfa and other hays are consumed directly on the farm by cattle, the major hay-producing areas coincide with the dairy-cattle-producing areas. Nevertheless, there are some differences where alfalfa and other hay acreage is located. The leading counties of alfalfa production, in order of importance, are Lancaster, Bradford, Franklin, Berks, Bedford, Washington, Chester, Westmoreland, and Centre. In many of these counties limestone soils dominate. The leading counties in other hays are Bradford, Tioga, Susquehanna, Crawford, Wayne, Somerset, Lancaster, and Adams. With the exception of Adams and Lancaster, these counties are in the Northeast, Northwest, and Southwest dairy regions where the soils are less fertile.

Wheat

Pennsylvania was one of the major wheat producing states in the nineteenth century. In 1899 some 1,514,000 acres of wheat were harvested. By 1990 the acreage had declined to about 215,000. In 52 of the counties wheat acreage declined from 80 to 100 percent.

With the decline in acreage the spatial pattern of production has experienced significant changes. At the turn of the twentieth century, wheat production was widely distributed, with a major belt extending across the southern half of the state, but by 1990 only southeastern Pennsylvania had a remnant of a wheat region. York was the leading county, with 23,200 acres, followed by Lancaster, Berks, Franklin, Adams, Lehigh, and Cumberland counties. These seven counties had about 46 percent of the wheat acreage of the state (Fig. 12.12).

The significance of the decline in acreage has been somewhat lessened by an increase in the yield per acre. In the early 1900s the yield per acre varied from 13 to 17 bushels, and in 1990 it was 50–60 bushels. Limiting wheat production to richer soils, new higher-yielding varieties, and greater fertilization raised yields. The variations from year to year are largely due to weather conditions. While the total acreage has declined by 85 percent, production has declined from 17 to 21 million bushels annually in the early 1900s, to 10.5 million bushels in 1990.

A number of interrelated factors are responsible for the decline in wheat acreage. As competition developed from farms in the Midwest and the Great Plains, Pennsylvania's wheat was placed at a cost disadvantage. The rugged areas with thin, acidic soils on the Appalachian Plateau were least able to compete, and wheat production declined and even disappeared in some counties.

In the general farming system of the nineteenth and early twentieth centuries, wheat played a major role in the crop rotation of wheat, oats, corn, and grass. With the decline of general farming and the growth of the dairy and poultry industries, crop rotation has declined and even disappeared in some areas of the state. There has also been a decline in the need for small grain nurse crops in starting alfalfa or grass production. Most farmers now seed alfalfa or other hays directly on plowed fields. Use of no-till methods to plant hay crops is also increasing. Emphasis is now on crops consumed on the specialized dairy and poultry farms, such as corn and hay. Corn commands the land, for compared with wheat it is a superior feed for livestock and poultry. It also has a higher yield per acre than wheat.

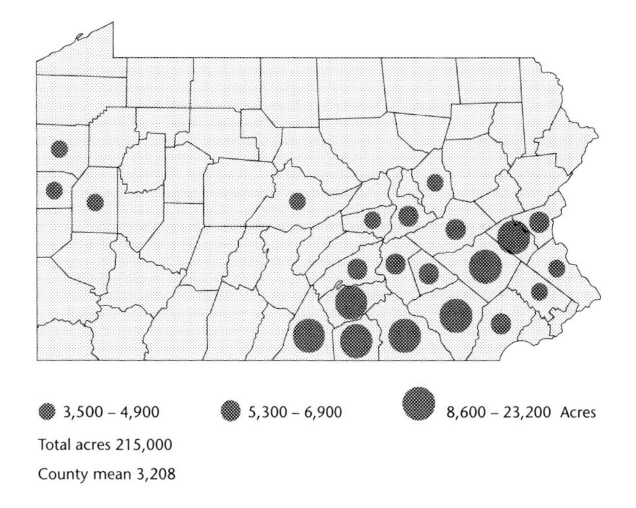

Fig. 12.12. Wheat acreage, Pennsylvania, 1990.

Wheat continues to be produced because it is a cash crop providing an immediate income on the farm. As the production of wheat declined, the availability of straw also declined. Consequently, straw prices increased relatively to corn and hay, and the comparative position of wheat improved slightly. This is not very significant because on many farms straw has been replaced by chopped cornstalks, paper, or rubber mats.

Oats

Oats was a major crop in the general farming system of the nineteenth century. From a peak in the 1890s of about 1,330,000 acres, the crop has experienced a consistent decline to about 270,000 acres in 1990. Yields per acre have increased, however, from about 17 to 28 bushels per acre in the 1890s to 50 to 60 bushels in the 1980s. The wide fluctuations in annual yields are due primarily to variations in weather conditions. Production in 1990 was 15.8 million bushels.

In the 1890s oats production was widely distributed, with major concentrations in southeastern, northeastern, and western Pennsylvania. As acreage declined, the importance of oats in the farming system was greatly reduced in southeastern Pennsyl-

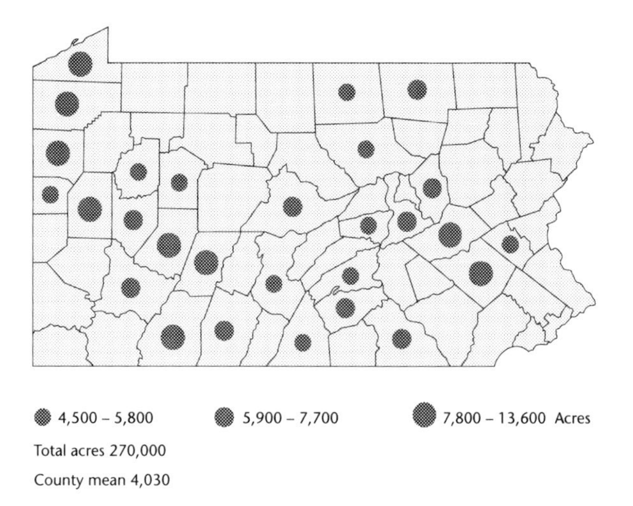

Fig. 12.13. Oats acreage, Pennsylvania, 1990.

vania. In 1990 Somerset was the leading county, with 13,600 acres planted in oats, followed by Crawford, Mercer, Cambria, Erie, Columbia, Indiana, Berks, Butler, and Schuylkill (Fig. 12.13). As a livestock and poultry feed, oats must compete for land with corn. On the average, the pounds of oats harvested per acre is about one-third less than for corn. Oats competes with corn only on the poorer upland, heavier acidic soils, where corn yields are the lowest.

A number of factors have maintained the small oats production. Because oats has a higher protein value than corn, many horse owners prefer to feed oats. With a declining horse population, even on the farms of the "plain people," the demand for oats is decreasing. Some farmers continue to produce some oats because they need straw, and some farmers use oats as a nurse crop in establishing alfalfa or other hay crops. Tradition may also play a role in that some farmers still follow rotation systems that include oats. As demand for oats as a feed grain wanes, however, acreage is likely to continue to decline in the future.

Barley

Barley was not one of the traditional grains in the farm economy of the nineteenth century. There was a short period from 1930 to 1955 when the acreage in barley increased from 45,000 to 245,000. During this period a number of farmers attempted to replace corn with barley as an animal feed, but crop failures caused barley acreage to decrease rapidly, and by 1990 barley acreage totaled only about 65,000. Yields have, however, increased from about 25 bushels per acre in the 1930s to between 55 and 70 bushels per acre in recent years.

Barley production is concentrated in the southeastern counties of York, Lancaster, Franklin, Cumberland, Berks, and Lebanon. The cold winters of northern and western Pennsylvania are likely to cause winter kill. Even in southeastern Pennsylvania barley can be damaged by cold temperatures. Barley has a higher grain yield than oats and provides an excellent livestock feed. Because it furnishes straw for bedding, it competes with corn on poorer soils. Barley is declining as a field crop and will continue to be of little importance to the farm economy of the state.

Rye

Rye was an important cash crop on the general farm of the nineteenth century. Its acreage has steadily decreased from a high of 390,000 in 1880 to about 15,000 in the 1980s. Because rye is neither a competitive livestock or poultry feed crop nor a profitable cash crop, it has essentially disappeared from Pennsylvania farms.

Soybeans

The soybean was introduced into Pennsylvania about 1924. Production grew slowly to 24,000 acres in 1950 but nearly disappeared in the next decade because of low yields. With an increase in the yield from about 20 bushels per acre in 1960 to more than 34 bushels per acre in the 1990s, acreage has increased to about 280,000. Soybean production is concentrated in southeastern Pennsylvania and the middle Susquehanna River lowlands. York County, with 32,000 acres, is the major county for soybeans, followed by Lancaster, Chester, Bucks, Lebanon, Adams, and Northumberland.

A number of factors have encouraged the modest growth in soybeans. It is one of the few cash crops on Pennsylvania farms, but the demand for livestock and poultry feed is so large that most farmers use cropland for feed rather than for a cash crop.

Furthermore, soybeans must compete with both corn and alfalfa for the best farmland in southeastern Pennsylvania. Production of soybeans may increase somewhat, but for the immediate future soybeans will remain a relatively minor field crop.

Tobacco

Tobacco has been grown in Pennsylvania since colonial days. Production was initially widespread, to satisfy local demands, but after 1870 it was concentrated in southeastern Pennsylvania, particularly in Lancaster County. Tobacco reached its peak acreage between 1910 and 1920, when 43,000 acres were devoted to the crop. As late as 1940 tobacco was the main cash crop in Lancaster County. It followed corn in the crop rotation scheme on about three-quarters of the farms.

Since 1930 tobacco has experienced a steady decline, and only about 10,000 acres were planted in 1990, of which more than 90 percent were in Lancaster County. A number of factors have adversely affected the industry. Tobacco requires a soil of high fertility, preferably limestone, and also heavy fertilizer with applications of commercial fertilizer with a high potash content. Because tobacco is a soil-depleting crop, and as a result of greater understanding of soil fertility, many farmers are turning to other crops. Tobacco requires not only more manual labor than any other field crop but also a supply of labor that is highly specialized in production, curing, and marketing. The cost of farm labor is rising, which means the cash income from tobacco declines. Tobacco culture in Lancaster County was almost universal among the German sectarian groups. The availability of family labor throughout the year made tobacco an attractive crop. The use of tobacco among the Amish, the Dunkards, and the Mennonites was forbidden by their religious beliefs, and many farmers are now producing other types of cash crops. The Pennsylvania tobacco industry has been experiencing increased competition from other areas, particularly Florida, in recent years. Because the cigar-producing industry has nearly disappeared from Pennsylvania, the demand for broadleaf Pennsylvania tobacco has declined greatly. Finally, tobacco production is no longer compatible with the specialized dairy and poultry industry that has evolved in southeast Pennsylvania.

Specialized Agricultural Regions

A number of small regions have developed specialized products within the agricultural regions of Pennsylvania. These regions exist because of a unique physical environment combined with excellent market locations.

Mushroom Culture

Mushroom culture in Pennsylvania is localized in Chester County in the vicinity of Kennett Square. The crop was introduced from France around 1885 to provide a gourmet food for the wealthy residents who lived in the "mainline" suburban towns west of Philadelphia. The first mushroom culture was associated with the production of cut flowers largely by Italian growers. The mushrooms were grown in dark areas beneath the benches and wooden frames of the greenhouses.

Although the industry began when mushrooms were an exotic food grown by only a few producers for a restricted market, the industry prospered. The early growth was fostered by two favorable factors. First, one chief hazard was eliminated: the uncertain results from imported spawn, the mycelium of the fungus that was dried on manure bricks—some beds yielded a crop, others did not. About 1910, scientists at the Pennsylvania State College developed a pure culture spawn that provided the basis for reliable production. Second, Chester County was in a major populated region with a high market potential, which was important because mushrooms are extremely perishable and at that time required a nearby market.

As the demand for mushrooms grew, a unique farming landscape developed. Because mushrooms can be produced only in darkness, clusters of long windowless sheds lying side by side under a continuous roof dot the region. Each shed has a ventilator and cooling system along its gable. The acreage devoted to the crop is negligible, and there is no climatic or soil requirement for the growing of mushrooms. Originally the season for mushroom production was from October to May, for during the summer the growing sheds became too warm. The temperature in the sheds is now controlled with air conditioning, and the growing season extends throughout the year. Mushrooms were once grown in a compost made from horse manure collected from barns in the Philadelphia area. By 1930 the replacement of horses by motor vehicles made it impossible to have an adequate supply of horse manure. As a result, a synthetic compost that controlled disease and ensured the quality of the mushroom crop was developed.

Mushroom production in Pennsylvania has grown steadily from 145.6 million pounds in 1972–1973 to 336.5 million pounds in 1990. A number of factors have affected the modern industry both positively and negatively. The development of high-speed refrigerated trucks has greatly enlarged the market for fresh mushrooms. Pennsylvania mushrooms are now marketed throughout the nation. By 1973 some 24 percent of the mushrooms were marketed fresh, while 76 percent were processed. By 1990 some 58 percent were marketed fresh and 42 percent were processed. This trend is a response not only to high-speed marketing but also to increased foreign competition and a change in food habits. In the early 1980s competition from low-priced imported canned mushrooms, particularly from China, depressed the market for Pennsylvania-processed mushrooms. Mushrooms are a labor-intensive industry because the crop must be picked by hand each day. Although transportation costs are high for shipment of canned mushrooms from foreign countries, low labor costs have made it possible for foreign canned mushrooms to compete on the American market. In spite of the decline in the processed market, total mushroom production has increased because of the rise in the fresh market. The development of the salad bar in restaurants introduced the mushroom to a new and larger segment of the population, and as the taste for mushrooms grew, their use increased in home cooking prepared in a variety of ways. Demand for mushrooms in natural-food diets has also grown in recent years. As the demand has increased, new centers of mushroom cultures have developed, particularly in the western United States, and the relative importance of Pennsylvania has declined from 57 percent of total U.S. production in 1972–1973 to about 46 percent by 1990.

South Mountain Fruit Region

In the foothills of South Mountain in Adams, Franklin, and Cumberland counties lies the most important fruit-producing district of the state. Apples are the most important crop, with Adams County producing 215 to 300 million pounds annually, about half the state's crop. Franklin County provides about 12 percent, and Cumberland 2 percent. Other important apple-producing counties are Lehigh, Allegheny, Bedford, and Berks. Peaches are the second most important fruit crop. In the South Mountain area, Adams County led in 1990 with 37 percent of the state's peach output of 76 million pounds, followed by Franklin with 15 percent and York with 8 percent. Other peach-producing counties are Lehigh and Berks. Cherries, plums, and pears are important fruit crops in the South Mountain area.

The fruit industry began as a supplement to general farming after 1870, both to supply the farm family with fruits and to serve a growing local market. As late as 1900, however, commercial orchards were essentially lacking in the South Mountain area, but in the early twentieth century a change in American eating habits caused the demand for fruits to grow rapidly.

The South Mountain area provided a desirable physical environment for fruit-growing near the major eastern United States market. The commercial orchards are planted in the foothills of South Mountain on slopes with good air drainage so that the orchards are protected against late spring frosts. The same type of site provides excellent soil water drainage conditions, a prerequisite for fruit growing. Most fruit orchards are planted at elevations of 600 to 800 feet, and a few are as high as 1,000 feet. The soils of the foothills are also favorable to fruit production. The orchards seek the well-drained gravelly loams, which permit deep penetration of the roots. In some places the orchards have been planted on steep slopes so that a ground covering must be maintained to prevent erosion.

Many orchards were originally planted near railroads, which provided the only means of transportation. The eastern slopes of South Mountain possessed a locational advantage in reaching the eastern market, so the industry had its greatest concentration in Adams County. This concentration of production continues, although trucks have now replaced railroads in transporting fruit to market. The change in transportation that occurred in the 1930s greatly affected the varieties of apples grown. Because the apples were now shipped in bushel baskets instead of large barrels, diversification in apple varieties was encouraged. Major varieties produced today include Delicious, McIntosh, Rome, York Imperial, among other types.

As the apple industry grew rapidly, production exceeded the fresh market demands. In the 1930s about half the apple crop was canned or made into cider, vinegar, and other apple products. As the fresh-fruit markets expanded and quality controls reduced the culls and low-grade apples, emphasis has been on the fresh-fruit market, where profits are greatest. Nevertheless, a processing industry persists, with Biglerville the principal center. Smaller centers include Orrtanna, Chambersburg, Aspers, Peach Glen, and Newburg.

As general farming declined, the percentage of land in farms has also decreased. In 1880 some 85 percent of the total land in Adams County was in farms. By 1990 the land in farms had declined to 55 percent. Similar trends occurred in Franklin and Cumberland counties. In value of production, field crops, vegetables, and fruits account for more than 50 percent of the total in Adams County.

Lake Erie Fruit and Truck Farm Region

The Lake Erie Fruit and Truck Farm Region is a part of a larger area that extends from Buffalo to Toledo along the shore of Lake Erie. The Pennsylvania portion is particularly noted for its grape vineyards, which are concentrated in the two distinct areas of Girard and North East. The Pennsylvania Lake Erie Plain produces about 85 percent of the state's grape crop. Besides grapes, the lake plain possesses orchards of apples, cherries, peaches, plums, and other fruits. This area is also the leading vegetable-producing region of the state. The wide variety of vegetables grown include tomatoes, sweet corn, cabbage, carrots, beets, and potatoes. Of the crops produced in Erie County, fruits make up about 27 percent of the total, vegetables 23 percent, and field and forage crops the remainder.

The physical environment is highly favorable for the production of fruits and vegetables. Climate is the most important physical factor. Lake Erie has a notable influence on the climate of the two- to five-mile wide lake plain. In the spring the cool waters of the lake retard the blossoming of the fruit trees and vineyards until danger of frost is past. By contrast, the warm lake waters in autumn delay the first frost until about October 31. The long growing season of 190 to 195 days encourages grape production, for it ensures a high sugar content, which makes the grapes particularly desirable for juice production. The lake effect moderates the temperature during the summer months so the fruit and vegetables are not subjected to extremes in heat, which lower both yield and quality. The soils of the lake plain are weathered lacustrine deposits laid down during the Pleistocene glacial period. They vary from sandy and gravelly loams to heavier clay soils. The brown to grayish-brown soils have medium to poor drainage. Most of the soils are satisfactory for fruit and vegetable production, but yields vary somewhat, based on texture. The highest-quality grapes are produced on the heavier soils, but the highest yields come from the lighter loams. The highly sandy loams require the most fertilizer.

Grape production is a year-round industry. In the autumn after the first hard frost, when the vines are bare, they are pruned. When the freezing and thawing of winter has passed in early spring, the stakes supporting the trellises are driven firmly into the ground, and the wires supporting the new growth are stretched anew. As the vine grows, it is tied to the wire by twine. The vineyards are fertilized several times during the growing season, and the ground between the rows of grapes is shallow plowed to control weed growth. During the summer the grapevines must be sprayed to control insects and blight. Grapes are harvested by machine when the grapes are to be used for juice and jellies. For grapes to be sold fresh, special grape shears are used to cut each cluster from the vine. Government inspectors examine all grapes and establish their grade as No. 1 table, No. 1 juice, or unclassified.

More than 95 percent of the grape types are the Concord variety. Niagara, Worden, and other types are of limited importance. The varieties other than Concord are for local consumption. Grape production has been increasing in the region and in recent years has been between 53,000 and 65,000 tons annually. Between 85 and 90 percent of the output is used to produce juice. Welch Foods, with a plant at North East employing about 400 workers, is the principal producer of grape juice. The emphasis on juice production has greatly retarded the development of the wine industry. Only two tiny wineries exist at North East. Two to 3 percent of the grapes are sold fresh.

Besides the production of grapes, cherries and apples are the most important fruits. The region produces between 7 and 8 million pounds of apples annually. Erie County ranks tenth in the state. Sweet and tart cherries are grown in the orchards. Production of sweet cherries has varied from 300 to 800 tons annually, depending on growing conditions. The region has long been noted for the preparation of maraschino cherries from its red tart-cherry crop. Production has, however, declined from about 24 million pounds in 1965 to 3.5 million pounds in 1990. This specialized industry has long experienced competition from foreign sources. Although total production has decreased, superior quality has maintained the industry.

Of the vegetables produced, cabbage is of major importance, and Erie outranks any other county in the state in this crop. Tomatoes are grown chiefly for juice production. Erie County is the state's largest potato producer, with 17 percent of the total. In recent years frozen vegetables have increased in importance. The two largest processors are Keystone Foods in North East and the Heinz Company in Lake City. Besides the processing of fruits and vegetables, there is a considerable local and regional market for fresh products. A large number of farms sell a part of their produce at roadside stands. The region is also noted for its dairy and poultry products.

The fruit and vegetable industries of the lake plain are labor intensive. The demand, however, is highly seasonal, reaching a peak at the height of the fall harvesting period. The harvest of most fruits and vegetables must occur within a very limited period in order to maintain the highest quality. Consequently, migrant workers are a fundamental part of the labor supply during the intense harvesting period. The migrants are primarily Puerto Ricans and Mexicans. Many of the farmers have built permanent quarters to house these migrant workers during their several-week stay each year.

Bibliography

Dahlberg, R. E. 1961. "The Concord Grape Industry of the Chautauqua-Erie Area." *Economic Geography* 37: 150–169.

Department of Agriculture Annual Report. Annual. Harrisburg: Pennsylvania Agricultural Statistics Service.

Dum, S. A., W. F. Johnstone, and B. J. Smith. 1980. *Pennsylvania's Food and Agricultural Industry Through the 1980s: Dairy Industry.* University Park: The Pennsylvania State University, College of Agriculture.

Fletcher, S. W. 1949. *Pennsylvania Agriculture and Country Life*. Vol. 1, *1640–1840,* vol. 2, *1840–1940*. Harrisburg, Pa.: Historical and Museum Commission.

Gasteiger, E. L., and D. O. Boster. 1954. *Pennsylvania Agricultural Statistics, 1866–1950*. Harrisburg: Pennsylvania Federal State Crop Reporting Service.

Miller, E. W. 1986a. "Agriculture in the Industrial Age." In *Pennsylvania: Keystone to Progress,* by E. W. Miller. Northridge, Calif.: Windsor Publications.

———. 1986b. "Agriculture: The Era of Specialization." In *Pennsylvania: Keystone to Progress,* by E. W. Miller. Northridge, Calif.: Windsor Publications.

———. 1988. "Evolution of Agriculture in Southwestern Pennsylvania." *Pennsylvania Geographer* 26 (Fall-Winter): 11–18.

Partenheimer, E. J. 1980. *Pennsylvania's Food and Agricultural Industry Through the 1980s: Field Crops*. University Park: The Pennsylvania State University, College of Agriculture.

Pennsylvania Abstract: A Statistical Fact Book. Annual. Middletown: Pennsylvania State Data Center.

Rizza, P., and S. Haylett. 1977. "An Overview of Pennsylvania's Changing Rural Land-Use Patterns." *Pennsylvania Geographer* 15 (December):1–33.

U.S. Census of Agriculture. Various years. Washington, D.C.: Government Printing Office.

13

MINERAL RESOURCES

E. Willard Miller

In the nineteenth century Pennsylvania's vast energy resources were the foundation for the state's rise as an industrial power. Bituminous coal, anthracite, petroleum, and natural gas provided the energy for an industrial economy. The importance of minerals was so great that large areas of the state were referred to as bituminous, anthracite, and petroleum regions. Thousands of people were attracted to these areas for the single purpose of developing the natural resources, and from this exploitation of the primary wealth, permanent secondary and tertiary economies have evolved.

Mineral Position in the Nation

In the nineteenth century Pennsylvania was the leading mineral-producing state in the nation, and it held a preeminent position for more than a century. For example, in 1909 Pennsylvania produced 28 percent of the nation's mineral wealth, while Illinois (in second place) produced about 6 percent.

As the nation's economy developed, mineral exploitation in other states grew too. While coal was the basic source of energy in the nineteenth century, petroleum became increasingly important as a supplier of energy in the twentieth century. Texas, with its huge petroleum production, became the nation's leading mineral producer in 1935 (Pennsylvania dropped to second place), a position it held until the early 1950s. By 1960 Pennsylvania had fallen to fourth place, dropping to seventh place in 1970 with an output of only 3.6 percent of the nation's minerals. By 1990 Pennsylvania had declined to tenth place, providing only 2 percent of the nation's minerals. Fig. 13.1 shows the comparative gains and losses in coal production among the states in the most recent decades.

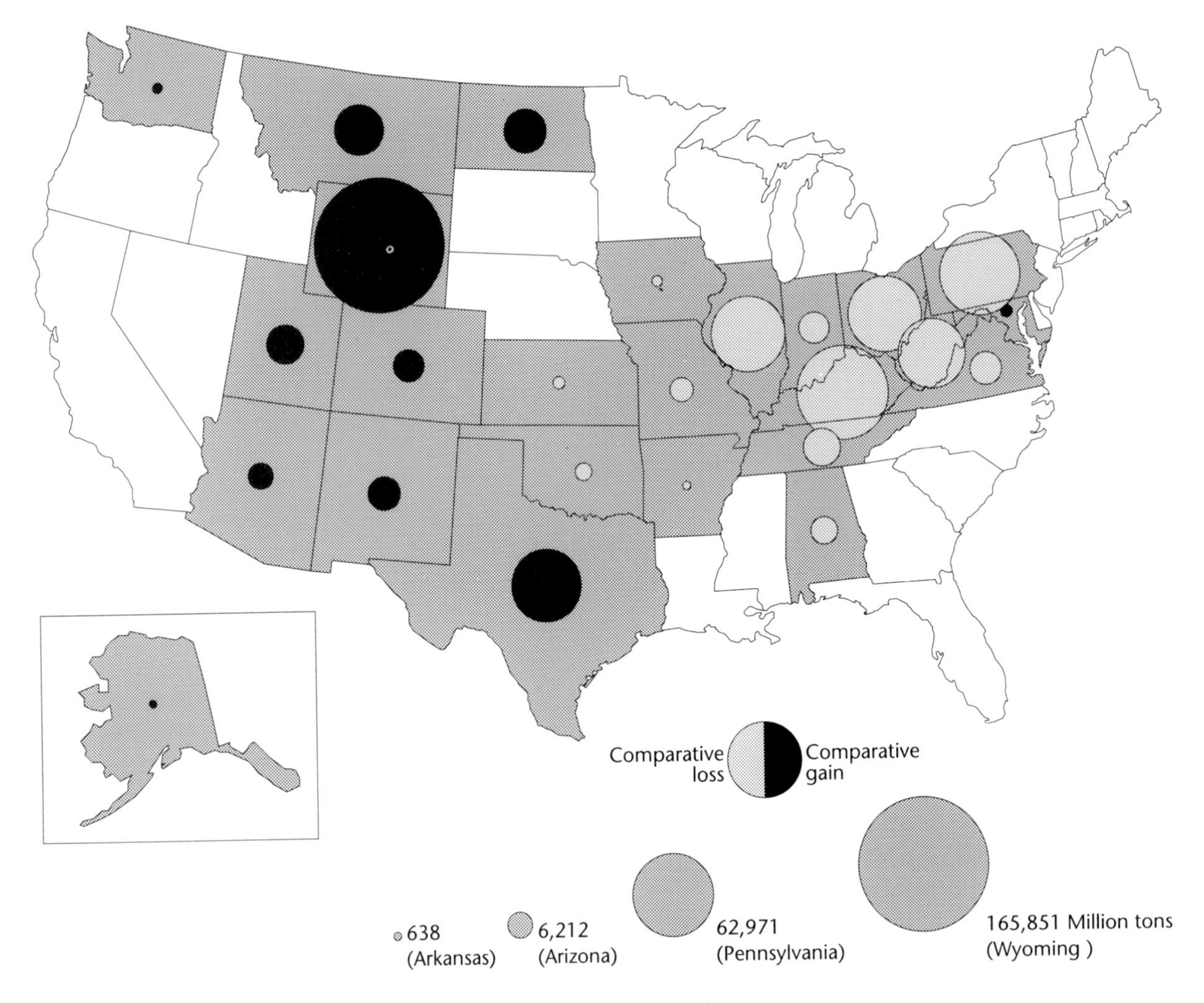

Fig. 13.1. Gain and loss in coal production, United States, 1972–1990.

The Structure of Pennsylvania's Mineral Economy

The energy minerals—bituminous coal, anthracite coal, petroleum, and natural gas—have traditionally provided 85 to 90 percent of the mineral wealth of the state. The relative importance of the fuel minerals has changed significantly since about 1960. With the decline of the anthracite industry, its value has fallen from 21 percent of total value in 1960 to about 3–5 percent in recent years. In contrast, the value of bituminous coal has risen from 50 to 75 percent of the total. During the same period the value of petroleum fell from 4.1 to 2.7 percent and natural gas went from 5.2 to 2.8 percent.

The nonmetallic minerals have provided 10 to 15 percent of the state's mined output. Of these minerals, stone is most important, with 7 to 10 percent of the total, followed in value by lime and sand and gravel. Metallic minerals are of no importance in the state's mineral output.

The Bituminous Industry

Bituminous coal seams that outcropped on the hillsides in western Pennsylvania were mined by the earliest settlers. Used initially to heat homes, bituminous coal became the leading industrial fuel by the early nineteenth century.

Location and Geology

The bituminous coal fields of Pennsylvania lie at the northernmost extension of the Appalachian coal fields that extend from Pennsylvania to Alabama. Coal is found in portions of 33 counties, and the total area underlain exceeds 13,000 square miles. The bituminous coal beds were formed in great swamps that range in age from Pennsylvanian to early Permian. Deposition of the coal-bearing sequences persisted uninterrupted for a period of 50 to 60 million years. The coal-bearing measures are thus from 260 to 310 million years old. The geologic structure throughout the coal fields is characterized by nearly horizontal to gently undulating broad northeast-southwest trending folds.

The coal-bearing Pennsylvanian and Permian stratigraphic section ranges between 1,200 and 3,000 feet thick. Major rock types are shale, sandstone, limestone, claystone, and of course coal. There are approximately 50 named coal beds that are recognizable in the coal field. A majority of the coal beds have been mined locally where seam thickness and quality enable profitable operations. Only 8 of the 50 named coal beds, however, maintain sufficient continuity and quality to be of economic importance throughout the entire coal region. These, in order of economic importance, are Pittsburgh; Upper and Lower Freeport; Upper, Middle, and Lower Kittanning; Waynesburg; and Sewickley. Most of the coal beds average less than 5 feet in thickness, although some attain a local thickness of 7 to 10 feet.

Coal quality throughout the bituminous coal field of Pennsylvania varies both locally and regionally. The coal beds exhibit a progressive change in fixed carbon, volatile matter, and calorific value from east to west. Because regional metamorphism was greatest on the eastern margins of the field, the rank of coal changes from east to west, from low-volatile bituminous (fixed carbon between 78 and 86 percent) to medium-volatile bituminous (fixed carbon between 68 and 78 percent) and finally to high-volatile bituminous (fixed carbon less than 68 percent). The majority of Pennsylvania's coals are moderate to high sulfur, 1 to 3 percent, and have a moderate to high ash content from 8 to 15 percent. The heat value of the coal normally ranges between 10,000 Btu/lb to 14,000 Btu/lb on an as-received basis.

Numerous estimates have been made of the original bituminous coal reserves in Pennsylvania. These vary from 75 to 90 billion tons, depending on the method of calculation. Between 10 and 20 billion tons of coal have been mined, or lost in the mining process. Estimates of recoverable reserves indicate that about 20.4 billion tons of underground reserves or 93 percent of the total and 1.5 billion tons of surface reserves, or 7 percent of the total remain. The ten counties with the largest total recoverable reserves are Greene, Washington, Fayette, Westmoreland, Indiana, Somerset, Armstrong, Cambria, Jefferson, and Butler. These counties account for about 87 percent of the total recoverable reserves. Greene County has the most recoverable reserves of deep coal, about 3.9 billion tons, and Butler County has the most recoverable reserves of strip-mineable coal, about 233 million tons. At current rates of production, 300–350 years of bituminous coal production can be realized from Pennsylvania's reserves.

Bituminous Coal Production Trends

Bituminous coal production can be divided into two distinct periods (Table 13.1). The initial period, extending from the first mention of coal production in western Pennsylvania in 1751 to 1918, was characterized by steadily rising production. The second period, extending from 1918 to the present, is characterized by fluctuations in production with a basic downward trend.

Table 13.1. Bituminous coal production, 1877–1990.

Year	Total Production	Stripping Production	No. of Employees
1877	1,305,932	—	16,627
1880	16,564,440	—	33,391
1890	40,884,103	—	67,383
1900	79,318,362	—	108,735
1910	148,770,858	—	193,488
1918	177,217,294	—	181,678
1920	166,929,002	—	184,168
1930	123,417,850	—	133,703
1940	111,416,916	2,808,607	117,832
1947	144,761,964	35,946,734	109,202
1950	103,439,887	25,460,343	94,514
1960	65,595,999	21,028,887	33,396
1970	80,091,922	24,161,015	24,667
1980	87,068,738	43,447,368	35,071
1990	67,947,083	27,066,611	14,083

SOURCE: *Annual Reports on Mining Activities*, Pa. Dept. of Environmental Resources.

Initial Period of Growth, 1751–1918

Coal production began with the pioneer settlement. The early coal mines supplied fuel for domestic heating and cooking and for such pioneer activities as blacksmith shops. The first important growth in demand came when railroads began to be built in western Pennsylvania in the 1850s. Nevertheless, production remained small until coal was used to produce coke for the rapidly expanding iron and steel industry. In 1877, bituminous production totaled only 1,305,932 tons, but it rose to 12,500,741 in 1878. From that year production increased rapidly to an all-time peak of 177,217,294 tons in 1918. Employment in the bituminous fields reached its peak in 1914, when 196,038 miners produced coal. During the nineteenth and early twentieth centuries, coal was the major source of energy for manufacturing, transportation, and residential heating.

In western Pennsylvania, coal production was initially widely dispersed. The near-surface coal seams were readily accessible, and transportation of a bulky, relatively low-cost fuel favored local developments to satisfy market demands. With the growth of iron and steel and related manufacturing in southwestern Pennsylvania, Fayette, Westmoreland, and Allegheny counties became the major areas of production. Between 1881 and 1917, Pennsylvania mines produced 3,111,250,000 tons of coal. Of this amount, Fayette County produced 18.6 percent, Westmoreland 18.5 percent, and Allegheny 12.8 percent of the total. In this 36-year period, these three counties produced 49.9 percent of the state's coal.

The concentration of the coal industry and its allied coke industry in Fayette and Westmoreland counties was due to the location of the "Pittsburgh seam." At an early date, coal from this seam was recognized as being nearly physically and chemically perfect for producing beehive coke. With a carbon content of 57 to 60 percent, volatile material at 30 to 35 percent, moisture content under 2 percent, average ash content less than 8 percent, and, most important, sulfur content under 1 percent, it was an ideal coking coal.

Besides the excellent quality, accessibility and a large reserve were important features localizing the industry. The Pittsburgh seam outcrops in Fayette and Westmoreland counties and underlies all of southwestern Pennsylvania. The reserves were huge. In Fayette County, for example, the Pittsburgh coal seam was 8 to 11 feet thick, and the original reserve was 1,900,000,000 tons.

The coke was produced at the mine, for in the beehive process the gaseous liquid hydrocarbons were burned off so that only the carbon remained as coke. From 1,200 to 1,300 pounds of coke were produced from a ton of coal. The weight and bulk were thus reduced at the mine, greatly lowering the cost of transportation to the blast furnaces. This region had more than 90 percent of the coke production in the United States. The peak output of the region was reached in 1916, when 40,000 coke owners consumed 33,792,000 tons of coal to produce 22,496,000 tons of coke.

While the growth of the beehive ovens was spectacular, their decline was even more striking. With the development of the by-product coking process, beehive coke oven production declined rapidly. This new process prevented the great waste of liquid and gaseous hydrocarbons inherent in the beehive process. The change in coking processes resulted in a complete change in the location of the coke industry. When using the by-product method, the coal is processed at the place where the coal gases and liquids can be utilized. Because coal gases were used for heating blast furnaces, open hearths, power stations, and other plants, by-product ovens were built at the site of the primary iron and steel industry. Coal was thus shipped out of the mining region to the consuming centers.

Modern Trends

Although the general trend in Pennsylvania's coal production has been downward since 1918, there have been major fluctuations in output. Sandwiched between three distinct periods of decline (1918–1939, 1947–1961, 1980–1990) are two periods (1940–1947, 1961–1979) when the coal industry experienced a recovery (see Table 13.1).

Periods of Decline

Coal production reached its all-time peak in 1918, when 177,217,000 tons were produced. Although the 1920s were a period of economic prosperity, coal production declined steadily to 142,351,000 tons in 1929. A number of factors adversely affected coal production. Foremost of these was the rising competition from petroleum and natural gas. The availability of low-cost alternative fuels drastically affected a large number of coal markets, and some disappeared entirely. When coal was the dominant fuel at the turn of the twentieth century, domestic heating consumed from one-quarter to one-third of bituminous production, but by the World War II period bituminous coal as a residential heating fuel had declined to insignificance. Coal was the leading fuel for locomotives well into the twentieth century, but development of the diesel engine meant that diesel fuel replaced coal in operating locomotives. In industry, petroleum and natural gas have made massive inroads replacing coal. Metallurgical coke can be made only from bituminous coal, but this market also declined as blast furnaces became more efficient and required less fuel for each ton of ore smelted.

Besides competition from petroleum and natural gas, the growth of new coal-producing states in the 1920s adversely affected the Pennsylvania coal industry. Pennsylvania was one of the first states where the vast majority of the coal miners were members of the United Mine Workers of America. As a result, union demands increased the price of coal, making it less competitive with petroleum and natural gas. To combat the rising costs, coal companies began to exploit the coal resources in such nonunion areas as West Virginia and Kentucky. The middle and southern Appalachian fields, although farther from the large markets, possessed high-quality coals and enjoyed a lower ton-mile freight rate to markets. Most important, however, was the lower wage scale. By 1927 the expansion in the nonunion fields had been so great that these fields could supply the entire nation's needs without the output of the unionized areas. Although the middle and southern Appalachian fields were gradually unionized, excessive production capacity remained and excessive competition continued.

During the early 1930s, as a response to the world economic depression, production declined drastically, to a low of 74,162,000 tons in 1932. During the remainder of the 1930s output varied considerably, depending on annual economic conditions. For example, production recovered to 110,417,000 tons in 1937, but as the depression deepened in 1938 production declined to 76,881,000 tons.

The second major period of decline occurred between 1947 and 1961, when production fell from a high of 144,761,000 tons in 1947 to a low of only 63,171,000 tons in 1961. During this period petroleum and natural gas replaced coal in railroad transportation, in large segments of manufacturing, and in residential heating. Technology for the use of petroleum or natural gas increased, and little effort was made to use coal. By 1960 the only remaining coal markets were in the public utilities and for production of coke in the metallurgical industries.

The last period of decline began in 1980. Between 1979 an 1983 production declined from 89,166,000 tons to 68,732,000 tons but has remained at that level, with an output in 1990 of 67,947,000 tons. As the price of oil declined in the 1980s, the competitive position of Pennsylvania coal decreased. Particularly significant has been the decline of the European export market, where petroleum and nuclear energy are providing a larger share of the energy needs. The U.S. Clean Air Act continues to depress Pennsylvania coal production, for low-cost removal of sulfur is still in the future.

Between 1979 and 1990 U.S. coal production increased by about 28 percent. During the same period Pennsylvania coal production declined by 21,219,000 tons, or 23.8 percent. If production had grown at the same rate as the nation, the output would have been about 114 million tons. Thus Pennsylvania has both a comparative and an absolute loss. Nevertheless, the coal industry has survived at a lower level of production, due to coal consumption by the electric utilities in areas of low population in order to conform to the state's air standards, the production of coking coal, and a small export of coal to foreign countries, particularly Canada.

Periods of Growth

Between the three periods of decline there were two periods when coal production grew, reversing the basic downward trends. These were periods when demand for all types of energy was high and economic considerations as well as national policy encouraged development of all domestic energy resources. In the first period, 1940–1947, during World War II, the coal industry made a remarkable recovery. Output increased from 89,397,000 tons in 1939 to a peak of 144,761,000 tons in 1947. Petroleum was required in this war effort, and gasoline was rationed. The increase in coal production met domestic energy needs, and in the immediate postwar period coal exports to the devastated warring nations rose significantly. By 1948, however, petroleum and natural gas were once again available, and coal continued its decline.

The second period of modest growth occurred between 1961 and 1979, when total production rose from 63,595,000 tons in 1961 to 89,166,000 tons in 1979. This time span can be divided into two periods: 1961–1972 and 1972–1979.

In the early 1960s the general economic expansion created a demand for increased energy production. For the Pennsylvania coal industry, particularly significant was the increase in demand for metallurgical coal due to the growing iron and steel industry, and there was an increasing demand for steam coal for an expanding production of electricity. Between 1961 and 1966 coal production increased by 18,278,000 tons to a total output of 81,449,000 tons. Between 1967 and 1972 production fluctuated between 75 and 81 million tons, reflecting economic conditions.

From 1972 to 1979 production increased by 13,293,000 tons to a modern high of 89,166,000 tons. Because strip mines can be brought into production quickly, strip production increased by 19,399,000 tons. In contrast, underground production declined by 6,106,000 tons. The retarding influence of high-sulfur coals that could not be burned before sulfur removal, and the increased cost of labor for workers who did not mine coal, reduced the growth of the Pennsylvania coal industry. A significant export market was an important factor in the growth of the industry.

Conflicting Factors in Coal Production

Until the 1970s, domestic economic conditions were the most significant factors influencing coal production and consumption. In the 1970s political factors—both domestic and international—came to the forefront and influenced production trends The actions of OPEC in the early 1970s increased the price of oil and changed the energy structure of the world. On the domestic scene, the Clean Air Act of 1970 and the Federal Health and Safety Act of 1969 were passed. These resulted in conflicting trends: OPEC's actions encouraged coal production to reduce dependence on oil, while the federal legislation retarded development of the Pennsylvania coal industry.

The OPEC Factor

In the early 1970s OPEC was able to raise the price of a barrel of oil from about $2 in 1972 to more than $30 by the late 1970s. In response the United States attempted to develop domestic resources in order to reduce its energy dependence on the world. It could do this most rapidly by replacing oil with coal. In the mid-1970s there was high hope that the Pennsylvania coal industry would increase greatly. As a result, as noted above, coal production in Pennsylvania increased to its modern high.

The Clean Air Act

The basic purpose of the Clean Air Act was to control the pollution of the atmosphere. On December 23, 1971, the Environmental Protection Agency (EPA) established basic national standards for the emission of sulfur dioxide (SO_2), oxides of nitrogen (NO_x), and particulates (fly ash). Concern is greatest for sulfur dioxide because it requires a complex and costly cleaning process before burning, or scrubbers to remove it during the burning process. The standards have been raised, and additional legislation was enacted in 1979, 1983, and 1991.

The Clean Air Act has been a major factor in reducing Pennsylvania coal output since 1970. Between 1970 and 1979 bituminous coal production in the United States rose 127.3 percent, from 602,932,000 to 767,856,000 tons. In order to measure the dynamics of change, the net shift analysis can be utilized. Thus, if a state were to experience a comparative gain, its production had to be 127.3 percent or greater than in 1970, or less than 127.3 percent, to experience a comparative loss. Pennsylvania's coal production rose from 80,491,000 tons in 1970 to 89,166,000 tons in 1979, an increase of 8,665,000 tons, or 9.8 percent. This increase, however, was far below the national average growth. The comparative loss totaled 13,300,000 tons. The states with comparative losses were the high-sulfur-coal states east of the Mississippi, and the states with comparative gains were the low-sulfur states in the West. In essence, the Clean Air Act created two regional markets for steam coal—high sulfur versus low sulfur coals.

A formula of SO_2 emission devised by the EPA was expressed in terms of weight of SO_2, relative to the Btu present in the coal. The calculations are as follows:

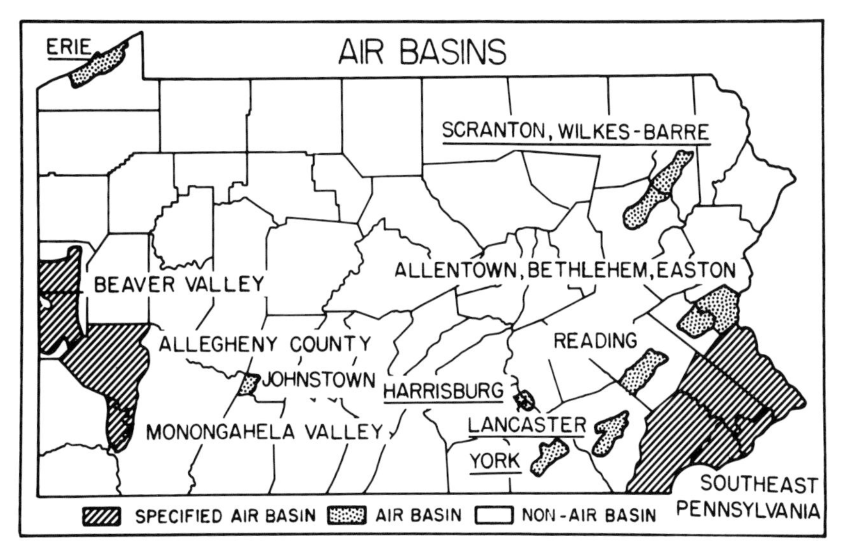

Fig. 13.2. Pennsylvania's air basins.

Assume coal with properties of

3 percent sulfur
10,000 Btu/lb, 20 million Btu per ton
(In order to understand the relationships of SO_2 to sulfur, it must be remembered that the weight of sulfur is twice that of oxygen. Thus SO_2 formed in combustion is equal to twice the weight of sulfur contained in the coal.)

The sulfur content of the coal is:

.03 × 2,000 lb = 60 lb of sulfur emission when coal is burned
2 × 60 lb per ton = 120 lb of SO_2 per ton. When related to heat content 120 lb SO_2 per ton/20 million Btus per ton = 6 lb SO_2 per million Btus.

EPA regulations classified coal into three categories of SO_2 contents, each subject to a different level of SO_2 reduction:

Category 1: All coals with SO_2 content equal or less than 2 lb/million Btus must remove 70 percent of all SO_2.

Category 2: All coal with an SO_2 content between 2 lb/million Btus and 6 lb/million Btus must reduce the SO_2 content so that final emissions do not exceed 0.6 lb/million Btus. This requires a variable percent reduction of SO_2 by 70 to 90 percent.

Category 3: All coals with an SO_2 content between 6 lb/million Btus and 12 lb/million Btus must reduce SO_2 emissions by 90 percent. The maximum final emissions level, or ceiling, after 90 percent reduction is 1.2 lb/million Btus.

To be in compliance with the Clean Air Act, Pennsylvania established a three-class level emission standard for SO_2. To implement the plan, the state was divided into air basins and non–air basins (essentially a function of urbanization) with a discrete SO_2 emission level described for each air basin class (Fig. 13.2). Emission levels were expressed as amounts of sulfur dioxide that may be released per quantity of energy consumed, the units being pounds of SO_2 per million (10^6) Btu. On August 1, 1979, the emission standards were revised (Table 13.2). The changes permit use of higher-sulfur coal over shorter time periods, but a more stringent longer time average in most areas. In general, the standards were relaxed slightly. To illustrate the 1979 standards, Allegheny County lies in a specified air basin, and the pounds of $SO_2/10^6$ Btu allowable in this region range from 0.6 to 1.0 lb/10^6 Btu.

Because Pennsylvania coal varies considerably in both thermal (Btu) and sulfur content, it is necessary to determine emission standards as a percentage of sulfur relative to the Btu content of the coal. Fig. 13.3 graphically displays the relationships between SO_2, the heating value of coal, the unit size of boiler, and the percentage of sulfur for each air basin type and facilitates the recognition of sulfur levels in relation to Btu content necessary to meet the limits for each air basin class. As the Btu content of coal increases, so does the percentage of sulfur in the coal

Table 13.2. Sulfur dioxide emission standards in Pennsylvania, August 1, 1979.

Air Basin Type	Sulfur Dioxide Emission Limits (#10^6 BTU) by Combustion Unit Size (10^6 BTU/hour)		
	<250	<250 with Permit	>250
Non-Air Basin[a]			
1-hour average	4.0	—	—
Daily average maximum	—	4.8	4.8
Daily average, 2 days in 30-day period	—	4.0	4.0
30-day running average	—	3.7	3.7
Annual average[b]	—	3.3	3.3
Air Basins			
1-hour average	3.0	—	—
Daily average maximum	—	3.6	3.6
Daily average, 2 days in 30-day period	—	3.0	3.0
30-day running average	—	2.8	2.8
Annual average[b]	—	2.5	2.5
Specified Air Basins			
Allegheny–Beaver Valley–Monongahela Valley	0.6–1.0[c]	—	0.6[c]
Southeast Pennsylvania Inner Zone			
1-hour average	1.00	—	—
Daily average maximum	—	1.20	0.72
Daily average, 2 days in 30-day period	—	1.00	0.60
30-day running average	—	0.75	0.45
Annual average[b]	—	0.67	0.40
Southeast Pennsylvania Outer Zone			
1-hour average	1.20	—	—
Daily average maximum	—	1.44	1.44
Daily average, 2 days in 30-day period	—	1.20	1.20
30-day running average	—	0.90	0.90
Annual average[b]	—	0.80	0.80

SOURCE: Applications of the Pennsylvania Coal Model.

[a]Including Erie; Harrisburg; York; Lancaster; and Scranton/Wilkes-Barre air basins.

[b]Calculations based on statistical distributions assumed for coal sulfur content by Pennsylvania Department of Environmental Resources.

[c]2.5–50 million Btu/hr: 1.0
50–2000 million Btu/hr: $A = 1.7E^{-0.14}$ where E = heat input
>2000 million Btu/hr: 0.6

that can be burned legally. In Fig. 13.3 the left-most vertical line for each air basin represents the SO_2 level for large combustion units. The right-most vertical line represents the less restricted SO_2 levels. The smaller boilers have fewer restrictions on SO_2 levels, so that they can utilize higher-sulfur coals.

The Federal Mine Health and Safety Act

Since the beginning of the coal-mining industry, technological progress has attempted to make coal mining a less hazardous occupation. National standards were, however, not established until the passage of the Federal Coal Mine Health and Safety Act of 1969. The purpose of this act was "to provide more effective means and measures for improving the working conditions and practices in the Nation's coal and other mines in order to prevent health and other serious physical harm and in order to prevent occupational disease originating in such mines." The act established mandatory health and safety standards and required each coal mine operator and every miner to comply with the established standards. The act also contained provisions to expand research and training programs aimed at preventing coal mine accidents and occupationally caused diseases in the industry. The mandatory health and safety regulations require inspection of all aspects of the mining operation.

In order to enforce the regulations, inspection of underground and surface mines is required on a regular basis. If a mine inspection finds violations, the operator receives a citation. To continue operation, the violation must be removed within a specified time. If the violation continues, penalties,

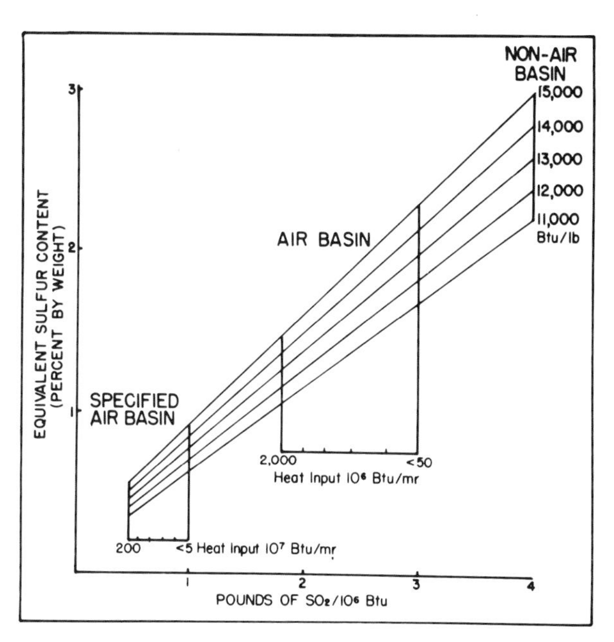

Fig. 13.3. Sulfur content equivalent to Pennsylvania's sulfur dioxide limitations.

including work stoppage in the mine, are imposed. Although it is difficult to question the health benefit of the act, many workers who do not mine coal are necessary to implement this legislation and secure these benefits. Consequently, the price of mining increases but the output of coal per worker has declined.

Recent Structural Changes in the Coal-Mining Industry

The coal-mining industry of Pennsylvania has experienced numerous structural changes in production, employment, productivity, and the number and types of mines (see Fig. 13.4). Federal Health and Safety Act standards have had an influence on these changes.

Underground and Surface Production

Traditionally bituminous mining occurred underground. Only in the past half-century has strip mining become a significant factor in production with the development of equipment that can move massive amounts of rock. In 1947 some 76 percent of bituminous coal produced in Pennsylvania was from underground mines. By 1961, although total production had declined, strip production had risen to 33 percent of the total. The importance of strip mining remained relatively stable between 1961 and 1972, at 27 to 33 percent of the total. With the development of the energy crisis in the 1970s, strip mining increased its importance in Pennsylvania, reaching a peak between 1977 and 1982, when 50 to 57 percent of the coal mined was from stripping operations. Since then the importance of strip mining has declined, and in 1990 strip products totaled 40 percent of the Pennsylvania output. Strip mining responds quickly to rises in market demand for coal and declines quickly as the market is depressed. In contrast, underground mining with more stable long-term markets has a greater stability in production rates.

Employment

From 1947 to 1961 total employment declined from 109,202 to 29,632, or 73 percent. Underground employment declined from 96,171 to 22,950, a 76 percent drop, and employment in surface mining dropped from 13,031, to 6,682. This decline was primarily a response to declining production, but rising productivity in a period of declining production was also important. In the 1960s the decline in employment continued, even with increased production due to increasing productivity. Underground mining reached a low of 17,711 workers in 1970 and a low of 4,132 for strip mining in 1969. The long-time trend of decline in employment was halted after 1970. Underground employment rose to 23,535 in 1979, an increase of 32 percent, and employment in strip mining rose to 6,057, an increase of 46 percent. These changes reflect the influence of the Federal Health and Safety Act in the need to hire people to implement mandated federal mining standards, and the state and federal environmental requirements for reclaiming surface-mined land.

After 1979 the employment trends were once again reversed. By 1990 underground employment totaled only 8,300 and surface mining 4,690. These declines reflect not only declining production but also the disappearance of low-productivity underground mines with high labor costs.

Productivity

Productivity in both underground and surface mining rose steadily from 1947 to 1971, with increases in

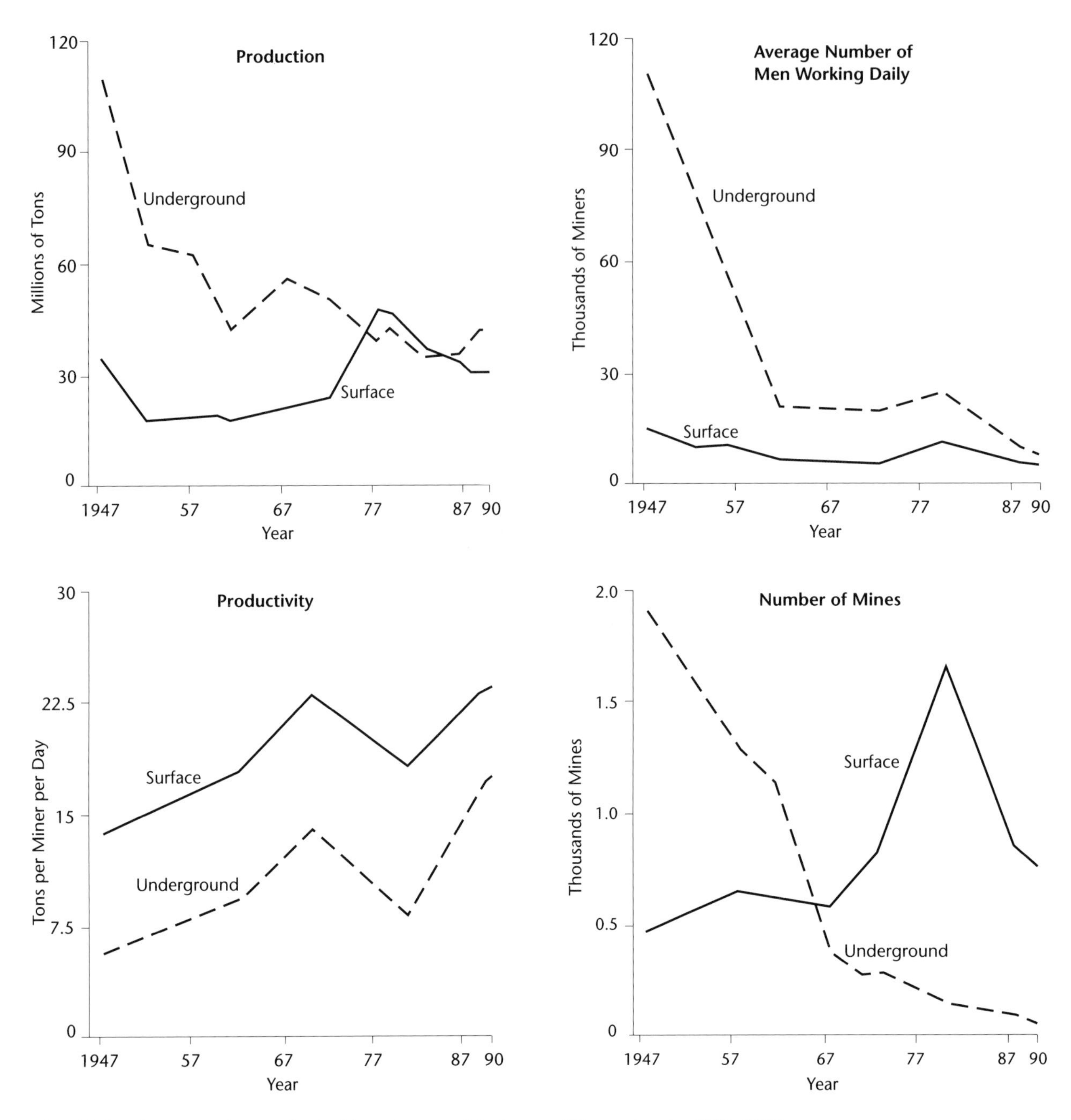

Fig. 13.4. Structure changes in the bituminous coal industry of Pennsylvania, 1947–1990.

underground mining from 6.3 tons per worker per day to about 14 tons and in surface mining from 14 tons per worker per day to a peak of 24.9 tons. In order to reduce the cost per ton of coal and make the industry competitive with other energy sources, technological advances were significant during the decline in total output. For example, continuous mining machines were not available in 1947, but by 1960 some 70 percent of underground coal was mined by continuous cutting machines. Huge draglines that could move more than 100 feet of overburden to mine a two-foot coal seam were developed.

From 1970 to 1979 productivity declined in underground and surface mining. In 1979 an underground miner produced about 7.8 tons a day, a decline of 44 percent from 1970, and a strip miner produced 18 tons a day, a decline of 27 percent. The decline in productivity reflects the increase in workers who were engaged in associated activities, not in producing coal.

In the 1980s productivity increased once again, particularly in the underground mines. Only the highly mechanized mines remained, and output averaged about 22 tons per worker per day. Strip-mine productivity also increased to 27 tons per worker per day.

Number of Mines

The trends in the number of underground and surface mines reflect the changing production patterns. Since 1947 there has been a nearly continuous decline in the number of underground mines, from about 1,800 in 1947, to 245 in 1970, to only 88 in 1990. Only the most modern and highly productive underground mines have been able to survive economically. The decline in number of many older underground mines reflects the economic problems of meeting the federal health and safety standards. Furthermore, with a depressed market for coals and the high cost of developing a new mine, few new, highly mechanized underground mines have been developed in recent years. The capital outlay for a modern-size underground mine is now about $100 million. In contrast, a small surface mine can be put into operation for a few million dollars. An underground mine must have an operating life of about 25 years or more to be profitable. A surface mine may have an operating life of only a few months to at most a few years. Although the initial cost of new underground mines and the increased cost of maintaining the old underground mines have caused this decline, the low cost of developing and maintaining health and safety standards of surface mines, plus their higher productivity and consequently profitability, has favored the continuation of surface mining.

As the market for strip-mined coal in the 1970s rose, the number of strip mines increased from 908 in 1972 to a peak of 1,738 in 1979. With the rapid decline in oil prices after 1980, the market for strip-mined coal was depressed and the number of strip mines declined to a low of 766 in 1990.

Spatial Changes

The bituminous industry has been subject to rapid spatial changes due to rise or fall of production in recent years (Fig. 13.5). This has created instability in employment, resulting in the economic decline of many coal areas.

1947–1961

Although coal was mined in 29 counties of western Pennsylvania between 1947 and 1961, production was concentrated in a few counties. In underground mining, eight counties, each with a minimum of 1,000,000 tons of output in 1947, provided 87 percent of the total. These counties still led in production in 1961, with 91 percent of the total, although production had declined. The greatest declines were in the older counties, where production costs were high because easily mined coal sources had been depleted. For example, in Fayette County output declined from 16,731,000 to 873,000 tons, a drop of 94 percent between 1947 and 1961. Other counties with declines from 4,000,000 to 15,000,000 tons include Greene, Somerset, Allegheny, Westmoreland, and Cambria.

In surface mining, the condemnation was not quite so great, with 10 counties producing 75 percent in 1947 and 72 percent in 1961. Strip-mining production increased on the northern periphery in Clearfield, Clarion, Butler, Mercer, and Armstrong counties, where the coal seams were shallow. These coals were less costly to mine and remained relatively competitive with other fuels. In contrast, major declining counties, including Allegheny, Fayette, Washington, and Westmoreland, were concentrated in southwestern Pennsylvania. These counties had few shallow seams, and the deep seams were costly to strip mine.

1961–1979

The spatial pattern between 1961 and 1979 reflected changing market and political conditions. The number of underground-producing counties declined from 25 to 12. Production was concentrated in 7 counties, each with more than 1,000,000 tons production. These counties had 88 percent of the output in 1961 and 83 percent in 1979. While underground production increased in 4 counties and declined in 3, only Indiana experienced a major growth, from 4,295,000 to 10,112,000 tons, due to a new electric utility facility. Most significant was the disappearance of the traditional counties of Fayette and Westmoreland as major coal-producing counties. The production of metallurgical coal in Greene and Washington counties maintained production in those counties. In the counties with smaller produc-

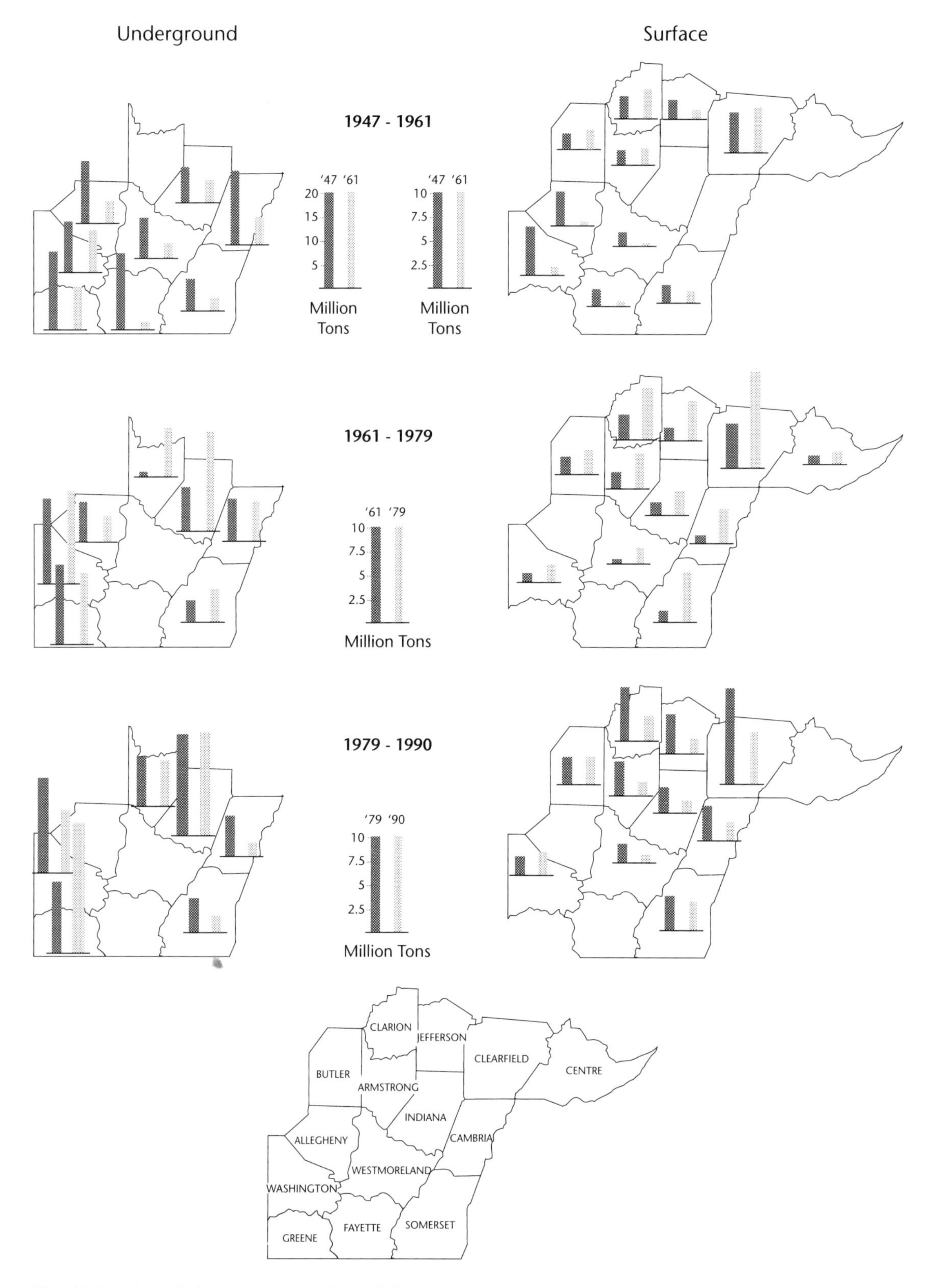

Fig. 13.5. Spatial changes in Pennsylvania's bituminous coal production, 1947–1990.

tion, the small mines could not comply with the new federal regulations and were abandoned.

In contrast to the shrinking pattern of underground production, strip mines operated in 24 counties. Of these, 12 counties had production totals from 1,137,000 to 9,081,000 in 1979. All major counties experienced growth in output between 1961 and 1979. While growth was widely dispersed, the greatest gains in production were on the eastern and northern periphery of the coal region, where the coal seams were near the surface. In the period 1970–1979 Clearfield County increased its output by 3,756,000 tons, and 8 other counties—Armstrong, Cambria, Clarion, Fayette, Indiana, Jefferson, Somerset, and Westmoreland—increased production by more than 1,000,000 tons. The larger earth-moving equipment made it possible to renew strip mining in Fayette and Westmoreland counties. The increase of strip mining, particularly in the 1970s, reflects the ease of bringing a strip mine into production when market demand increases.

1979–1990

Although coal production declined, the total number of coal-producing counties was stable, with 13 having underground production and 25 counties with strip mining in 1990. Although the number of counties was stable, there were significant changes in production within the counties. Seven of the underground producing counties increased production and six decreased. In 1979 the seven counties—Allegheny, Armstrong, Cambria, Greene, Indiana, Somerset, and Washington—with more than 1,000,000 tons of underground production, produced 40,697,000 tons (93 percent) of the total. By 1990 these seven leading counties produced 39,045,000 tons of coal (95 percent of the total). However, within the major counties wide variations were evident. For example, production in Greene County increased from 7,830,409 to 18,430,843 tons, but in Washington County production declined from 9,681,856 to 5,041,079, and in Cambria County from 4,407,109 to 2,177,914. These spatial changes created unemployment in areas of declining production and worker demand in other areas.

Between 1979 and 1990 strip-mine production declined from 45,166,917 to 26,616,353 tons. Twenty counties experienced declining production, while only 5 increased production. In 1979 the 11 counties with a production of 1,000,000 tons or more had an output of 32,091,000 tons (72 percent of the total). By 1990 only 8 counties—Armstrong, Butler, Cambria, Clarion, Clearfield, Indiana, Jefferson, and Somerset—produced 1,000,000 tons or more. The total for these counties was 20,241,000 tons (76 percent of the total). Major declining counties included Clearfield, where production dropped from 9,041,407 to 5,831,437 tons, and Armstrong County, with a decline from 3,253,204 to 1,875,820 tons. Strip mining remained concentrated in the northern and eastern periphery of the coal fields, where seams were closer to the surface and mining was less costly. The largest declines were in the counties that had the largest increase in the previous period. This illustrates how rapidly operations respond to the cost of production and to market demands.

The Anthracite Industry

The northeastern Pennsylvania antracite coal deposits constitute one of the few areas in the world where anthracite is found. Because of the massive deposits lying close to the huge residential market of eastern United States, few areas of the world have been dominated so long by mineral exploitation. During the nineteenth century a total mineral economy evolved.

Location and Geology

The surface area of the anthracite fields is remarkably small, covering only 484 square miles in a 10-county area (Fig. 13.6). Three of the fields occupy valleys or basins (Northern, Western Middle, and Southern fields), whereas the Eastern Middle Field occupies a plateau-like tableland. The Northern Field of 176 square miles extends through Luzerne, Lackawanna, and small portions of Susquehanna and Wayne counties. The largest cities of the Northern Field are Scranton and Wilkes-Barre. The Eastern Middle Field, centering on Luzerne County with extensions into Schuylkill, Carbon, and Columbia counties, has an area of only 33 square miles. Hazleton is the urban center of this field. The Western Middle Field occupies 94 square miles in Northumberland, Columbia, and Schuylkill counties. The largest urban center is Shamokin. The Southern Field is the largest

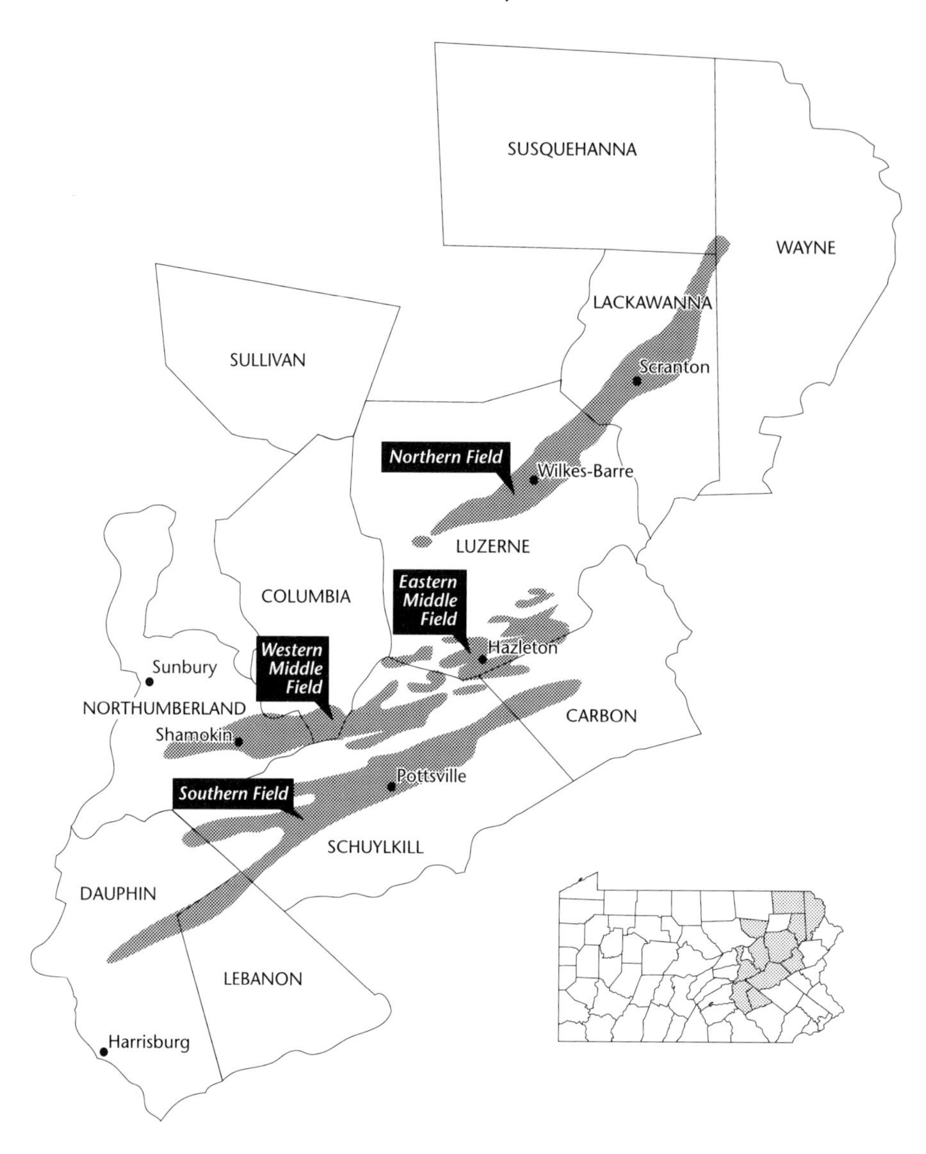

Fig. 13.6. Anthracite fields of Pennsylvania.

in area, occupying 180 square miles. It extends northeast-southwest in Schuylkill, Carbon, Dauphin, and Lebanon counties. Pottsville is the largest settlement.

The Pennsylvania anthracite fields, lying in the folded and faulted Ridge-and-Valley province, are geologically complex. The coal fields have been subjected to intense folding into synclines and anticlines in which thrust faulting is common. The structural complexity increases from north to south, so that in the north the folds are open and symmetrical, while in the south there is tight folding with associated high-angle faults.

More than 200 coal seams, ranging from 1.1 to 19.3 feet in thickness, have been mined in the anthracite fields. Most seams are less than 5 feet thick. The original reserve of coal is estimated at 24.4 billion tons, of which about 9 billion tons have been mined or lost in the mining process. Of the 15.4 billion tons that remain, about 7 billion could be recovered by modern mining technology. Of the reserves, about 6.9 billion tons can be recovered by underground mining. The reserves are, however, not evenly distributed. Mining was initially concentrated in the less geologically complex areas, so that reserves in the Northern Field are largely depleted or have been flooded after mine abandonment. Of the total remaining reserves, about 9.5 billion tons are

Table 13.3. Anthracite production, 1970–1989.

	Total Production	Culm Bank Production	Strip Production	Employees
1870	14,172,004	—	—	35,600
1880	27,974,532	—	—	73,373
1890	44,906,206	—	—	119,989
1900	57,363,396	1,818,170	—	143,824
1910	83,603,994	5,412,167	—	168,175
1917	100,445,299	6,408,227	—	156,148
1920	89,636,032	5,435,697	—	149,117
1930	68,776,559	1,468,572	—	151,171
1940	51,526,404	2,599,280	6,010,364	90,790
1944	64,112,589	10,527,481	10,925,619	78,145
1950	46,339,255	4,027,249	12,291,325	75,231
1960	17,721,113	3,006,803	7,138,743	20,269
1970	9,248,011	2,956,707	4,624,610	6,286
1980	5,983,149	1,945,054	3,454,821	3,429
1989	3,375,315	576,850	2,247,395	2,443

SOURCE: *Annual Reports on Mining Activities*, Pa. Dept. of Environmental Resources.

located in the Southern Field, and an additional 3.5 billion are in the Middle Fields. These two fields have approximately 80 percent of the remaining anthracite reserves.

Anthracite Coal Production Trends

The anthracite industry has experienced two distinct production trends (see Table 13.3). The first period, extending from initial exploitation in the eighteenth century, was characterized by a steady growth in production to a peak of 100,445,000 tons in 1917. Since then, production has declined steadily from 46,335,000 tons in 1950 to 9,248,000 tons in 1970 and 3,414,000 tons in 1990.

Period of Growth

The first recorded use of anthracite was in 1769, when Obadiah Gore used it in his blacksmith shop in Wilkes-Barre. The use of anthracite increased slowly for several reasons: it was difficult to ignite, there was a relative abundance of firewood, and many problems had to be solved in transporting the fuel to market. Annual production did not exceed 1,000,000 tons until 1837.

The smelting of iron ore was the first major market for anthracite. The anthracite furnace was larger than the charcoal furnace, so it assumed a major role in supplying the growing American iron market. By 1841 a dozen anthracite furnaces were in operation in eastern Pennsylvania, and four in New Jersey. Because the market for iron in the anthracite region was limited, the coal moved southeastward toward the iron ore and the ultimate market.

The Lehigh Valley became the center of the new industry. By 1860 more than 1,000,000 tons of iron were produced using anthracite, with more than half of it in Pennsylvania. Anthracite was beginning its 20-year reign as the chief furnace fuel. By the late 1870s coke produced from bituminous coal assumed leadership, although as late as 1915 a score of blast furnaces were still using anthracite mixed with coke.

In the last half of the nineteenth century the use of anthracite for home heating became the dominant market. As the forests were depleted in New England and the East Coast, a new domestic heating fuel was sought. Anthracite was the fuel nearest these expanding markets and, furthermore, was considered a nearly perfect home heating fuel. Because of its high carbon content of 92–98 percent fixed carbon and 2–8 percent hydrocarbons, it produces a large amount of heat (averaging 12,695 Btu/lb) and is nearly smokeless. Its ash and sulfur content is low, making it a clean fuel. Because bituminous was nearly twice the cost of anthracite on the East Coast due to transportation costs, it was eliminated from these market areas.

In order to develop the anthracite market, the first prerequisite was a transportation system, which was initially achieved by a canal system. With the completion of the Schuylkill Navigation Canal and the Delaware & Hudson Canal in 1827, and the

Lehigh Canal in 1829, anthracite became the most important commodity moving on these waterways. By 1846 the canal system of the anthracite region consisted of 643 miles of waterways. Canals greatly reduced shipping costs, giving anthracite a competitive advantage over other fuels in eastern cities.

The canals were superseded by the railroads. The canal companies themselves were among the earliest investors in the anthracite railroads. The first railroads in the anthracite region were built to carry coal to the canals. By the Civil War period, railroads had largely replaced canals for shipping anthracite. Railroad trackage increased from more than 1,000 miles in 1863 to 2,290 in 1873. In 1900 there were 10 railroads providing a dense network in the coal fields, each of them serving a specific East Coast market.

Period of Decline

Many reasons have been offered to explain the decline of the anthracite industry. Of these, the depletion of reserves has often been suggested, because anthracite is a nonrenewable resource and billions of tons have been mined. This type of statement applies only to depletion of reserves at a particular mine or small district, however. Although the effect of local depletion of reserves is severe, for a specific settlement might not have an alternative viable economic base, production of coal from other areas is available to serve all market demands. Local depletion of reserves cannot explain the drastic decline in production, so that other factors must be sought.

It has also been suggested that because anthracite mining is technically difficult due to geological conditions, mining problems have become too great and costs have become prohibitive. Because the anthracite region is an old mining area, with many of the more accessible underground coal seams depleted, and stripping operations have mined the beds nearest the surface, there is little doubt that mining is becoming more difficult. However, a general rising curve of production per worker per day from 2.83 tons in 1950 to about 7.20 in the 1980s indicates that technical problems are being solved. Increasing technological problems do not appear to provide a satisfactory explanation for the decline in production.

A more plausible explanation for the decline in the use of anthracite lies in the increasing availability and consumption of petroleum and natural gas. In 1920 anthracite supplied more than 95 percent of the home heating needs of the area east and north of the Pennsylvania anthracite fields. As competition from petroleum and natural gas increased, anthracite's share of the market declined sharply. By the early 1950s fuel oil and natural gas supplied about 66 percent, anthracite 32 percent, and coke and briquettes about 2 percent of the home heating fuel needs of this area. By the 1980s the demand for anthracite as a home heating fuel had declined to insignificance.

The question immediately arises whether this loss of market in the traditional anthracite marketing area was due to a price differential between competing fuels. A study by the U.S. Bureau of Labor Statistics in 1952 revealed that the price of anthracite for most grades was considerably lower in eastern cities than that of fuel oil or natural gas for an equal amount of heat. In New York, for example, the heat equivalent of fuel oil from one ton of pea-grade anthracite costs $23.49, while the cost of pea-grade anthracite was $19.90. Natural gas had an even greater cost disadvantage over anthracite. For the equivalent of one ton of pea coal, natural gas in New York cost $58.74 and in Baltimore $37.73. When cost differentials did exist against anthracite for selective grades, they were slight. Because cost differentials have changed little over time, cost has not been a major factor when a residential heating fuel was selected.

Two principal factors appear to explain the change in fuel use in heating homes. The first is the convenience in the use of fuel oils and natural gas, the second is the psychological determinants that evolved after 1920. The trend toward the use of more and more laborsaving devices in the household was of the greatest importance. The American home could be heated through no more effort than the delicate adjustment of a thermostat; tons of anthracite did not have to be stored in a basement area, coal shed, or unsightly open pile in the backyard of a home; and homeowners no longer had to shovel coal into the furnace or a stoker and remove the ashes after burning. The anthracite industry, largely through research conducted by the Anthracite Institute, did attempt to develop and promote the automatic stoker well after the decline had begun, but these efforts were totally unsuccessful. An additional factor discouraging the use of anthracite in home heating was the difficulty of igniting the hard coal in the furnace. Because anthracite is nearly pure carbon with little hydrocarbons, it requires a high ignition temperature.

The psychological factors, though more elusive, were no less real. The use of anthracite was no longer considered fashionable. It had become an outdated fuel. When economic factors were of little importance, other considerations played a crucial role. In the mode of the times, "keeping up with the Joneses" required that a "modern" fuel be used to heat a home.

Besides these major factors, a number of lesser causes contributed to the decline of anthracite. Several labor strikes, particularly one lasting 170 days in the winter of 1925–1926, gave the general public the impression that anthracite might be in short supply, or not even available, during the cold winter season. Many anthracite users therefore turned to the use of fuel oil or natural gas. To implement this change, the coal furnace was discarded and a new furnace purchased, or the old furnace was converted at considerable cost to the new fuel. While anthracite was never in short supply due to company stockpiling, fear on the part of the general public created a fear that caused a shift to other energy sources.

In mining anthracite there was frequently a large amount of slate in the original tonnage. The breakers removed some of this slate, but a large quantity remained. When the producing companies had little or no competition from other fuels, the general public was forced to purchase coal in which there were hundreds of pounds of slate in each ton. For many years the coal companies did not recognize the seriousness of the competition from fuel oil and continued to exercise poor quality control in the industry. As a result, the general public gradually lost confidence in the industry's ability to provide a quality product. This sapped the marketing potential of anthracite.

As the market for home heating declined, a number of smaller, auxiliary markets were affected. Originally, essentially all of the coal was marketed by rail transportation. The anthracite railroads used anthracite as a fuel. Because their freight consisted predominantly of coal, railroad consumption of anthracite declined as the domestic market vanished. To aggravate this situation, trucks gradually replaced the railroads in marketing coal. Consumption of coal also declined at the mines as deep mines were abandoned.

Anthracite has not been able to expand its market beyond home heating. A market in manufacturing and public utilities has not developed. In fact, anthracite's use as a fuel to produce cement has disappeared. In electric-power generation, anthracite requires equipment entirely different from that used in bituminous-fired plants. The higher ignition temperatures and longer burning time necessitates a much larger boiler to provide the longer flame path. The boiler sizes for anthracite must be about twice the size of the equivalent-capacity bituminous boilers. Because anthracite is much harder than bituminous coal, the grinding is more costly. While there is little sulfur emission, the control of particulates (fly ash) requires an expensive bag filter system, so the capital expenditure for an anthracite power facility is much greater than that for a plant using bituminous coal. To date, an anthracite plant has not been built, nor is one contemplated for the future. The Sunbury plant of Pennsylvania Power & Light has boilers with a capacity to produce 50,000 kilowatts and does use some anthracite recovered from culm banks, which does not require elaborate crushing equipment. Although billions of tons of anthracite remain, anthracite has played no role in modern-day attempts to develop an energy policy for the United States. Anthracite is not recognized as one of the nation's energy resources.

A Mineral Economy in Transition

The decline of anthracite production has brought great changes, not only in the mineral economy—such as locational shifts of production, decline of employment, corporate changes in ownership, changes in the mode of transportation, and the importance of the export trade—but also in the social and economic conditions of the region.

Locational Shifts in Production

In addition to the decline in total production since 1917, there have been major locational shifts in production (Table 13.4). During the period of peak production, the Northern and Middle Eastern fields, lying in Luzerne and Lackawanna counties, accounted for about 66 percent of production, while the Southern and Middle Western fields, lying predominantly in Schuylkill and Northumberland counties, produced about 34 percent. By 1960 the Northern and Middle Eastern fields were producing

Table 13.4. Anthracite production by Pennsylvania county 1917 and 1990.

County	1917	1990
Carbon	3,687,302	49,329
Columbia	1,563,200	60,286
Dauphin	967,704	113,134
Lackawanna	23,072,582	121,061
Luzerne	37,580,170	876,307
Northumberland	7,889,237	434,736
Schuylkill	23,707,223	1,731,500
Sullivan	598,859	28,614
Susquehanna	652,504	—
Wayne	726,518	—
Total	100,445,299	3,414,967

SOURCE: *Annual Reports on Mining Activities*, Pa. Dept. of Environmental Resources.

only 40 percent of the total, and by 1990 only 22 percent of the total. In contrast, the Southern and Middle Western fields had increased their share to 70 percent, with Schuylkill County producing 51 percent of total production.

The locational shift is due to the changing importance of deep-mine, strip-mine, and culm (refuse) bank production. In 1917 about 94 percent of anthracite production came from deep mines. Deep-mine production was initially concentrated in the Northern and Western Middle fields because of the less complex geology. Deep mining of anthracite has declined as a result of increased costs from deep seams, as compared with the lower costs of strip mining and other types of mining. Consequently, deep-mine production declined steadily, to 63 percent in 1950 and to only 15 percent in 1990. By the early 1980s deep mine production had disappeared from the Northern and Middle fields, and in 1990 only 529,000 tons produced from deep mines came from the Southern and Western Middle fields, with 82 percent from Schuylkill County alone.

Strip mining occurs throughout the anthracite fields, but because of greater accessibility about 50–70 percent of production is concentrated in the Southern and Middle Western fields and only 30–50 percent in the Northern and Middle Eastern fields. Because strip mines can be relatively short-lived, the percentage of total production from a particular field can change significantly from year to year. In 1983, for example, Luzerne County had 46 percent and Schuylkill County had 34 percent of strip production, but in 1990 Schuylkill County had 60 percent and Luzerne 33 percent of total strip production. Because of rapid fluctuations in production, strip mining does not provide a sound economic base for a region.

Culm-bank production became significant when it became economically sound to recover the coal remaining in refuse piles. Because the breakers were inefficient in sorting the coal from the slate, many culm banks are 20–30 percent coal. The remining of the culm banks began after 1900, and by 1930 about 1.5 percent of the anthracite came from this source. Coal from this source increased to about 17 percent in 1960 and to about 40 percent of total production in 1975, but has decreased to 28 percent in 1990. The richest culm banks have been remined, so their importance in total coal production is declining. Because these refuse piles are found throughout the coal fields, production is widely distributed. However, mining operations have depended not so much on the location of the culm banks as on the acquisition of a site that a coal-processing operator believes will be economically feasible. Locational shifts are thus common. Over the years Schuylkill, Luzerne, and Northumberland counties have had the largest bank production, with the percentage from each county varying considerably from year to year.

A minor source of anthracite production has been from river dredging. In the coal fields the proportion of coal in breakers and washeries, and the natural erosion of culm banks, has washed coal into the Susquehanna River system, providing another source of anthracite. The Susquehanna has a number of places where deeply pitching rocks outcrop or are near the surface in the riverbed. Behind these outcrops are calm pools of water where the sediment laden anthracite settled out of the water. The bottom of the river was thus covered by fuel rich deposits, in places several feet thick. Small dredges scooped the anthracite deposits from the bottom of the river. Mining in the Susquehanna occurred as far south as Harrisburg. The maximum production occurred in 1941, when 1,517,563 tons of anthracite were produced by river dredging. This was about 2.7 percent of total anthracite production. In 1960 river-dredged coal totaled 3.8 percent of anthracite production, but output has declined to insignificance with the depletion of the river deposits.

Employment in Mines

The number of miners is closely related to production. The peak employment year was 1914, when 179,679 miners were employed. There has been a steady decline, with a few periods of drastic reduction when production dropped precipitously. Between 1931 and 1932, for example, employment dropped from 139,431 to 121,243. By 1990 total employment had declined to 2,258 miners.

The employment decline was greatest in the underground mines. Between 1914 and 1990 employment in the deep mines declined from 180,899 to only 453. Strip mining's labor force in 1990 was only 784 miners (34 percent of the total). Culm-bank employees totaled 368. In addition, there were 653 employees in breakers and washeries.

Until the 1930s, mining provided the primary basis of employment in scores of settlements that had been built around a deep mine, but this has completely changed. There is not a single settlement in the anthracite region where mining remains the dominant activity. These traditional mining settlements are now primarily "bedroom communities," with the workers commuting to the large urban areas. Settlements have not developed at strip-mining sites. Strip mines exist for short periods of time and employ relatively few miners. When coal is exhausted at a site, equipment is moved to a new location. Strip mining is thus migratory, and the miners travel to each new operation.

Corporate Reorganization

With the decline of production, a new corporate structure has evolved. During the nineteenth and early twentieth centuries large-scale corporations, in which mining and transportaton were interrelated, controlled coal production. These corporations had headquarters in such cities as Philadelphia and New York. In 1917 eight companies, each with a production of more than 5,000,000 tons, produced 71,369,000 tons, or 71 percent of the total for the year. The Philadelphia Reading Coal & Iron Company produced 14,946,000 tons of coal, followed by the Delaware, Lackawanna & Western Railroad, with 12,885,000 tons. Of a total of 128 anthracite companies, only 35 produced more than 300,000 tons each. The small companies sold their coal locally or in nearby markets.

As coal production declined, the old-line corporations drifted into receiverships, and small owner-operators bought the coal lands and equipment. By the 1970s not a single corporation of the early twentieth century produced coal in the anthracite fields. A U.S. Department of Energy official has characterized the present ownership situation in the anthracite region as mere cottage industries. In 1990 there were 248 mines operating, of which 87 were deep mines, 101 were strip mines, and 60 were bank mines. In addition, there were 70 breakers and washery plants.

Production of the companies in 1990 varied from about 100 tons to a maximum of 328,970 by the Veddo Highland Coal Company of West Pittston. Because the companies are small, there are not only many failures but also many new companies organized each year. For example, the four leading companies in 1987 were no longer producing in 1990. Most companies do not have the resources necessary for modern-day operations.

Transportation Evolution

In 1917 some 88,147,050 tons of coal, 100 percent of the anthracite marketed outside the region, were transported by railroads. The remainder, 1,228,249 tons, was consumed in local trade or used at the colliery. The railroads remained the principal carriers of coal until the 1950s, when trucks became increasingly important. By 1990, of a total of 3,268,000 tons marketed, outside the region trucks carried 84 percent, railroads carried 14 percent, and barges transported 2 percent on water routes. About 4 percent of production was consumed in local trade and at mines for steam and heat.

A number of factors contributed to the changes in the method of transporting coal. As deep-mine production declined, there was not sufficient production to make up a train load of coal. Haulage of other types of freight did not develop in the coal settlements. As a result, the railroad network deteriorated and the tracks to the mines were abandoned. As strip mining increased in importance, railroads did not build lines to these mines. The life of these mines was too short to justify the capital outlay. Consequently, the coal had to be hauled from the mine by trucks. In the initial period the trucks moved some of the coal a short distance to railroad loading terminals, but as the road system improved and trucks increased in size, producers found that a cost advantage existed when trucks moved the coal to its final market destination.

Export Markets

Anthracite has had a traditional market in eastern Canada. The cold winters and lack of an adequate supply of domestic fuel provided the basis of the trade. In 1917 some 5,917,000 tons of anthracite, or

98 percent of total exports—about 6 percent of the 1917 production—were shipped to Canada.

As the domestic market declined, the export market became more significant, rising from about 8 percent in 1960 to a high of about 45 percent in 1981. As oil prices declined in the 1980s, anthracite export has also declined. In 1990 only 8 percent of anthracite production was exported. Canada remains the largest importer, but its importance has declined. Canada now takes about 50 percent of the exports. The remainder is largely sent to the western European countries, and a small percentage goes to Latin America and Japan. Much of this coal is consumed in the American bases overseas. But exports are so small that they are of no significance in maintaining anthracite production.

Economic and Social Conditions

With the decline of the anthracite economy, the economic and social structure of the region is in transition. These changes have occurred relatively gradually, for the decline of coal production did not occur precipitously. Miners and coal companies long shared the hope that the decline was temporary and that the coming years would be better, but this has not occurred.

A major indicator of change is the decline in population. The population of the four major anthracite counties—Lackawanna, Luzerne, Northumberland, and Schuylkill—reached a peak in 1930 with a total of 1,119,515. In each of the censuses since, the population of the region has fallen, reaching a low of 796,544 in 1990. Many of the small mining settlements have virtually disappeared, but the decline in the major cities has been most significant. The population of Scranton, the largest city, has declined from 143,433 in 1930 to 81,805 in 1990, and that of Wilkes-Barre during the same period declined from 86,626 to 47,523. The decline of population reflects the erosion of economic opportunities over an extended period of time.

The lack of economic activity is reflected in many measures in the census reports. In 1986 the U.S. Department of Commerce revealed that the average per capita income in the anthracite counties was among the lowest in the state. The average per capita income in 1984 for Schuylkill County was $5,172, for Luzerne County $6,609, for Northumberland $5,683, and for Lackawanna $7,045. By contrast, average household income in the noncoal counties of southeastern Pennsylvania was: Lancaster, $7,954; York, $8,231; Cumberland $10,492; and Montgomery, the highest in the state, $13,270.

The houses of the anthracite region are considerably older, on the average, than those in the rest of southeastern Pennsylvania. In the four anthracite counties, 60 percent of houses were built before 1939, compared with 35–40 percent in other southeastern counties of Pennsylvania. The median value of the houses is also among the lowest in the state. In 1990 the median value of a house in Northumberland County was $39,500, in Schuylkill $38,200, in Luzerne $56,000, and in Lackawanna $68,900. In nearby noncoal counties the median values were $89,400 in Lancaster County, $85,000 in Cumberland County, and $155,900 in Chester County. About 20 percent of the houses in the anthracite counties have one or more individual air coolers, compared with about one-third of the houses in nearby noncoal counties.

When anthracite production was rising, the economy was overwhelmingly dominated by coal. In 1924, for example, the total value of production from mining and manufacturing in Luzerne County was $277,058,000, of which anthracite accounted for $198,501,000; in Lackawanna County the figures were $173,348,000 and $109,780,000, respectively. Anthracite did not provide the basis for a manufacturing economy similar to that of the bituminous coal industry.

The male population worked in the mines. By tradition, no woman could work in a mine. To utilize the available female labor, textiles became the first major industry in the region. It was a "parasitic" industry, taking advantage of the unused labor supply paying minimal wages. Girls and women provided between 80 and 90 percent of all textile employees, which represented 50 to 60 percent of total manufactural employment. Tobacco and leather, also parasitic industries, were much less important, employing from 3 to 8 percent of the region's workers. A number of other industries existed to serve the coal industry. Powder and explosives were the most important products of the chemical industry, and the fabricated metal product industries were primarily engaged in producing equipment for the mines.

With the decline of mining, manufacturing was not attracted to the region. Despite high unemployment, coal miners were reluctant to accept other jobs. The miner was a highly skilled individual, and factory work did not provide equally skilled opportunities. The miners were unionized, and average wages were higher in mining than in manufacturing. Manufacturers were not attracted to a high-cost labor area. As the miners lost their jobs, the women of the region earned sufficient money to maintain the household, and a large number of miners remained outside the labor force. However, as anthracite production declined, the texitle industry also declined. Many textile plants closed or migrated to lower-cost production areas in the southeastern states. As a result, for example, the textile industry, which had provided 57 percent of all manufacturing employment in Luzerne County in 1924, declined to 40 percent in 1950, and to less than 5 percent in 1990. Similar declines occurred in all the anthracite counties.

With the decline in textile manufacturing, the employment of women shifted to the apparel industry. Because this industry required more labor and thus had a higher value-added in the process of manufacturing, it could absorb slightly higher labor costs. The apparel industry provided increasing employment from the 1930s until the mid-1960s, when the peak was reached. During that period employment in the apparel industry in Luzerne County increased from 6 percent to about 40 percent of the total. The industry concentrated on such low-fashion items as overalls, shirts, and blouses.

Since the mid-1960s the apparel industry has declined in importance with the growing importation of clothing from Taiwan, Hong Kong, South Korea, and many other foreign sources. Between 1969 and 1989 employment in the apparel industry in the four leading anthracite counties—Lackawanna, Luzerne, Northumberland, and Schuylkill—fell from 45,538 to 17,073. The decline of the apparel industry, along with the continued decline of the textile industry, has been the major factor in the decrease in total manufacturing employment in these four counties from 124,961 in 1969 to 83,327 in 1989.

Although the textiles and apparel industries declined from 48.4 percent of manufactural employment in 1969 to 27.2 percent in 1989, no new significant industries have entered the anthracite region in the past two decades. Food and kindred products is the third largest industry, with 9.1 percent of employment, followed by fabricated metals, with 7.7 percent of employment, and electric and electronic products, with 6.6 percent. In spite of considerable regional efforts to attract industry through special inducements, the results are not encouraging.

While employment in mining has nearly disappeared and manufacturing has declined significantly, the tertiary industries—transportation and other public utilities, wholesale and retail trade, finance and insurance, and services—have exhibited a dynamic growth in employment. Employment in the tertiary industries has risen steadily since 1965, from 96,988 to 127,130 in 1980, to 152,300 in 1989 in the four leading anthracite counties. Between 1965 and 1989 employment in service industries has more than doubled, rising from 25,329 to 59,585. The greatest growth was in the health services.

The tertiary industries have brought a considerable degree of stability to employment in the anthracite counties. The population now appears to have stabilized, unemployment has declined, and job opportunities are expanding. A region normally progresses from a primary to a secondary and finally to a tertiary economy. It appears that the secondary phase of manufacturing has been partially bypassed in the anthracite region, with the economy evolving from one dominated by primary activities to one in which tertiary activities provide the major growth potential in the future.

The Coal-Mining Industry and the Environment

The major goal of coal mining has traditionally been to meet the energy requirements of the nation. Little consideration, if any, was given to addressing and solving the environmental problem created by the mining process. When Pennsylvania was striving to become an industrial giant, there was no attempt to reckon the environmental costs of economic progress. In the nineteenth century, many modern signs of environmental degradation that are no longer tolerated were considered signs of economic prosperity. As a result, the vast problems of today were spawned. Four environmental problems—reclamation of strip-mined land, acid mine drainage, mine fires, and surface subsidence—are critical in the mining of coal.

Strip-Mine Land Reclamation

Strip mining, the process of removing an overburden of rock from an underlying coal seam, began about 1915 in Pennsylvania. Because tremendous quantities of earth must be moved to reach the coal, strip mining became important only after the development of large power shovels in the 1930s. In the mining process, first bulldozers remove the weathered surface materials, and then power shovels or draglines remove the bedrock. The largest of these draglines in Pennsylvania has a capacity of more than 85 cubic yards and can move overburden to a depth of several hundred feet. After the overburden is removed, the coal is mined by smaller shovels and loaded into trucks or, rarely, into railroad cars.

The advantages of strip mining over underground operations are many. In strip mining, the accident rate is much lower than in underground mining, especially with regard to fatalities. As a result of minimum danger in open-pit operations, insurance for a surface miner is significantly lower than for underground miners. From a coal recovery perspective, strip mining is far more efficient. One acre of coal 3 feet thick will produce by stripping operations approximately 5,000 tons, whereas in deep mining the average recovery will be 3,300 tons or less. About 80 to nearly 100 percent of the coal can be recovered by surface mining, whereas underground mines recover only 40 to 60 percent of the seam being mined. The profits of a strip-mining operation will depend on the size of the equipment used and the amount and type of overburden to be moved. In strip mining the most expensive piece of equipment is the dragline, which will cost more than $10 million, but the salvage value is greater than for underground equipment. In addition, several million dollars' worth of other equipment, such as small shovels, bulldozers, and trucks, are required. In strip mining the time between initial investment and full production is comparatively short. An underground mine may require two to four years of preparatory work before production begins.

The amount of stripped land varies greatly from county to county (Fig 13.7). In the early 1970s aerial photos of Pennsylvania were taken and tabulated by the U.S. Geological Survey's Geographic Information Retrieval and Analysis System. At that date, stripped land constituted only 1.3 percent of the state's total area. In the anthracite fields the largest amount of stripped land was in Schuylkill County, where 10.2 percent of the area had been stripped, followed by Lackawanna County, with 5.4 percent (Fig. 13.8). In the bituminous fields, Clearfield County led with 9.8 percent of its area strip mined

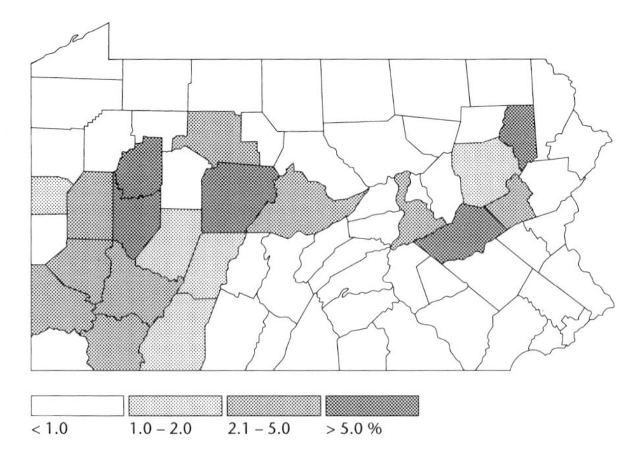

Fig. 13.7. Stripped land as a percentage of all Pennsylvania land.

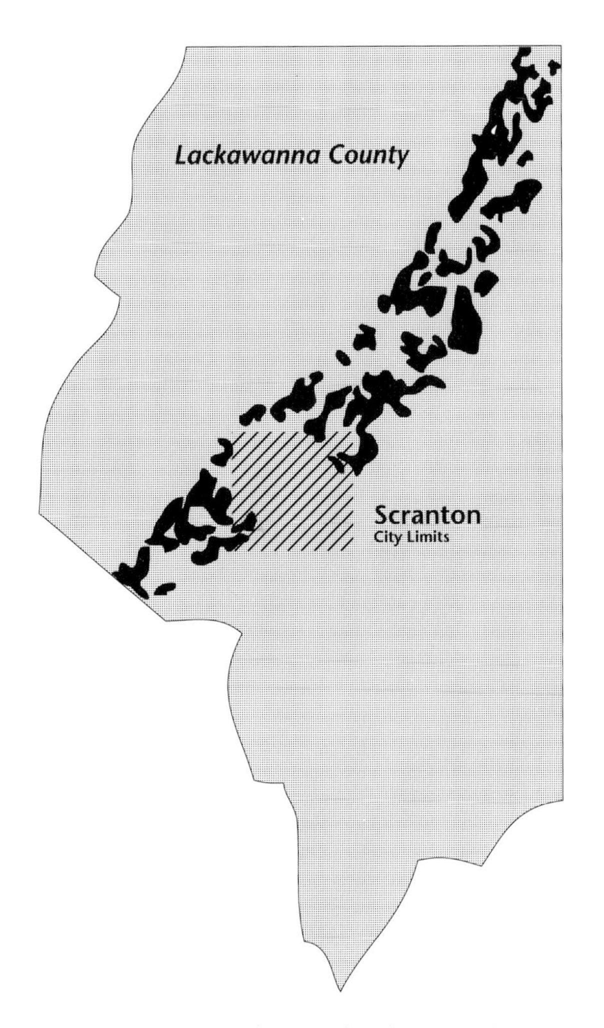

Fig. 13.8. Strip-mined areas of Lackawanna County.

followed by Clarion with 6.8 percent, Armstrong 5.4 percent, Lawrence 3.5 percent, Somerset 2.7 percent, and Indiana 2.5 percent. Since the 1970 land-use survey, strip mining (and the resulting stripped land) has increased greatly, but much of the land has been reclaimed as mandated by law.

While the total percentage of land stripped in any country is relatively small, large tracts of land in close proximity to the area stripped are affected. In the hill country of western Pennsylvania, the coal seam frequently outcrops on the side of a hill. Strip mining follows the contour of the hill and proceeds into the hillside until the overburden becomes too thick for economical operations. Vast areas are thus scarred by these narrow, mined strips of land.

To the general public, the most striking feature of strip mining is the unsightly spoil banks that mar the scenic beauty. During the early days of strip mining, the prevailing attitude among landowners was that land in Appalachian Pennsylvania had little economic value and therefore did not justify the cost of reclamation. This attitude was fostered by a number of considerations. Because the abandonment of farmland began in western Pennsylvania about 1890, long before strip mining began, there were no demands for agricultural land. As farmland was abandoned, farmers began to allow their land to be strip-mined in order to secure a windfall profit, frequently at the time the farmer retired. Although trees are the natural vegetation of western Pennsylvania, there was little interest in reforestation. The strip-mined lands were dispersed and occupied little area so that a large forest stand was difficult to achieve and large lumber companies were not attracted to the region. Finally, the rugged coal-mining regions were isolated, so the tourist industry was of minor economic importance. As a consequence, few outsiders complained about the loss of the aesthetic value of the region.

The first legislation to force strip-mine reclamation in Pennsylvania was enacted in 1945. The Pennsylvania Bituminous Coal Open Pit Mining Conservation Law, one of the earliest reclamation laws in the nation, required that a coal-mining company deposit a filing fee of $100 for each stripping operation and in addition post a bond of $300 per acre to be stripped, with a minimum of $3,000. This legislation required each strip-mine operator to cover the exposed face of the unmined coal within one year after completion of mining and to level and round off the spoil banks sufficiently to permit the planting of trees, shrubs, or grasses to the specifications of the state Department of Forest and Waters. If the operator did not comply with these regulations, the posted bond was forfeited. Although the law appeared to provide the basis for regulating the reclamation of strip-mined land, it was essentially ineffective. Forfeiture of the bond was not a sufficient penalty to encourage land reclamation—land reclamation cost an estimated two to six times the posted bond. Furthermore, the strip-mine operator had little interest in land reclamation, and there was no legal means of preventing the mining company from securing another concession to strip a new area.

As a response to strong environmental pressure, the Pennsylvania Surface Mining Conservation and Reclamation Act was passed in 1963 and revised in 1968, 1971, 1972, and 1974. The present act is recognized as a model law for the control of strip mining and was used to formulate the national strip-mining law of 1977. The present Pennsylvania law requires that the mining company secure an operator's license and a mining permit, post a bond, and provide a reclamation plan before strip mining can begin. If the stipulations of the law are not fulfilled, the bond is forfeited and the company cannot obtain a permit to strip another area.

Reclamation of the stripped areas in Pennsylvania is difficult. In hill country when the land is restored to its original slope, erosion can be a serious problem before a vegetation cover develops. Much of the overburden is shale or sandstone. In the weathering process, sandstones are mechanically much stronger than shales and in the overburden form a jumble of boulders that resist weathering. When limestone or limey shales occur in the overburden, they tend to neutralize the acidic waters and thus aid vegetational growth.

The spoil banks usually contain some materials that retard plant growth. Pyrite, an iron sulfide, is commonly found near the top of a coal seam, so when the earth is overturned in the stripping operation, pyrite is frequently found near the surface of the spoil bank. When exposed to air and water, these particles ultimately form iron oxides and sulfuric acid. The acid, though hastening the weathering process, hinders plant growth. Soluable alumina, formed by the reaction of acid on shales, likewise is harmful to plants. When pyrite is abundant, it requires a much longer period of weathering to obtain a satisfactory vegetational cover.

The quantity, quality, and distribution of the spoil-bank waters are also important factors in revegetation. Acidic waters will slow growth and may kill vegetation. The irregular surface of the leveled areas may have numerous clay depressions when water collects for long periods after rains. In these places newly planted trees may be covered with water for several weeks and are usually drowned. During dry periods these tracts become hard pans, and trees are killed by drought conditions.

In both the anthracite and the bituminous regions, revegetation of the spoil banks has traditionally been achieved by planting trees. The red, shortleaf, and pitch pine, Norway and white spruce, and hemlock are commonly used. The shortleaf and pitch pines are particularly suitable for sandstone and acid shales. If the soil is highly acidic, Japanese larch is one of the more acid-resistant trees to plant. If a quick vegetational covering is desired on moderately acid soils, red oak and black locust are particularly satisfactory. In Pennsylvania most plantings have been of a single species.

Acid Mine Drainage

The presence of acid water draining from coal seams was recognized as early as 1803 when T. M. Morris noted that springwater issuing from hills with coal seams was "so impregnated with bituminous and sulfurous particles as to be frequently nauseous to the taste and prejudicial to health." Acid mine drainage began with the first mining operation and today an estimated 80 percent of the acid mine drainage pollution comes from abandoned mines. Within the bituminous region, acid mine water is coincident with the mining of coals from the Allegheny, Conemaugh, and Monongahela groups of the Pennsylvania and the Waynesburg formations of the Permian. These formations contain the major coal seams mined in Pennsylvania.

Pennsylvania has the most widespread acid drainage problems of the Appalachian coal fields. An estimated 2,750 tons of acidic waters enter the state's streams daily (1,003,750 tons annually), or about 45 percent of the estimated 6,000 tons of mineral acidity entering Appalachia's waters daily. In Pennsylvania about 2,300 miles of streams are polluted with acid waters. These include major portions of such rivers and their tributaries as the Monongahela, the Allegheny, the Conemaugh, the Kiskiminetas, and the Clarion of the Ohio River system and portions of the Susquehanna and Schuylkill rivers.

Acid mine drainage results in the deterioration of surface and ground water systems due to a lowering of the pH, reducing the natural alkalinity and increasing the total hardness with the presence of iron, manganese, sulfates, alumina, and other dissolved and suspended solid concentrations. The quantity of mine-drainage pollutants produced from either surface or underground mining depends on such conditions as whether a mine is active or inactive; the hydrologic, geologic, and topographic features of the surrounding terrain; the type of mining method employed; and the availability of air, water, and iron sulfide minerals.

Acid water from underground mines is normally at a constant flow. In contrast, the discharge from surface mines is often intermittent, generally occurring during and after periods of precipitation. Runoff in strip-mined areas may flow directly into surface streams or may be trapped in a depression and reach a stream only during high water periods. These concentrated "slugs" of acid are particularly damaging to aquatic life in streams. Between flush-out periods, the pools often drain slowly into the backfill to emerge as acid seepages downslope from the mining operations. Another major source of acid drainage is from culm banks and refuse from coal preparation plants.

The problems of acid water in rivers and lakes have been widely studied. The flora and fauna are reduced and sometimes eliminated by acidic conditions. Each fish species is vulnerable to a different level of acidity. When the pH level falls below 6.5, brook and rainbow trout disappear. The females lay eggs, but they do not hatch in acid water. When the acid level reaches pH 5.0, smallmouth bass, walleye, and lake trout have no future, and finally at about pH 4.0 (acidity 1,000 times stronger than neutral water) only chubbs, rock bass, and herring survive. Besides the depletion of the fish community, acid water affects the entire stream ecosystem. Algae and higher plants are reduced to a few acid-tolerant forms. Sensitive macroinvertebrates, including the majority of mayflies, stoneflies, and caddisflies, are unable to tolerate the altered chemical environment.

The quality of water is of major importance in its utilization. Acid water restricts recreational uses. A number of communities depend on streams in the mining region. Acid drainage can destroy this source of water. Water quality is also a factor in industrial location. Acid conditions reduce the usability of water in many manufacturing processes. Acid pollution of streams is one of the major problems in terms of severity of damage for recreational, human, and industrial use.

The attempts to control acid mine drainage was limited until the passage of the Pennsylvania Clean Stream Act of 1966 and the Federal Water Pollution Control Act of 1972, and as amended in 1977. Commonly known as the Clean Stream Act, the federal act has as its basic objective "to restore and maintain the chemical, physical, and biological integrity of the nation's waters." The major goal is to eliminate the discharge of pollutants into streams.

The Pennsylvania Clean Stream Act requires that discharged mine waters have a pH of not less than 6.0 or greater than 9.0, and that the water not contain more than 7.0 mg of dissolved iron. The law also requires mine operators to submit an application for a drainage permit for their proposed operations to the Department of Environmental Resources indicating the method of controlling acid drainage during mining and how acid discharges will be prevented after completion of mining. Penalties, such as forfeiture of bond funds, can be applied if acid mine waters are discharged. There has thus been fairly effective control of acid water discharges from operating mines in recent years, but there is no control of acid discharges from long-abandoned mines. Although technology is available to control acid waters by such technologies as neutralization, reverse osmosis, or ion exchange, the processes are costly. A number of pilot plants have been built in Pennsylvania to test each of the technologies. Modern estimates indicate that several billion dollars are required to clean up the Pennsylvania streams alone. The original goal of the Clean Stream Act to have pure streams within 15 years of its passage was not realistic. It will be many decades before the problem of acid mine drainage is solved due to technological and economic difficulties.

Mine Fires

Coal-mine and culm-bank fires are a major problem in the coal regions, but the magnitude of the fire problem in the coal regions is not easy to assess. In 1977 the U.S. Bureau of Mines recorded 261 known fires burning in virgin coal outcrop and abandoned underground mines in 16 of the coal-bearing states. Of the 42 fires reported in Pennsylvania, 12 were in the anthracite region and 30 were in the bituminous fields. In 1964 the Bureau of Mines reported 495 culm-bank fires in the United States, of which 26 were in the Pennsylvania anthracite region and 48 in the bituminous fields.

A number of these fires were started by spontaneous combustion, but most resulted from the burning of rubbish near mines and culm banks. Once a coal seam or culm bank has been ignited, it is extremely difficult to extinguish. Sufficient ambient oxygen is available for smoldering and burning to continue until all the fuel is exhausted. Underground mine fires sometimes smolder for years unnoticed, feeding on vestigial coal seams. Eventually these noxious and poisonous gases break through the surface.

Since 1965, when the program Operation Scarlift began in Pennsylvania, there have been serious attempts to extinguish the fires in culm banks. To put out a fire in a surface culm bank requires the removal of the burning hot spots. High-power water guns have become the principal equipment in extinguishing the blazes. To reach the burning materials, thousands of tons of material must be moved by draglines.

The largest burning culm banks in the United States are in the anthracite region and the oldest mining areas of Appalachia. In the past 20 years at least 33 burning, abandoned refuse piles have been eliminated in Pennsylvania. The Huber Bank at Ashley covered 97 acres, and the Glen Burn Bank at Shamokin was hundreds of feet high and a mile long. In the Morvine Bank north of Scranton, nearly 2.5 million tons of smoldering, fiery culm were removed over four years of reclamation operations.

Control of underground fires is more difficult. The normal technique is to seal the mine, shutting off the supply of oxygen, then fill the mine with refuse materials. This process has proved successful in controlling a number of fires, including one of the largest underground fires, known as the Cedar Avenue project in Scranton. Operation Scarlift extinguished the fire, and Cedar Avenue has once again become a major thoroughfare and rehabilitated area of the city.

The largest uncontrolled mine fire in Pennsylvania exists at Centralia, near Mount Carmel, in the southern anthracite field. It started in 1962 when a refuse fire ignited underground coal seams. The fire was ignored for many years in the belief that it would burn itself out, but the fire persists under

Centralia in an area of 14 or 15 abandoned coal seams. The old mine shafts and tunnels act as perfect chimneys, and the supply of coal and mine timbers may be sufficient to maintain the fire for hundreds of years.

For more than two decades, the fire has burned beneath the 1,200-inhabitant town. This situation has gradually created devastating conditions. Smoke rises ominously from natural vents over the burning shafts and tunnels, and the stench of sulfur dioxide permeates the air. The ground temperature in some places exceeds 1,000 degrees Fahrenheit, so hot that thermocouples burn off and paper falling on the ground turns instantly to white ash. Unexpected cave-ins occur throughout the area, causing walls and backyards to collapse. At night the town becomes a Dantesque landscape of red-glowing hot spots. Highly combustible hydrogen gas and deadly carbon monoxide gases are a constant problem. Many homes were equipped with carbon monoxide detectors that sound a warning when the CO level reaches 35 parts per million. In other homes, residents resorted to the traditional miners' warning of potent gases by keeping a canary.

All modern technology has failed to control the fire. In the more than two decades the state and federal governments have spent more than $7 million in 16 attempts to contain or extinguish the fire. In 1983 there was a proposal to tap the energy created by the fire to produce electricity. It was estimated that by using 17 different sites in Centralia, 430 billion kilowatt hours of electricity could be produced. Accelerating the fire to such a magnitude would increase the likelihood of cave-ins and underground explosions. The technological dangers were considered too great and the plan was abandoned. The plan now is to isolate the fire by building a fire wall 450 feet deep, 500 feet wide, and three-quarters of a mile long. This plan abandons Centralia but protects such nearby communities as Mount Carmel.

The removal of the families from the town began in 1981 when the U.S. Bureau of Mines, responsible for mine fires and other problems resulting from bad mining practices, purchased 27 homes that were located in the "impact area." In October 1983 Congress allocated $42 million to move the residents of Centralia, but no definite timetable was established. By August 1987 only 37 families remained of the 500 originally in the fire impact area. In 1992 final notices were given to abandon the town.

Although the removal of people from their homes has been difficult, the lives of the people of this community have been disrupted in many other ways. Families and friendships have been torn apart by arguments over moving. As one longtime resident put it, "We want to be buried here. There is no amount of money they could give me for the blood, sweat, and tears that went into building this house." Many believed that their health was affected, with symptoms ranging from allergies, nausea, dizziness, bronchitis, and heart disease to a drug dependency. Psychological problems are increasing. A state-financed "stress center" does a brisk business at its counseling sessions. Regardless of the reluctance to move, the town will disappear.

Mine Subsidence

In underground mining, the removal of vast quantities of coal and rock creates a void that can under certain conditions result in subsidence at the surface. The problem is exacerbated if several seams of coal that lie on top of each other are mined. As the roof of the mine falls into the void, cracking and caving of the overlying rock progresses upward. Earth movement at the surface may result in many different types of damage. Building foundations and walls may be cracked or displaced. Railroad tracks and roads may subside or shift out of alignment.

Subsidence damage is normally greater in urban areas than in rural areas. In the anthracite region, many towns were built on top of underground mines. In this area, entire blocks of buildings have been affected. Forestlands are little affected by mine subsidence, and damage to croplands is minimal. The greatest effect may be the loss of groundwater sources. Water from wells and springs may be diverted, and sources could disappear. The water table may be lowered, affecting the supply of water. There may be permanent changes that require adjustment to a different water regime.

Subsidence may begin as soon as deep mining occurs, so the problem in Pennsylvania is long-standing. To provide some relief to today's resident in the coal regions, the Anthracite and Bituminous Coal Mine Subsidence Fund was established by the Pennsylvania legislature in 1961. Any individual can purchase insurance giving protection against mine subsidence. In 1982 some 1,650 new policies were written, bringing the total policyholders to 17,500 providing total insurance coverage to $645 million. Forty-eight damage claims were paid in 1982 totaling $281,707.

To control mine subsidence, the Pennsylvania legislature passed the Mine Subsidence and Land Conservation Act in 1966, which provided protection for public buildings, churches, hospitals, schools, any dwelling used for human habitation, and all cemeteries and burial grounds. The Bureau of Mining and Reclamation has established guidelines to control subsidence. In underground mines, 50 percent of the coal must be left in place in uniformly distributed pillars that cannot be smaller than 6.1 by 9.1 meters (20 by 30 feet). No mining can occur if the overburden is less than 30.5 meters (100 feet), and pillars cannot be extracted between two support areas when the distance between them is less than the support cover.

This act has largely controlled subsidence in modern-day underground mining. However, billions of tons of coal were mined before the passage of the 1966 act. To solve the traditional problems, some attempts have been made to fill the underground voids. Red ash, a nonflammable residue from culm banks, is commonly used, as is fly ash, the waste from electric-power generating stations. Sand is also utilized. These types of filling processes are not only costly but also reduce subsidence no more than 50 percent. In the late 1960s 11 mine filling projects in the anthracite fields cost more than $7 million. Because of prohibitive costs and limited success, this procedure has been essentially abandoned in Pennsylvania—it is less costly to pay damages through the Subsidence Fund as subsidence occurs.

The Petroleum and Natural Gas Industry

The world's petroleum industry began along Oil Creek at Titusville, Pennsylvania, on August 27, 1859, when "Colonel" Edwin L. Drake drilled a well and discovered oil at a depth of 69 feet. Because the discovery came at a time when industry was seeking a better lubricant than whale oil and there was a growing world demand for a better illuminant than candles, the early oil industry grew rapidly. In the period from 1859 to 1874 an estimated 10,500 wells were drilled in Pennsylvania, of which only 3,250 were productive. Within this 15-year time span the oil region of Pennsylvania was defined, extending from the western New York border to West Virginia and Ohio on the southwest. Pennsylvania's oil production rose to a peak of 31,500,000 barrels in 1891 and then gradually began its initial decline.

With the discovery of oil, a totally new industry came into existence. Not only did drilling techniques have to evolve, but new transportation, refining, and marketing systems had to be created. By the 1890s it was recognized that the oil fields were being depleted and oil men, who a few years before produced oil with no regard for conservational practices, now realized that the future looked bleak. One of the greatest contributions from Pennsylvania to the oil industry was the realization that an oil field could be rejuvenated by means of flooding the oil sands.

Because most of the oil fields were extremely small, they did not lend themselves to secondary recovery of oil by water flooding. Only the Bradford field in McKean County, with its large area of uniform oil-producing sand, lent itself to this method. The introduction of water into one well that increases the production of other wells was probably done accidentally. In some depleted wells the iron casing was removed without proper plugging of the water horizons, and in others the casing and tubing corroded, admitting fresh water from shallow horizons to the producing sand. By 1905, however, it was recognized that water flooding was an effective way to increase production. At this time the laws of Pennsylvania required that all abandoned wells be plugged to prevent water from entering the oil sands. As a result, producers who practiced flooding did so illegally. By 1921 flooding had become so prevalent, and so much political pressure was exerted, that the Pennsylvania legislature passed a special act legalizing the practice when applied to the Bradford Third and certain other specified sands.

After experimentation, it was found that a five-spot system of flooding in which the center well was the water injection well, and four wells around the injection well were producers, gave the highest yield of petroleum. Production in the Bradford field rose from about 3,000,000 barrels in the early 1920s to a second peak of 17,000,000 barrels output in 1937. This was more than 90 percent of the state's total production. Petroleum production fell steadily, to a low of 2,564,485 barrels in 1978. Due to the high price of oil new wells were drilled, and production increased to 4,241,673 barrels in 1984 but declined significantly to 3,419,667 barrels in 1986.

A number of regional trends are occurring in the oil region. The Bradford field of McKean County long dominated production due to secondary recovery effort. As the Bradford field oil sands near exhaustion, production in this field has declined from 1,224,384 barrels in 1977 to 664,717 in 1986. During the same period the number of producing wells decreased from 12,734 to 6,894. In contrast, production in nearby Warren County rose temporarily, from 458,641 barrels in 1977 to 1,295,296 in 1984, but declined rapidly to 759,919 in 1986, and in Venango County from 903,016 to 577,474. The largest percentage increase occurred in Elk County, where production rose from 19,226 barrels in 1977 to 710,545 barrels in 1986. Fifteen additional counties had small increases in production. The average production per well increased from 0.26 barrels to 0.59 barrels per day in 1984. Production per well ranges from about 0.1 barrel to rarely more than 3.0 barrels per day. The smallest "stripper wells" are being abandoned, as shown by the decline in total number of wells from 28,400 in 1977 to 14,271 in 1986.

A number of factors have favored continuation of the tiny Pennsylvania oil industry. Because Pennsylvania crude oil produces the finest lubricating oil, it demands a premium price. Such trade names as Quaker State, Pennzoil, Kendall, and Valvoline are known throughout the world. Small refineries exist in the region to process local production, but because of small production, refineries must import oil from other oil areas. Once a well is drilled and placed in production, maintenance costs are minimal and production can continue even if it is only a tiny output each day. The landowners also encourage continued production, for most receive royalties of one-eighth of production. In a depressed economic region, the small oil income is a welcome bonus. Finally, proved reserves total more than 55,000,000 barrels. Small oil production will continue for many decades.

Natural gas is more important than oil production to the local economy. As demand for natural gas has increased, production has risen from an average of 85 billion cubic feet per year in the late 1960s to about 120 billion cubic feet per year in the early 1990s. Because most natural gas deposits are deep—some more than 10,500 feet—drilling costs are high and exploration and exploitation are essentially limited to large companies with major assets. Unfortunately, many natural gas deposits are found in "pockets," which means that although initial production may be significant the decline in output is rapid. Furthermore, a consequence of the erratic distribution of natural-gas-producing sands is that one well may be a major producer while another well drilled only a short distance away may have no production at all.

Nonmetallic Minerals

Pennsylvania is a major producer of nonmetallic minerals, with a value in 1989 of $671,906,000 (Table 13.5). The nonmetallic mineral deposits are widely scattered (Fig. 13.9), and because they are heavy, bulky, and relatively low in value they are usually marketed within 50 miles or less of the production site.

Of the nonmetallic minerals produced, crushed stone has about 67 percent of the total value. About 80 percent of the crushed stone product is limestone and dolomite. Other types include granite, sandstone, traprock, and marble. Most of this product is used for road surfacing, road bases, and cement manufacture. Although production takes place in more than 50 counties, about 70 percent of the state's output is concentrated in southeastern Pennsylvania. In 1989 about eight tons of stone per person was produced in Pennsylvania, compared with a national average of five tons per person.

Of the dimension stone quarried in Pennsylvania, slate is most important. Commercial-quality slate is

Table 13.5. Nonmetallic mineral production, 1989.

	Quantity (000)	Value ($000)
Stone		
Crushed stone (short tons)	93,123	$455,004
Dimension stone (short tons)	44,000	10,032
Sand and gravel (short tons)	19,500	94,600
Lime (short tons)	1,660	92,600
Clays (metric tons)	1,049	4,936
Combined value of kaolin, mica, industrial sand and gravel, crushed granite, tripoli	—	14,734
Total		$671,906

SOURCE: Pa. Dept. of Environmental Resources.

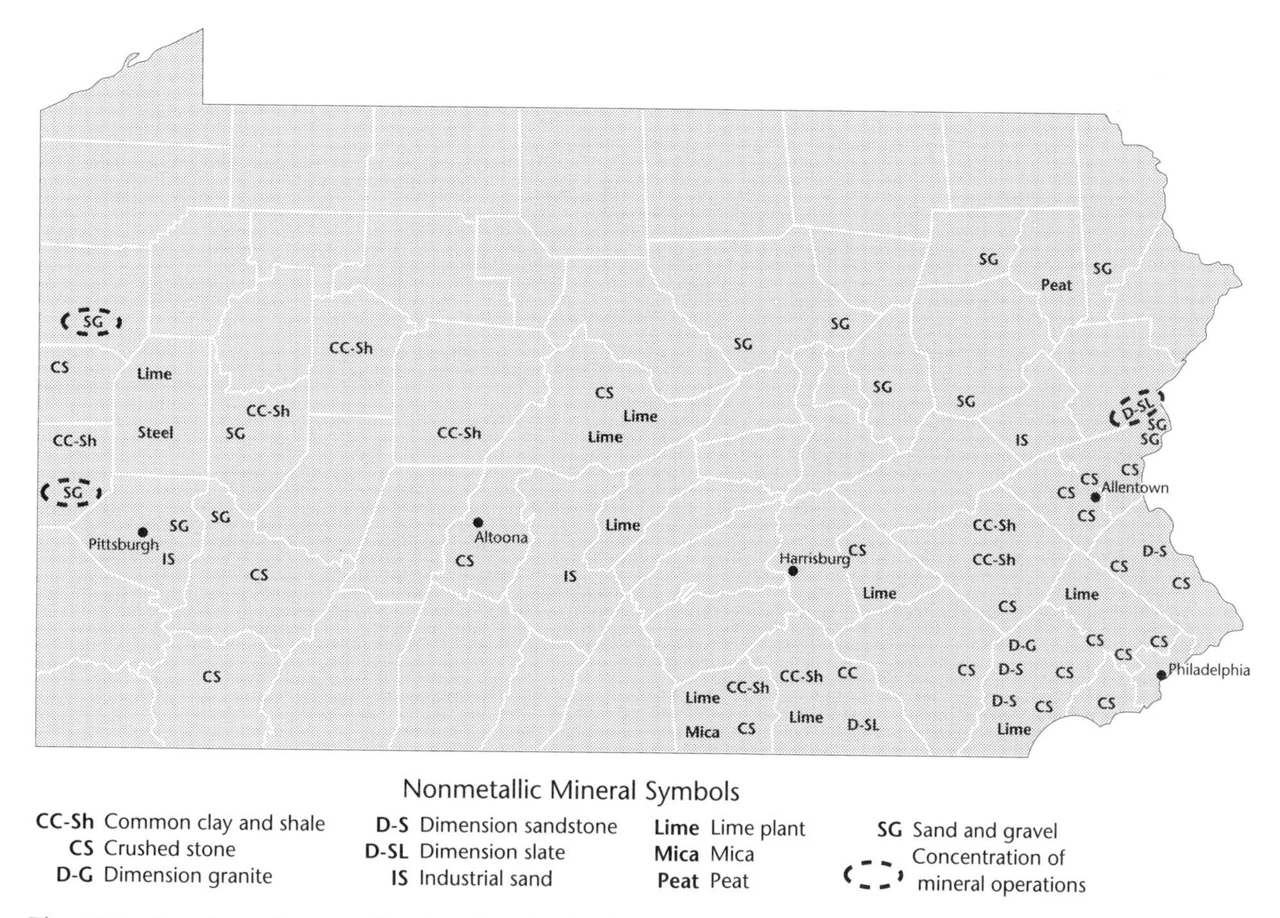

Fig. 13.9. Locations of nonmetallic mineral production in Pennsylvania.

quarried in two regions in the state. In the Lehigh Valley of Northampton and Lehigh counties is the most important area, centering on the towns of Bangor, Pen Argyl, Slatington, Windgap, Walnutport, and several smaller boroughs. The other center of output is the Peach Bottom area of southeastern York County. In the nineteenth and early twentieth centuries a major mining area developed in the Lehigh Valley to produce slate for roofing, structural and sanitary uses, blackboards, electrical uses, and other purposes. With the introduction of less-expensive substitutes, particularly for roofing and blackboards, demand for slate has plummeted. A remnant industry producing electrical, structural, and sanitary slate remains in Bangor and Pen Argyl.

Sand and gravel deposits are widely distributed in Pennsylvania. These nonmetallics are heavy, bulky, and low in value, so transportation costs are high and local markets prevail. The number of sand and gravel operations varies considerably from year to year, depending on market demand. Little capital is needed to open a new or old quarry site, so that many are sporadic. Because the Allegheny and Susquehanna river headwaters are located in glacial deposits, the rivers have been a major source of high-grade aggregates. Environmental constraints have reduced the dredge operations in the rivers.

Production of industrial sand is much more limited than that of sand and gravel. Because more than half the industrial sand is used in the manufacture of glass, it must be free of impurities that would color the finished product. Industrial-sand production has been limited to Allegheny, Huntingdon, and Carbon counties in recent years. The glass container industry has been adversely affected by plastic and cardboard containers, resulting in a decline of industrial-sand production.

Lime is produced in some 80 plants in seven counties. Centre County is the leading county, followed by Butler, Adams, Chester, York, Montgomery, and Mifflin. The market for lime has been undergoing significant changes. In the past, half the lime manufactured was used in the basic oxygen blast furnaces. As steel output declines, a greater percentage of the output is used for water purification, pulp, and paper, and as an agricultural fertilizer.

Clay production has declined in Pennsylvania, from a record high of 3,000,000 tons in 1973 to about 900,000 tons in 1989. The decline is largely due to the drop in demand for bricks, particularly firebrick used in lining blast furnaces. Clay is widely distributed, but production is concentrated in Berks, York, Adams, Clearfield, and Jefferson counties. There is a single kaolin mine in the state in Lancaster County, where output is used for fertilizer and paint.

Metallic Minerals

Pennsylvania's metallic mineral wealth is extremely limited, but small and widely distributed deposits of iron ore have played a significant role in establishing an early iron industry. The first iron ore mined in Pennsylvania was in 1716, when Thomas Ritter erected a blooming forge, known as Pool Forge, on Manatowny Creek in Berks County. Local iron ore deposits were mined throughout the state until the end of the nineteenth century, providing the raw material for the widely dispersed iron industry.

Of the deposits, the Cornwall ore hills in southeastern Pennsylvania were the most productive in the state. Located in Lebanon County, they consisted of three deposits of magnetite iron ore. Although the ore averaged only 35 to 40 percent iron, it could be concentrated so it was profitable to mine. Production was continuous from about 1740 until the mines were flooded during Hurricane Agnes in 1972 and never reopened. Production peaked in 1889, when 769,200 tons of ore were mined. Cumulative production from 1853 to 1907 was more than 21,000,000 tons.

In the 1950s a new, underground magnetite mine was opened at Morgantown in Berks County. The ore is crushed, concentrated by magnetic separation, and pelletized at agglomerating plants at the mine site. Most of the iron ore pellet production was shipped to the Bethlehem Corporation's iron and steel plants in Pennsylvania and Maryland for consumption in blast furnaces. The Morgantown mine was closed in 1986.

Zinc, principally sphalerite, has been mined and concentrated at the Friedensville mine in Lehigh County for decades. Production has been sporadic depending upon demand, but in the 1980s annual totals were from $18 million to $25 million. Since the early 1900s, the ore was smelted and refined at Palmerton, where the New Jersey Zinc Company consolidated its smelting and refining operations into one large plant. Most of the ore came from the Franklin Furnace district of northern New Jersey to supplement the Pennsylvania ore.

A number of other metallic minerals are processed in Pennsylvania. Beryllium is processed in Hazleton. Cerium is produced in a solvent-extraction plant in York from ore shipped from California mines. Small quantities of copper, cobalt, gold, and silver were obtained from the magnetite ores of the Lebanon and Berks county mines.

Bibliography

Annual Report on Mining Activities. Annual. Harrisburg, Pa.: Department of Environmental Resources.

Bise, L. 1983. "Mining Technologies and Development Underground." In *Pennsylvania Coal: Resources, Technology, and Utilization,* ed. S. K. Majumdar and E. W. Miller. Easton: Pennsylvania Academy of Science.

Deasy, G. F., and P. R. Griess. 1959. *Atlas of Pennsylvania Coal and Coal Mining, Bituminous Coal,* Part 1. Bulletin of the Mineral Industries Experiment Station, Mineral Conservation Series No. 73. University Park: The Pennsylvania State University.

———. 1965. "Effects of a Declining Mining Economy on the Pennsylvania Anthracite Region." *Association of American Geographers Annals* 55:239–259.

Elliott, J. M., S. L. Friedman, and S. C. Braverman. 1983. "An Overview of Pennsylvania Coal Law." In *Pennsylvania Coal: Resources, Technology, and Utilization,* ed. S. K. Majumdar and E. W. Miller. Easton: Pennsylvania Academy of Science.

Hoffman, J. N. 1978. "Pennsylvania Bituminous Coal Industry." *Pennsylvania History* 45:351–363.

Majumdar, S. K., and E. W. Miller. 1983. *Pennsylvania Coal: Resources, Technology, and Utilization.* Easton: Pennsylvania Academy of Science.

Miller, E. W. 1953. "Connellsville Beehive Coke Region: A Declining Mineral Economy." *Economic Geography* 29:144–158.

———. 1955. "The Southern Anthracite Region: A Problem Area." *Economic Geography* 31:331–350.

———. 1980. The Bituminous Coal Industry of Pennsylvania: Localization and Trends." In *Energy, Environment, and the Economy,* ed. S. K. Majumdar. Easton: Pennsylvania Academy of Science.

———. 1986a. "Exploitation of a Rich Natural Heritage."

In *Pennsylvania: Keystone to Progress,* by E. W. Miller. Northridge, Calif.: Windsor Publications.
———. 1986b. "Mineral Resources: Modern-Day Trends." In *Pennsylvania: Keystone to Progress,* by E. W. Miller. Northridge, Calif.: Windsor Publications.
———. 1988. "Spatial Changes in Bituminous Coal Mining in Pennsylvania: 1947–1986." *Pennsylvania Geographer* 26 (Fall-Winter): 19–26.
Minerals Yearbook. Annual. Bureau of Mines, U.S. Department of the Interior. Washington, D.C.: Government Printing Office.
Phelps, L. B. 1983. "Mining Techniques and Development Surface." In *Pennsylvania Coal: Resources, Technology, and Utilization,* ed. S. K. Majumdar and E. W. Miller. Easton: Pennsylvania Academy of Science.
Schnell, G. A., and M. S. Monmonier. 1983. "Coal Mining and Land Use–Land and Cover: A Cartographic Study of Mining and Barren Land." In *Pennsylvania Coal: Resources, Technology, and Utilization,* ed. S. K. Majumdar and E. W. Miller. Easton: Pennsylvania Academy of Science.

14

TRANSPORTATION

E. Willard Miller

Since the colonial period, when waterways and trails provided the earliest means of travel, modes of transportation have evolved to serve a changing economy. The first major effort to improve transportation came with the building of a network of improved turnpikes in Pennsylvania. The turnpike era was followed by the canal era, which dominated transportation from around 1830 to the 1850s, when the canals were superseded by railroads as the primary means of transportation until early in the twentieth century. During the railroad era, the canals were abandoned and roads were neglected. The modern period of transportation began with the widespread use of motor vehicles. As highways were built and improved, the importance of the railroads declined. Most recently, the airplane has added an additional dimension to the transportation system of Pennsylvania.

Trails and the Turnpike Era

The transportation system of the pioneer era, which extended from 1682 to about 1830, was based on the needs of commerce and the military. The wooded land of southeastern Pennsylvania was so difficult to travel across that early patterns of travel were determined by natural waterways. As settlement spread inland from the rivers, Indian paths served as the first land routes. The packhorse trails were especially important in the movement of settlers and military expeditions to western Pennsylvania. By the 1750s the route across the Allegheny Mountains was crowded with packhorse trains.

Pennsylvania's first roads were built in the more densely populated areas, their extent and direction influenced by topography and settlement patterns. As settlement spread from Philadelphia, it became the hub of transportation. Gradually routes penetrated to the interior of the southeast. In 1706, for example, the Queen's Path was established from Philadelphia to Chester. The Old Conestoga Road from Philadelphia to Lancaster, the first major route to the west, was begun in 1721 but not completed until 1733. These were earthen roads, which had no firm base and were virtually impossible to travel during the wet seasons of the year.

William Penn established the road-building policy that persisted for more than a century. This policy made local governments responsible for building and maintaining the roads within their respective borders. Unfortunately, local governments lacked the capital and the ability to plan a coordinated system for the colony. Furthermore, the local governments did not have the authority to levy or collect road taxes. The provincial government appropriations for roads were small and irregular and normally only for military routes.

In 1784 the burden of maintaining roads was partially removed from local governments, as the state assumed the cost of building long-distance roads. The first state appropriation, in 1785, was for the Western Road that extended from Cumberland County to Pittsburgh. This road, not completed until 1810, followed the low valleys and the wind and water gaps to Bedford. From Bedford westward to Pittsburgh, the route followed the military trail cut through the forest by General John Forbes in 1758 in the British expedition to drive the French from Fort Duquesne. Forbes, in turn, had followed an old Indian trail known as the "Trading Path."

While a number of other roads were constructed in the state, most of the existing roads, including the state-owned roads, were in deplorable condition. To solve this problem, the Society for Promoting the Improvement of Roads and Inland Navigation was formed in 1789 in Philadelphia. This organization laid the foundation for the turnpike era, which began on April 9, 1792, when the General Assembly passed an act incorporating the Philadelphia & Lancaster Turnpike Company.

The Lancaster Turnpike, financed by sale of stock, was completed in 1794. A toll was collected to use this stone-based road, acclaimed a "masterpiece" of its kind. The road was an immediate economic success and was the foundation for expanded turnpike construction. Although most of the turnpikes were in southeastern Pennsylvania, other parts of the state benefited. The major long-distance turnpikes included two turnpikes from Philadelphia to Pittsburgh, one from Philadelphia to Erie by way of Sunbury, Bellefonte, Franklin, and Meadville; two roads from Philadelphia to the New York border, one through Berwick the other through Bethlehem; and a stone-based road from Pittsburgh to Erie through Butler, Mercer, and Meadville.

By 1832 the toll turnpikes reached their zenith. The system consisted of 2,400 miles of completed highways, and more than 600 miles were projected (Fig. 14.1). After this date more roads were abandoned than were newly built. A number of factors caused the decline. Except for the Lancaster Turnpike, most turnpikes were financial failures. Further, as the economy developed, state and local governments assumed greater responsibility for building toll-free roads through public funding. Finally, the development of the canals resulted in cheaper transportation, followed soon by the greater speed and efficiency of the railroads.

The Canal Era

The nineteenth century witnessed a revolution in the development of transportation. The improvement of waterways and the building of canals was the initial step in this transportation advancement. Although canals in Pennsylvania were considered as early as 1762, the Union Canal between Reading on the Schuylkill River and Middletown on the Susquehanna River was not completed until 1827.

With the building of the Erie Canal across New York, it became evident that the economy of Pennsylvania would suffer if a similar water route was not provided. But the rugged topography of central Pennsylvania provided a physical barrier that was not present in the lowland of northern New York State. Nevertheless, a legislative commission appointed in 1824 reported that a lock system of canals from Philadelphia to Pittsburgh was feasible, except for thirty-six miles between Hollidaysburg and Johnstown—the stretch over the Allegheny Front and the eastern edge of the Allegheny Plateau. Although railroads were technologically primitive at this time, it was decided that the only solution was to construct a railroad over the mountainous area to haul passengers and goods between the two canal

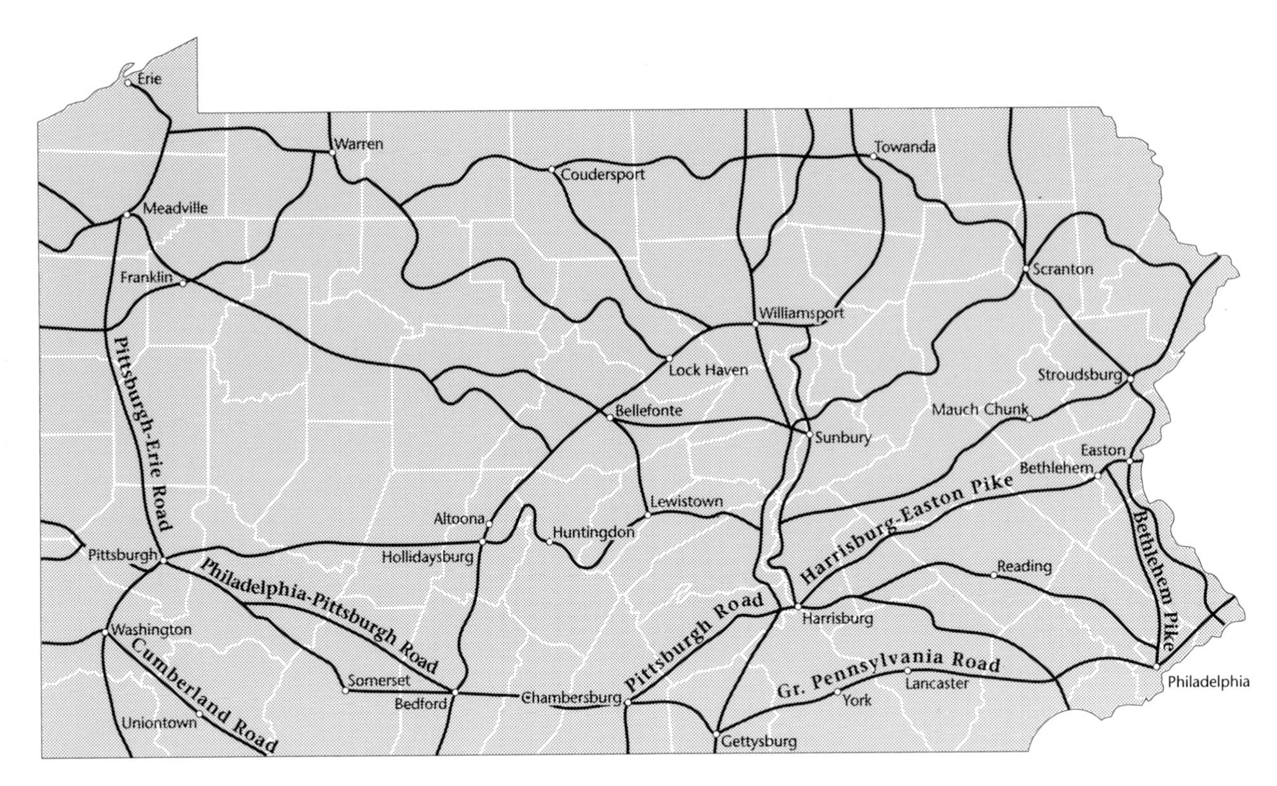

Fig. 14.1. Major Pennsylvania turnpikes of the eighteenth and early nineteenth centuries.

points. Construction of the Pennsylvania Canal began in 1826, a year after the Erie Canal was completed, and finished in 1834.

This transportation route west from Philadelphia combined railroads and waterways. A railroad was built from Philadelphia to Columbia on the Susquehanna River. The central section, from Columbia to Hollidaysburg extending along the Susquehanna and Juniata rivers, consisted of 15.8 miles of slackwater navigation and 156 miles of canals. The third section was the Allegheny Portage Railroad between Hollidaysburg and Johnstown. To extend the canal over the Allegheny Front would have required 100 locks, each with a 14-foot lift. Traversing the Allegheny Plateau would have been equally difficult. Consequently, an incline railroad to raise boats 1,398 feet from Hollidaysburg to the top of the Allegheny Front was built. The railroad was then extended to Johnstown. Finally, the western portion of the canal extended 104 miles from Johnstown to Pittsburgh, following the deep, narrow, meandering valleys of the Conemaugh, Kiskiminetas, and Allegheny rivers. This section required the construction of 10 dams and 61 locks.

The Pennsylvania Canal provided the foundation for the canal system of Pennsylvania. From this route, one canal extended northward along the Susquehanna from the Juniata to Northumberland, connecting with a railroad to Elmira, New York. From Northumberland another canal was built along the West Branch of the Susquehanna to Lock Haven with an extension to Bellefonte. Along the Delaware and Lehigh rivers a canal extended from Bristol to Easton and to Mauch Chunk (Jim Thorpe) and Stoddardville to serve the southern anthracite fields. The Schuylkill Canal extended from Philadelphia along the Schuylkill River to Pottsville in the southern anthracite fields. Many canals were projected in western Pennsylvania, but few were completed. The most important of these was the canal that extended northward from Beaver on the Ohio River to New Castle and on to Erie.

Although major canal projects developed later in Pennsylvania than in most other states, Pennsylvania became a leader in canal construction. By 1838 Pennsylvania had a total of 788.5 miles of canals completed and 133.75 miles under construction. At the peak of the canal era, about 1852, there were 1,024 miles of canals and 314 miles of unfinished improvements in the waterway systems of Pennsylvania.

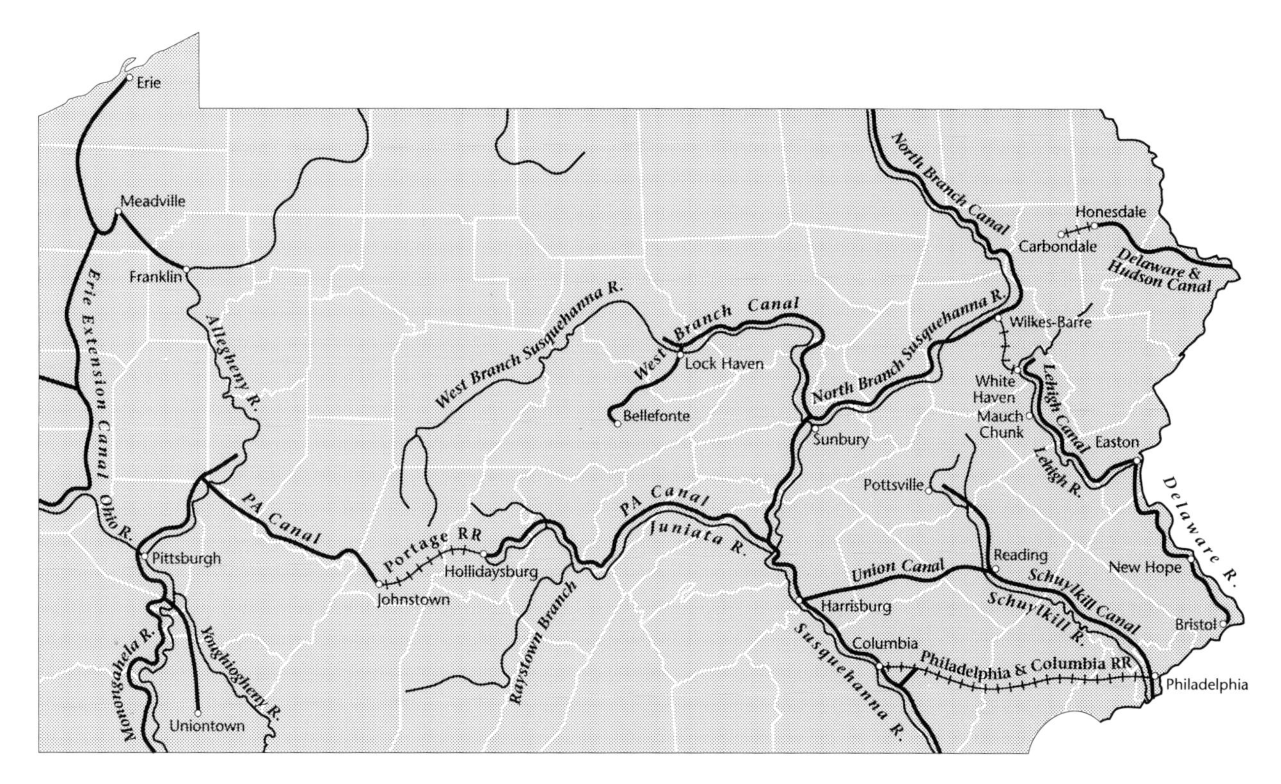

Fig. 14.2. Canal network of Pennsylvania, 1840.

Although the canal system was extensive (see Fig. 14.2), the waterways failed to achieve the movement of traffic expected. It soon became evident that the Pennsylvania Canal, the main link in the system, would never become the profitable route that had been anticipated. The weak link was the Portage Railroad. Because the goods and passengers had to be moved in boats or transferred from boats to rail cars and back again, costs were high and the speed of movement greatly reduced. Local traffic was normally more important than through traffic.

The canals of Pennsylvania experienced significant problems. Many were in a poor location for getting a large volume of traffic. Although the eastern and western sections of the Pennsylvania Canal passed through developed economic areas, the central section was in an area of limited development. Similarly, the canals of the north and west branches of the Susquehanna served thinly populated areas. The regions best served by canals were those that produced bulky, low-value commodities, such as anthracite, where canals provided low-cost transportation to distant markets. Canals serving southeastern Pennsylvania had peak loads during harvest periods followed by long stretches of little activity. Unlike the railroads, canals could not adjust quickly to new technological developments. Finally, canals were closed by ice during a considerable portion of the year, and at times floods or lack of water made navigation difficult.

At the height of the canal era, the railroads began an intensive construction program that provided stiff competition because of their speed and efficiency. The Pennsylvania Railroad was incorporated in 1846. Between 1846 and 1854 the railroad was built between Philadelphia and Pittsburgh, essentially following the route of the Pennsylvania Canal. To avoid competition, the Pennsylvania Railroad purchased the Pennsylvania Canal and operated it for some time. By 1864, however, the western section from Pittsburgh to Johnstown was abandoned. The central section in the Juniata Valley discontinued operations in 1899, and the Susquehanna River section was discontinued in 1900.

By the early 1870s most canals had been purchased by railroads. By 1906 some 908 miles of the 1,200 miles of improved rivers and canals that the state and private corporations had built were abandoned. In eastern Pennsylvania by the early twentieth century, only the Lehigh and Schuylkill canals and the improved portions of the Delaware River provided water transportation. In western Pennsylvania all canals had been abandoned and only the natural waterways of the Ohio River system carried traffic.

The Railroad Era

The first railroad was built in Pennsylvania in Delaware County in 1809 to haul stone from a quarry. Other short lines were built to haul commodities, such as anthracite, to canals. By 1840 there were more than 36 lines varying from 7 to 50 miles in length, mostly in the east. As railroad lines were being built from Baltimore, New York, and Boston, in order to supplement canal systems to the Midwest, Philadelphians recognized the need to have access to the western markets not provided by the Pennsylvania Canal.

The railroad era in Pennsylvania began in 1846 when a group of Philadelphia merchants formed a company to build a railroad from Philadelphia to Pittsburgh, to be known as the Pennsylvania Railroad Company. This transportation company established the major east-west railroad route across the state. The railroad was completed to Hollidaysburg by 1850 and reached Pittsburgh in 1854. Philadelphia merchants now had a route they could use to compete successfully with other eastern cities for the growing Midwestern market.

The Pennsylvania Railroad not only extended its system westward to Chicago and St. Louis and eastward to the Atlantic Coast, but also developed a network of routes in Pennsylvania. In order to develop the state's economy, the Pennsylvania Railroad system extended into the bituminous coal region, the oil fields of the state, the forest areas, and the anthracite fields.

The network of railroads of Pennsylvania was built between 1860 and 1900, blanketing the states (Fig. 14.3). By 1900 only a few areas were more than 10 miles from a railroad. The counties of Fulton and Bedford in south-central Pennsylvania, and Pike and Wayne counties in the northeast, had limited railroad accessibility.

Track mileage increased dramatically each decade. In 1860 the state had 2,598 miles of railroads. In

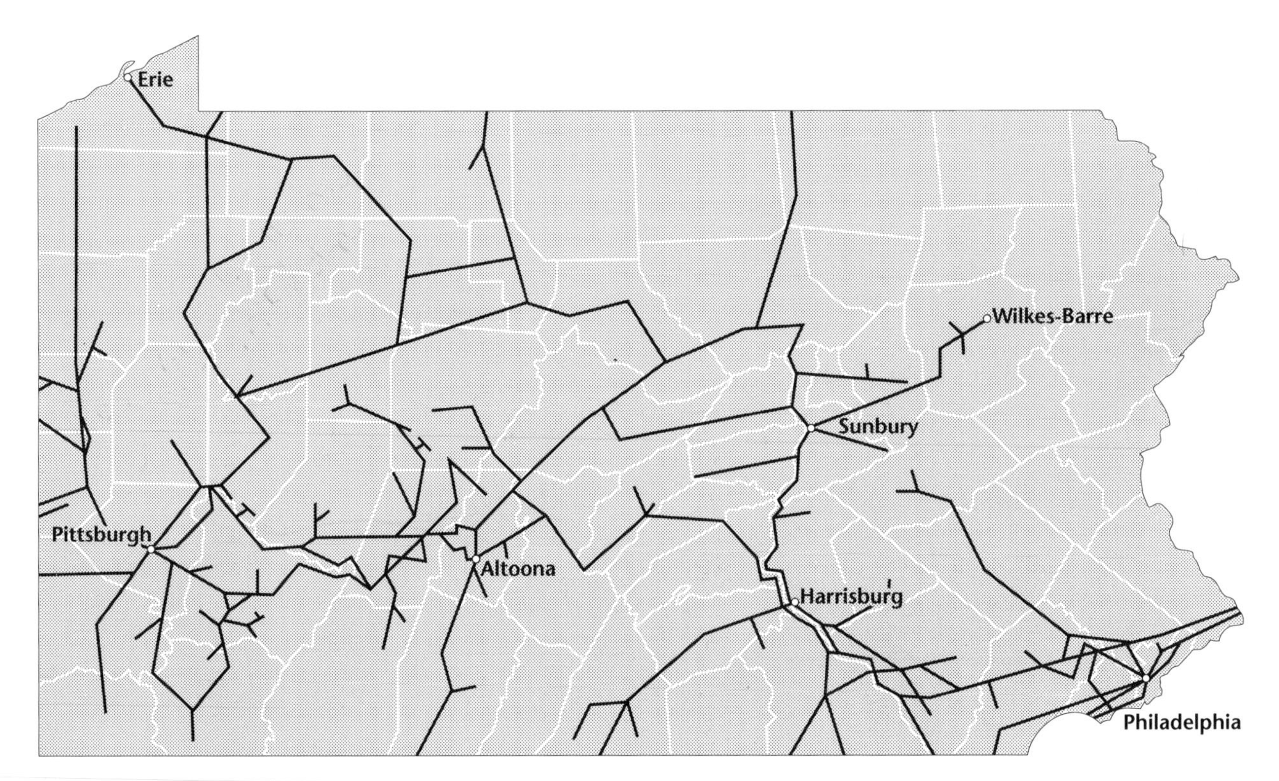

Fig. 14.3. Pennsylvania railroad network, 1900.

1874–1875 the Department of Internal Affairs reported a total trackage of 8,960, of which 1,806 miles were double track. In 1901 the Annual Report of the Bureau of Railways listed 10,697 miles of main-line railroads. Including sidings, spurs, and other lines, the grand total was about 25,000 miles of track.

In the latter part of the nineteenth century the railroads provided the means to move people and freight. In 1900 Pennsylvania railroads carried 478,684,683 tons of freight. More than half of this tonnage consisted of mineral products, primarily anthracite and bituminous coal. Railroads were also important movers of iron ore, iron and steel, and manufactured products. Passenger service was available to every county in the state except Fulton County.

The topography of Pennsylvania played a major role in establishing the railroad pattern. Because railroads sought low-level routes, the rail lines have followed valleys wherever possible. Only in the southeastern plains could routes ignore the lay of the land. In the Ridge-and-Valley region the routes follow the lowlands, crossing from one valley to another at wind and water gaps. When these were absent, tunnels had to be built through the ridges. In the Appalachian Plateau the major routes are confined to the river valleys, and the spurs rise over the steep slopes. The meandering streams lengthened the mileage of many railroads by as much as 50 percent. The topography is so rugged in many sections of the state that the number of potential routes was quite limited. The Pennsylvania Railroad achieved a dominant position in the state by obtaining rights-of-way along the most low-level routes. Once a right-of-way had been obtained, a railroad network could be developed over a large area with negligible competition from competing railroads.

By 1900 the railroad system was complete. The system was dominated by the Pennsylvania Railroad (Fig. 14.3). The anthracite mining region, however, had developed the densest network of railroads with such major railroads as the Lehigh Valley; the Delaware, Lackawanna & Western; the Erie; and the Central Railroad of New Jersey. Anthracite railroads were primarily carriers of coal to the northeastern United States and Canada. The Baltimore & Ohio served western Pennsylvania, centering on Pittsburgh. The New York Central Railroad served the northern portions of the state. The Erie Railroad also extended into northern Pennsylvania from New York. In addition, a large number of small railroads connected the isolated towns and regions to major routes. Examples of these railroads included the Susquehanna & New York Railroad, the Bloomsburg Sullivan Railroad, the St. Mary & Emporium Railroad, the Cumberland Valley Railroad, and the Bellefonte Central.

After 1900 essentially no new railroad lines were laid, and for several decades the track mileage remained virtually unchanged. There were a number of reasons for the stoppage of railroad-building. First, the fine network of trunk lines plus short spurs to isolated areas provided a complete network, with frequent passenger and extensive local freight service. Second, the advent of motor vehicles focused attention on improving the state's road network instead. As the road system expanded, the need for passenger services declined and trucks began to carry freight. The short lines that served local communities became uneconomical, and abandonment began in the 1920s and continued for decades.

The Modern Era

In the twentieth century a transportation system has evolved in which highways, railroads, and air routes each play a role in providing passenger and freight services.

Highways

Highways provide the major means of moving people and freight. As a consequence, the dense network of railroads of the nineteenth century has been replaced by a dense highway network throughout the state. The twentieth-century development of Pennsylvania's highways can be divided into three stages: establishment of the state road system; creation of a system of U.S. highways extending across the state and connecting with adjacent states; development of the interstate system. Pennsylvania today has about 108,600 miles of highways, of which the state has primary responsibility for 43,095 miles. The remainder are roads controlled by townships, counties, boroughs, and cities (Fig. 14.4).

Evolution of the Modern Highway System

During the railroad era of the nineteenth century, trains provided the passenger and the freight services

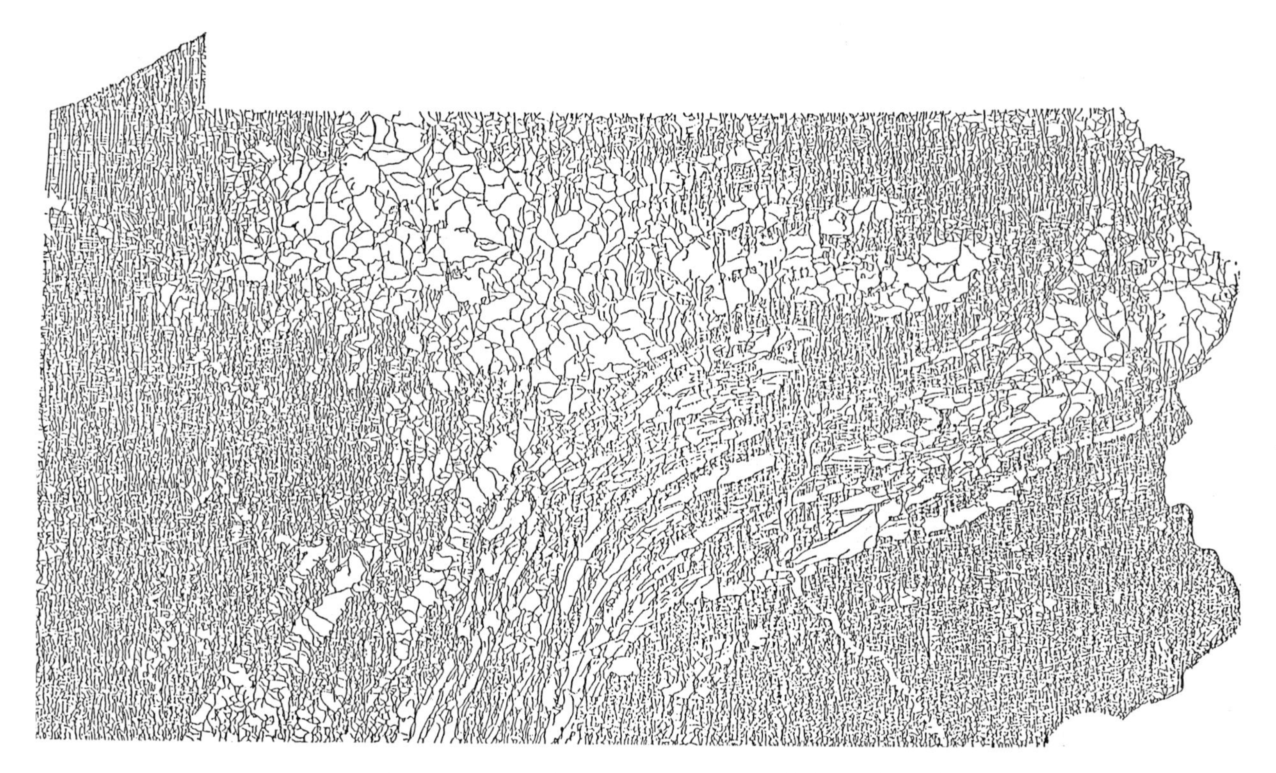

Fig. 14.4. Total road patterns of Pennsylvania. This unique map shows all the roads and main streets in the state but purposely omits any features that indicate road importance.

the developing industrial society required. As a result, the road system of the state was neglected. Before 1900 less than 5 percent of the state's 80,000 miles of roads could be classified as "fair," and there were no paved roads in rural areas. In the 1890s the need for better highways became evident, and in 1897 the Hamilton Road Bill provided $1 million to repair the state's roads. A general report of road conditions to the General Assembly in 1899 led to the passage of the Sproul-Roberts Act of 1903. This legislation created the Department of Public Highways and was the initial step in developing a state highway system. To demonstrate the importance of paved roads, 64 short, object-lesson macadam roads were built throughout the state. When requested by local road commissioners, the Department of Public Highways was required to prepare plans for construction of main highways in the district. Two-thirds of the cost was paid by the state, and one-third by the local government. Initially the urban areas were the first to develop paved routes, but after 1910, with the introduction of the Model-T Ford, farmers began to use automobiles and thus added their voice to the swelling chorus demanding better roads.

In the evolution of the state highway system the next step was the passage of the Sproul Road Act of 1911. Senator William C. Sproul, who later became governor, is known as the "father of good roads" in Pennsylvania. The 1911 act expanded the operations of the State Highway Department and increased state aid. The state took over 8,855 miles of main highways connecting the principal urban areas and assumed responsibility for their reconstruction with "durable material" and their maintenance. The state also purchased all toll roads that were a part of the state highway system. The Sproul Act initiated a complete survey of all roads of the state, which became the basis for the evolution of the state highway system. To secure funds for highways, a legislative act in 1913 required that all funds obtained from the licensing of motor vehicles were to be used exclusively by the state for highway development.

By the 1920s funds were growing rapidly for road improvements. During the administration of Governor Sproul (1919–1923) more than $120 million was expended, and in the administration of John D. Fisher (1927–1931) the amount exceeded $235 million. By 1929 the state had assumed responsibility for all borough streets that were part of the state highway routes and for county bridges on such routes. By 1930 the state was responsible for maintaining 13,500 miles of highways.

Most of Pennsylvania's main routes had been paved by 1930, but most of the rural roads remained gravel or dirt. In 1931 Gifford Pinchot campaigned for governor on the issue "Take the farmer out of the mud." During the first year of his administration, more than 1,500 miles of township dirt roads were lightly surfaced with bituminous material at a cost of about $5,000 per mile, in contrast to $40,000 a mile on primary roads. In the 1930s the state acquired about 20,000 miles of township roads, providing a sound financial base for maintenance. By 1940 the state's modern highway network had been established.

U.S. Highways

Federal aid to improve Pennsylvania's roads was initiated by an act of Congress in 1916. This act stipulated that a state could designate not more than 7 percent of its total highway mileage as part of a federal highway system, thereby creating a federal-state highway partnership. Under this agreement the states were responsible for choosing the routes for development, for planning projects, and for establishing priorities. As the project developed, the Federal Highway Administration's Bureau of Public Roads guided, reviewed, and approved each step. Federal funds, which had to be matched by state funds, were apportioned to each state by legislated formulas. Between 1917 and 1971 more than $3 billion in federal highway funds were apportioned to Pennsylvania.

In the development of U.S. highways in Pennsylvania, five east-west routes (U.S. routes 6, 22, 30, 322, 422) and four north-south routes (U.S. routes 11, 19, 219, 220) were selected, as well as a number of shorter routes. These routes, established by 1940, provided a loose grid of roads in the state (see Fig. 14.5). Except for a few short sections, they were two-lane highways, but provided ready access to

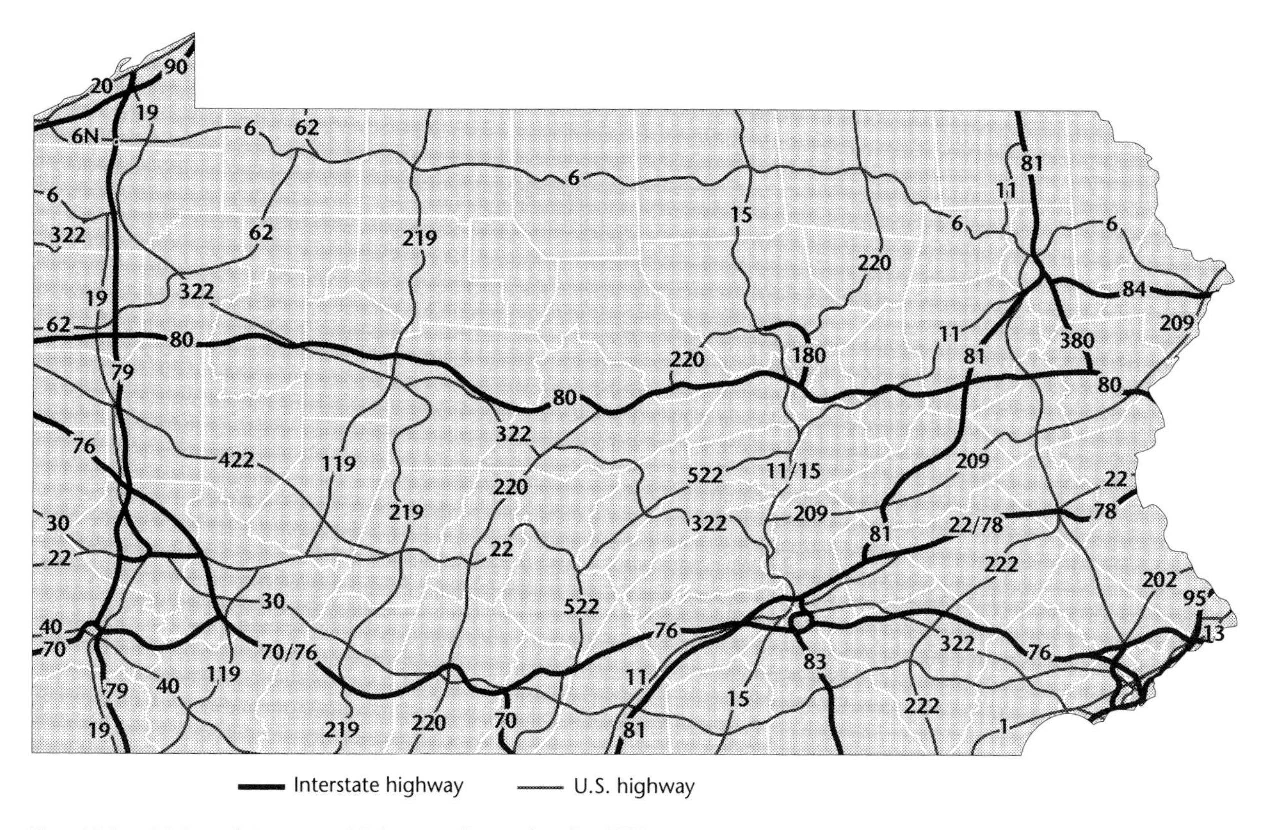

Fig. 14.5. U.S. and Interstate highways, Pennsylvania, 1992.

large regions of the state. Many of the Interstate Highways such as routes 70, 76, 78, 79, and 81 parallel the U.S. routes, and in a few places the U.S. routes have been rebuilt to Interstate Highway standards. To recognize the importance of through routes, many of the U.S. highways were given specific names. U.S. 6 was named the Grand Army of the Republic Highway; U.S. 322 is the Twenty-Eighth Division Highway; U.S. 30 is the Lincoln Highway; U.S. 422 is the Benjamin Franklin Highway; U.S. 22 is the William Penn Highway; and in recent years U.S. 220 has come to be known as the Appalachian Thruway. Many of the U.S. highways pass through many states, such as U.S. 6 and U.S. 30, which extend across the United States.

The Pennsylvania Turnpike

The problem of crossing the rugged Pennsylvania topography from east to west was not solved until the Pennsylvania Turnpike was built. This turnpike was not only the first four-lane, long-distance highway in the nation but also provided a low-grade route through the mountains of central Pennsylvania. The turnpike extends for 327 miles, from Philadelphia on the east through Pittsburgh to a point on the Ohio line a few miles south of Youngstown.

In the 1880s some of the greatest empire-builders of the day, including William H. Vanderbilt, J. Pierpont Morgan, Jay Gould, and others, formed a company to build a railroad across central Pennsylvania. After two years of construction work on the tunnels and right-of-way, the Pennsylvania Railroad bought the South Pennsylvania Railroad to control competition and the rail route was abandoned. In the 1930s, survey-crew reports indicated that the old South Pennsylvania Railroad route was "the best ever devised between the Ocean and Ohio." In order to stimulate the economy and provide jobs for the unemployed in the 1930s, the federal Works Progress Administration did surveys to determine the cost of building the central mountainous section. The Pennsylvania Turnpike Commission was created in 1937 and instructed to build a highway from Middlesex in Cumberland County, about 18 miles west of Harrisburg, to Irwin, about 24 miles east of Pittsburgh, a distance of 160 miles.

At first, private funds were sought to build the highway, but no buyers could be found for $60 million of bonds sold through a New York investment company. To salvage the project, President Franklin D. Roosevelt, who saw the value of a high-speed route for military purposes if war developed in Europe, ordered two federal agencies to provide the funds. The Reconstruction Finance Corporation, headed by Jesse Jones, underwrote a $35 million issue of turnpike bonds, and the Public Works Administration, under Harold Ickes, provided an outright grant of $29.1 million to the commission.

In September 1937 the commission secured title to the nine tunnels of the South Penn Railroad, and on October 27, 1938, construction of the road began. Within 23 months, on October 1, 1940, the 160-mile-long road was completed "Across the Barrier." The highway was a marvel of construction. Most of the route was free of sharp curves, and the maximum grade was 3 percent. To achieve this low-grade route, 53.3 million cubic yards of earth and rock were moved, cutting off ridges and filling in valleys. Six of the nine original railroad tunnels were completed, and one new tunnel was built.

To complete the turnpike, the 100-mile Philadelphia Extension to the east was begun in 1948, and the 67-mile westward extension in Ohio was begun in October 1949. The Northeast Extension from Philadelphia to Scranton was opened to traffic in 1957.

With increasing traffic, the Turnpike has been continually improved. In the 1960s three tunnels were eliminated, and the remaining four on the original route were given an additional tunnel to provide a four-lane route. Because this original highway did not have divider strips, a median guard rail has been installed along the entire route to provide additional safety.

Traffic on the turnpike has grown steadily. In 1989 a total of 78,000,000 passenger vehicles and 12,000,000 commercial vehicles entered the 30 exchanges from Philadelphia to the Ohio border. An additional 7,006,000 passenger cars and 1,000,000 commercial vehicles used the Northeast Extension from Philadelphia to Scranton. This forerunner of all superhighways continues to be a major traffic route in Pennsylvania.

Interstate Highway System

In the mid-1950s, President Dwight D. Eisenhower recognized the need for a national system of superhighways. The plan was not to improve existing roads but to build an entirely new network of highways, with federal funds providing 90 percent of construction and right-of-way costs and the states providing the remaining 10 percent. The federal funds came from the Highway Trust Fund created by the Highway Reserve Act of 1956, which was a depository for the taxes collected on motor fuels and other highway-related levies. These funds were used solely to finance the federal share of the interstate system. No funds were provided from the general budget of the U.S. Treasury.

The basic goal of the system was to provide a modern network of highways in the state connecting regions within the state and ultimately to the nation. Within Pennsylvania, two east-west routes were designated. The Pennsylvania Turnpike was labeled Interstate 76, one of the few highways in the United States not specifically built for the network. In central Pennsylvania, Interstate 80 (better known as the Keystone Shortway) was built as part of the major route from New York to Chicago and extending to San Francisco on the Pacific Coast. Interstate 80 in Pennsylvania was built through an area of limited economic development and did not follow any previously built highway. The physical obstacles to construction made it one of the most difficult building projects in eastern United States. Particularly spectacular, for example, was the "Big Rock Cut" in Centre and Clinton counties. In a stretch of 4.9 miles, rock excavations totaled almost 7 million cubic yards. Nearly 1.4 million square feet of presplit blasting was necessary, creating a nearly vertical face standing 240 feet high. In order to bring the roadway grade from the adjacent valley to the proper elevation, a 200-foot-high embankment was built using 6 million cubic yards of blasted rock.

Interstate 80 is fundamentally a transit area through Pennsylvania. At many of the interchanges, services such as motel lodging, restaurants, and gas stations have been developed, and these services have poured millions of dollars into the local economy. The route has also spurred the development of new industrial projects in many adjacent small cities, such as Sharon, Franklin, and Hazleton.

Two major north-south routes extend across the state (see Fig. 14.5). In eastern Pennsylvania, Interstate 81 extends from the New York border south of Binghamton through Scranton, Wilkes-Barre, Hazleton, and Harrisburg and leaves Pennsylvania at the Maryland border at Hagerstown. This new highway parallels U.S. 11. At Scranton two routes branch to the east. Interstate 380, known as the Pocono Extension, goes through the Poconos and connects with Interstate 80. The other route is Interstate 84, extending to Port Jervis on the New York border and providing access to New England. Branching off from Interstate 81 at Harrisburg are Interstates 78 and 83. Interstate 78 extends to Allentown-Bethlehem at the New Jersey border, and 83 stretches southward to York and on to Baltimore, Maryland. The East Coast Interstate 95 follows the Delaware River from Trenton through Philadelphia and on to Wilmington, Delaware.

In western Pennsylvania, Interstate 79 extends from Erie to Pittsburgh and leaves the state at the West Virginia border north of Morgantown. Interstate 70 enters Pennsylvania north of Hancock, Maryland, and connects with Interstate 76 at Breezewood. It follows 76 west to New Stanton, then branches off and leaves the state at the West Virginia border east of Wheeling. Interstate 90 borders Lake Erie in the northwest.

The Interstate system was developed as through routes. Every effort was made to see that the highway did not pass through urban areas—not only to reduce costs but also to eliminate congestion and heavy traffic in built-up areas. The Interstate system of Pennsylvania as originally visualized is now complete. There are, however, vast areas in central, western, and northern Pennsylvania that are not served by a modern highway network.

Accessibility

The linkage of one region with another is of fundamental importance in a modern industrial economy. There must be continued efforts to minimize distances traveled in the completion of required interactions. The minimizing of distance has both a physical and a time dimension. Thus, directness measures not only the amount but also the quality of the connections between places. The spatial allocation of transportation routes has a major influence on economic development in a region. For example, investment in through routes to the neglect of local access routes is likely to result in economic decline of the isolated areas. At the same time, the reverse may foster inefficient provincial regionalism because of limited access to national markets.

The density of state highways per square mile in Pennsylvania varies tremendously (Fig. 14.6). Within the state the mean density is 1.04 miles of highway per

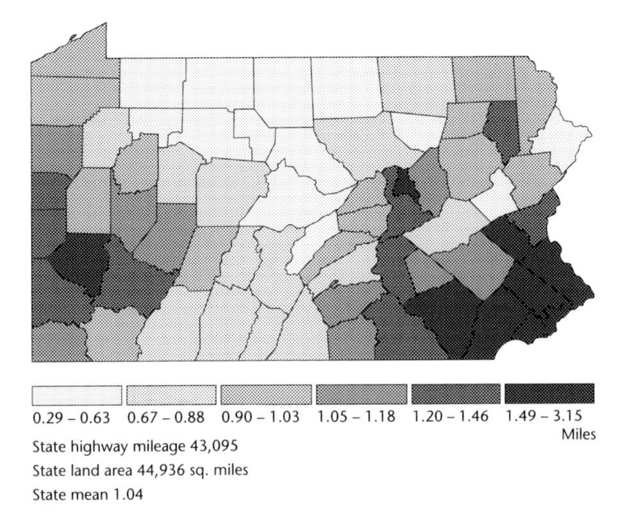

Fig. 14.6. Miles of highway per square mile, Pennsylvania, 1990.

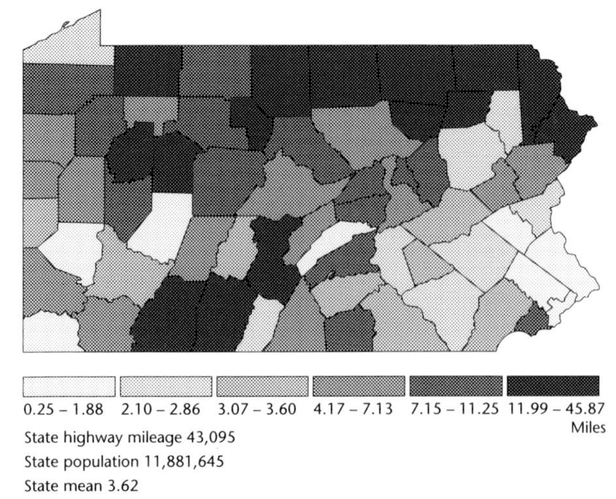

Fig. 14.7. Miles of highway per 1,000 persons, Pennsylvania, 1990.

square mile, with 27 counties above and 40 below the mean. Philadelphia has the greatest density, with 3.15 miles of highway per square mile, while Cameron County has the lowest density, with only 0.29 miles of highway per square mile of area. Many counties in central and northern Pennsylvania have a low density of highways in relation to area. In southeastern and southwestern Pennsylvania the density of highways in relation to area is heavier, but the highways are also more heavily used due to population density.

Another measure of accessibility is revealed by the miles of highway per 1,000 persons (Fig. 14.7). These figures contrast sharply with the data showing miles of highway per square mile. The state mean is 3.62 miles of highway per 1,000 persons. However, there are wide variations within the state. Philadelphia County has only 0.25 miles of highway per 1,000 persons, while Sullivan County has 45.87 miles of highway per 1,000 persons. Forty-three counties are below the mean, and only 24 counties are above. The counties below the mean are those that have the largest total population. In contrast, the counties that have the highest mileage per population are the lowest populated counties of central and northern Pennsylvania.

Pennsylvania, with 43,095 miles of state highways and 7,986,207 passenger cars and trucks, has about 185 vehicles on the average per mile of state highway (Fig. 14.8). The variation in number of motor vehicles to highway mileage is much greater than the number of motor vehicles per population. The number varies from 47 vehicles per mile of highway in Sullivan County to 1,518 in Philadelphia County. The counties with the lowest densities of vehicles are located in the northern and central portions of the state. The highest density patterns are found in the southeast and in other densely populated counties.

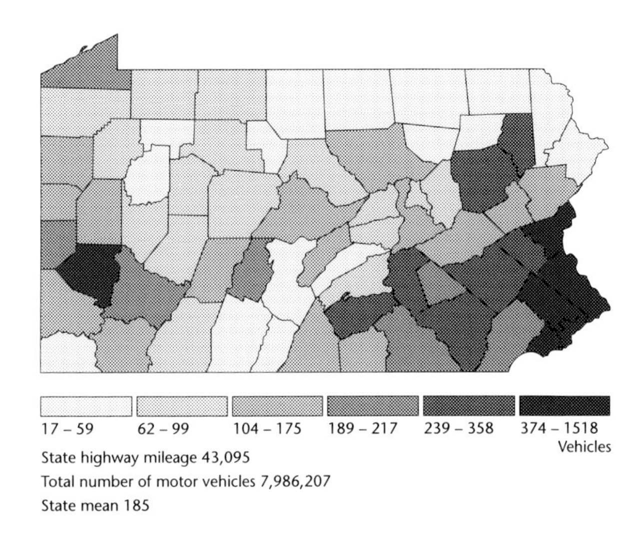

Fig. 14.8. Number of vehicles per mile of highway, Pennsylvania, 1990.

Air Transportation

Air transportation is a twentieth-century phenomenon. As early as 1908, aircraft were being designed and constructed in the state, and by 1909 air shows were being held throughout the state. The Aero Club of Pennsylvania was organized in 1909 with 10 charter members and was incorporated in 1910. Today, it is the oldest active incorporated organization of its kind in the state.

The Age of Barnstorming

The first airfields of Pennsylvania were developed to serve the early aviation enthusiasts. By 1920 Pennsylvania was one of the leaders in "permanent" flying fields, with ground facilities at Altoona, Bellefonte, Philadelphia, Chester, Clarion, Clearfield, DuBois, Essington, Everett, Gettysburg, Huntingdon Mills, Johnstown, Lebanon, Lehighton, Lewistown, Ligonier, Mount Union, Philadelphia (Island Road), Pittsburgh, Stroudsburg, and Wilkes-Barre, plus the Essington seaplane base. In the early days of flying there were no air charts except those published by local airplane clubs and by Rand-McNally. These maps normally contained place names but no terrain features. In flying, pilots usually followed roads, and if they became lost they landed in an open field and asked directions to an airport.

In the 1920s many pilots sought the thrill of flying. Air shows were spectacular events, including parachute jumps and walking on the wing of a plane. These pilots played a major role in developing future air transportation by testing and improving planes and encouraging the improvement of airports.

The Development of Commercial Aviation

The first important commercial development began with the establishment of airmail service by the federal government. In 1918 airmail service was established between New York and Washington with a stop in Philadelphia. At this time, none of the three cities had improved airports. In Philadelphia the planes landed in a farm field near the Lincoln Highway. In 1920 Philadelphia was eliminated from the route.

There was also a route from New York to Chicago via Cleveland, crossing Pennsylvania with stops every 20 to 50 miles, but no airmail service was provided for cities along the route in the state. It was a hazardous flying route, and the 20-mile stretch across the Bald Eagle Ridge at Bellefonte was known as the "Hell Stretch." There were no radio aids, and the rotating beacons were not much help because on "clear nights you did not need them and on bad nights you could not see them."

These early endeavors proved, however, that airplanes could provide a fast and reliable airmail service. This service was placed on a sound basis when the 1925 Air Mail Act was passed, authorizing the Postmaster General to designate airmail routes and contract with private companies for their operation. With the establishment of commercial airmail routes, the vast system of scheduled airlines, not only in Pennsylvania but also throughout the nation, began.

A second piece of legislation, known as the Air Commerce Act of 1926, regulated the efficient and safe use of air space and airports. The new law provided for the licensing of new pilots and aircraft, established rules of the air, and included enforcement provisions.

In the initial establishment of the airmail system, Pennsylvania was served by one route from Pittsburgh to Cleveland. Airport facilities were being expanded, and in 1928 the Avey Air Corporation listed 183 landing strips in Pennsylvania, including 148 intermediate fields with improved landing strips, and 5 federal, 15 municipal, and 15 commercial fields. By the early 1930s the major cities and some smaller cities were served by a growing network of routes. During this period the state encouraged the development of air transportation by creating an advisory State Aeronautics Commission under the Department of Internal Affairs and the Pennsylvania Bureau of Aeronautics. In 1929 these were combined to form the State Aeronautics Commission.

As late as the 1930s, however, many Pennsylvania communities were without airmail service, for small airfields had not been developed. Dr. Lytle S. Adams, a dentist in Irwin, near Pittsburgh, developed a solution to this problem by inventing a device that permitted a plane to pick up a sack of mail without landing. In 1939 Postmaster James Farley gave a contract to All American Aviation (AAA) to link 54 cities and towns in Pennsylvania, West Virginia, Ohio, and Delaware with the national airmail system. The system was successful, and by 1941 there were 92 pickup and delivery points in Pennsylvania alone, with major terminals in Philadelphia and Pittsburgh. the pickup points were 5–20 miles apart. Although flying under instrument conditions was impossible, the pilots completed an average of 94 percent of their flights each year, a remarkable record considering the terrain and weather conditions. The airmail pickup and delivery system was terminated in 1949, by which time it had become outmoded.

Establishment of Air Passenger Service

Air passenger service developed slowly in Pennsylvania. The first scheduled commercial service in the United States began in July 1926 in conjunction with the Sesquicentennial Exposition in Philadelphia between Philadelphia and Washington, and later on to Norfolk. The service lasted until November 1926 and was then discontinued when the fair ended.

In 1930 the New York–Philadelphia–Washington Airline was founded by C. Townsend Ludington. Its slogan was "Every Hour on the Hour." During 1931 Ludington had 6,300 miles of flying scheduled daily, out of a national total of 152,000. Because operations were successful, the airline hoped to receive an airmail contract on the New York–Atlanta–Miami route in 1931, but the contract was awarded to Eastern Air Transport. The Ludington line floundered due to the depression in 1932 and was sold to Eastern Air Transport.

During the 1930s many airmail carriers supplemented their income by carrying passengers. Without an airmail subsidy, airlines did not succeed as passenger carriers alone. Not until after World War II did passenger carriers become important. Airplanes had become larger and safer, and the potential for a network of passenger routes became evident. The transition from airmail to passenger airlines required vast changes in personnel and operating techniques. A major economic change has been the development of independent passenger airline companies. Today the Pennsylvania network consists of three types of carriers: the national carriers, such as United, American, TWA, and U.S. Air; regional carriers, such as Allegheny Commuter, Air Atlanta, American Eagle, and Piedmont Commuter; and local nonscheduled airlines that act as feeder lines.

Airport Network

A network of airports has developed in Pennsylvania (Fig. 14.9). The two major hubs consist of the Philadelphia and Greater Pittsburgh international airports. In addition, 18 regional airports have scheduled airline services. Of these, three (Erie, Harrisburg, Wilkes-Barre/Scranton) are also small international airports. The 18 regional airports are served by one or two airlines that provide commuter air service to the larger terminals within the state and to major airports immediately outside the state, such as Kennedy or Dulles. The Allegheny Commuter Airline, which is one of the largest regional airlines, serves Johnstown, Altoona–Blair County, Chess Lamberton (Franklin), Allentown-Bethlehem-Easton, DuBois-Jefferson, Harrisburg, Reading and Wilkes-Barre/Scranton. Other regional commuter airlines include the Pennsylvania Commuter, Allegheny Airlines, Crown Airways, Altair, Freedom, Holiday, Vec Neal, Wings Airways, Pocono Airways, and Meridian Airways.

There are, in addition, about 140 public airports that do not have scheduled services but have landing strips, some navigational, fueling, and repair facilities, rental cars, and food services. They are frequently operated by an individual, but many are associated with a local company. The airports are suitable only for small planes, and many have obstructions, such as turns at the end of the runways.

In addition to the public-use airports, there are about 368 for personal use. These vary greatly in quality, from an unpaved landing strip with few or no facilities to a paved landing strip with advanced navigational facilities. There are also 6 public-use and 10 personal-use seaplane bases, and 13 public-use and 45 personal-use heliports.

The greatest density of airports is located in the southeastern and southwestern portions of the state. The least density is found in the rugged, low-population areas of north-central Pennsylvania.

Major International Airports

As airplane traffic increased, the international airports of Philadelphia and Pittsburgh became the major hubs of the Pennsylvania system. The advent of the jet age required enlargement of these airports. Philadelphia and New York were selected to demonstrate the importance of the Boeing 707 in carrying large passenger loads, for these two cities had the only runways on the East Coast capable of having the 80-ton, 140-passenger airplanes land at a speed of nearly 130 miles an hour. To accommodate the larger planes, Philadelphia's main instrument-landing runway was extended from 5,000 to 7,500 feet in the late 1950s, and in the 1970s the airport added additional waterfront acreage to extend the runway to a length

Fig. 14.9. International and regional airports, Pennsylvania, 1990.

of 10,499 feet. An additional runway of 5,460 feet has been added. This airport services 10 national airlines and is a terminal for Lufthansa, British Airways, Air Jamaica, and Mexicana. A number of commuter routes feed into the airport. Philadelphia International now has direct service to more than 100 cities and nonstop service to more than 50 cities. The airport handles more than 14 million passengers annually.

Pittsburgh's first major airport was completed in 1931. After a bond issue was approved, hills were leveled and some 50,000 square yards of macadam were laid to develop the Allegheny County Airport, which served for more than 20 years as Pittsburgh's major air terminal. With increasing traffic and larger planes, it became evident that a new airport was needed. In 1946 the county commissioners secured a World War II military stopover airport 12 miles west of downtown Pittsburgh. The Greater Pittsburgh Airport opened in 1951 and in its first year served more than one million passengers. Both American and United airlines immediately recognized Pittsburgh as a prime new interim service point.

The airport was renamed Greater Pittsburgh International Airport in 1977, when its international terminal was opened. It was enlarged in 1992. The airport has four runways, 8,000, 9,500, 10,000, and 10,500 feet in length. The facility handles more than 10 million passengers annually and is served by nine national American carriers, of which U.S. Air has its headquarters at the terminal. Of the international carriers, British Airways and Nordair have service from the airport. Allegheny Commuter is the principal regional air carrier serving the airport.

Railroads

Railroad transportation dominated in the nineteenth century, but the twentieth century witnessed the decline of railroads as roads and air routes evolved. The place of railroads in the modern scheme of transportation has been dominated by two crucial themes. First is the persistent problem of excess capacity, stemming first from overbuilding in the nineteenth century and then from the competition of motor vehicles and air carriers. Second, in the nineteenth century railroads were the exemplification of private initiative, but in the twentieth century the railroads moved steadily toward ever-greater reliance on government intervention to deal with growing economic problems.

The Pennsylvania and the New York Central

The Pennsylvania Railroad remained the dominant railroad in the state in the first two decades of the twentieth century, not only serving a larger area but also carrying more passengers and freight than any other railroad. By the 1920s, however, the Pennsylvania Railroad began to experience serious economic problems.

As automobiles increased in number, travelers deserted the rails for long-distance travel. While overall passenger service declined, a strong demand continued for rush-hour suburban service into and out of the central cities, particularly Philadelphia and Pittsburgh. This created an excessive economic burden, for the railroads were forced to maintain a large pool of equipment and a substantial labor force to operate a service that handled passengers only in the early morning and late afternoon. For the rest of the time the equipment and labor sat idle. The result was an increasing deficit that the railroad could ill afford.

Competition with trucks was also developing for hauling freight. Unlike trucks, railroads have high terminal costs, as they are forced to maintain large marshaling and freight yards in many locations. Terminal costs remain the same no matter how far a commodity is shipped, a factor that works against short-distance rail shipments. Truck transportation is much more flexible. Trucks can pick up freight at its origin, reaching every part of the city or town, and thus can move small shipments easily.

With the decline of passenger and freight service, the economic problems of the Pennsylvania Railroad became acute in the late 1920s. The economic depression of the 1930s aggravated the situation. Because of lack of funds, modernization programs were not implemented and the railroad system deteriorated badly.

Finally, in the early 1950s the railroads launched a campaign to gain public support for their struggle against trucking firms. They hired a high-pressure relations firm, Carl Byoir & Associates of New York, as lobbyists. Initially, they appeared to be successful. When the Pennsylvania legislature passed a Big Trucks Bill, raising from 50,000 to 60,000 pounds the maximum gross weight of truck and load, Governor John S. Fine vetoed the legislation. Fine's veto meant a reported $5 million worth of freight traffic in the state would be retained by the Pennsylvania Railroad.

In 1953 lawsuits were filed by both the truckers and the railroads, claiming that each was vilifying and slandering the other. After a long series of court battles, in 1957, the courts declared for the truckers. The decision stated that the railroads had made "a deliberate attempt to injure a competition for an illegal purpose by destroying public confidence in it." The Pennsylvania Motor Truck Association was awarded $852,000 in damages.

By 1957 the Pennsylvania Railroad and the New York Central had become so economically distressed that they began to consider cooperation in order to survive. It was initially envisioned that the Pennsylvania would serve the core of one area and the New York Central the core of another area. Thus the historic competition would be preserved, while the chances of survival would be strengthened.

This cooperation did not develop. Instead, on November 1, 1957, a joint committee of the two railroads was appointed to study a possible merger. The negotiations did not succeed, and on January 12, 1959, the New York Central discontinued discussions. As the economic situation of the railroads worsened, the committees were reconvened on October 25, 1961, and agreement was quickly reached on key issues. The New York Central would sell its 20 percent interest in the Baltimore & Ohio, and the Pennsylvania Railroad would divest itself of its one-third interest in the Norfolk & Western. The New York Central would be merged into the Pennsylvania Railroad, and the system would be called the Penn Central.

On January 12, 1962, the boards of both railroads approved the merger plan, and in May the stockholders overwhelmingly endorsed it. In August 1962 the Interstate Commerce Commission (ICC) began hearings that would continue for about four years. A major concern of the ICC was that the Penn Central's size would be unmanageable. The railroad executives stated that because many of the routes had been abandoned, the merged railroad would be considerably smaller than the Pennsylvania alone had been in its heyday. By merging, it was hoped that the railroads would be able to modernize and become profitable. On April 27, 1966, the Interstate Commerce Commission finally authorized the merger, providing the new Penn Central organization would take over the bankrupt New York, New Haven & Hartford Railroad. However, the U.S. Justice Department continued to oppose the merger until November 1967. Finally on February 1, 1968, the merger was concluded.

With the merging of the two railroads, problems immediately became evident. The computer systems of the railroads had not been made compatible in advance, so data from the old New York Central could not be sent directly to Philadelphia, headquarters of the old Pennsylvania and now of the Penn Central. The merged railroads were soon confronted with lost waybills, missing freight cars, clogged yards, and dissatisfied customers, resulting in a sharp decline in business. The equipment of both lines had deteriorated badly, but with continued losses there were no funds to develop the hoped-for modernized electronically operated system.

The Penn Central continued to lose money. Its operating losses for the first quarter of 1970 were probably the greatest in American railroad history. Receipts from rail operations totaled about $5 million a day, but operating expenses were more than $6 million a day. In June 1970 the railroad had liabilities of $748,974,320 and assets of $426,472,382, for a capital deficit of $286,501,938. The balance sheet showed cash on hand of only $7,308,130. Although desperate efforts to secure funds were made, the Penn Central was forced to declare bankruptcy on June 21, 1970.

The Penn Central was the largest business failure in the history of the United States. If it had discontinued service, large areas not only of Pennsylvania but also of the northeastern United States would have had to operate under the bankruptcy laws of the United States.

Consolidated Rail Corporation (Conrail)

Besides Penn Central, five other railroads with lines in Pennsylvania—the Reading, the Erie, the Lackawanna, the Lehigh Valley, and the Central Railroad of New Jersey—declared bankruptcy in the early 1970s, as did the Boston & Maine and the tiny Ann Arbor Railroad. Despite strong opposition to nationalization of those seven lines, no other solution was found, and in January 1974 Congress passed the Regional Rail Reorganization Act. This new act established the United States Railway Association, out of which the Consolidated Rail Corporation (Conrail) was formed.

Beginning in 1974, Conrail provided service on more than 80 percent of the rail lines in Pennsylvania. Private railroads no longer provide a statewide regional system, but instead serve only regional areas. The Baltimore & Ohio has the most extensive network in western Pennsylvania, with tracks from Pittsburgh to Wheeling in the south and to Buffalo in the north. The Pittsburgh & Lake Erie and the Bessemer & Lake Erie are specialized freight lines in western Pennsylvania. Short routes entering Pennsylvania but originating outside the state are the Norfolk & Western, the Western Maryland, and the Delaware & Hudson.

In order to revitalize the system, the new law provided an initial outlay of $2.2 billion. Between 1976 and 1981 the government spent $3.3 billion to modernize the system. At the same time, there was also a major effort to reduce the miles of track. Between 1974 and 1982 the entire mileage of Conrail was reduced from about 30,000 to 15,000 miles. In Pennsylvania the railroad mileage was reduced from 8,064 to 6,961 and employment from 44,000 to 33,000. The lines that have been abandoned were branch lines with low traffic density, many of them in the anthracite and bituminous regions where truck traffic had replaced the railroads.

Conrail was originally responsible for numerous commuter trains, and it was required to serve Amtrak under the same general terms as other railroads. Initially Amtrak had to buy or lease all passenger equipment from Conrail. However, this divided responsibility for passenger service did not function well and Conrail has become a freight service organization only.

The Conrail system lost money until 1981. The streamlined Conrail has received no government revenues since 1980. In 1983 the railroad had a profit of $250 million. As soon as Conrail became profitable, the federal government wanted to return the railroad to private ownership. According to federal stipulation, Conrail could be sold to a corporation, to the public in a stock offering, or to railroad employees.

In June 1983 the railroad employees—members of 17 unions plus nonunion workers—made an offer to purchase Conrail for $2 billion, including $500 million in bank loans. The employee group planned to sell 20–30 percent of the stock to the public. The U.S. Department of Transportation, concerned that the employees did not have sufficient funds to maintain the railroad and that it would once again

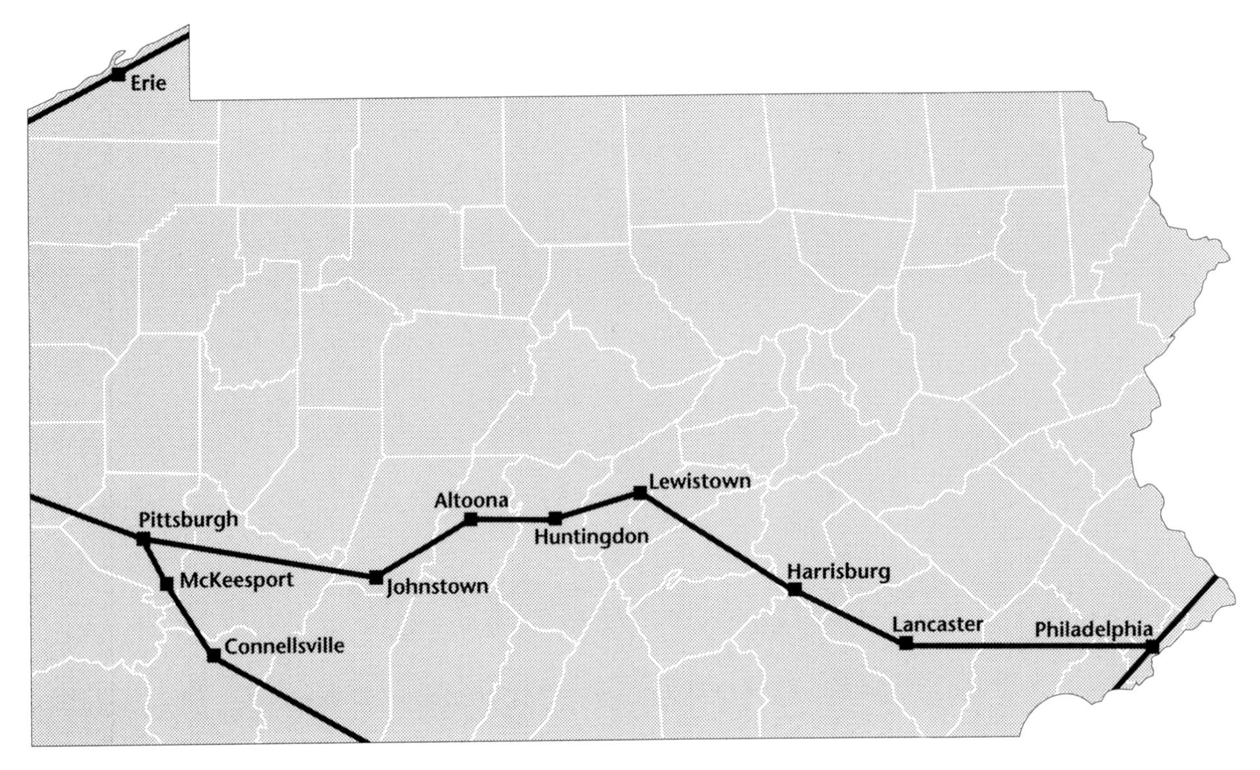

Fig. 14.10. Amtrak routes and stations, Pennsylvania, 1990.

need government aid, rejected the offer. In 1985 the Norfolk & Southern Railroad prepared offers for Conrail, but opposition rose in Pennsylvania out of fear that the repair shops would be removed from the state, and the offer failed. Finally, in early 1987 Conrail became a public corporation with a stock offering to the general public.

Amtrak

Passenger service on railroads was strongly neglected as it became increasingly unprofitable. In 1970 the private railroads lost $400 million operating what was left of their passenger fleet. The rail system had deteriorated to such an extent that on one stretch of the Penn Central system in Pennsylvania unwary passengers were sometimes thrown out of their dining-car seats.

To salvage the nation's passenger routes, the federal Rail Passenger Service Act of 1970 was passed, creating a national integrated system of passenger service that came to be known as Amtrak. It was thus conceived to provide passenger service between the largest American cities.

Within Pennsylvania, the Amtrak system centers on Philadelphia and Pittsburgh (Fig. 14.10). The major East Coast corridor route extends from Boston to New York, Philadelphia, and Washington. This is the major passenger route in the nation. The major east-west route across the state from New York through Philadelphia and Pittsburgh to Chicago, stopping at Lancaster, Harrisburg, Lewistown, Huntingdon, Altoona, and Johnstown. This route has consistently had two trains daily: the Pennsylvania from New York to Pittsburgh, and the Broadway Limited from New York through Philadelphia and Pittsburgh to Chicago. Service varies, however, depending on demand.

A third route, originating in Washington, enters Pennsylvania in the southwest, stopping at Connellsville and McKeesport, and connects at Pittsburgh to the New York–Chicago route. A fourth route crosses northwestern Pennsylvania on the Buffalo-Chicago route, stopping at Erie.

In the late twentieth century vast areas of the state are not served by passenger service. Only the highways and automobiles serve the many local communities that had railroad passenger service in the nineteenth century.

Bibliography

Alexander, E. P. 1971. *On the Main Line: The Pennsylvania Railroad in the Nineteenth Century.* New York: C. N. Potter.

Bogen, J. T. 1927. *The Anthracite Railroads: A Study in American Enterprise.* New York: Ronald Press.

Daughen, J. R., and M. C. Kennedy. 1971. *The Wreck of the Penn Central.* Boston: Little, Brown & Co.

Faris, J. T. 1927. *Old Trails and Roads in Penn's Land.* Philadelphia: Lippincott.

Hilton, G. W. 1980. *Amtrak, the National Railroad Passenger Corporation.* Washington, D.C.: American Enterprise Institute for Public Policy Research.

Jones, P. R. 1950. *The Story of the Pennsylvania Turnpike.* Mechanicsburg, Pa.: P. R. Jones and E. N. James.

Jordan, P. D. 1948. *The National Road.* Indianapolis, Ind.: Bobbs-Merrill.

MacAvoy, P. W., and J. W. Snow, eds. 1977. *Railroad Revitalization and Regulatory Reform.* Washington, D.C.: American Enterprise Institute for Public Policy Research.

McLean, H. H. 1980. *Pittsburgh and Lake Erie Railroad.* San Marino, Calif.: Golden West Books.

Miller, E. W. 1992. "The Pennsylvania Turnpike." In *Geographical Snapshots of North America,* ed. D. J. Janelle. New York: Guilford Press.

Miller, E. W., and R. M. Miller. 1985. *Pennsylvania Transportation: A Bibliography.* Monticello, Ill.: Vance.

Rhoads, W. R. 1960. "The Pennsylvania Canal." *Western Pennsylvania Historical Magazine* 43:203–238.

Saunders, R. 1978. *The Railroad Mergers and the Coming of Conrail.* Westport, Conn.: Greenwood Press.

Smith, F. K., and J. P. Harrington. 1981. *Aviation and Pennsylvania.* Philadelphia: Franklin Institute Press.

Soel, R. 1977. *The Fallen Colossus* (*Penn Central R.R.*). New York: Weybright & Talley.

Stuehldreher, M. 1974. "Automobiliousness." *Western Pennsylvania Historical Magazine* 57:275–288.

15

MANUFACTURING

E. Willard Miller

In the nineteenth and early twentieth centuries the growth of manufacturing played a dominant role in the expanding economy of Pennsylvania. Although growth in the tertiary industries now exceeds that of manufacturing, Pennsylvania continues to be a major manufacturing state, with 5.5 percent of the manufacturing workers of the nation, exceeded only by California and New York. The annual value of manufactured goods exceeds $150 billion and generates a payroll of more than $30 billion. Pennsylvania ranks second in the nation in employment in primary metals and in stone, clay, and glass products, and third in employment in apparel and other related products, paper and paper products, food and kindred products, and leather or leather products.

Employment Trends

Employment in manufacturing rose steadily in the nineteenth century as Pennsylvania created an industrial society. By 1850 there were 146,766 workers in manufacturing industries. Between 1850 and 1880 the number grew to 387,072, a growth of 240,306. Between 1880 and 1890 the manufactural work force increased by 60 percent, the largest percentage increase in a single decade, to a total of 663,960. The work force continued to increase to an initial peak in 1919 of 1,135,837. Between 1909 and 1919, 261,294 workers were added to this work force (Table 15.1).

The growth of manufacturing ended in 1919. During the 1920s, employment was essentially stable, and in 1929, with the beginning of the Great Depression, employment fell to 1,014,046, some 121,791 lower than in 1919. In the early 1930s employment continued to decline to a low of about 800,000 employees in 1933–1934, and only at the end of the 1930s, when the nation began to rearm, did the employment rise to 964,273 in 1939. The demand for manufactured goods rose rapidly during World War II and in the early postwar period, when America supplied the world with manufactured goods. Pennsylvania's manufacturing thrived

Table 15.1. Employment in manufacturing, 1899–1992.

Year	Employment
1899	663,960
1909	877,543
1919	1,135,837
1929	1,014,046
1939	964,273
1949	1,439,534
1959	1,386,000
1969	1,579,000
1979	1,407,000
1989	1,048,000
1992	950,000

SOURCE: U.S. Department of Commerce.

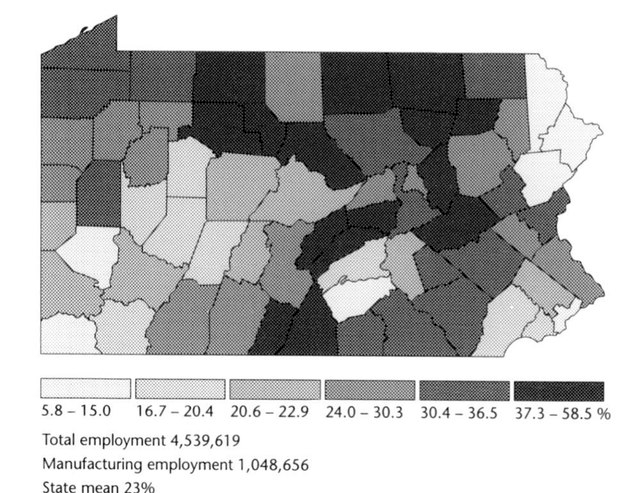

Fig. 15.1. Manufactural employees as a percentage of total employment, Pennsylvania, 1989.

during this period, and employment in manufacturing rose to an all-time peak of 1,618,000 in 1953.

Between 1953 and 1969 manufactural employment was nearly stable. Yearly fluctuations reflected general economic conditions. In 1969 employment stood at 1,579,000. Since then, manufactural employment has steadily declined. In the 1970s the decline was a gradual dropping to a total of 1,407,000 in 1979.

The early 1980s experienced a period of rapid decline in employment in manufacturing. The traditional industries of textiles, iron and steel, and leather products declined most rapidly. By 1984 manufacturing employment totaled only 1,127,344, a decline of 284,000 jobs since 1979. It remained stable until the economic depression of the early 1990s and has once again decreased to a modern low of 950,000 workers in 1992. In 1992 Pennsylvania had a total employed work force of about 5,460,000, of which less than 3 percent were engaged in primary activities and about 25 percent in secondary activities (22 percent in manufacturing and 4 to 5 percent in construction). Tertiary activities now provide about 72 to 73 percent of the state's employment, or about 4,040,000 jobs.

Of the 20 two-digit Standard Industrial Classification industries, 16 experienced declines in employment from 16.6 to 92.7 percent between 1969 and 1989. Tobacco products had the greatest decline (92.7 percent), followed by leather and leather products (70.0 percent), primary metals (63.3 percent), textile mill products (62.3 percent), apparel and other textile products (53.6 percent), electronics and other electronic equipment (47.7 percent), petroleum and coal products (43.5 percent), transportation equipment (40.6 percent), industrial machinery and equipment (31.1 percent), furniture and fixtures (27.3 percent), food and kindred products (22.4 percent), instruments and related products (22.0 percent), chemicals and allied products (19.5 percent), paper and allied products (18.1 percent), fabricated metals (17.2 percent), and stone, clay, and glass products (16.6 percent). Three two-digit industries experienced employment growth ranging from 21.0 to 120.7 percent. The greatest growth occurred in lumber and wood products (120.7 percent), followed by rubber and miscellaneous plastic products (34.1 percent) and printing and publishing (21.0 percent).

Manufacturing Employees as a Percentage of Total Employees

Although manufacturing employs about 23 percent of the total work force of 4,539,619, there is tremendous variation in the state (Fig. 15.1). The range is from 5.8 percent of the workers in Greene County to 58.5 percent in Elk County. In 1989 there were 44 counties above the mean and 23 counties below. Although the service economy has come to dominate employment in the state, there are a large number of counties, possibly 20, where manufacturing remains the major source of employment. These are largely rural counties where a single industry dominates the industrial structure. The economy of the county is thus based on the health of a single establishment. If this industry flourishes, the economy of the county remains strong. If, on the other hand, employment at this plant declines, the entire economy of the county suffers.

In the past 30 years the total work force in the counties of Pennsylvania has been quite stable. In most of the manufacturing counties, employment has declined and service industries have grown. Thirty years ago economists assumed that the vitality of a region's economy was inextricably linked to the health of the region's manufacturing sector. According to this view, only manufactured goods could be exported, and hence only manufacturing enjoyed a multiplier effect in employment, meaning that jobs in export-oriented industries lead to additional jobs in the firms providing local consumer goods. In Pennsylvania, however, while there was a loss of jobs in manufacturing, the overall economy continued to grow. The increase in jobs in the service industries compensated for the manufacturing job losses, so there was little change in the total employment.

This suggests that the health of the economy is not necessarily tied to manufacturing. In fact, services as well as manufactured goods can be exported, so the multiplier effect applies to services as well as manufacturing. It is now well accepted that such services as insurance and financial services can be exported.

In order to understand why employment in both manufacturing and services in many counties in Pennsylvania has changed, it is necessary to look at the causes of changes in employment and their multiplier effects. Three major factors account for changes in employment: productivity (changes in the amount of output produced with each unit of labor), demand (changes in the nature and location of demand for the products or services produced), and business practices (changes in the ways goods and services are produced). Each factor contributes to the potential changes in the multiplier effect associated with the major sectors in the economy, creating fewer job losses or gains.

The causes and consequences of replacement of manufacturing jobs by service jobs has been the subject of a great deal of conjecture. The average wage levels in manufacturing are higher than those in the service sectors, but within the different manufacturing sectors there are significant variations in wage levels, such as between food and kindred products, and instruments and related products. The difference in wages appears to vary with the size of the establishment, higher wage levels being found in the larger establishments. Furthermore, in determining the effect on the economy, the degree to which a firm sells or purchases goods is a more important factor than whether the firm is in the service or manufacturing sector. In particular, the percentage of purchases made within the region has a major influence on the size of the multiplier effect.

One of the negative effects of the change from manufacturing to service-dominance is that in many cases those who lost manufacturing jobs do not find comparable jobs in the growing service sectors. Instead, new jobs in the service sector were taken by new entrants into the labor force at lower wage rates.

The loss of manufacturing jobs did not reduce the volume of production in Pennsylvania. In fact, in 1989 real output (adjusted for inflation) was approximately the same as in 1969, although this production did come from a smaller number of establishments. There is also evidence that Pennsylvania firms secured more of the intermediate inputs from outside the state. A significant consequence of this process has been the decline in the multiplier effect.

International competition, the attraction of new investment opportunities elsewhere in the United States, and the belief that wage costs were higher in Pennsylvania caused many manufacturing businesses to close or relocate in another state. In the twenty-two-digit manufacturing sector in Pennsylvania in 1987, wages accounted for 6.6 to 20.3 percent of total value of shipment, compared with a variation of from 9 to 26 percent for the United States as a whole. But wage rates alone may be a misleading indicator of the productivity of a region's labor force. For example, in the food and kindred products sector the average wage and salary in larger manufacturing firms in 1987 were almost twice that paid in smaller firms. However, larger firms had a significant advantage in sales generated for each dollar spent on wages. Within all two-digit manufacturing sectors in the state, sales of larger firms were 30–50 percent higher than in smaller firms for every dollar spent on wages and salaries. While higher wages may have contributed to the decline of manufacturing, attention must be focused on the productivity of expenditures of wages and salaries, and not just on the levels. When Pennsylvania is compared to the United States as a whole, its manufacturing productivity is equal to or exceeds the national average. It

Table 15.2. Manufacturing in Pennsylvania, 1987 (in millions of $).

	Employee Payroll	Production Worker Wages	Value Added by Manufacturing	Cost of Materials	Value of Shipments
Food and kindred products	$1,774.0	$1,107.4	$7,039.3	$9,586.5	$16,588.3
Tobacco products	13.3	9.4	34.8	80.0	115.7
Textile mill products	465.8	336.1	1,047.6	1,219.5	2,238.8
Apparel and other textile products	1,091.9	816.5	2,198.9	1,816.7	4,019.1
Lumber and wood products	518.7	373.8	1,177.3	1,430.7	2,601.3
Furniture and fixtures	427.9	276.6	927.8	775.4	1,688.6
Paper and allied products	1,018.5	697.6	2,980.5	3,639.8	6,608.0
Printing and publishing	1,812.5	927.4	4,710.2	2,639.0	7,334.2
Chemicals and allied products	1,142.4	527.2	5,471.9	3,930.1	9,361.1
Petroleum and coal products	245.1	148.6	1,204.3	6,271.9	7,402.4
Rubber and misc. plastic products	861.3	554.0	2,079.5	1,971.6	4,027.6
Leather and leather products	134.1	106.3	265.4	371.4	629.1
Stone, clay, and glass products	979.4	724.1	2,643.8	2,123.1	4,769.1
Primary metals industry	2,236.8	1,636.8	4,305.8	7,679.8	11,908.7
Fabricated metals products	2,288.7	1,469.8	4,696.4	4,773.3	9,503.4
Industrial machinery and equipment	2,408.8	1,340.9	5,183.2	3,762.6	8,812.3
Electronic equipment	1,794.6	1,047.5	4,332.2	3,468.0	7,751.0
Transportation equipment	1,504.8	862.2	3,637.6	3,548.4	7,114.7
Instruments and related products	1,061.4	496.7	2,484.0	1,592.2	4,065.8
Misc. manufacturing industries	416.4	272.3	1,185.6	926.1	2,112.0
Total	25,301.6	13,731.1	57,605.2	61,606.1	118,651.3

Source: U.S. Department of Commerce.

then appears that the remaining establishments are able to compete effectively at the national level. Regrettably, the rise in productivity has reduced the number of jobs.

Changes in business practices have also been an important component of the change in manufacturing employment. In fact, all sectors of the economy have reduced their purchases of manufactured products since 1969. Similarly, improvements in labor productivity have contributed to the reduction in employment demand in manufacturing. By contrast, there has been a significant increase in demand for services. There is no doubt that increases in female labor force participation changed not only the structure of employment but also the volume and composition of demand generated by the larger number of two-earner households.

Until the late 1980s there was little evidence of improvements in labor productivity in services. However, the same phenomena that affected manufacturing 20 to 30 years ago (pressure to contain labor costs and to improve labor productivity) appear to have become a factor in the service sector. Here, labor costs are an even higher percentage of total costs (ranging from 22 to 55 percent), and in the recession of the early 1990s there was a substantial reduction in service employment in order to gain an edge in an increasingly competitive environment.

Production Workers in Relation to Other Measures

Ratios of different measures of manufacturing (Table 15.2) provide information on regional variations within the state. The state mean ratio of production worker wages to cost of materials reveals that production wages are 22.2 percent of the cost of materials (see Fig. 15.2). Wide variations occur within the state, from a maximum of 54.2 percent in Cameron County to 8.8 percent in Delaware County. Forty-one counties were located above the mean, 21 counties were below, and 5 counties had no data. The large range in values illustrates the great variations in the importance of wages to cost of materials in the state.

"Value added" is a standard measure of the importance of manufacturing in a region. At the state level, production worker wages in 1987 were 23.8 percent of the value added (Fig. 15.3). There

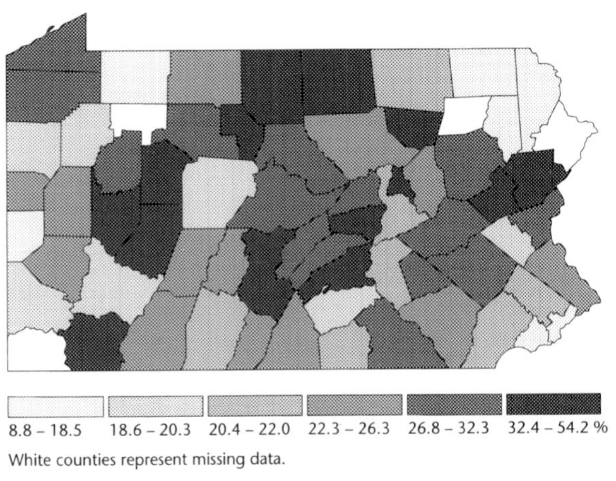

Fig. 15.2. Production worker wages as a percentage of cost of material, Pennsylvania, 1987.

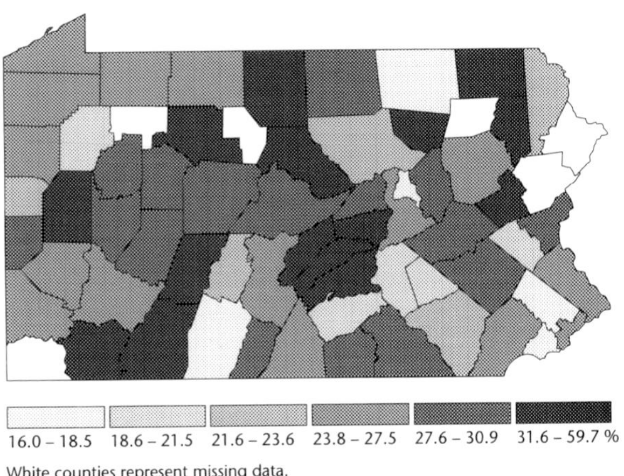

Fig. 15.3. Production worker wages as a percentage of value added, Pennsylvania, 1987.

were wide variations in the state, from 59.7 percent in Susquehanna County to 16.0 percent in Delaware County. Forty-seven counties were above the Pennsylvania mean, 15 counties were below the mean, and 5 counties had no data. In general, the higher the value added, the greater the value of labor input.

In 1987 production worker wages averaged 11.5 percent of the value of manufacturing shipments in the state. The county range was from 23.3 percent in Cameron County to 5.7 percent in Delaware

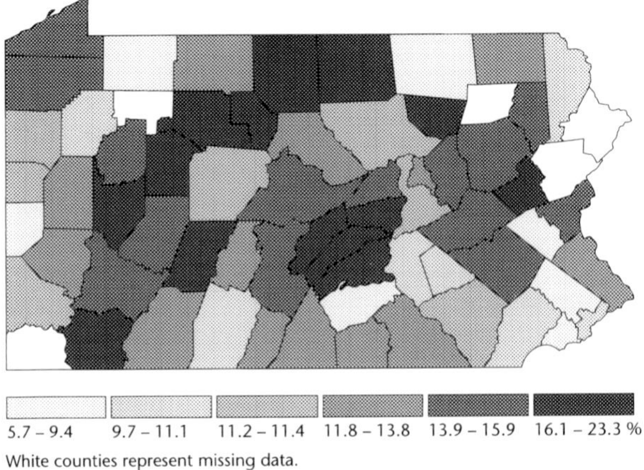

Fig. 15.4. Production worker wages as percentage of shipment value, Pennsylvania, 1987.

County (Fig. 15.4). There were 41 counties above the mean, 21 counties below, and 5 counties had no data. Note the large number of counties in northern and western Pennsylvania, where the wages were a high percentage of the shipment value. In southeastern Pennsylvania the percentage of wages was low in relation to the value of shipments. A lower percentage indicates the lesser importance of production worker wages in the manufacturing process; in contrast, the higher the value the greater the significance of wages.

Exports and Imports

Exports are important to the manufactural economy of the state. In 1987 manufactured exports totaled $16,964 million, or about 14.3 percent of the total manufactural shipment of $118,651 million. Primary metals led the exports with 17.5 percent of the total, followed closely by electronics and other electronic equipment (17.1 percent), industrial machinery and equipment (11.9 percent), fabricated metal products (8.6 percent), and chemicals and allied products (7.5 percent) (see Table 15.3).

When manufactured exports of a particular industry are related to the total manufactural shipments of that industry, a number of industries are strongly related to its export markets. Exports are most important to electronics and other related equip-

Table 15.3. Manufactural exports from Pennsylvania, 1987 (in millions of $).

	Value of Manufactural Products	Value of Manufactural Exports	Total Exports as % of Total Shipment
Food and kindred products	16,588.3	568.5	3.4
Tobacco products	115.7	29.8	25.8
Textile mill products	2,238.6	188.9	8.4
Apparel and other textile products	4,019.1	147.6	3.7
Lumber and wood products	2,601.3	158.6	6.1
Furniture and fixtures	1,688.6	44.7	2.6
Paper and allied products	6,608.0	763.3	11.6
Printing and publishing	7,334.2	429.0	5.8
Chemicals and allied products	9,361.1	1,277.3	13.6
Petroleum and coals products	7,402.4	787.7	10.6
Rubber and misc. plastic products	4,027.6	578.0	14.4
Leather and leather products	629.1	95.2	15.1
Stone, clay, and glass products	4,769.1	466.7	9.8
Primary metals industries	11,908.7	2,979.8	25.0
Fabricated metal products	9,503.4	1,471.3	15.5
Industrial machinery and equipment	8,812.3	2,033.2	23.0
Electronic equipment	7,751.0	2,913.3	37.6
Transportation equipment	7,114.7	1,067.2	15.0
Instruments and related products	4,065.8	741.5	18.2
Misc. manufacturing industries	2,112.0	221.5	10.5
Total	$118,651.3	$16,964.1	14.3

SOURCE: U.S. Department of Commerce.

ment, with 37.6 percent of production exported. This is followed by the primary-metals industry and tobacco products (25 percent each), industrial machinery and equipment (23 percent), and instruments and related products (18 percent). Exports have increased in importance, particularly since 1980.

The leading manufacturing counties in the state are also the leading manufactural exporting counties. There were 18 counties above the mean, with 83 percent of the total exports, and the top one-third above the mean had 33 percent of the exports. Ten of the counties were concentrated in the southeastern manufactural belt of Pennsylvania, three were centered in the Pittsburgh region, and the others were scattered (Fig. 15.5).

The port of Philadelphia has well over 90 percent of the total exports from the state. In 1990 imports totaled $15.2 billion while exports totaled $2.9 billion. Of the imports, petroleum and petroleum products totaled 51 percent in value, followed by road vehicles (6 percent), fruit and vegetables (4.9 percent), animals (4.6 percent), and iron and steel (3.8 percent). Nearly 100 types of products were imported. Of the exported products, coal and coke had the greatest tonnage, with 3,213,923 tons in 1990. This was about 4.4 percent of the total export value. When based on value, road vehicles had 21 percent of the value, followed by chemicals and pyrotectonic materials (8.6 percent), metalliferous ores, and scrap (5.4 percent).

In 1990 some 170 countries carried on trade with Pennsylvania. The countries with major petroleum exports to the state included Nigeria, Venezuela, Saudi Arabia, Mexico, and Tunisia. The industrial nations of Europe and Japan provided a wide variety of commodities.

The Pennsylvania Industrial Development Authority

The Pennsylvania Industrial Development Authority (PIDA) was created in 1956 by the state legislature to stimulate the construction and acquisition of industrial buildings to increase employment in Pennsylvania. There are three kinds of PIDA projects: an

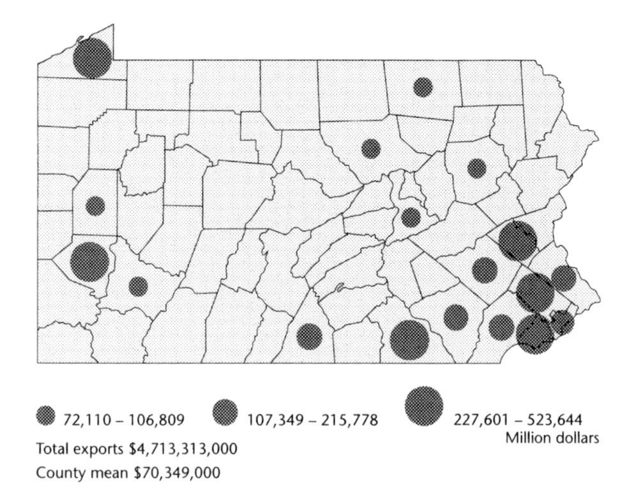

Fig. 15.5. Value of exports of manufactural production, Pennsylvania, 1987.

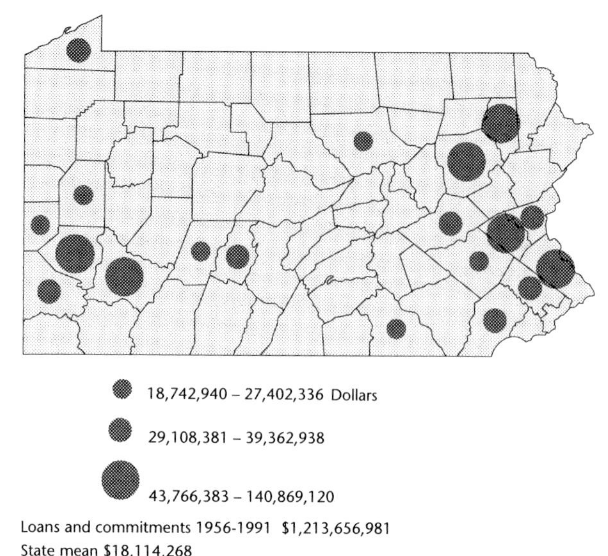

Fig. 15.6. Loans and commitments of the Pennsylvania Industrial Development Authority (PIDA), 1956–1991.

industrial development grant, an industrial park project, and a multiple-tenancy building project.

The industrial development grant may consist of a manufacturing, research, or agricultural enterprise established by an industrial development agency in a critical economic area. The Act stipulates that PIDA may participate in financing the cost of establishing an industrial development project, which includes the cost of land and buildings and related costs, but does not include the cost of machinery and equipment and related installation and maintenance costs. PIDA has defined an industrial enterprise as an enterprise (other than a mercantile, commercial, or retail establishment) that requires substantial capital (no less than $200,000) and that creates employment opportunities of not less than 25 jobs, existing or to be created.

A second kind of PIDA project is an industrial park, which consists of land and buildings required by an industrial development agency for establishment of two or more industrial development projects. The third kind of PIDA project is a multiple-tenancy building project, defined as a facility or undertaking acquired or constricted for occupancy by two or more industrial, manufacturing, or research and development enterprises.

Between 1956 and June 1991 some 2,939 projects were funded. Every county in Pennsylvania had at least one project, and the top 20 counties above the mean had 78 percent of the loans and commitments (Fig. 15.6). Loans and other commitments totaled $1.2 billion, and project costs exceeded $3.3 billion. The planned employment totaled 257,353, with an annual payroll of $2.5 billion. From 1956 to 1991 the state appropriations to PIDA totaled $393 million. Annual appropriations varied from zero dollars to a maximum of $35 million in 1991–1992. The annual number of projects varied from 24 between 1955 and 1957 and 185 in 1983–1984.

Changing Factors of Localization

The factors that attracted industry to Pennsylvania in the nineteenth century have changed dramatically in the twentieth. A major factor in the development of industry in the state was the availability of low-cost energy, primarily from bituminous coal. The primary factor in the growth of the greatest heavy industrial center in the nation in southwestern Pennsylvania was based on low-cost available energy. The smoky environment was a sign of prosperity, and many industries, with their extravagant consumption of energy, were known as "smokestack" industries.

As fuel costs have risen in the twentieth century, plants have become more energy efficient, and total consumption of energy has declined. For example, coal consumption in the processing of a ton of iron ore has declined from about two tons to one ton. Blast furnaces no longer need to be oriented to a specific fuel source. Because fuels are no longer a critical cost factor, present-day manufacturing is not tied to an energy site. It would be difficult to find a single manufacturing company that has developed in Pennsylvania in the past quarter-century specifically because of a local power source.

Nineteenth-century industry was labor intensive. Industry was at a much lower technological level, and unskilled and semi-skilled workers performed numerous industrial tasks. The growing population of Pennsylvania, supplemented by vast numbers of immigrants, provided a pool of low-cost labor. Female labor was particularly important in establishing the textile and apparel industries of eastern Pennsylvania. In the twentieth century, as labor costs rise, there is a growing efficiency in the use of human labor, and machines have replaced humans in many plant operations. It is now recognized that robots can produce a better product than human labor. Over the decades wages have risen as greater skills have been acquired and as the influence of unions in the industrial regions of Pennsylvania has increased. As a consequence many plants have migrated to the newer industrial areas of the nation, seeking lower-cost labor. Competition has also arisen from foreign countries that have an abundant supply of low-cost labor. A critical factor for the future of manufacturing in Pennsylvania is the growing importation of industrial goods from low-cost labor areas.

In the nineteenth century most industrial raw materials were obtained locally or regionally. As manufacturing became more complex and diversified, the sources of raw material are now worldwide. This factor releases industry from localization to a raw material source.

A century ago it took only a relatively small amount of capital to start an industry. A limited amount of low-cost equipment was needed, and an industry could thrive with a small output of goods. As industry became increasingly complex, more capital was needed, and today high capitalization characterizes most modern industrial plants. Because advances in modern industry are extremely rapid, a large capital investment must be made in research and development to remain competitive. Small firms rarely have these resources and either disappear or are purchased by larger companies if they have a unique product with market potential.

Originally industry was limited to satisfying local and possibly regional markets. Cost of transportation was only a minor factor. Because many industries now seek national, or even world markets, low-cost transportation is often an important consideration. In the nineteenth century much of the industrial output was heavy and bulky, whereas many twentieth-century growth industries, such as electronic equipment and instruments, produce items that are small and extremely costly. These products can withstand the cost of distant transportation.

In the nineteenth century the market for manufactured goods was relatively small. In the twentieth century, with an increasing population with high buying power, the market for manufactured products increased greatly. To remain competitive, an industry must operate at its optimum economy-of-scale. The emphasis in the twentieth century is to produce each unit of a product at the lowest possible cost. All industry is migratory in seeking the region where it can produce most efficiently.

Regional Patterns of Manufacturing

In 1969 the county manufactural employment in Pennsylvania (Fig. 15.7) varied from a low of 129 employees in Pike County to 265,764 in Philadelphia County, showing that counties in Pennsylvania experienced differential rates of manufactural growth. The Pennsylvania county mean was 23,570 employees. Eighteen counties were above the mean and 49 were below. The 18 counties above the mean had 77.6 percent of total employment. Two major and two minor regional concentrations of manufacturing are readily identifiable. Southeastern Pennsylvania, with 11 of the counties, contains the largest concentration, centering on Philadelphia. The second major concentration of three counties, centering on Allegheny County and the city of Pittsburgh, is in southwestern Pennsylvania. Two minor concentrations consist of three counties in northeastern Pennsylvania and an isolated county, Erie, in northwestern Pennsylvania.

Between 1969 and 1989 manufactural employment declined from 1,579,213 to 1,048,656 (see Fig. 15.8). As a result, the county mean declined to 15,651. In 1989 some 20 counties were above the mean, and these counties continued to dominate,

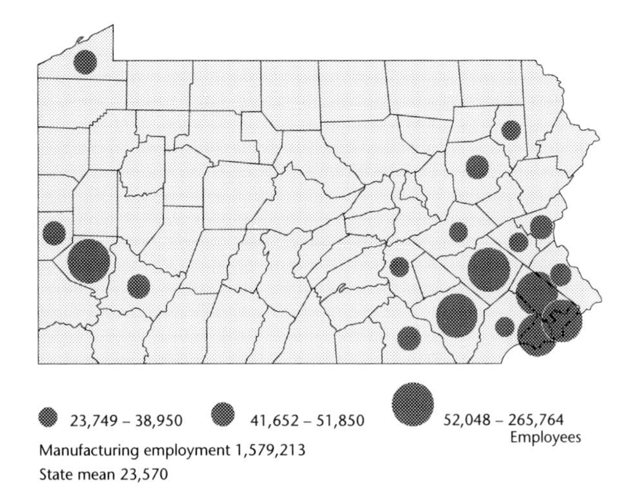

Fig. 15.7. Employment in manufacturing, Pennsylvania, 1969.

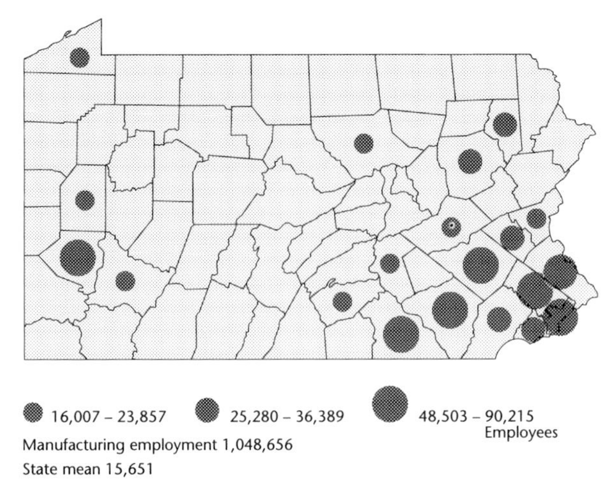

Fig. 15.8. Employment in manufacturing, Pennsylvania, 1989.

with 73.7 percent of the state total. Most significant is that the 7 counties in the top third above the mean had 45.1 percent of the manufactural employment of the state. Within the counties above the mean, the rank changed slightly between the two base years. Most striking was the rise of Montgomery County as the leading manufactural county of the state, replacing Philadelphia, which had been the leader since its settlement in 1682. Also significant is the rising importance of Lancaster, York, Cumberland, and Lehigh counties on the western and northern borders of the region. This reflects the westward expansion of manufacturing in southeastern Pennsylvania. Equally significant was the decline in importance of the heavy-industry counties of Allegheny, Beaver, and Northampton.

A basic question now is whether, although the spatial pattern of the major manufacturing counties was relatively stable between 1969 and 1989, there have been differential rates of growth within the leading counties that lie above the mean. The growth differential of counties can be measured when compared to change at the state level. The net-shift technique is used to measure these changes in counties. Manufactural employment declined by 33 percent, so in net-shift analysis to experience a comparative gain a county had to have employment in 1989 greater than 67 percent of that of 1969. If employment is less than 67 percent of the 1969 total in 1989, the county experienced a comparative loss (Fig. 15.9).

Comparative Growth Regions

In southeastern Pennsylvania nine counties—Bucks, Montgomery, Chester, Berks, Lancaster, York, Lehigh, Dauphin, and Cumberland—form the largest region in the state where manufacturing experienced a comparative gain in employment. These nine counties in 1989 had 411,889 employed in manufacturing, or 40 percent of the manufactural employment of the state. Of these counties, Lancaster and Montgomery had the greatest comparative gains, of 24,876 and 23,549. Montgomery County (with 90,215 employed in manufacturing) had about 8.6 percent, and Lancaster (with 61,845) had 5.8 percent of the manufactural employment of the state.

The industrial structure of the southeastern Pennsylvania comparative growth region consists of industries that have experienced their greatest growth in recent years. The instruments and related products, an example of a high-value-added industry, increased its employment in the counties from 10,150 in 1969 to 20,264 in 1989. During the same period employment in electric and electronic products grew from 29,055 to 32,467, and in printing and publishing from 11,774 to 34,927. The industries that experienced great growth in the nineteenth and early twentieth centuries—for instance, textiles, apparel, paper, chemicals, leather, primary metals (particularly iron and steel), and glass products—have exhibited little growth potential.

A series of factors have encouraged growth in this

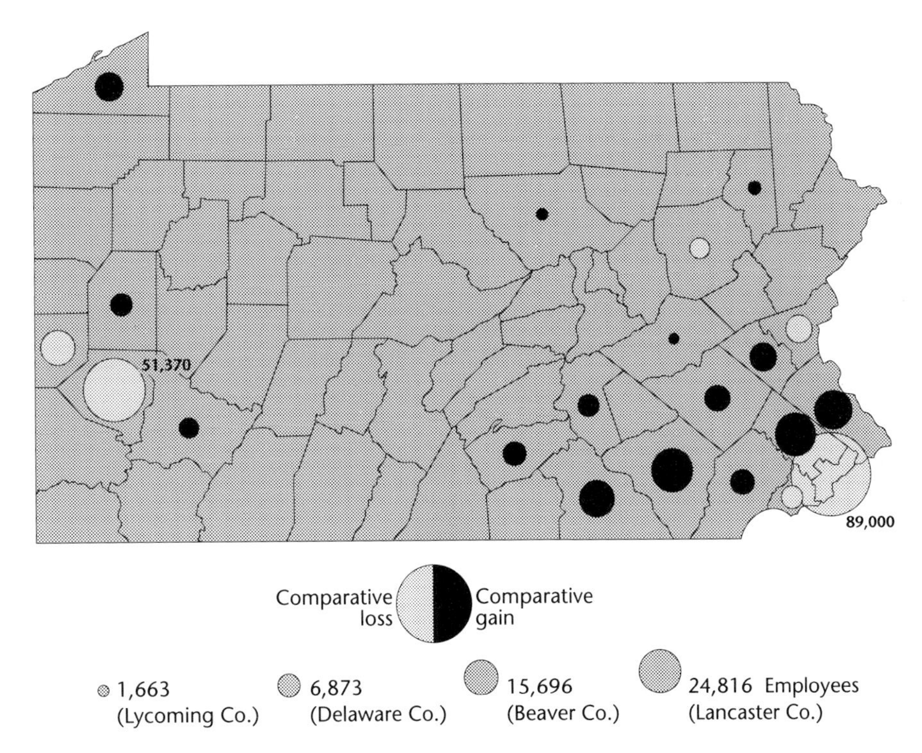

Fig. 15.9. Comparative loss or gain in manufactural employment, Pennsylvania, 1969–1989.

region. The migration of skilled labor, including engineers and scientists, to the suburban areas has been most important. There, industrial plants are modern with efficient operations, and the availability of relatively low-cost land for industrial plants, including large parking areas for employees, has been an additional factor. The emphasis on high-value-added industries ensures that products can seek national and world markets. The development of a modern transportation network has been crucial for rapid assembly of raw materials and distribution of finished products. Finally, development of such amenities as parks, golf courses, riding academies, and cultural centers has provided a gracious living environment.

Outside the southeast, Erie County, with a comparative gain of 10,425 employees, is the only large county with a significant gain. Erie is part of the lower Great Lakes corridor and has developed such modern industries as electric and electronic products, instruments and related products, and transportation equipment. Other counties with small comparative gains include Lycoming, Westmoreland, Schuylkill, and Lackawanna.

Comparative Declining Regions

The major absolute losses and comparative declines in manufactural employment between 1969 and 1989 occurred in Philadelphia, Delaware, and Northampton counties in the southeast, and in Allegheny and Beaver counties in the southwest. Absolute employment in Philadelphia and Delaware counties declined from 327,235 to 123,293, a loss of 203,942 and a comparative loss of 95,923. During the same period employment in manufacturing in Philadelphia and Delaware counties dropped from 20 percent of the state's total to 11.7 percent. The first industries to decline in Philadelphia and Delaware counties were those that had dominated in the nineteenth century, most notably the textile and apparel industries. In the period 1969–1989 textile-mill employment in Philadelphia and Delaware counties dropped from 18,314 to 2,679, and the apparel industry has experienced an employment decrease from 45,219 to 12,000. Other older declining industries include paper, printing and publishing, tobacco, leather products, lumber and wood products, fabricated metals, petroleum refining, and food and kindred products. These declining older industries have not been replaced by modern, twentieth-century industries. The older industries, have not only experienced a decline in employment but the modern twentieth-century industries have also been experiencing a decline. For example, employment in electric and electronic equipment dropped from 29,502 to about 400, machinery from 30,352 to 4,646, transportation equipment from 25,639 to 1,648, and instruments from 7,181 to 2,069. In fact, every major category of industry (the 20 two-digit Standard Industrial Classification) declined between 1969 and 1989.

The decline in manufacturing in Philadelphia and Delaware counties is attributable to a wide range of factors. The nineteenth-century industries have not been able to compete economically with other regions. At the early stages of this change in locational evolution, new regions developed in the United States, such as the development of the textile industry in the Piedmont, but in the migrational process new developments have frequently been in foreign nations, particularly Third World countries where low-cost labor exists.

The twentieth-century industries grew significantly until the mid-1950s but have since experienced a significant decline. The general economic environment of decline has been caused by many factors, including old and antiquated plants for which renovation is too costly, lack of space for new plants, migration of high-quality labor to the suburbs, high wage rates, high cost of land, high taxes in the urbanized areas, a high crime rate in the industrial districts, transportation congestion due to an antiquated urban road system, and a lack of modern amenities, such as parks, recreational facilities, and cultural events. A modern, complex industry requires continual product advancement. In the older regions inventive abilities decline.

In southeastern Pennsylvania, Northampton County employment declined by 27,138 from 52,043 in 1969 to 24,807 in 1989. The comparative loss was 9,932. The employment in Northampton County was dominated by the iron and steel industry, where major declines occurred. There were also major declines in the textile and apparel industries.

In southwestern Pennsylvania the comparative loss centered on Allegheny and Beaver counties. Employment in these two counties declined from 242,799 in 1969 to 91,588 in 1989, an absolute loss of 151,211 employees. The comparative loss was 67,066. These southwestern counties have long been dominated by the primary metals, where employment declined from 84,327 to 13,780. Other major industries with declining employment included food and kindred products, stone, clay and glass, fabricated metals, machinery, and electric and electronic equipment.

In northeastern Pennsylvania, Luzerne County's total manufactural employment between 1969 and 1989 decreased from 52,042 to 29,418, with a comparative loss of 5,454. The apparel industry that dominated employment in Luzerne County dropped from 18,258 to 5,789. Other major industries that experienced declines included textiles, leather, and electric and electronic equipment.

Manufacturing in Rural Pennsylvania

In 1989, of the 47 rural counties that were located below the county mean, 31 experienced comparative gains in employment and 18 experienced comparative losses. The comparative gains ranged from 4,033 in Centre County to 36 in Venango County, and the comparative losses ranged from 7,684 in Cambria County to 15 in Perry County. Diverse economic forces were shaping the manufactural trends in the smaller counties.

There are two broad regions where the comparative-growth counties dominate, and two regions where comparative loss dominates. A major comparative-growth area centers on south-central Pennsylvania, with Centre County at the core. The second comparative-growth region consists of the northern tier of counties (except McKean) from Crawford to Wayne County. The major comparative loss exists in southwestern Pennsylvania in the triangle—Mercer, Cambria, and Fayette counties. The second region of comparative loss centers on the anthracite region, with an arm extending westward through Clinton, Cameron, and McKean counties.

No single factor or even group of factors appears to explain growth or decline in the rural counties. In most counties the growth or decline could be attributed largely to changes in employment in one or two industries. In 28 of the counties the comparative change—gain or loss—was less than 1,000 employees.

In most of the counties that experienced a comparative loss, the major industries were the traditional industries of the nineteenth and early twentieth centuries. In southwestern Pennsylvania the primary metal and allied industries have spilled over into the rural counties. These counties, such as Cambria, have experienced the same trends as the major steel counties of Allegheny and Beaver. Similarly, in the anthracite region the textile and apparel industries developed adjacent to the major counties are now in decline. In a number of counties the decline of individual industries provided the basis for the comparative losses. In Armstrong County the decline of the glass and clay industries was most significant. The migration of an aircraft company to Florida was a major factor in the loss of employment in Clinton County. In McKean County the declining petroleum industry led to the decline in petroleum field equipment, and in Cameron County the decline of the electric and electronic equipment industry was significant.

A wide variety of factors had important influences on the counties when comparative growth occurred. In south-central Pennsylvania the westward spread of Megalopolis provided the incentive for growth of industry in Adams, Franklin, Juniata, Snyder, and Union counties. In Centre County the presence of The Pennsylvania State University provided a scientific and technological environment attracting such high-value-added industries as electronics, instruments, and specialty metal products. The growth of the Northern Tier counties is strongly tied to that of southern New York. For example, in Bradford County the employment growth in instruments and machinery appears to be related to that of Binghamton, New York. The growth in Wyoming County was due to a single new sanitary paper plant. Crawford County is noted for its growth in light chemicals, plastics, and machinery.

Because most rural Pennsylvania counties have only one or two manufacturing centers, a single industry frequently dominates manufacturing. For example, in 1989 the machinery industry accounted for 36 percent of all manufacturing employment in Venango County; electric and electronic equipment accounted for 34 percent in Elk County; stone, clay, and glass accounted for 42 percent in Armstrong County, and paper production for more than 90 percent in Wyoming County. These single dominant industries may utilize most of the labor resources of the area. If employment declines in the major plant, economic consequences are severe. Dominance of a single industry has discouraged the diversification of manufacturing in a number of counties. Although the company town as such has disappeared, the economy of many towns remains tied to a single industry.

Of the factors discouraging industrial growth in rural Pennsylvania, lack of a specifically trained labor supply is paramount. The rural Pennsylvania population is relatively small, and scientifically and technically trained workers are in short supply. In rural Pennsylvania many workers travel 50 or more miles to their place of employment. Many sections of the state are poorly served by transportation facilities. The sparse rural population provides only a limited market, and when goods are marketed outside the region the lack of adequate transportation facilities—both roads and railroads—presents a severe problem. Capital is also limited in rural Pennsylvania, so that outside resources are required to establish a plant.

The future of manufacturing in rural Pennsylvania is a major issue in the total economy of the state. The economic degeneration of some urban areas has created a need to reappraise how economic activities might be better distributed. The fundamental importance of manufacturing to the economic stability of a large portion of rural Pennsylvania makes these areas a key element in the total industrial structure. In most rural areas manufacturing employment has been more stable than in the urban counties. Nevertheless, present-day trends give little indication that major new manufacturing centers will evolve in rural Pennsylvania. Future growth in manufacturing is most likely to occur in industries that require a limited supply of semi-skilled to skilled labor producing products of low bulk and relatively high value that can be marketed nationally.

Industrial Structure

The traditional industrial structure of Pennsylvania was established in the nineteenth and early twentieth centuries. The dominant industries were primary metals, apparel, machinery, textiles, electrical machinery, fabricated metals, and food and kindred products. These industries employed more than 70 percent of the workers into the 1950s, at which point the traditional industries began their decline in

Table 15.4. Structure of manufacturing in percentage of employment, 1969 and 1989.

	1969	1989
Food and kindred products	7.080	8.360
Tobacco products	0.006	0.066
Textile mill products	4.720	2.700
Apparel and other textile products	12.280	8.640
Lumber and wood products	0.090	3.150
Furniture and fixtures	1.900	2.090
Paper and allied products	3.930	3.900
Printing and publishing	4.810	8.860
Chemicals and allied products	3.410	4.100
Petroleum and coal products	0.090	0.070
Rubber and misc. plastic products	2.160	4.400
Leather and leather products	1.990	0.090
Stone, clay, and glass products	3.770	4.210
Primary metal industries	15.580	8.540
Fabricated metal industries	7.880	9.920
Industrial machinery and equipment	9.590	10.040
Electronic equipment	9.110	7.240
Transportation equipment	6.030	5.500
Instruments and related products	2.250	4.170
Misc. manufacturing industries	1.720	2.130

SOURCE: U.S. Department of Commerce, *County Business Patterns*, 1969 and 1989.

employment. This trend has continued to the present. Primary metals have experienced the greatest employment losses, going from 18.0 percent of the total in 1949 to 14.5 percent 1969 and only 8.0 percent in 1989. Textiles and related products have experienced a similar trend, dropping from 9.3 percent of the total in 1949 to 4.4 percent in 1969 and 2.4 percent in 1989. In contrast to the declines in the traditional industries, a number of industries have grown in importance. For example, employment in the printing and publishing industries has risen from 3.9 percent of the total in 1949, to 4.4 in 1969 and 8.1 percent in 1989. A number of industries have maintained about the same relative share of manufacturing: apparel and related products, industrial machinery, and fabricated metals (Table 15.4).

Food and Kindred Products

Food processing began in the home. As the industrial age advanced, the demand for commercially prepared foodstuffs grew. An increasingly urban population could no longer depend on home-grown produce and home-processed foods. Mills for preparation of flour were developed throughout the state as the settlement pattern evolved. By the early nineteenth century small meatpacking establishments were widely distributed, for fresh meat could not be transported great distances without refrigeration. With the development of refrigeration, however, the small packing plant was doomed. The first shipment of meat in refrigerated cars from Chicago to the East occurred in 1869. Because Midwest meat was lower in cost than meat that could be produced in Pennsylvania, the small meatpacking plants based on locally produced meat disappeared. By 1910 the meatpacking industry in the state was concentrated in Philadelphia, Pittsburgh, and Lancaster, where the largest markets existed and where economies of scale could be achieved. A small meatpacking industry persists in these three centers.

Of the processors of food products in the nineteenth century, Henry J. Heinz of Pittsburgh became the Pennsylvania leader. As a boy in the 1870s Heinz grew horseradish in his mother's garden, grated it, and sold the product in the neighborhood. By the time he was 16 he was growing and canning vegetables and selling them in the Pittsburgh market. His prepared horseradish was the first specialty product of the famous "57 varieties" that became the slogan of the Heinz Company.

In 1888 the H. J. Heinz Company was formed to increase production and reach a wider market. By the early 1900s Heinz had built one of the nation's biggest and most profitable businesses, manufacturing more than 200 products. At the turn of the century it was the nation's largest producer of pickles, vinegar, and ketchup; the largest grower and processor of cabbage, onions, and horseradish; the second largest producer of mustard; and the fourth largest packer of olives.

Today, with headquarters in Pittsburgh, the company has nine branch factories in six states, a branch company in London, and agencies around the world. It is the only company of its kind in the world that manufactures its own bottles and has its own private tank cars. In 1990 the company employed about 1,600 workers in its Pittsburgh plant. As the food processing has become increasingly mechanized, the number of workers has declined from about 2,800 in the last 50 years. The Heinz Company was built without mergers and with the purchase of only two or three small companies.

The H. J. Heinz Company contributed greatly to the technology of producing pure foods in quantity. Advances in the canning process were noteworthy. The Heinz laboratories devised a process in which a machine deftly fashioned a cylinder open at both ends, the sides being rotated together and crimped in a lapped double-lock seam. The bottom was then crimped and double-locked onto the can. After the can was filled, the top was put on in the same manner and the can was placed in a pressure cooker. One machine could turn out 40,000 of these airtight cans in a 10-hour day. Heinz believed that for the food-processing industry to produce pure foods and earn the confidence of the public there had to be a partnership with a federal regulatory agency. The company thus became a major force in the passage of the Pure Food and Drug Act of 1906 and other legislation that followed.

Another famous producer of specialized food products in Pennsylvania was Milton S. Hershey. Hershey was born on a farm in Dauphin County in 1857 and learned to make candy in small plants in Lancaster, but for many years he was unsuccessful in establishing his own business. Finally, at the Chicago World's Fair in 1893, he was impressed by a German exhibit of chocolate-making machinery, bought the machine, and began making caramel candy in Lancaster. It was a most successful venture, and in 1898 Hershey sold the caramel business for $1 million.

Hershey and his major partner, William F. R. Munie, then decided to erect a plant to produce chocolates. They situated the plant in Dauphin County—in the heart of a dairy region—so that the most perishable of the raw materials, milk, could be obtained locally. The first chocolate was produced in 1905, and the present Hershey Chocolate Company was founded in 1910. By 1920 the Hershey Company was using one-tenth of the world's crop of cocoa. In order to reach a world market, the candy bars were wrapped in a foil wrapper to maintain their freshness. No candy bar became better known than the Hershey chocolate bar.

The town of Hershey is a model industrial settlement. The Hershey Company controlled the physical plan of the town by developing residential areas and providing the utilities and services until the Derry township government assumed responsibility. The homes were always privately owned, but permits were required to conform to an established building code. Present growth is still largely controlled, for the Hershey Trust still owns the majority of the undeveloped land. In 1907 Hershey Park was opened as an amusement center for Hershey employees. It has become a major amusement center in eastern Pennsylvania. The M. S. Hershey Foundation operates the Hershey Educational and Financial Center. Within the area, the Milton Hershey School, located on a 10,000 acre tract of land, is an institution that houses and educates 1,250 boys and girls from broken and needy families. The Hershey Medical Center of The Pennsylvania State University opened in 1970 as a response to a $50 million contribution from the M. S. Hershey Foundation.

The processing of foods and kindred products has grown tremendously in the twentieth century. In 1987 this industry processed raw materials valued at $9.5 billion. The value added in the manufacturing process exceeded $7.0 billion, and the total value of output was more than $16.5 billion.

A wide variety of food items are produced in the state, with no single branch dominating the industry. In 1989 bakery products, including bread, cake, cookies, crackers, and related items, accounted for the largest share of employment, with about 18 percent of the total. Other important branches included meat products, sugar and confectionary products, preserved fruits and vegetables, and beverages.

Although the food and kindred products industry was located in 46 counties in 1989, the industry is concentrated in the 20 counties that lie above the county mean of 1,208 employees (Fig. 15.10).

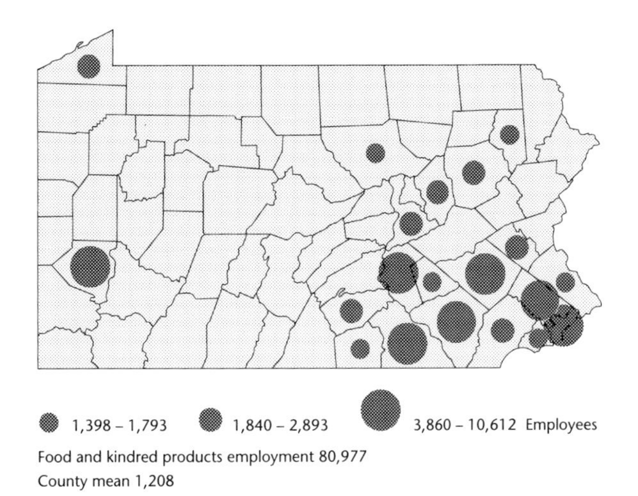

Fig. 15.10. Employment in food and kindred products, Pennsylvania, 1989.

These 20 counties have about 85 percent of the state's employment, and the top seven counties—Philadelphia, Lancaster, Allegheny, Montgomery, Dauphin, Berks, and York—have 52 percent of the employment.

Two factors are important in the localization of this industry. First, because many of the food products are perishable, a market orientation is of the utmost importance. Second, because many food products are bulky and relatively low in value, transportation costs must be kept to a minimum. In a number of counties, such as Lancaster, Erie, Cumberland, and Columbia, food-processing industries have developed because of the availability of fruits and vegetables.

In the period from 1969 to 1989 employment in the food and kindred products industries declined from 104,375 to 80,977. As total employment has declined, the number of food-processing establishments has decreased even more. In 1947 there were 2,789 establishments, with an average of 37 workers. By 1989 the number of businesses had decreased to 1,081, but the average number of employees had increased to 74. The industry has become highly mechanized, so that the small processors have disappeared, but as the factory size increased, production has been maintained and even increased in most types of food processing.

Employment in the food industries experienced diverse trends in the state. Some counties had major decreases while other counties grew modestly. The county of greatest decline was Philadelphia, where employment decreased from 26,691 to 10,612, or 60 percent—about 68 percent of the total state loss. The small bakeries, such as the neighborhood bakery, where employment declined from 8,610 to 3,382, were unable to survive in Philadelphia because name-brand products forced the high-cost specialty shops out of business. The meat and dairy products industries suffered the same fate, with employment declining from 7,820 to 2,094. All branches of the food industry experienced significant employment decreases. The large food producers could achieve economies of scale, producing products of high quality at a lower cost with far fewer employees.

A number of other counties besides Philadelphia County have experienced significant decreases in employment in the food industries. Allegheny County had the second largest decrease between 1969 and 1989, with employment declining from 12,248 to 5,045. The same factors influenced the employment trends in Allegheny County as in Philadelphia. In Chester County, employment between 1969 and 1989 declined from 3,558 to 2,876, due primarily to the decrease in the canning of mushrooms.

In a number of counties employment in the food and kindred products industries increased modestly. In Lancaster County employment grew from 4,585 to 7,461 due primarily to an increase in meat processing, grain milling, and sugar and confectionary products. Other southeastern Pennsylvania counties to experience growth include Montgomery, York, and Lehigh.

Textile Mill Products

The textile industry was one of the earliest industries established in Pennsylvania. The industry was centered in southeastern Pennsylvania, especially the Philadelphia area. Germantown was noted for the knitting of stockings. One report states that in 1768 the area produced 6,000 dozen pairs of woolen stockings. Much wool was manufactured into clothing, and linsey-woolsey, a cloth of linen and wool, was much in demand for women's dresses. Small craft mills produced serges and laces. During the pioneer period, textile manufacturing was second only to iron production.

In the early nineteenth century the textile industry changed from a craft to a factory operation. The industry grew steadily, from about 10,000 workers in 1810 to 79,500 in 1890 to 154,000 in 1919, when Pennsylvania produced about 15 percent of all textiles manufactured in the United States. Although the textile industry as a whole grew slowly after 1900, a number of branches—such as woolens and worsteds—initiated the long decline that has characterized the entire textile industry in the twentieth century.

Of the various branches of the textile industry, manufacture of knit goods (in which hosiery dominated) was preeminent. By 1810, when the state's output of hosiery totaled 107,508 pairs, Philadelphia accounted for half of the nation's total. Throughout the nineteenth and early twentieth centuries, Pennsylvania's hosiery output placed it first in the nation. In 1919 the state contributed 34.4 percent of the quantity and 40.5 percent of the value of all hosiery made in the United States. In the nineteenth century the industry concentrated on the production of cotton hose, but early in the twentieth century the demand for silk and woolen hosiery grew. Between 1909 and 1919 the value of silk hosiery grew from 4 to 60 percent of the total. Although hosiery still constituted about 66 percent of the value of knit goods in 1919, outerwear and underwear, fancy knit goods, and knitted cloth were growing in importance. The knit-goods industry was concentrated in eastern Pennsylvania in Philadelphia, Berks, Lackawanna, Luzerne, and Schuylkill counties. The major factor in its localization was plentiful low-cost labor, primarily women.

The manufacture of silk cloth was another early endeavor in the textile industry. Silk culture started in 1750, when a filature for reeling silk from cocoons was established under the patronage of Benjamin Franklin and others. In 1793 the making of fringes, laces, and tassels began in Philadelphia, expanding by 1815 to include trimmings, naval sashes, and ribbons. Power looms were introduced in 1837 for wearing ribbons and narrow goods. In 1880 Pennsylvania ranked fifth in the United States, accounting for 8.5 percent of the total value of silk textiles produced. The growth of the industry was particularly noteworthy between 1880 and 1900, during which period many towns offered financial and other inducements to attract the industry. By 1900 Pennsylvania's silk industry stood in second place, with 29 percent of the nation's total, and by 1919 Pennsylvania was the leader among the states, with 33 percent of total output by value. The industry was localized in Philadelphia, Scranton, Allentown, Easton, Reading, and a number of smaller centers. The introduction of synthetic fibers had a devastating effect on the silk industry, particularly hosiery. Women preferred synthetic hosiery to that produced from silk. Within a few years silk hosiery had essentially disappeared.

The woolen industry was a traditional branch of the textile industry. During the pioneer period the women of the family produced woolen cloth for family consumption in most homes in the state. The manufacture of woolen goods was first established by English millers in the Schuylkill Valley, in small factories using hand looms, and by 1850 Pennsylvania produced more yarn than any other state. With the development of factory production, the industry concentrated in Philadelphia. By 1870 the state ranked second in the manufacture of both woolen and worsted goods, with about 18 and 36 percent, respectively, of the nation's output by value. The woolen and worsted industry reached its peak about 1909 and then began to decline as competition from other areas increased. High-cost production in Pennsylvania, due primarily to higher wages and declining sources of local wool, placed the state at a cost disadvantage, and the industry began its migration to newer centers.

In the nineteenth century, Pennsylvania became the carpet and rug center of the nation. In 1850 the state ranked third, with production valued at 21 percent of the national output. By 1870 the state ranked first, with 45 percent of the nation's total, and this share increased to 48 percent in 1900. Philadelphia dominated the industry. After 1900 the industry began to decline, and by 1919 production (measured in square yards) was less than half the state's peak output. A number of factors adversely affected the industry. The decline of the woolen industry, which supplied the basic raw material for carpets and rugs, affected the industry adversely. Also, and more important, Philadelphia had gained its reputation in the production of ingrain carpets, and as demand for Brussels carpet grew, the Philadelphia manufacturers were slow to follow the fashion of the day. Producing Brussels carpet required a technological change as well as a change in design, and the industry developed in new centers.

Of the various branches of the textile industry, the production of cotton fabrics has been of relatively little importance in Pennsylvania. In 1900 the state ranked fifth in the nation, producing only 7 percent of the total output. The cotton fabric industry was concentrated in the New England states. Pennsylvania was, however, noted for its specialty items. The use of cotton for decorative fibers originated in Pennsylvania, and its cotton tapestries and chenille curtains gained national markets.

The textile industry of Pennsylvania has experienced a long decline in the twentieth century, particularly sharp since the 1940s. Employment has dropped from 139,533 workers in 1950, to 69,537 in 1969 and only 26,161 in 1989. During this period the number of establishments declined from about 740 to 341, of which 143 have fewer than 20 employees and only 5 have more than 500. Nonetheless, the industry provided to production workers $336 million in wages in 1987 and used $1.2 billion of material, with an output valued at more than $2.2 billion.

The industry is dominated by employment in knitting mills. In 1989 these mills had 11,439 workers, or 48 percent of the total. Of the textile knit products, outerwear mills employed 39 percent of the total, followed by knit underwear mills, knit fabric mills, and hosiery.

The localization of the industry has changed little since the nineteenth century. The industry is concentrated in a few counties in eastern Pennsylvania (Fig. 15.11). In 1989 some 29 counties had some textile production, but the 18 counties above the county mean of 390 employees had 78 percent of the employment. The six top counties were Schuylkill, Columbia, Philadelphia, Berks, Lehigh, and Luzerne.

Of the six leading counties, five experienced significant declines in employment from 1969 to 1989. In only one county—Schuylkill—did employment increase slightly, from 2,928 to 3,009. Philadelphia County experienced the greatest loss, with employment declining from 16,749 to 2,679. The knitting mills had the greatest decline, from 7,297 to 1,135 employees. Other significant declines were in weaving mills, yarn and thread mills, and miscellaneous textiles. Employment in production of narrow fabrics nearly disappeared.

The decline in knitting mills was nearly universal. In a number of counties that specialized in a particular type of textile product, the declines were devastating. In Berks County the employment in women's hosiery declined from 2,209 in 1969 to zero in 1989; in York County employment in weaving mills declined from 1,167 to about 300.

Since the beginning of the Industrial Revolution the textile industry was traditionally migratory, always seeking areas of lower-cost production. The first textile industry was localized in the nation's Northeast. In the late nineteenth century, however, the industry began to develop in the Piedmont region of North and South Carolina and in Georgia. One commonly cited factor was that the industry was seeking the source of its raw materials.

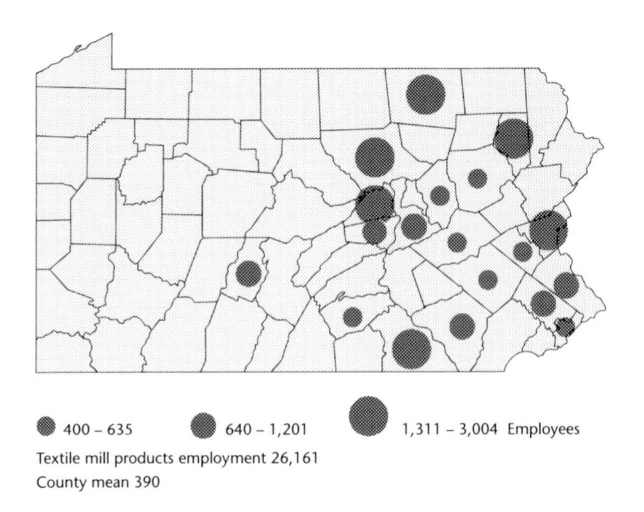

Fig. 15.11. Employment in textile mill products, Pennsylvania, 1989.

In the early development of the industry the Piedmont was undoubtedly stimulated by accessibility to cotton, but other factors were soon more important. One estimate is that in the 1890s the Carolina Piedmont mills had an advantage in the transport of cotton of between 0.8 and 1.5 cents per pound of coarse products over mills in New England. An even more significant transportation-cost advantage was provided at that time, when southern railroads offered special low shipping rates on finished cotton products to offset their competitive disadvantage in terms of distance to the northern market. These two advantages were important as an impetus to the early Piedmont industry, but they did not last long. With the attack of the cotton beetle, production declined in the Southeast and a large percentage of the cotton had to be imported from areas west of the Mississippi. Also, as the industry grew, the railroads reverted to a normal transportation rate structure. Therefore, transportation-cost differentials on either cotton or finished goods do not provide a sound reason for the continued growth of the cotton textiles in the Piedmont.

A more significant factor in explaining the growth

of the southern textile industry and the inability of the Northeast, including Pennsylvania, to compete is the sizable differential in labor costs between the northern and southern mills. In the highly unionized textile industry of the Northeast, wages and salaries represent the largest share of the mill costs, averaging 25 percent of the value of the product. Lesser costs include raw materials, depreciation, fuel and power, taxes, marketing, and management.

Because of the high proportion of labor costs to the value of the products, and, even more significant, because these costs constitute the major share of the mill costs, textiles are labor-cost oriented. The favorable differential in labor costs for the South was and is due not only to lower wages but also to the higher productivity of textile workers. In the 1920s it was estimated that the wages of the textile workers in the Carolinas were 20 to 30 percent lower than in Pennsylvania. In 1972 the average hourly earnings for cotton-textile workers were $2.62 in North Carolina, compared with about $3.00 in Pennsylvania. Since then the differentials have narrowed because of the increasing unionization of southern mills, government regulations pertaining to minimum wages, and the general increase in the level of the South's economy.

A number of minor factors have encouraged the growth of the textile industry in the South. Power and fuel costs are substantially lower in the South, partly due to climate and partly due to the relatively low-cost hydroelectric power and lower-cost coal. The South has also provided other inducements to attract the textile industry—lower taxes, capital at low interest rates, and physical inducements such as free plant sites and buildings. Although individually these are minor factors, collectively they are factors to be considered when new locations are chosen.

Just as the cost differential between Pennsylvania and the South was narrowing, competition began with foreign producers in Hong Kong, Singapore, Korea, and many other developing nations. Labor costs in these areas are as low as $1.00 per hour for textile workers. In recent years an increasing flow of foreign textiles into the United States has been devastating to the nation, including Pennsylvania.

Even if wages were comparable in the South and in foreign countries, the Pennsylvania textile industry would still be at a disadvantage because of its generally lower productivity. The work assignments are larger for Southern workers, and particularly for foreign workers, partly because of less union control and—in some factories, particularly in foreign countries—because plants and machinery are more modern. Foreign workers are less conscious of delegated workloads and have been more amenable to scientifically designed job assignments. In addition, the tempo of technological change has been rapid, and management in the South and in foreign countries has been more than willing to experiment with new technologies.

All branches of the textile industry in Pennsylvania are at a critical stage of their existence. Plants are old and not being modernized. There are essentially no funds for research and development. The Pennsylvania textile industry must adjust to both domestic and foreign competition if it is to survive.

Apparel and Other Textile Products

The apparel industry became an important factory industry in Pennsylvania late in the nineteenth century. The first center of production in the state was Philadelphia. This industry was normally established in large urban areas because it was labor intensive and because manufacturers sought areas where huge numbers of immigrant women skilled in needle handicrafts were willing to work for low wages. New York was the principal garment center on the East Coast, but other large cities also developed apparel industries, particularly producing low-style garments. Besides the availability of labor, Philadelphia was well situated in relation to the East Coast market. A general rise in the standard of living in the nineteenth century provided further impetus to the purchase of manufactured clothing.

A second center of the apparel industry developed in the twentieth century in the anthracite region. As the textile industry declined in this region, the apparel industry developed and grew. Textile establishments were readily converted to the production of clothing, and textile workers were easily trained to manufacture low-style garments. The apparel industry was a higher-value-added industry than textiles, so wages were higher. The anthracite region was also well situated to serve the large East Coast market.

After World War II the apparel industry began to migrate away from the major East Coast urban centers, where the industry had become highly unionized, in search of lower-cost labor and lower taxes. This movement favored not only the anthracite counties but also other rural areas of southeastern Pennsylvania.

The apparel industry reached its peak of employment about 1969, when 180,888 workers were employed. Since then employment has experienced a drastic decline, to 83,753 in 1989. The greatest decline in employment occurred in Philadelphia County, where it dropped from 44,048 in 1969 to 12,000 in 1989. Philadelphia was the only major center in Pennsylvania in the production of suits and coats for men and boys. This single branch declined by 15,095, dropping from 19,511 to 4,416. Other major branches of the industry to decline were women's and misses' outerwear, men's and boys' furnishings, children's outerwear, and miscellaneous fabricated textile products.

The anthracite counties experienced major declines in employment too. In Luzerne, Lackwanna, and Schuylkill counties, employment between 1969 and 1989 declined from 40,849 to 15,778. Women's and misses' outerwear was the largest branch of the industry and experienced the greatest declines.

The counties where the apparel industry had grown most rapidly from 1945 to 1965 have experienced the least declines in employment since 1969. These include Lancaster, Montgomery, Lebanon, and Lehigh counties. The apparel factories in these counties are modern and more efficient and have been better able to produce a marketable product at a lower cost.

The drastic decline in employment can be largely attributed to the growing importation of foreign clothing. Materials and labor, which make up about 75 percent of the cost of a garment, are much lower in Hong Kong, South Korea, Taiwan, Singapore, and many other Third World countries than in the United States. As Pennsylvania apparel becomes less competitive with foreign imports, the plants are abandoned.

Although there was a drastic decline in employment, the spatial pattern of apparel production changed little between 1969 and 1989 (Fig. 15.12). In 1969 the industry was found in 52 counties, and in 1989 in 54 counties. With a mean county employment of 1,250 in 1989, there were 18 counties above the mean and 39 below the mean. The 18 counties above the mean had 75 percent of the employees. In 1969 there were 13 counties above the mean with 83 percent of total employment. The same counties were above the mean in both 1969 and 1986, with Franklin County added in 1989. While the spatial pattern was stable, the counties in the top third above the mean declined slightly in importance.

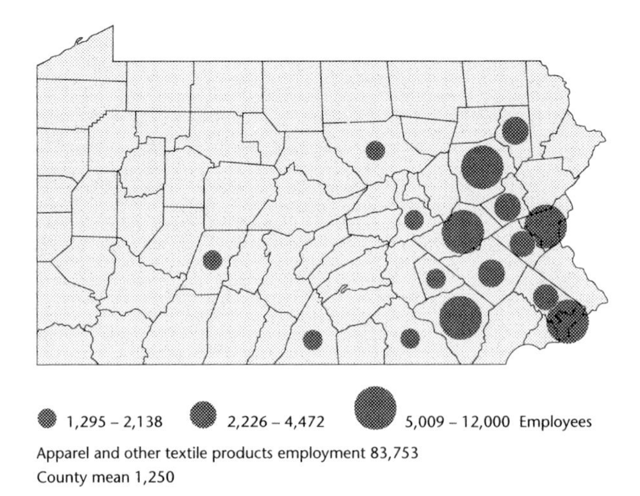

Fig. 15.12. Employment in apparel and other textile products, Pennsylvania, 1989.

With the decline in employment some changes have occurred in the type of clothing produced. Employment in women's and misses' outerwear, primarily dresses, increased from 36 to 45 percent of the total. In contrast, employment in men's and boys' suits and coats declined, from 17 to 13 percent of the employment, and men's and boys' furnishings also declined, from 18 to 14 percent of the total. There was little change in the other branches of the clothing industry. In 1969 and 1989 other branches of the industry included women's and misses' clothing, children's outerwear, and miscellaneous apparel and accessories.

The apparel industry in 1987 generated a payroll of about $1.0 billion, consumed material worth $1.8 billion, and had shipments valued at $4.0 billion. The industry is characterized by small plants. In 1989 some 534 of the 1,318 establishments had fewer than 20 employees. Only 15 plants had more than 500 employees. In 1982 new capital expenditures totaled only $36.3 million or $312 per employee. This is an extremely low capital expenditure so that most of the plants lack modernization. Mechanization is limited so that hand work is required for most operations. The future of the apparel industry in Pennsylvania is bleak.

Paper and Allied Products

The making of paper dates from the early colonial period of Pennsylvania. The first paper mill in the United States was built in Germantown in 1693. As the market increased, there was a continued incentive to develop new technologies, to utilize a greater variety of raw materials, and to improve the quality of the paper in Pennsylvania. For example, in 1816 the first steam-powered paper mill in the United States began operation in Pittsburgh. In 1830 Wooster & Holmes, of Meadville, secured patents for making paper from whitewood and hemlock. The process for making white paper from straw was developed at Flat Rock Mills in 1854. J. A. Roth of Philadelphia patented a process in 1857 to treat wood fiber with a combination of sulfuric acid and chlorine bleaching agents. In 1863 M. L. Keen of Royersford patented a process for making pulp from wood, and in the following year Richard Magee of Philadelphia patented a method for making quality writing paper. The industry flourished, and in 1880 Pennsylvania ranked third among the states. The leading counties were Philadelphia, York, Blair, Elk, and Erie.

The type of paper produced in Pennsylvania changed in response to new technology, the availability of raw material, and market demands. In the early part of the nineteenth century the mills of Pennsylvania produced all types of paper, including that required by the newspapers of the state. By 1900 there was essentially no newsprint made in the state. This change was largely due to the lack of coniferous forests, the wood best suited for newsprint manufacture. There was a growing concentration on quality paper, including book paper.

Employment in the paper and allied products industries has declined from 46,206 in 1969 to 37,797 in 1989. In both years about 25 percent of the employment was in paper mills and paperboard mills, and 75 percent in plants producing miscellaneous converted paper products and paperboard containers and boxes.

In 1989 the industry was located in 43 counties, but the 18 counties that were located above the county mean (Fig. 15.13) of 564 employees had 82 percent of the employment, and the top 6 counties—Philadelphia, York, Wyoming, Bucks, Delaware, and Montgomery—had 50 percent of the total. Philadelphia County continues to have the largest employment in paper products. However, employment declined from 8,820 in 1969 to 4,892 in 1989. The decline was largest in paperboard containers and boxes. Wyoming County had the greatest gain in employment, where a single new plant employs about 3,500 workers.

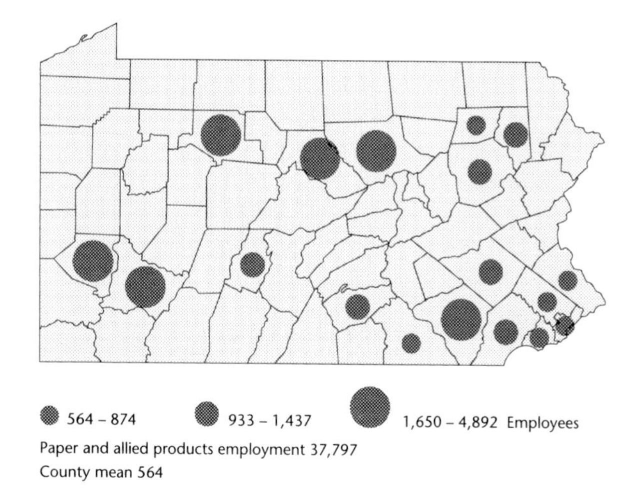

Fig. 15.13. Employment in paper and related products, Pennsylvania, 1989.

Most of the plants that produce paper are oriented to raw material sources. These include the plants at Lock Haven in Clinton County, at Erie in Erie County, at Roaring Spring in Blair County, at Spring Grove in York County, and at Mehoopany in Wyoming County. Trucks transport the raw materials for these plants, moving pulpwood sometimes more than 50 miles. The Erie plant of Hammermill Bond receives some of its raw materials by water across Lake Erie.

In contrast to the raw-material orientation of the paper-producing plants, the plants producing converted paper products and paperboard containers and boxes such as envelopes, bags, die-cut paper, and paperboard boxes are oriented to market locations. These plants are widely distributed, with concentrations in southeast and southwest Pennsylvania. These products are bulky and have a low value, so that they are normally marketed locally and regionally.

The future of the paper and allied products industries is bright because there is an increasing supply of raw materials from Pennsylvania's forests and because a major market for paper products exists not only within the state but also in adjacent areas. Because many of the paper mills are old, modernization of the industry is essential to its future productivity.

Printing and Publishing

The printing and publishing industry is one of the oldest in the state. The first printing press was erected in Philadelphia in 1686, only four years after the first English settlement. Publication of the first magazine was attempted by Benjamin Franklin as early as 1741. The first English Bible printed in the United States was from the press of Christopher Sauer of Germantown in 1743. A daily paper was printed as early as 1784, and by 1810 seventy-three newspapers were published. By the early nineteenth century an estimated half a million volumes were printed annually in Philadelphia.

By 1900 Pennsylvania firms accounted for 10 percent of the nation's printing and publishing output. Among the leading book publishers were Winston and Lippincott. The J. B. Lippincott Company, founded in 1792, published a wide range of books including fiction, nonfiction, reference books, children's books, and books of religious interest. The Winston Company, founded in 1884, developed an international reputation for its Bibles, children's books, dictionaries, and other reference books. The Curtis Publishing Company, organized by Cyrus H. K. Curtis of Philadelphia in 1891, ranks as one of the foremost publishers of magazines. In addition to the *Ladies' Home Journal* and the *Saturday Evening Post,* one of its early publications was *The Farm Journal,* the leading rural periodical in the nation.

The printing and publishing industries have experienced rapid growth in Pennsylvania in the twentieth century. Employment grew from 70,871 in 1969 to 85,821 in 1989. The structure of the industry has changed slightly. In 1969 newspapers had 34 percent of the employment but declined to 31 percent in 1989. In contrast, employment in commercial printing increased from 35 to 37 percent and in books from 7 to 10 percent of the total. In both 1969 and 1989 periodicals had 7 percent of employment, and blankbook and bookbinding and printing trade services each had 5 percent.

In 1989 the printing and publishing industries were located in 57 counties. The county mean was 1,280 employees, with 16 counties above the mean having 82 percent of the employment (Fig. 15.14).

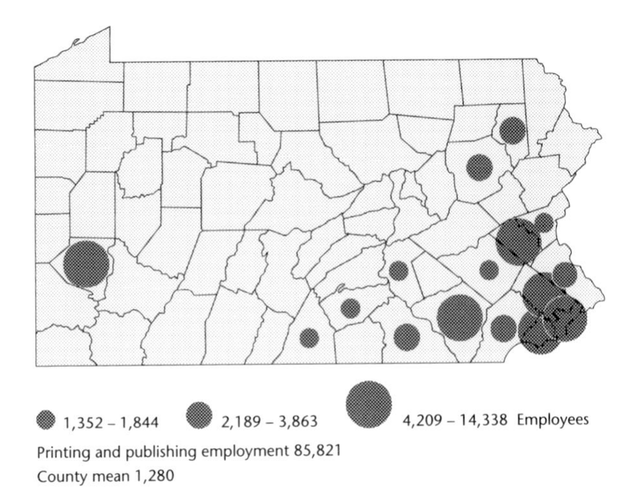

Fig. 15.14. Employment in printing and publishing, Pennsylvania, 1989.

The top six counties—Philadelphia, Montgomery, Allegheny, Lancaster, Lehigh, and Delaware—had 53 percent of the employment in the state.

Although employment has increased in essentially all counties, Philadelphia County has experienced a decreasing trend, from 28,282 in 1969 to 14,338 in 1989. Every branch of the industry experienced a decline in employment, but the drop was precipitous in newspaper employment—from 8,499 to about 4,000—and in commercial printing from 10,268 to 4,849. The development of modern newspaper printing methods reduced labor demands and as the local economy declined the demands for commercial printing decreased.

The principal growth counties in printing and publishing were in southeastern Pennsylvania. Counties with major increases include Montgomery, Lancaster, Lehigh, Bucks, and Chester. The printing and publishing industry closely reflects the economy of an area. When an area is prospering, demands for printing and publishing are high. Conversely, in areas suffering economic problems the demand is less. Furthermore, the printing and publishing industry is closely connected to population size.

Chemicals and Allied Products

Of a wide variety of chemicals produced in Pennsylvania, prescription drugs are most important, accounting for about 45 percent of the employment and about 45 percent of the $9.3 billion worth of

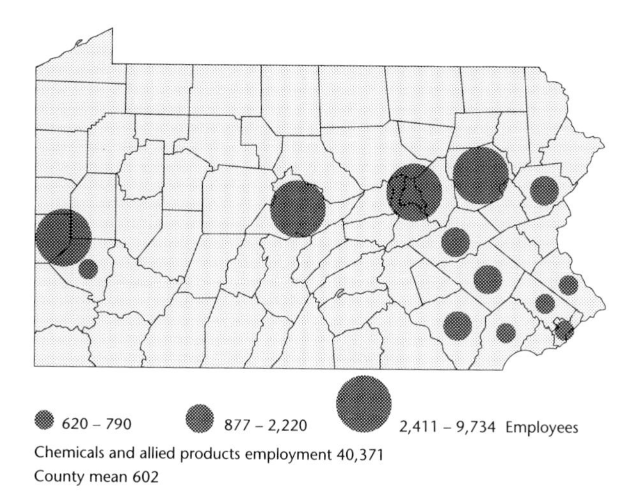

Fig. 15.15. Employment in chemicals and allied products, Pennsylvania, 1989.

chemicals produced in 1989. Of the drugs produced, pharmaceuticals make up about 88 percent of the total. Other important chemicals include industrial organic chemicals, with 8 percent of the value of shipments, followed by plastic materials and synthetics and miscellaneous chemical products.

Statewide employment in the chemical industries declined from 50,186 in 1969 to 40,371 in 1989. The greatest loss occurred in Philadelphia County, where employment declined from 12,882 to 6,607. In 1969 Philadelphia was the leading county in employment, but by 1989 it had dropped to second place, with Montgomery County assuming the leadership.

In 1989 the Pennsylvania mean employment for the 35 counties having a chemical industry was 602 (Fig. 15.15). The top five counties above the mean had 62 percent of total employment, and the top three counties—Montgomery, Philadelphia, and Allegheny—had 35 percent. Diverse employment trends are evident, however. Two counties, Philadelphia and Bucks, had a decrease of 7,871 jobs between 1969 and 1989. In contrast, employment grew by 3,994 in Allegheny, Chester, Lancaster, Montgomery, and Berks.

The industry possesses a large number of small plants. In 1989 of a total of 558 establishments only 249 had 20 employees or more. The size of the plant varied greatly for different types of chemicals produced. The plants producing drugs were among the largest. In contrast, a large number of plants producing soaps, cleaners, and toilet goods are extremely small. The drug factories averaged 273 employees each, while the soap, cleaners, and toilet goods plants averaged 29 employees. As a response to a growing plastics industry in the state, the production of plastic raw materials and resins has grown significantly since 1969.

The chemical industry of Pennsylvania is characterized by production of high-value-added products. The industry is not oriented to raw materials. Coal, the basis of a wide variety of chemical products in a number of European countries, has not been an important raw material for the production of chemicals in Pennsylvania. Throughout the nineteenth and well into the twentieth century, coal was coked in beehive ovens, and the coal tars and gases, the basis of the chemical industry, were lost in the coking process. When the by-product coke ovens came into major use about 1920, the chemical industry using coal tars and gases localized in other states.

In no group of industries is change more prevalent and more necessary than in the chemical field. New methods, new equipment, and new products are keys to survival in this highly competitive industry. In order for individual companies to survive, there must be an active research program, and after research has developed a new product, there must be adequate funds to achieve commercial production. Because of frequent changes in process and product, new equipment and continuous plant renovation are required. Because many of the companies in Pennsylvania are small, research and development funds are limited.

The growth areas of the chemical industry of Pennsylvania lie in the industrial growth areas of southeastern Pennsylvania. The industry is dependent on a highly skilled work force and research scientists to develop new products. Production of specialty chemicals, such as pharmaceuticals and plastic resins, will continue to dominate the industry in the future. The products will be of high value, but with greater efficiency the work force will probably decline slowly during the next decade.

Rubber and Miscellaneous Plastic Products

The rubber and plastics industry developed in the twentieth century and is a growth industry, with employment rising from 31,806 in 1969 to 42,683 in 1989. About 80 to 85 percent of the workers produce

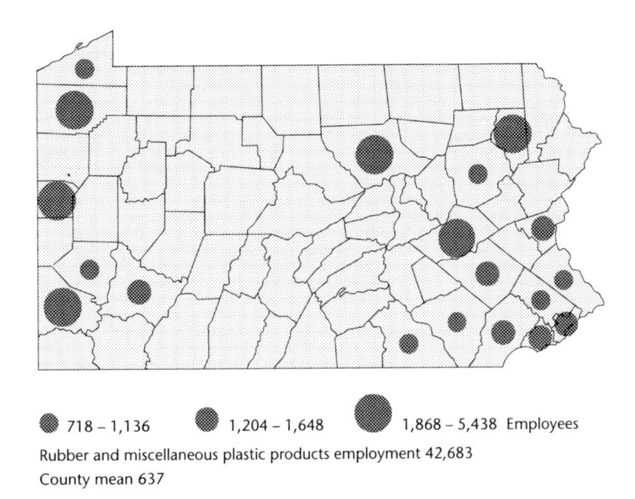

Fig. 15.16. Employment in rubber and miscellaneous plastic products, Pennsylvania, 1989.

plastic items. The remainder produce rubber footwear, fabricated rubber products, and a small output of tires and inner tubes. In 1989 the rubber and plastics industry consumed $1.971 billion of raw materials, with a payroll of $861 million, and had a shipment valued at $4.027 billion.

The plastics industry has expanded most rapidly in recent years. The number of counties possessing the industry increased from 32 to 48 between 1969 and 1989. In 1989 the 19 counties above the county mean of 637 employees had about 81 percent of total employment (Fig. 15.16). Furthermore, the top 6 counties—Erie, Bucks, Montgomery, Luzerne, Lancaster, and Allegheny—had 48 percent of the total. Employment in individual counties remains relatively small, with the leading county, Erie, having only 5,438 employees. Of the 710 plants in 1989, 317 had fewer than 20 employees. Only 7 establishments had more than 500 employees.

The plastics industry has had little development in Philadelphia and Allegheny counties—traditional centers of Pennsylvania's industry. And Philadelphia County employment decreased from 3,270 in 1969 to 1,239 in 1989, while Allegheny County employment was steady at about 1,860 workers. The greatest concentration of the industry is in southeastern Pennsylvania, but even here there are conflicting trends. Counties that have grown between 1969 and 1989 include Bucks, Lancaster, Delaware, Berks, and York. Counties that have declined include Montgomery and Chester. Entrance into the industry is easy because many of the companies are small, but lack of marketing skills makes failures frequent.

Many small towns in both rural and urban counties have attracted a plastics industry. Because plastic raw materials and resins are readily transported and energy demands are low, they are of little importance in localizing the industry. Plant and equipment needs are modest, so that capital investments are low. A labor force can be rapidly trained to operate the machinery. The product is light, though sometimes bulky, so it can be transported to distant markets. The industry therefore seeks sites where production costs are minimal. Because the quality of plastic articles has improved, the demand is increasing.

Stone, Clay, and Glass Products

Stone, clay, and glass industries were among the dynamic growth industries of the nineteenth century. In the twentieth century this growth has slackened, and between 1969 and 1989 employment declined from 55,562 to 40,842. There are a number of reasons for the decline of these industries in Pennsylvania. Because these products are heavy, bulky, and sometimes fragile, a market orientation in other parts of the nation is important to reduce transportation costs. Many of the plants are antiquated and as efficiency is reduced the operations are transferred to newer plants. As the economy of the United States has grown, these industries have become increasingly decentralized.

A wide variety of items are produced. Pressed and blown glass and glassware had the largest employment in 1989 with 23 percent of the total, followed by miscellaneous nonmetallic mineral products such as abrasives, gaskets and mineral wool, cement and concrete, structured clay products, and pottery. The stone, clay, and glass industries are widely dispersed in Pennsylvania, with 51 counties having some type in 1989 (Fig. 15.17). With a mean county employment of 609 some 26 counties were above the mean, with about 70 percent of the total employment. The employment in individual counties was quite small (Fig. 15.17). Allegheny County, with about 2,786 employees, and Montgomery County, with 2,415, had the largest numbers.

As in many other industries, the traditional and largest counties of production have experienced the greatest dip in employment. Between 1969 and 1989 major declines in employment occurred in Alle-

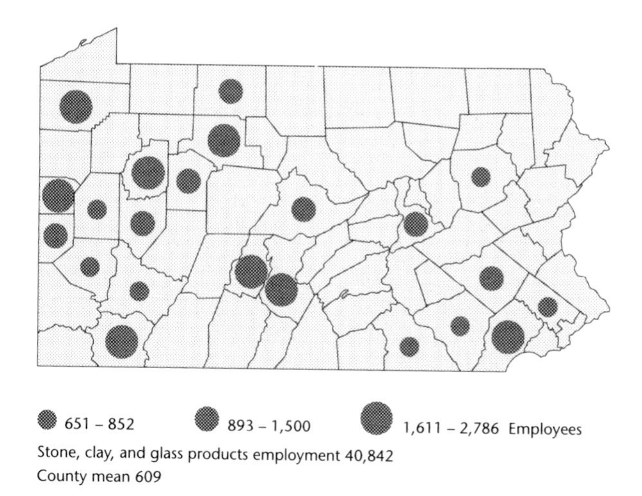

Fig. 15.17. Employment in stone, clay, and glass products, Pennsylvania, 1989.

gheny, Westmoreland, Washington, and Northampton counties, where the industry lost 8,990 workers. Counties experiencing employment declines were those that developed early glass and cement industries. In many other counties employment remained nearly stable or rose slightly. In 1987 the clay, stone, and glass industries together had an annual value of output of $4.76 billion and provided $724 million in wages.

The Portland Cement Industry

The natural cement industry was developed in the United States about 1820. The greatest impetus to the expansion of natural cements was the construction of canals. Because these waterways provided the only means of transporting the heavy, bulky cement long distances, the industry developed along canal routes. The industry was widely dispersed in the nation. A major center developed in the Lehigh Valley along the Lehigh Coal Navigation Company Canal. The availability of limestone, fuel, and a large market provided the foundation for the industry.

The first portland cement was produced commercially in the United States in the Lehigh Valley in 1875. For the next 25 years this area maintained a near monopoly in the production of this cement. In 1897 the year of its greatest dominance, this single region produced 74.8 percent of the nation's total output. The early importance of this district was due largely to the availability of large quantities of natural cement rock and its position near the important eastern market. During this period the natural cement rock in this region was believed to be superior to any other limestone in the United States.

The portland cement industry grew slowly in the Lehigh Valley, and two factors were important in that retarded growth. The natural cements had gained a national reputation as quality products, and many builders preferred to use them. European portland cement was also imported in this country before the American industry was established. Until about 1890 these cements could be brought to the United States and sold at a price below that of American portland cement. The foreign cements also gained a good reputation, which gave them a preferred market.

By the 1890s Lehigh Valley cement was, however, gaining wide acceptance. It was being sent to every section of the country and even developed a small export market. As a result, production increased from 57,150 metric tons in 1890 to 1,447,000 in 1900.

Between 1900 and 1905 there was a phenomenal decentralization of the cement industry. In 1900 the East Coast states of Pennsylvania, New Jersey, and New York produced 90 percent of the nation's total, but by 1905 these three states produced only 55 percent. By 1920, although the Lehigh Valley continued to be the major center of production, with 26 percent of the U.S. total cement, production was found in 27 states.

A number of factors caused the widespread growth of the industry. In the two decades from 1900 to 1920 complete mechanization of the industry occurred, the introduction of the rotary kiln being the most significant technical advancement. The rotary kiln revolutionized the making of cement while at the same time greatly widening its market. Cement made in rotary kilns was much better in quality and cost 25 percent less. With the introduction of the rotary kiln, it was soon realized that a wide range of raw materials could be used. The cement industry was no longer restricted to areas where natural cement rock was found. A standard, uniform quality of cement could now be produced. In 1904 the American Society for Testing Materials established standards so that all cement would be of the same quality. Thus, cement became a noncompetitive product.

Besides mechanization, the decentralization and expansion of the portland cement industry was encouraged by the rapidly expanding market after

1900. Until about 1915 the greatest market for cement was for public and commercial buildings, bridges, levees, and specialty uses. Between 1910 and 1920 the concrete highway was developed, and this single use soon became the most important market. Because cement was bulky, heavy, and had a relatively low price, and because the raw materials and fuels were widely available, plants were dispersed over a large area.

In 1989 the major concentration of cement plants in Pennsylvania remained in the Lehigh Valley, in Egypt, Nazareth, Bath, Whitehall, and Stockertown in Northampton and Lehigh counties, with a secondary concentration in western Pennsylvania in Lawrence County at Wampum. Cement production has steadily declined in Pennsylvania from 8,438,000 short tons in 1969 to 5,735,000 tons in 1986. Pennsylvania now produces about 8 percent of the nation's portland cement.

The Glass Industry

The glass industry was one of the earliest industries in western Pennsylvania, filling a fundamental need for containers not only for Pennsylvania but also for the settlers migrating west. The first glass plant west of the Alleghenies was built by Albert Gallatin in New Geneva on the Monongahela River in 1797. By 1810 there were at least eight glass factories in western Pennsylvania. As the industry developed, Allegheny, Westmoreland, Washington, and Armstrong counties became counties of major production.

The industry was oriented to raw materials and fuel supplies, and its growth depended on the large regional market. Bituminous coal was the first major source of energy to produce glass, but with the discovery of natural gas it became an important source of energy. The first successful use of natural gas to manufacture glass took place at a large glass factory at Creighton, near Pittsburgh, in 1883. High-quality glass sands were available, particularly in Juniata, Huntingdon, and Fayette counties. There was also an abundance of fireclay suitable for making bricks to line the furnaces.

By 1870 Pennsylvania was the leading glass-producing state, with about 52 percent of the nation's output, based on the value of the production. Since then the relative position of Pennsylvania has declined as other states developed new glass industries. The localization of the industry has been remarkably stable, with the industry remaining concentrated in the traditional counties of southwestern Pennsylvania. Employment in the industry has, however, experienced a significant decline. Between 1969 and 1989 employment in the production of glass and glassware declined from 7,023 to 2,500 in the four counties of Allegheny, Westmoreland, Washington, and Clarion. Plants are becoming antiquated and are not being renovated. Production has been discontinued at some centers, such as at Clarion in Clarion County. The region was noted for the production of bottles, but with the development of plastic and paper containers a major market has essentially disappeared. The largest producer is the Pittsburgh Plate Glass Industries, with major plants at Ford City, Creighton, East Deer Township in Allegheny County, and Greensburg. Other major glass companies in southwestern Pennsylvania include Brockway Glass Company, Corning Glass, Owens-Illinois, and the Jeannette Corporation.

Primary Metals Industries

In the nineteenth and early twentieth centuries, Pennsylvania was the national leader in primary metals, primarily iron and steel production. The industry was oriented to the fuel resources of the state—first wood for charcoal, then for anthracite, and finally for the massive bituminous deposits of southwestern Pennsylvania. Although new centers have arisen outside the state, the industry in 1987 still had 10.8 percent of the nation's primary metals employers. Ohio had risen to first place, with 11.7 percent of the nation's employers, followed by Indiana, Michigan, and Illinois. The Pennsylvania primary metals industries have experienced many problems recently, including the antiquity of the plant facilities and competition from imports from the low-cost producing countries of the world. The iron and steel industries are treated in detail in Chapter 16.

Fabricated Metal Products

The iron industry of the colonial period provided the raw material for making simple fabricated metal products such as horseshoes, cutlery, hand tools, nails, and hundreds of other items. In the nineteenth century the heavy castings and forgings produced by

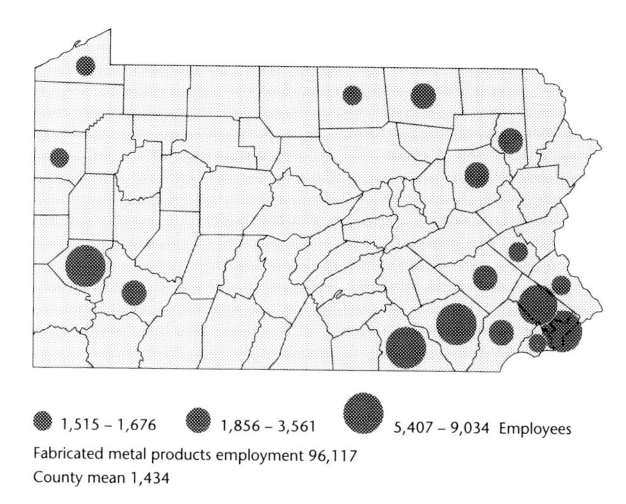

Fig. 15.18. Employment in fabricated metal products, Pennsylvania, 1989.

Pennsylvania's iron and steel industry provided the foundation for the machine age.

Pennsylvania today produces a wide variety of fabricated metal products. The structure of the industry has changed little in recent decades. Most notable has been the increase in miscellaneous fabricated products. Fabricated structural metal products account for about 32 percent of the total employment, followed by miscellaneous products, metal forgings and stampings, cutlery, hand tools and hardware, and screw machine products.

The fabricated metal industry produces products valued at $9.5 billion annually, using $2.1 billion worth of raw materials and paying wages of more than $1.2 billion. Between 1969 and 1989 employment in fabricated metals dropped from 116,114 to 96,117, a decline of about 17 percent.

In 1989 some 57 counties produced fabricated metals, of which 6 counties were above the county mean of 1,434 employees. These 6 counties—Philadelphia, Allegheny, Montgomery, Lancaster, Erie, and York—had 46 percent of the state's employment (Fig. 15.18). Between 1969 and 1989 the spatial structure of the industry was quite stable. A skilled labor force plus an established market are important in maintaining the industry.

Major declines occurred in counties where the industry had initially been concentrated. Philadelphia, the leading employer in both 1969 and 1989, had a decline in employment from 23,144 to 9,034, or about 45 percent. Every branch of the industry experienced a decline, with major declines in cutlery, hand tools and hardware, metal cans and shipping containers, and fabricated structural metal products. The same trends were exhibited in Allegheny County, where employment decreased from 17,041 to 8,418, a drop of about 50 percent. In Beaver County employment declined from 4,397 to 665, reflecting the decline in raw material from the basic primary industries and loss of the market for fabricated metal products. Other counties with significant employment declines include Montgomery, Erie, York, Lackawanna, Delaware, Lehigh, and Westmoreland.

A number of counties experienced modest gains in employment. Employment in Lancaster County increased, as did employment in Berks, Luzerne, and Chester. In each of these counties employment in most branches of the fabricated metals rose slightly.

Industrial Machinery and Equipment

The development of the machinery industry played a key role in the industrial evolution of Pennsylvania in the nineteenth century. Because this was a period of great technological innovation, new advances in machinery were demanded. The heavy machinery Pennsylvania's iron and steel industry needed was of major importance for its advancement. The Mackintosh-Hemphill Company of Pittsburgh, founded in 1803, was an early leader in the field. Noted for its ability to build complete rolling mills, it also designed the machinery for the first large tin-plate mill. This company also produced the machines used in the first Garret rod mill, the first reversing bloom mill, and the first continuous hot-strip mill. Pittsburgh was also the headquarters of the United Engineering & Foundry Company, for decades the world's largest designer and maker of rolls and rolling-mill equipment for the steel industry. Without these and other companies, the iron and steel industry could not have been developed.

Development of the state's resources also required the manufacture of specialized machinery. Because the oil industry began in Pennsylvania, the state was the predominant leader in the output of oil-field machinery, such as engines, pipe-laying machines, and refinery equipment. The coal industry, as it became more highly mechanized, relied on Pennsylvania factories to produce the machinery needed in the mines. In the nineteenth century Pennsylvania was also an important producer of farm machinery as agriculture expanded in the state.

Pennsylvania was also a center of textile-machine production to serve the expanding textile industry of the nation as well as the state. H. W. Butterworth & Sons Company of Philadelphia, established in 1820, pioneered in the development of complete textile mills. Reading was a major center of textile machinery. In the 1890s two German immigrants, Henry Jonssen and Ferdinand Thun, began building braiding machines. This venture became the basis for the Textile Machine Works, which developed the first practical machine in this country for making full-fashioned hosiery. Many other types of textile machinery were developed in Pennsylvania to meet the needs of the industry.

The production of machinery remains one of Pennsylvania's leading industries. With an annual output of more than $8.8 billion, machinery is exceeded only by primary metals, fabricated metals, chemicals, and food and kindred products in overall value of production. The industry consumes annually $3.7 billion worth of raw materials and pays wages totaling more than $1.3 billion. Employment grew steadily in the twentieth century, totaling 141,416 in 1969. Since then employment has dropped steadily, to a low of 97,234 in 1989.

A wide variety of machinery is produced in Pennsylvania. In 1989 industrial machinery was most important, with 36 percent of the total employment. Other important branches include metalworking machinery, construction and related machinery, office and computing equipment, and refrigeration and service machinery. Between 1969 and 1989 the office and computing machinery employment rose from 3.5 percent of the total to 10 percent, and miscellaneous machinery went from 8 percent to 15 percent. In contrast, employment in the production of engines and turbines fell from 11 percent to 4 percent, and employment in special machinery declined from 14 percent of the total to 9 percent. There was little change in the percentage employment of the other branches.

In 1989 machinery was produced in 56 counties in the state. The county mean was 1,451 employees. However, the industry was concentrated in the 19 counties above the mean, which have 82 percent of total employment (Fig. 15.19). Furthermore, the seven major counties—York, Montgomery, Allegheny, Lancaster, Philadelphia, Franklin,

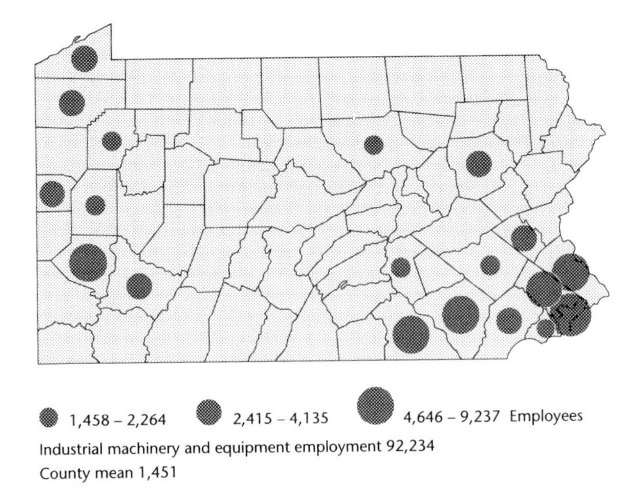

Fig. 15.19. Employment in industrial machinery and equipment, Pennsylvania, 1989.

and Bucks—in the top third above the mean had 48 percent of the state's total employees.

Employment trends have varied throughout the state, with some counties experiencing major declines and other counties having modest increases. The greatest declines in employment occurred in the largest producing counties. All branches of the machinery industry declined in Philadelphia, but the employment decreases were particularly large in construction machinery, special industrial machinery, and general industrial machinery. The machinery industry of Philadelphia County followed the decline of oil manufacturing in the county. In Allegheny County the decrease in employment in machinery reflected the decline in demand from the heavy industries of the area. Declines were greatest in special industrial machinery and metal-working machinery. In Berks County the decline of the textile industry, and in Venango County the decline of the petroleum industry, reduced the demand for specialized industrial machinery.

In a number of counties, machinery employment was stable or increased modestly. These include Chester, Bucks, Franklin, and York counties. Growth in these counties was normally due to increases in one or two branches of machinery. For example, in Chester County employment increased in office and computing machinery, in Bucks County growth occurred in general industrial machinery and miscellaneous machinery, in Franklin County in construction machinery, and in York

County by employment increases in refrigeration and construction machinery.

Although the 2,568 machinery establishments varied greatly in size in 1989, the industry is dominated by small plants. Of the 2,568 plants, 1,191 had fewer than 10 employees, and only 20 had more than 500 employees. In most of the small machinery-producing counties the establishments with fewer than 10 employees make up from 50 to 70 percent of the total. The small plants produce machinery usually for a local market. In an industry that has become increasingly complex, the small plant does not have sufficient capital for modernization, and certainly not for research and development.

The production of machinery remains a major industry in Pennsylvania. The state has a highly skilled labor force and its reputation for producing a quality product persists. On the negative side, new capital expenditures are minimal. In the 1980s about $320 million were invested annually in new capital expenditures—only about $2,600 per employee. This compares with $3,900 per worker of capital expenditure for each employee in electronics and electronic equipment, $3,200 per worker in the transportation equipment industries, $5,500 in the primary industries, and $14,000 in petroleum and coal products. The machinery industry of Pennsylvania will remain important in the state, but a modernization program is needed if the industry is to maintain its traditional importance.

Electronics and Electronic Equipment

Besides such basic industries as iron and steel, textiles, leather, and glass, many other industries of the nineteenth century thrived in Pennsylvania. Of these, none was of greater importance than the electrical machinery industry. By 1900 the Census of Manufactures listed nearly 100 manufacturers of some type of electrical apparatus. The manufacture of electrical machinery and apparatus was one of the state's fastest growing industries.

Of Pennsylvania's many leaders in the development of the electrical machinery industry, the most prominent was George Westinghouse of Pittsburgh, who was not only a major inventor of new electrical machinery but also a corporate leader who founded one of the nation's largest electrical-machinery companies. The original Westinghouse Electric & Manufacturing Company was founded in 1886 in Pittsburgh, with 200 employees producing 13 products.

The first electric light bulb was invented by Thomas A. Edison in his New Jersey Laboratory and test-proved on October 21, 1879. On July 4, 1883, in Sunbury, Pennsylvania, Edison conducted the first successful experiment in using a three-wire electrical circuit to light an entire building. The use of electricity was hampered, however, by its dependence on direct current transmission, which could not transport electricity more than a few feet. The inventive genius of Westinghouse solved this problem by developing the alternating current. In the late 1880s he secured patent rights for the alternating-current electrical transformer first invented in Europe but refined in Westinghouse's laboratory in Pittsburgh. Using this transformer, Westinghouse demonstrated how entire cities could be lighted.

Westinghouse made many other contributions to further the use of electricity. In 1890 he developed a steam turbine for producing electricity, and in 1898 he began using the turbines in his own factories. By 1905 Westinghouse had introduced the use of electricity in a steel mill, and in 1906 a railroad line was electrified. In 1920 the Westinghouse organization launched the world's first radio broadcasting station, KDKA, in Pittsburgh. No single person, including Thomas Edison, contributed more to the development of the electrical age through inventions and technological innovations and the manufacture of electrical machinery and apparatus than George Westinghouse. He made Pennsylvania the leader in electrical machinery.

In addition to the Westinghouse Company, Pennsylvania was the base of a number of other leading electrical-equipment manufacturers. The General Electric Corporation built major plants in Philadelphia and Erie. This giant had its origin in Philadelphia in the Thomson-Horston Electric Company, organized to supply electricity to Philadelphia. A group of Philadelphians in 1892 established the Philadelphia Storage Battery Company, later to become the Philco Corporation, one of the world's largest producers of electrical equipment. Another large manufacturer of electrical equipment was the Sylvania Electric Products Company, with original headquarters in Emporium.

The electronics and electronic equipment industries continue to be of major importance in the state. In 1987 the industry consumed $3.4 billion worth of raw materials, and in the manufacturing process produced products with a value added of $4.3 billion to produce shipments of goods with a value of $7.7 billion. From a peak employment in 1969 of 134,219 employees, it declined to 70,164 in 1989. As employment declined, the different branches of the industry experienced diverse trends. The only major branch that did not decline in employment was electronic computers and accessories. Major declines in employment were experienced by electronic testing and distributing apparatus, radio and television receiving equipment, and communications equipment. Lesser declines occurred in electrical industrial apparatus and in electric lighting and wiring equipment. Small branches of the industry that were stable included household appliances and miscellaneous electric equipment and supplies where employment rose slightly.

Diverse trends were also being experienced by the major producing counties. In 1969 some 13 counties were above the county mean employment of 2,796. These counties had 71 percent of the total employment. In 1989 there were 16 counties above the county mean employment of 1,047 (Fig. 15.20). These counties had 78 percent of the state's total employment. Eleven counties appear in both 1969 and 1989, two counties (Erie and Beaver) disappear in 1989, and four new counties (York, Lackawanna, Centre, and Cumberland) appear as important producers of electronics and electronic equipment.

A number of major counties experienced significant losses in employment. The greatest loss in employment occurred in Philadelphia, where employment dropped from 26,941 to 9,433. Many branches of the electronics and electronic equipment industries essentially disappeared in Philadelphia, including electric testing and distribution equipment, electrical industrial apparatus, and household appliances, radio and television receiving equipment, communications equipment, and miscellaneous electrical equipment. The electric lighting and wiring equipment industries had a small gain in employment, and electronic computing and accessories had a significant decline in employment.

Allegheny County was the second major employer of the electronics and electronic equipment industries in 1969, with 10,904 employees. By 1989 it had fallen to eighth place among the counties, with 2,719 employees. All branches of the industry declined significantly except electric industrial apparatus, where employment was at about 860 employees. Other counties with major declines include Erie, Elk, and Chester.

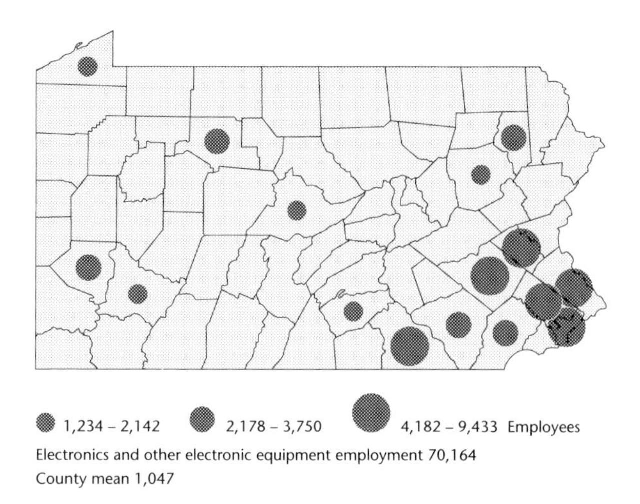

Fig. 15.20. Employment in electronics and electronic equipment, Pennsylvania, 1989.

In contrast, there was a slight increase in employment in a number of counties. Montgomery County was the leading county in both 1969 and 1989, with about 10,000 workers. The major branch of production is electronic components and accessories, with about 2,500 employees. Other counties experiencing small growth in southeastern Pennsylvania included Berks, Bucks, York, and Cumberland. The growth in these counties was largely attributable to the increased employment in electronic components and accessories. The only growth county outside southeastern Pennsylvania was Lackawanna, where electronic components also provided the impetus for growth. Lancaster and Lehigh counties were the only other counties to experience a decline in employment in southeastern Pennsylvania. While the electronic component employment grew, other branches declined more rapidly, causing a total drop in employment. Elk County developed a specialized branch of the electronic equipment industry in the production of carbon and graphite products used in electrical industrial apparatus. As this specialized industry grew in other states, employment has declined in Elk County.

In the late nineteenth and early twentieth centuries the electronic equipment industries were concentrated in Allegheny and Philadelphia counties. The major factor in this localization was the inventive genius and management skills of a few people. Westinghouse Electric in Pittsburgh and General

Electric in Philadelphia illustrate the growth of major companies that provided new concepts to industrial progress.

The old centers of Pennsylvania have not been in the forefront in developing new innovations. Within Pennsylvania the newer electronic component and accessories industries developed in the counties where industry was expanding in southeastern Pennsylvania and in Centre County, where contact with The Pennsylvania State University aided the industry.

Transportation Equipment

At the end of the nineteenth century the manufacture of locomotives and the building of railroad cars provided the fifth largest employment in the state, exceeded only by iron and steel, textiles, foundry products, and machinery. The tremendous growth of railroads in the nineteenth century provided the basis for the locomotive and railroad car industry. Because the Pennsylvania Railroad, with its headquarters in Philadelphia, served a large region in the northeastern quadrant of the United States, the Pennsylvania industry had a major regional market, and the quality of Pennsylvania locomotives and equipment provided the basis for a national market.

One of the first steam locomotives was built in Pennsylvania by Matthias Baldwin in Philadelphia in 1831 for the new railroad being constructed between Philadelphia and Germantown. The engine, built partly of iron and wood, became known as "Old Ironsides." Baldwin made Philadelphia the locomotive-building center of the nation, and by 1866, at the time of his death, he had built 1,500 locomotives for the American railroads. The Baldwin Locomotive Works, later becoming the Baldwin-Lima-Hamilton Corporation, was a major builder of steam locomotives. Many other companies built locomotives in Pennsylvania, including the Pennsylvania Railroad at Altoona.

A number of companies built railroad equipment too. The American Can & Foundry Company, with its main plant at Berwick, was an early builder of equipment for American railroads. Founded in 1840, it was the oldest company of its kind in the nation. When railroads developed distant travel the sleeping and dining cars became standard equipment. The Pullman Company, with plants in Butler, became a major producer of specialty equipment. The Edward G. Budd Company, incorporated in 1912, at Ardmore near Philadelphia, was a pioneer in the building and design of streamlined cars, including the high-speed cars used on the Pennsylvania Railroad between New York and Philadelphia. Budd built the pioneer Burlington Zephyrs running between Chicago, Omaha, and Denver.

In the twentieth century, when coal-fired engines gave way to the diesel and later the electric engine, Pennsylvania lost its advantage in locomotive production. Pennsylvania was also a major producer of Pullman cars, but as the railroads of the nation curtailed this type of service the demand for Pullman cars disappeared. Finally, in the decline of freight service the demand for rolling stock has been greatly reduced.

The modern transportation equipment industry is based on production of a wider range of products than that in the nineteenth century. Motor vehicles and parts are most important branches of the industry, with about 38 percent of the total, followed by aircraft and aircraft parts with 19 percent, railroad equipment with 18 percent, and ships and boatbuilding with 3 percent, and about 30 percent of the workers produce motorcycles, bicycles, guided missiles, space vehicles, trailers and campers, and tanks and their components. In 1987 the transportation equipment industries consumed raw materials costing \$3.548 million and had a value added of \$3.637 million in producing shipments valued at \$7.114 million.

Between 1969 and 1989 employment in the industry declined about one-third from 89,808 to 53,376. Although the industry is found in 38 counties, employment is concentrated in southeastern Pennsylvania and a number of isolated counties (Fig. 15.21). In 1989 the mean employment of the 38 counties was 796. Twelve counties were above the mean, with about 80 percent of total employment. All the major producing counties experienced employment declines from 1969 to 1989. For a few counties the decreases were extremely large—for example, employment in Delaware County dropped from 18,144 to about 7,599 because of declining boat and ship production.

A major characteristic of the modern transportation equipment industry is the dominance of one or two products in many of the counties. For example, in Delaware County the Sun Shipbuilding & Dry Dock Company and the Boeing Vertol Company, producer of aircraft parts, have more than 95 percent of the county's transportation equipment workers.

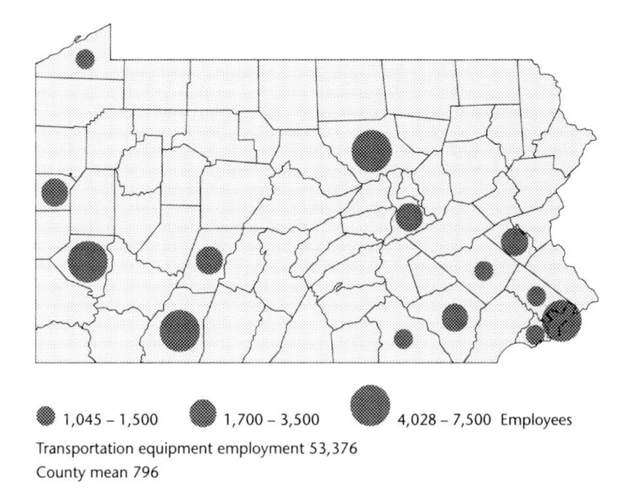

Fig. 15.21. Employment in transportation equipment, Pennsylvania, 1989.

In Erie County the General Electric Company, producing railroad equipment, has 95 percent of the total. Guided-missile production is concentrated in Philadelphia and Montgomery counties. A number of counties, particularly Berks, Philadelphia, and York, produce motor vehicle parts and equipment. In York County the AMF Corporation is noted for its production of motorcycles.

Pennsylvania is not one of the major states in the manufacture of transportation equipment. The early automobile industry concentrated in Michigan, and assembly plants were later built in other states. Pennsylvania was once noted for its heavy-truck production dominated by Mack Trucks, but in January 1986 Mack Trucks closed operations at one of its Allentown plants and transferred them to a new South Carolina plant. Friction between the company and the automotive union over wages and other benefits was primarily responsible for the closing. A significant percentage of the Pennsylvania workers were given the opportunity to move to the new location.

In recent years foreign automobile companies have been establishing branch plants in the United States. The first of these companies was the German Volkswagen Company, which began the production of cars at New Stanton. The German company was attracted to this site because of a vacant plant built by the Chrysler Corporation but never occupied because of a downturn in Chrysler sales. Unfortunately the Volkswagen Company discontinued operations in 1986 due to a declining market. The Japanese producers of automobiles have selected sites in other states, where costs, particularly labor, are lower.

The aircraft industry has been concentrated in other states, such as California, Washington, and Kansas. Although some manufacture of aircraft engines and parts remains in Pennsylvania, most of the small companies have left Pennsylvania, including the Piper Aircraft Company, which discontinued production at Lock Haven in Clinton County in 1983 to consolidate its operations in Florida. The anticipated increases in sales for small planes, such as the Piper Cessna, did not occur.

Although Pennsylvania does produce some missiles and space vehicles, the industry is centered in other states, such as Florida and Texas. Neither has Pennsylvania been a major producer of motor-vehicle parts. The dominance of nineteenth-century industries, such as iron and steel, textiles, glass, and leather, was so great in the state that the new twentieth-century industries sought new centers of development.

Instruments and Related Products

The instruments industry developed in the nineteenth century along with the machinery industry as demand for precision measurements increased. The Brown Instrument Company of Philadelphia, later an affiliate of the Minneapolis-Honeywell Company, illustrates the development of the industry. The company dates from 1859, when a young English engineer and draftsman, Edward Brown, migrated to the United States. The instruments used at that time were extremely crude, and Brown recognized the need for greater precision. During his lifetime Brown developed and improved many instruments, including the pyrometer, the thermometer, and pressure gauges. Scores of instrument companies were widely distributed in the state.

The production of instruments has continued as a major industry of the state. Employment grew steadily from 18,847 in 1950 to 33,129 in 1969 and reached a peak in 1980 with 41,992. Since then, however, employment has declined slightly, to 40,429 in 1989. In 1982 this industry consumed $871 million of raw materials and had a value added of $1,778 million to produce shipments valued at $2,647 million. Each employee in the industry produced an average of $70,034 in products. This

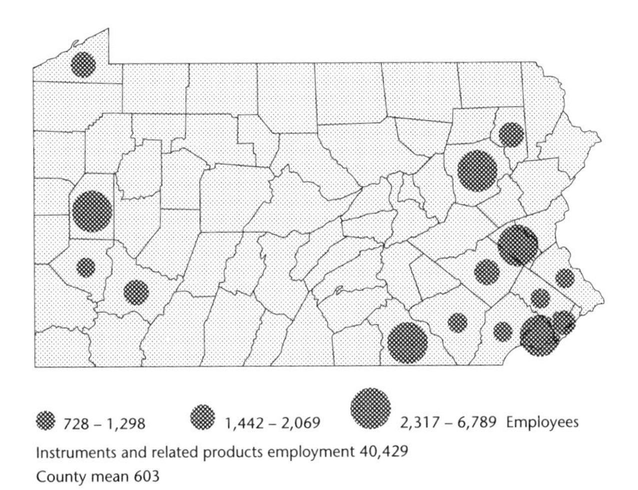

Fig. 15.22. Employment in instruments and related products, Pennsylvania, 1989.

compares with the output of each employee in machinery of $73,034, in fabricated metals of $79,687, and $114,489 in primary metals. The instrument industry is a skilled, high-cost, labor-intensive industry. The major branches of the industry include measuring and control devices, with 46 percent of employment, and medical instruments and supplies, with 29 percent.

In 1989 the county employment mean in the instrument industry was 603. The 15 counties above the mean (Fig. 15.22) had 77 percent of the total employment, and the top 5 counties above the mean (Montgomery, Bucks, Chester, Allegheny, and Lancaster) had 49 percent of the employment. The industry is therefore highly concentrated (Fig. 15.22).

The instrument industry has experienced a number of locational changes. Philadelphia County was a traditional center of instrument production, but between 1969 and 1989 employment declined from 6,795 to 2,069. The major change was in measuring and control devices, where employment dropped from 4,487 to 1,389. In 1969 Philadelphia County had 20.5 percent of the employment in instruments in the state, but by 1989 it had only 7.0 percent. Allegheny County, also an early center, experienced a decline in employment from 3,319 to 2,990. The older regions have antiquated methods and plants and have not been able to revitalize their old establishments.

The major growth in instrument production is in southeastern Pennsylvania in Montgomery, Bucks, Chester, and Lancaster counties—once again reflecting the growth of manufacturing in suburban counties around the old center of Philadelphia. Major new counties include Bradford (part of the Binghamton, New York, complex), and Westmoreland and Erie counties in western Pennsylvania. In these counties, growth in measuring and controlling devices has played a major role in the new growth pattern. The plants are new and efficient, producing instruments for modern industrial purposes.

Between 1969 and 1989 the number of counties with instrument plants grew from 22 to 34 (Fig. 15.22). There were in 1989 some 523 establishments producing instruments in the state. However, many of the plants are small, with 295 having fewer than 20 employees. The 19 plants with more than 300 workers produce a major share of the instruments of the state. In a complex industry where technological advances are rapid, the lack of capital in the smaller companies for research and development provides a limiting factor in the future growth of the industry.

Other Industries

A number of the two-digit industries have had a limited development or have experienced a significant decline in Pennsylvania. These include lumber and wood products, furniture and fixtures, petroleum and coal products, leather and leather products, and miscellaneous manufacturing. These five industries have a total of 88,000 workers or 8.3 percent of total industry.

Bibliography

Benhart, J. E. Fall-Winter 1988. "Manufactural Changes in Southwestern Pennsylvania: 1947–1985." *Pennsylvania Geographer* 26:27–35.

Bridenbaugh, C. 1966. *The Colonial Craftsman*. New York: New York University Press.

Epps, R. W. 1966. "Strategy for Industrial Development." *Federal Reserve Bank of Philadelphia Business Review,* November, 3–17.

Miller, E. W. 1943. "The Industrial Development of the Allegheny Valley of Western Pennsylvania." *Economic Geography* 19:388–404.

———. March 1966. "Trends in the Localization of Manufacturing in Pennsylvania, 1947–1964." *Pennsylvania Geographer* 5:3–17.

———. 1986a. "Manufacturing: Core of the Industrial System." In *Pennsylvania: Keystone to Progress,* by E. W. Miller. Northridge, Calif.: Windsor Publications.

———. 1986b. "Manufacturing in a Mature Economy." In *Pennsylvania: Keystone to Progress,* by E. W. Miller. Northridge, Calif.: Windsor Publications.

Robson, C., ed. 1875. *Manufactories and Manufacturers of Pennsylvania in the Nineteenth Century.* Philadelphia: Galaxy.

Roscoe, E. S. 1958. *The Textile Industry in Pennsylvania: Report of a Survey on Status of the Industry.* University Park: The Pennsylvania State University.

Stevens, S. K. 1955. "A Century of Industry in Pennsylvania." *Pennsylvania History* 22:49–68.

Swank, J. M. 1908. *Progressive Pennsylvania: A Record of the Remarkable Industrial Development of the Keystone State.* Philadelphia: Lippincott.

16

THE RISE AND DECLINE OF PENNSYLVANIA'S STEEL INDUSTRY

Allan L. Rodgers

For a long time, Pennsylvania was recognized as the largest steel-producing state in the nation. However, although it continues to be an important steel producer, production, share of national output, and employment in Pennsylvania's steel industry have sharply declined. These reductions will continue in the near future, at least. In this case, restructuring has meant the imminent fall of what was the state's leading industry.

This chapter reviews the historical geography of Pennsylvania's steel industry, including the massive changes that have occurred since the early 1970s. The focus is on locational change, or the evolution of spatial patterns in this industry, and on changing material supplies, technological innovations, and domestic and foreign competition.

For the earlier period through 1970, I have been guided by the laudable studies of Pounds (1959), Hogan (1972), and especially Warren (1973). Unlike Hogan, both Pounds and Warren were concerned with locational change. I, however, will also discuss the impact of locational shifts on the economies of several areas that were traditionally highly dependent on steel.

The iron industry of America in the postcolonial period was concentrated in the northeastern states, including eastern Pennsylvania. Its location reflected the presence of growing markets, seemingly inexhaustible forests, and scattered iron deposits both in bogs and in the hills. The technology was comparatively primitive, compared with western Europe and particularly Britain. It involved the use of charcoal furnaces to smelt the rather low-grade ores of the day. These furnaces had very low capacities, and the product was impure and lacked tensile strength.

The output of these early charcoal furnaces went in part to forges whose chief market was cannon production. The remainder was shipped to refineries and rolling mills designed to produce agricultural implements and hardware. Integrated plants including furnaces, forges, refineries, and rolling mills were rare. The charcoal furnaces were located in river valleys in rural areas or in small urban-type agglomerations. Proximity to raw materials and waterpower was essential. In contrast, forges and rolling mills tended to be relatively more concentrated in their locations.

We tend to think that those early furnaces were scattered—and indeed they were, for the population and market at the turn of the nineteenth century was dominantly rural. For example, less than 10 percent of Pennsylvania's population was urban in 1790. Yet within that scattering, there were distinct geographical clusters. In the case of Pennsylvania, furnaces tended to be arrayed along the major rivers east of the Alleghenies, such as the Schuylkill, the Delaware, the Juniata, and the Susquehanna, exploiting the timber and iron resources of their hinterlands. Warren (1973, 11) offers astute explanations for the low capacities, slow growth, and technical backwardness of the American iron industry before 1850, which are also applicable to Pennsylvania's iron industry. He ascribes the industry's problems to "competition from Britain, high labor costs, lack of mineral fuels and transport difficulties in a widely dispersed economy," and then deftly examines each of these explanations.

The transportation problem was clearly the most burdensome. Transportation innovations were needed to overcome the friction of distance from furnace to markets. This led to an era of transportation improvements east of the Alleghenies, such as road-building and canal construction. Canals, which ultimately were far more important, successfully reduced transportation costs, particularly on bulk goods.

As Pennsylvania's population moved westward, charcoal furnaces developed west of the Alleghenies. That development was also keyed to local iron deposits and forests. Like the pattern in eastern Pennsylvania, the furnaces were located along the river networks, particularly the Monongahela, the Youghiogheny, and the Allegheny. Pittsburgh became, over time, the chief western trading port. Its foundries and rolling mills, using pig iron from the scattered furnaces, cast and shaped nails, wire, bars, tools, stoves, and other metal products. Their output was designed largely to serve the needs of settlers moving to the frontier.

It took until 1852 for the Pennsylvania Railroad to reach Pittsburgh. The delay was attributed to extremely high construction costs. Questions about the railroad's economic feasibility were eventually overcome. Remarkably soon, however, Pittsburgh was outdistanced as a trading center by Cincinnati, St. Louis, and ultimately Chicago. In its place, manufacturing came to the fore.

Technological innovations soon began to reshape locational patterns. Charcoal remained the dominant fuel used in iron smelting in the United States until the middle of the nineteenth century. By that time, a railway network had spread over the eastern half of the nation. Contemporaneously, the sources of charcoal and iron ore were becoming exhausted, leading to abandonment of many charcoal furnaces. In addition, there was a rapid diffusion of technical knowledge, typically of British origin, on the use of coal, furnace design, and mill operations.

The result was the gradual decline of charcoal as a smelting fuel in favor of coal, the growth of the puddling and rolling of wrought iron, and, concomitantly, the concentration of the industry in a far more restricted number of centers, typically accessible by canals, waterways, and later railroads. At the same time, there was a change in scale. The size of the new furnaces had increased exponentially. The refining and rolling of iron was now far more oriented to markets. The period was notable for the rapid expansion of iron production. That growth was due to the adaption of the new technology of smelting iron with high grade coal—in this case, anthracite from northeastern Pennsylvania.

We now turn to what is usually termed the anthracite era, roughly a 40-year period dating from the 1840s that most concede was a major evolutionary step in the iron industry. The earliest anthracite furnace was built at Pottsville on the Schuylkill River. With improved transportation, particularly waterways, it was most economical to place the blast furnaces *south* of the anthracite district and closer to iron-ore sources and markets. More iron ore was used then per ton of finished product than coal, largely because of the lower metallic content of the product. It came from three districts: Cornwall in the Lebanon Valley, southwestern Pennsylvania, and the Middlesex region of northern New Jersey. With improved canal, waterway, and rail transport, the new hot blast anthracite furnaces spread outward to the Lehigh, Juniata, and Susquehanna rivers, all east of the Alleghenies. At its peak, almost 50 percent of Pennsylvania's iron was produced in these furnaces, but concurrently charcoal, raw coal, and later coke were also used for the production of iron.

Developments in western Pennsylvania were led by Pittsburgh, which, as noted earlier, was the leading foundry and rolling mill center west of the Alleghenies. Lesley (1859) estimated that by 1859 the region west of the Appalachians produced more than 40 percent of the rolled iron output of the nation. Pittsburgh's mills were supplied by charcoal furnaces scattered over its tributary region. However, during the decade of the late 1850s and early 1860s, there was a sharp drop in the price of charcoal iron and a rise in production costs, coupled with the near exhaustion of the surrounding forests. Thus, the number of charcoal furnaces rapidly declined, precipitating a shift to coal made possible by the introduction of large hot blast furnaces and by the adoption of coal in the form of coke as a smelting agent. Its technical origins are traceable to success in Britain a half-century earlier.

The utilization of raw coal was largely confined to furnaces in the Mahoning Valley of eastern Ohio and the Shenango Valley of western Pennsylvania. Its success was short-lived because pig iron smelted by this fuel could not compete with that produced by coke. The triumph of coke was in large measure due to the development of large hot blast furnaces. Coke, a semi-processed fuel, was the product of beehive ovens lined with refractory bricks where quality coking coals were carbonized. The waste gases were then removed, and the resulting product was used in the new blast furnaces to smelt iron ores. A good coking coal should be relatively low in volatile matter, and particularly low in sulfur, and the coke produced from that bituminous coal should have a physical structure strong enough to bear the burden of ore and limestone in the blast furnace.

The finest coking coals in the United States were found in a narrow 50-mile belt, near Connellsville, on the Youghiogheny River about 35 miles southeast of Pittsburgh. Its gently sloping seams were roughly 9 feet thick and readily and cheaply mined. The coke was then shipped by river and rail to the Pittsburgh district. That city and its hinterland now became a major coke iron producer. Ultimately, coking coal was also derived from the Pittsburgh seam. These were qualitatively inferior to those from Connellsville, but their reserves were huge and the new technology made their coke worthy of use in the new blast furnaces.

Turning to the availability of iron ore, until 1856 the western iron industry depended on coal measures and bog ores. Because of their proximity to the furnace area of western Pennsylvania, transportation costs were low. Their quality, however, was inferior to the ores of the east, and as the demands of the iron industry grew those supplies were progressively exhausted. Simultaneous with the growth of the production of Connellsville coal and coke, new supplies of iron were discovered near the shores of Lake Superior. These were first found in the Marquette Range of northern Michigan in the mid-1840s, but transportation to Lake Erie awaited the opening of the Soo locks in 1855. With the opening successively of the Gogebic Range of Wisconsin, the Menominee of Michigan, the Vermillion of Minnesota, and finally the largest and best of them all, the Mesabi of Minnesota in 1892, low-cost shipments by specialized ore carriers from Lake Superior to Lake Erie began. These carriers could then move a relatively short distance by rail to Pittsburgh. At the same time, coal and coke were transported to Pittsburgh by river barge carriers. Overall transportation costs were therefore far lower than those of its competitors east of the Alleghenies, and Pittsburgh's raw material supplies were qualitatively cheaper, better, and far more secure. The resulting pig iron found a ready market in the production of iron castings and the rapidly growing wrought-iron industry of the Pittsburgh region. That area had now become one of the leading iron centers of the nation.

By the 1870s, iron casting continued its importance, but wrought iron lost its way to steel. Industry now demanded a product essentially free of impurities and ultimately high tensile strength—thus the initial shift to Bessemer converters, which produced steel from low phosphoric iron ores like those from Lake Superior. These converters carbonized the impurities with hot air blasts, but the open hearths, which soon followed, use a mixture of pig iron and scrap along with small amounts of dolomite. In the long "cooking" process, the impurities could be removed from the mix, and strengthening alloys could be added. With these advantages, the number of Bessemer converters in the United States was soon matched and ultimately outstripped by open hearths. But in the case of Pittsburgh, the opposite was true, for its plants were using Bessemer-grade ores from the Lakes.

The new integrated plants now comprised coke ovens, blast furnaces, steel furnaces, and rolling mills. These facilities were far more clustered than had been the case earlier. Pittsburgh initially lagged in the adoption of the new technology, but by the 1870s Andrew Carnegie and his associates began to invest in steel production. Carnegie merged a

Table 16.1. Steel ingot production, by key district, 1890 and 1902 (000 of tons).

District	1890	%	1902	%
Lehigh, Schuylkill & Lower Susquehanna valleys	2,000	46.5	1,700	20.5
Allegheny County	1,500	34.9	4,700	56.5
Illinois and Indiana	800	18.6	1,900	22.9
Total	4,300	100.0	8,300	99.9

SOURCE: Modified from Warren 1973, 145.

number of rival producers into the Carnegie Steel Company. Then he acquired the lion's share of Connellsville coke production to feed his blast furnaces and gained control of a major share of the Lake Superior deposits to ensure a reliable supply of high-grade iron ore (50 to 60 percent metal). Finally, he created or bought a sizable number of steel fabrication firms to ensure a dependable market for his steel output. Pittsburgh's growth in steel production in the last quarter of the nineteenth century was extraordinary. According to Muller (1987), beginning with the Edgar Thomson works at Braddock in 1876, eventually 11 mills were constructed along Pittsburgh's waterways and those of its immediate hinterland. By 1908 that region produced nearly one-third of the nation's steel output. Its importance was derived from its high-quality low-cost raw materials, a growing local demand, and its access to expanding western markets.

Meanwhile, eastern Pennsylvania, in the last decades of the nineteenth century, lost ground to its western competitors. Charcoal and anthracite iron had declined drastically in importance, and eastern ores were deficient in both quantity and quality. Many small and outdated eastern mills closed. Yet a number of plants continued operations even at an expanded pace. Eventually, the survivors became competitive again. Some did so with coal deliveries from West Virginia using the railroads to transport their freight to Hampton Roads and by ship to eastern mills. Others bore the cost of direct rail shipments via the Pennsylvania Railroad. Then too there was a relative decline in the locational attraction of coal because of improved technology. The new by-product ovens of the eastern integrated mills produced a quality product at competitive prices. Ore supplies came in part from the Lakes, but increasingly also from foreign sources like Santiago in Cuba and later from the El Tofo mines of northern Chile and the Kiruna deposits of northern Sweden. The east had the advantage of large supplies of cheap scrap iron and of course market proximity. Table 16.1 demonstrates the absolute levels and shares of steel ingot output for several of the key producing districts from 1890 and 1902. Note that the volume of production in the East declined minimally, while that for Allegheny County, in the heart of the Pittsburgh district, tripled over those 12 years. Clearly, Allegheny County now accounted for the lion's share of the nation's steel output, yet one-fifth of the production was still centered in eastern Pennsylvania.

Outside the Pittsburgh district, the other Pennsylvania integrated mills were located in Johnstown (U.S. Steel and Bethlehem Steel) and Steelton and Bethlehem (both Bethlehem Corporation mills), and there were a number of smaller plants in the Schuylkill Valley near Philadelphia. U.S. Steel (and its predecessor, Carnegie Steel) dominated production in western Pennsylvania. Over time, steel was produced mainly by the open-hearth process, and nonintegrated mills became rarities.

By the turn of the century, the Pittsburgh district, Wheeling, and the Valleys (Youngstown in particular) were producing nearly 60 percent of the nation's steel output. Eastern Pennsylvania's share had declined sharply. Pittsburgh had the cheapest and best fuel, quality ores came by cheap lake transportation from Lake Superior, and short rail hauls went from there. For the time, Pittsburgh had one of the most developed transportation systems in the nation, and entrepreneurs of extraordinary ability. Meanwhile, consumption was spreading to the South, the Midwest, and even the Pacific. The new integrated mills required heavy investments in specialized and relatively permanent capital equipment. For this reason, the steel industry offered strong resistance to locational change when existing equipment was not fully depreciated. Yet the success of the established mills depended in large measure on their ability to sell at a profit in the new outlying market centers. To solve this problem, the industry adopted a basing-point system of pricing that enabled mills in the older established centers around Pittsburgh to penetrate the new market areas and yet postpone construction there almost indefinitely. This pricing practice was used for more than half a century, with major effects on the location of steel production and in turn on the nation's overall industrial location pattern.

The single-basing-point system, known as "Pittsburgh Plus," was in force during the first quarter of the twentieth century and was modified only in response to a Federal Trade Commission order in 1924. Under this system, steel was sold at the Pittsburgh base price plus the rail freight from Pittsburgh to the consuming center, regardless of where the producing plant was located. As a result, Pittsburgh producers were able to quote the same competitive prices as companies whose plants were closer to the new market centers. Geographical position was thereby eliminated as a substantial element in competition, and distant markets were opened to producers in the older districts on the basis of an approximate price parity with competitors elsewhere. Production tended to stabilize or expand in the older districts, and the industry was enabled to minimize the potential costs of building new mills in outlying market areas.

The higher prices prevailing at outlying producing points restricted demand in their areas. Although centers such as Chicago obtained a maximum yield from the so-called "phantom freight" they collected, which possibly protected their infant industry, once the regional demand was satisfied and they were forced to sell toward Pittsburgh, their returns were reduced both by the lower delivered prices and by rising transportation costs. Moreover, fabricators were attracted to the older districts, primarily because steel costs were lowest but also because location near Pittsburgh meant the widest possible radius of distribution for steel products. Thus the single-basing-point system favored the location of fabrication in the Pittsburgh district and in turn stimulated increased development of the basic steel industry there.

Table 16.2 is a comparison of regional capacity for producing hot rolled steel products and consumption of these products in 1920, 1937, and 1947. The data for 1920 confirm the conclusions stressed above, for although the first two decades of this century were characterized by continuing movement of population and markets westward, the steel industry did not reorient its productive capacity to serve these outlying markets. The degree of concentration of steel production as compared with consumption is best illustrated by the fact that in 1924 the Pittsburgh-Cleveland-Buffalo region produced about 60 percent of the national output yet probably consumed less than one-fourth of this total.

After "Pittsburgh Plus" was declared illegal in 1924, the steel industry adopted a multiple-basing-point system of pricing. Under the new system, the number of basing points was increased, and differentials above the Pittsburgh base price were established at these points.

The additional basing points undoubtedly diminished the advantages accruing to producers in the Pittsburgh district and provided a great incentive for both production and fabrication in the outlying market areas. The advantages that had rested solely with Pittsburgh and its satellites were now spread somewhat more evenly over a small group of centers. Nevertheless, although competition was intensified and the market areas of the older centers were contracted, the Pittsburgh district still remained in the distinctly favorable position of maintaining the lowest base price. Delivered prices were still kept high enough to make it profitable for Pittsburgh producers to compete in the outlying market areas at a profit without building facilities there.

Despite these changes, there was no general association of basing points and producing points, and Pittsburgh remained the starting point for most basing-point calculations. The delayed establishment of many producing points as basing points resulted in higher delivered prices in their areas, so that demand was restricted and the growth of new capacity in the outlying market areas was retarded.

By 1937 there had been significant shifts in the relative importance of the steel-producing districts (see Table 16.2). Certainly the most important change was the decline in relative importance of Pennsylvania. In 1920 this state produced about half the nation's steel products, and in 1937 it was less than one-third. Concurrently there was a rise in production in the outlying centers. However, although production in the deficit areas increased markedly, it barely kept pace with the rise in consumption in these areas, so that the overall pattern did not change greatly; there was still a disparity between production and consumption of steel.

Table 16.2. National production capacity for, and consumption of, hot rolled steel products, 1920, 1937, 1947 (in percentages).

Region and State	Capacity[a] 1920	Consumption[a] 1919–21	Surplus or Deficit
Mountain and Far West	*1.4*	*4.0*	*− 2.6*
California	0.0	2.3	− 2.3
Central	*38.9*	*49.5*	*−10.6*
Illinois	7.3	12.9	− 5.6
Indiana	10.7	5.7	+ 5.0
Michigan	0.0	8.5	− 8.5
Ohio	19.8	12.6	+ 7.2
South and Southwest	*2.4*	*5.9*	*− 3.5*
Alabama	1.4	0.8	+ 0.6
Middle Atlantic	*56.9*	*36.1*	*+20.8*
West Virginia	2.4	0.8	+ 1.6
Pennsylvania	49.0	22.3	+26.7
Maryland and D.C.	1.4	1.5	− 0.1
New York	3.7	7.6	− 3.9
New England	*0.4*	*4.5*	*− 4.1*

Region and State	Capacity[b] Jan.1, 1938	Consumption[c] 1937	Surplus or Deficit
Mountain and Far West	*2.4*	*6.9*	*− 4.5*
California	1.0	4.7	− 3.7
Central	*46.6*	*51.5*	*− 5.1*
Illinois	7.8	11.3	− 3.5
Indiana	11.9	4.5	+ 7.4
Michigan	5.1	16.2	−11.1
Ohio	20.3	12.1	+ 8.2
South and Southwest	*5.1*	*10.4*	*− 5.3*
Alabama	3.2	1.2	+ 2.0
Middle Atlantic	*45.3*	*28.2*	*+17.1*
West Virginia	3.5	1.4	+ 2.1
Pennsylvania	32.7	17.0	+15.7
Maryland and D.C.	3.6	1.8	+ 1.8
New York	4.7	5.7	− 1.0
New England	*0.6*	*3.0*	*− 2.4*

Region and State	Capacity[b] Jan. 1, 1948	Consumption[d] 1947	Surplus or Deficit
Mountain and Far West	*5.6*	*8.6*	*− 3.0*
California	2.6	5.4	− 2.8
Central	*46.3*	*51.2*	*− 4.9*
Illinois	8.2	11.6	− 3.4
Indiana	12.6	5.1	+ 7.5
Michigan	3.6	13.5	− 9.9
Ohio	21.0	10.9	+10.1
South and Southwest	*6.3*	*11.5*	*− 5.2*
Alabama	4.3	1.5	+ 2.8
Middle Atlantic	*41.2*	*25.1*	*+16.1*
West Virginia	2.8	1.4	+ 1.4
Pennsylvania	28.0	13.8	+14.2
Maryland and D.C.	5.1	1.9	+ 3.2
New York	4.6	5.3	− 0.7
New England	*0.6*	*3.6*	*− 0.3*

SOURCES:
[a]C. R. Daugherty, M. G. de Chazeau, and S. S. Stratton, *Economics of the Iron and Steel Industry*, 2 vols. (New York, 1937), 1:62; computed and corrected from chart 9.
[b]Directories of the iron and steel works of the United States and Canada, 1938 and 1948.
[c]Computed from data compiled by Marion Worthing from TNEC Schedule A, which was corrected by Worthing from data reported by the American Iron Steel Institute for shipments of steel products during 1937.
[d]Census of Manufactures 1947, Preliminary Report Ser. MC00-10, December 30, 1949. Compiled and corrected by the method described by Walter Isard and J. H. Cumberland, "New England as a Possible Location for an Integrated Iron and Steel Works," *Economic Geography* 26 (1950): 255.

But for the exigencies created by the last war, the production pattern might have continued relatively unchanged. The urgent need in the South and West for steel for ship construction, armaments, and so on forced an expansion of production there. About one-fourth of a program of more than $2 billion was devoted to new construction in these hitherto neglected areas. Large integrated establishments were created at Fontana, California, at Provo, Utah, and at Houston, Texas, and there were modest increases in capacity in the Birmingham district. As a result of these changes, plus postwar expansion programs, the regional patterns of production and consumption of steel altered significantly (see Table 16.2).

The Far West and the Mountain States, which had experienced a major boom in production during the war, in 1947 consumed almost 9 percent of the nation's output of rolled steel products and produced almost two-thirds of its own requirements. The South and Southeast, however, although their production had increased during the war, still produced only about half of their needs. The relative position of the Central district remained almost constant. There was still no significant output of steel in New England, which in 1947 was producing only one-sixth of its requirements. The Middle Atlantic district remained the only surplus area in the country, though its share of the nation's output had decreased at the expense of the South and the West. Note the surplus in Pennsylvania.

In 1948 the basing-point system was declared illegal by the Supreme Court, and subsequently the various steel companies adopted an f.o.b. mill system of pricing. Under the f.o.b. pricing system, each producing region has had a competitive advantage within a relatively limited area. Therefore, mills in areas of surplus capacity, such as Pittsburgh and Youngstown, have made determined efforts to cultivate and expand local markets. These markets, however, could never absorb the total output of such centers. Witness the Youngstown district, which normally produced about one-tenth of the nation's steel output yet absorbed only about one-eighth of its own production. Mills in such an area had to seek markets elsewhere, and to do so they had to absorb freight. However, freight absorption has become more difficult because of successive rises in freight rates during the postwar era, so that on many standard products the amount of freight that must be absorbed was probably too great to warrant penetration of distant markets. In addition, the growing markets of the East and South required expansion of capacity there and also led to the construction of a new mill at Morrisville, Pennsylvania, near Trenton, on the Delaware River. This was U.S. Steel Fairless Works. In addition, there was also a significant expansion of capacity at Bethlehem's Sparrows Point plant near Baltimore. As a result, steel production was in better accord with the market.

Dispersion was further stimulated by the spread of fabrication away from the older basing-point centers. In addition, the attraction of the market was augmented by the sharply increased use of foreign ores, particularly from Labrador and Venezuela, induced by the decline in the Lake Superior high-grade ore production. This, of course, was partially offset by the development of sintered and agglomerated low-grade ores, known as taconite, in that region. Coal quality deteriorated, and this led to heavy expenditures on washing and cleaning facilities, particularly for the production of coke.

At this point, a few words about new technology are in order. In the postwar era, there was a slow development of basic oxygen converters with their high capacities, and the use of oxygen in the older blast furnaces and open hearths also contributed to increased steel production per plant. There was also an increase in the building of electric furnaces, which led to a reduced consumption of coal. These furnaces used scrap as their basic raw material, leading to a rise in the consumption of this readily available material. This in turn encouraged a further dispersal of steel plants to growing market areas. Continuous-casting of steel, with its lower costs, became increasingly the mode of product conversion.

By the early 1970s, steel production had peaked (see Table 16.3), but Pennsylvania's share had declined from half to roughly one-quarter of the national output. Figure 16.1 demonstrates the locational pattern as of 1974 in terms of raw steel capacity. Because plants were operating at or near their potential, this was a reliable surrogate for production. Some 70 percent of capacity was found in western Pennsylvania, an area defined as locations from Johnstown westward. But Johnstown itself, with two mills—Bethlehem and U.S. Steel—had seen the latter essentially closed. Western Pennsylvania mills still used local coals, which now need beneficiation, in conjunction with taconite iron pellets derived primarily from Minnesota mines located near Lake Superior. In contrast, plants in

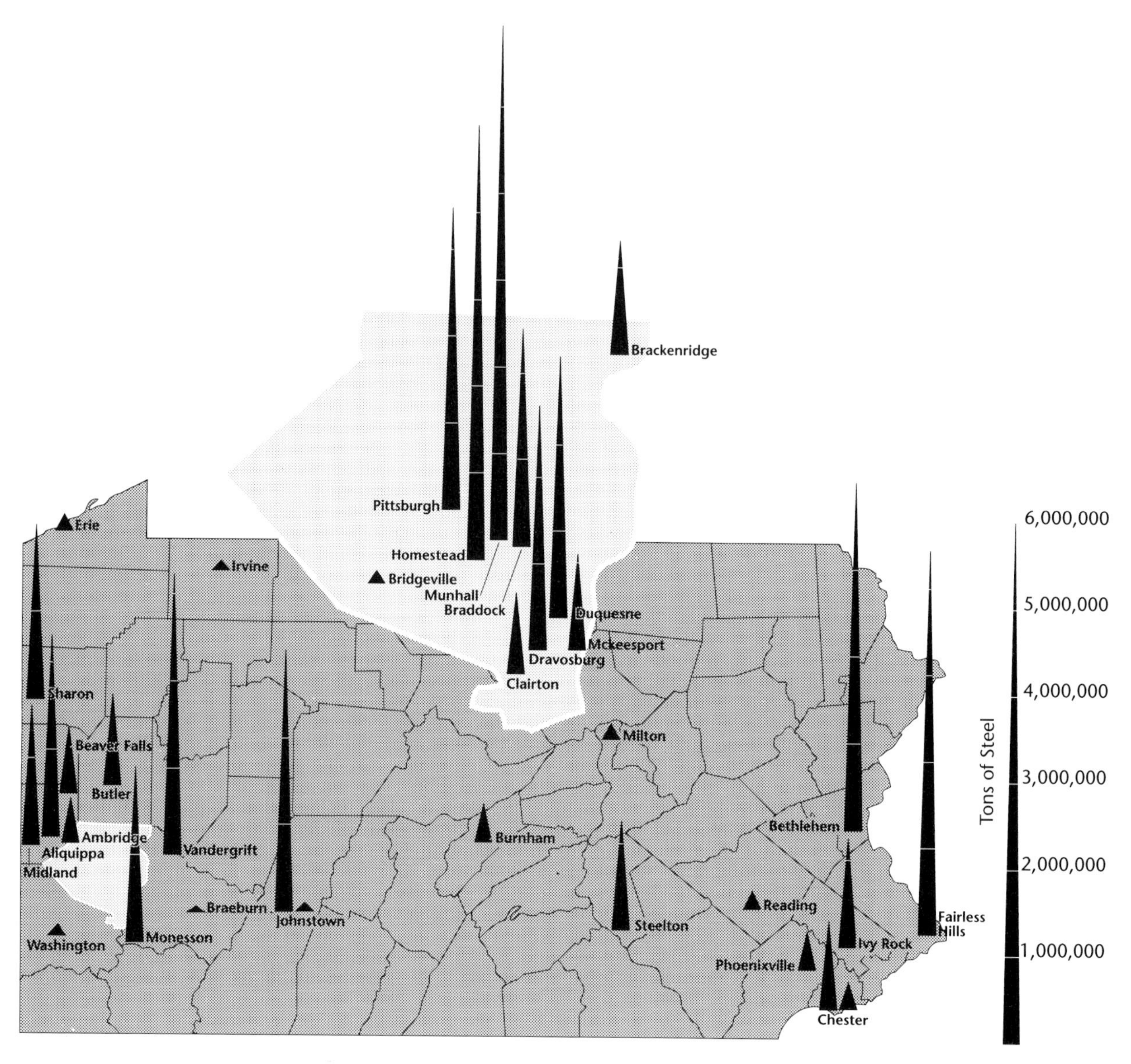

Fig. 16.1. Steel ingot capacity of Pennsylvania as of 1974.

eastern Pennsylvania relied heavily on imported high-grade ores from Labrador and Venezuela and on Appalachian coals shipped primarily from Norfolk. Capacity in the west still exceeded local consumption.

Students of the Pennsylvania steel industry still described it as "prosperous" in the early 1970s, yet in succeeding years the industry's fortunes waned and plant after plant closed—with powerful negative effects on local economies. There has been a plethora of bankruptcies and failures; huge numbers of workers were fired, and unemployment rolls have notably increased. Local communities suffered because of the loss in their economic base, and as a result government services were drastically reduced. Fire and police protection and garbage disposal were among the first to feel the cutbacks. These services were eliminated, severely reduced, or amalgamated with those of neighboring towns. It is also reported that educational budgets were cut as well. Outmigration became a serious problem, with the younger segments of the work force seeking sources of employment elsewhere. The plight of the middle-aged and elderly was sobering.

Table 16.3. Raw steel output (ingots and steel for castings), 1910–1991 (Pennsylvania as a share of the U.S., in millions of tons).

	Pennsylvania	U.S.	Share (%)
1910	13.2	26.1	50.6
1920	17.6	42.1	41.8
1930	14.3	40.7	35.1
1940	20.3	67.0	30.3
1950	27.3	96.8	28.2
1960	23.8	99.3	24.0
1974	33.5	145.7	23.0
1986	9.5	80.5	11.8
1987	11.6	89.2	13.0
1988	13.7	99.9	13.7
1989	11.9	97.9	12.2
1990	12.0	98.9	12.1
1991	9.4	87.3	10.8
1992	9.1	92.9	9.8

SOURCE: American Iron & Steel Institute annual statistical reports, 1912–1991.

By 1992 absolute production, both at national and at state levels, had shrunk drastically (see Table 16.3). Although shutdowns and layoffs have not been confined to Pennsylvania, cutbacks in the state have been more drastic than those in the largest steel centers of the Midwest. By 1992 Pennsylvania produced 9 million tons of steel, or less than one-tenth of the nation's output.

Because of the industry's reluctance to release geographic data, I have had to use data on employment by district in 1974 and 1991 as surrogates for production values. There is a significant but imperfect correlation between these two variables, but technological levels are equally important. Plants equipped with the latest basic oxygen converters and electric furnaces all tend to have very high outputs per employee. The reverse is true of the traditional plants. But technological levels vary from plant to plant and region to region. In the absence of a satisfactory measure of technology levels, by plant or county, employment data were the only alternatives. It must be stressed, however, that the correlation today is far lower than it was in 1986. Employment changes by district are shown in Table 16.4.

How can we explain the precipitous decline of Pennsylvania's steel industry? How too can we interpret spatial variations in the level of those reductions?

First, the obvious must be stressed: The steel industry both in the United States and in Western Europe in the 1980s and early 1990s was truly depressed—production levels were far below capacities. This can be attributed both to the sharply reduced market for steel products and to intense international competition, particularly from Japan and South Korea.

The industry and some scholars argue that high labor costs and low labor productivity were the main reasons for the steel industry's decline, but these are only partial explanations. I alluded earlier to the impact of new technology. In that regard, Japan and Germany were the first major producers to use advanced techniques. Of course, *their* old plants and equipment were largely destroyed during World War II, so they could start afresh, using the most modern technology. More recently, South Korea has developed a steel industry that is state-of-the-art. Its labor costs are the lowest in the world. American producers, on the other hand, delayed modernization because of their huge investment in existing steel-making facilities, their reluctance to adopt new technology until "proven," and corporate inertia. Their rationale was of course *buttressed* by their high labor costs and low productivity. Ultimately, many American plants were supplied with the latest equipment, but the industry as a

Table 16.4. Employment in Pennsylvania steel industry, by district, 1974–1991.

District	1974	%	1991	%	Absolute Change	Relative Change (%)
Eastern Pennsylvania[a]	35,301	24.2	21,521	41.8	−13,780	−39.1
Johnstown	12,490	7.2	1,821	3.6	−10,669	−85.4
Monongahela Valley	69,798	48.1	18,775	36.5	−51,023	−73.1
Beaver Valley	29,874	20.5	9,321	18.1	−20,553	−68.8
Total	147,463	100.0	51,438	100.0	−96,025	−65.1

SOURCE: Pa. Dept. of Industry.

[a]By 1992 the decline in eastern Pennsylvania had barely begun. Johnstown's two steel mills and the USX Fairless Works were closed then; Steelton had been shut down, and the major cuts at Bethlehem were in process.

whole was still unable to meet foreign competition and that of the new mini steel mills, which use electric-arc furnaces and continuous-casting equipment. These are the reasons I believe led to the industry's decline, and in turn to the sharp reduction of production in Pennsylvania. As of 1992 imported steel products and those produced by mini mills accounted for well over two-fifths of domestic consumption.

Some of the same reasons for decline also apply to Pennsylvania, where producers were among the last to adopt new technology. Most of the plants that were closed or driven into Chapter 11 bankruptcy were antiquated. Productivity in Pennsylvania has also been among the lowest of the steel-producing states. The development of mini mills has been far slower than the national average—perhaps because these mills are typically nonunion, while the strength of the United Steel Workers, though significantly diminished, is greatest in Pennsylvania, its headquarters. That could be a discouraging factor. As for competing imports, the data cloud the issue, for Pennsylvania itself is not a large importer. But its customers outside the state may be important users of foreign steel products. This is only surmise, however, for data that would support this thesis are not available.

It is clear, however, that all regions in Pennsylvania suffered employment declines in their steel industries. Nevertheless, by the late 1980s it was only the western part of the state that witnessed shutdowns and bankruptcies with attendant loss of jobs (see Table 16.4). These closings are permanent, and the key issue is now the loss of pensions—particularly true of LTV in Aliquippa. By 1991, the eastern plants had also sharply reduced their labor forces. All the major plants there had closed or were in the process of closing. These included Bethlehem's plants at Steelton and the planned drastic shutting down of parts of its major works at Bethlehem itself. USX's plant at Fairless, near Philadelphia, has been closed for several years. Then too, in the west both the USX and Bethlehem plants in Johnstown are no longer working.

The day of the open-hearth furnace has ended. Today, 62 percent of our raw steel is produced by basic oxygen converters and the remainder by electric furnaces that use scrap as their basic raw material. There is also the beginning of direct reduction of iron ore pellets into steel, which will further hasten the decline of coke ovens and blast furnaces. Just as significant has been the spread of continuous casting of steel, which until recently was confined to rods, bars, and plates, but the newest mills will produce strip and sheets for the appliance sector and ultimately quality steel for the automobile industry. Some 60 percent of American rolled steel is now produced by this process, still lower than the level in Japan (93 percent) and West Germany (85 percent). There has also been the rise of the mini mill, typically with a capacity of 1 to 4 million tons, accounting for 20 percent of the nation's steel production.

It must also be stressed that the man-hours to produce a ton of steel are now notably fewer than in Japan, West Germany, Britain, and South Korea. Concomitantly, labor costs per ton of output have been significantly reduced, but are still higher than those of chief competitors.

On the negative side, there has been a vast reduction of the industry's labor force. The largest share of these workers, as noted earlier, cannot readily be reskilled. There has been a wholesale reduction of 160,000 workers, or 40 percent of the labor force, in the industry. There has also been a parallel loss of jobs in many of the industries using steel.

The impact of plant closings on the economy of Pennsylvania has already been noted. Employment in steel had declined to 49,000 by 1992, compared with 160,000 in 1974, and there will be a further loss of jobs with the most recent and planned closings. Although the role of mini mills in the United States has been on the increase, there are to date only two or three mini mills in Pennsylvania. Mini mills tend to be dispersed more throughout the South, the Midwest, and the Pacific Coast, catering to nearby markets.

Other negative elements of the restructuring of the steel industry in Pennsylvania and the nation are the large amount of new equipment, particularly for continuous casting, that has been purchased abroad (particularly from Japan), and the minimal industry expenditures for research and development which does not bode well for the future of steel production in Pennsylvania and in the United States.

Much has been written about economic changes in the Pittsburgh region, which was once the heart of the state's steel industry. The fact is that currently there is *no* true steel production in Pittsburgh itself, and even in the metropolitan area steel employment is down to 3,000 workers. Between 1979 and 1988 some 67,000 steel jobs were lost in the Pittsburgh metropolitan area, coupled with another 63,000 jobs in other branches of heavy industry. In 1950 metropolitan Pittsburgh had a population of 650,000; as of 1990 the number was 330,000. Yet unemployment in 1989 was 5.4 percent below the state average. Income is clearly up. The growth of jobs in services and technical industries has to a degree counterbalanced the job losses in heavy industry. There has been an increase in the number of small and medium-size firms producing computer software, robotics, and biotech products. The old mill sites are now the homes of industrial parks and research centers. Pittsburgh has strengthened its role as a headquarters town with the gleaming skyscrapers of Westinghouse, Alcoa, and Mellon dominating the urban landscape. Population decline has largely been halted, moving from a high of 35,000 in 1985 to less than 10,000 in 1989. But the old one-industry steel towns have not recovered, nor is there any evidence of future resurgence. Philadelphia, which was never a "steel town," has lost thousands of jobs in other sectors.

Bibliography

Alexandersson, G. 1961. "Changes in the Locational Pattern of the Anglo-American Steel Industry, 1948–1959." *Economic Geography* 37:95–114.

American Iron & Steel Institute. Annual. *Annual Statistical Reports*. New York and Washington, D.C.

———. 1974 and 1986. *Iron and Steel Works: Directory of the United States and Canada*. New York and Washington, D.C.

Benhart, J. E., and M. Dunlop. 1989. "The Iron and Steel Industry of Pennsylvania: Spatial Change and Economic Evolution." *Journal of Geography* 88:173–183.

Heykman, J. S. 1979. "An Analysis of the Changing Location of Iron and Steel Production in the Twentieth Century." *American Economic Review* 68:123–133.

Hogan, D. 1972. *The 1970s: Critical Years for Steel*. Lexington, Ky.

Isard, W. 1948. "Some Locational Factors in the Iron and Steel Industry Since the Early Nineteenth Century." *Journal of Political Economy* 56:203–217.

Isard, W., and W. Capron. 1949. "The Future Locational Pattern of Iron and Steel Production in the United States." *Journal of Political Economy* 57:118–133.

Karlson, S. 1983. "Modeling Location and Production: An Application of U.S. Fully-Integrated Steel Plants." *Review of Economics and Statistics* 65:41–50.

Lesley, Peter. 1859. *The Iron Manufacturer's Guide to the Furnaces, Forges, and Rolling Mills of the United States*. New York.

Markusen, A. 1956. "Neither Ore, nor Coal, nor Markets: A Policy-Oriented View of Steel Sites in the U.S.A." *Regional Studies* 20–25:449–462.

Muller, E. K. 1987. Historical Aspects of Regional Structural Change in the Pittsburgh Region." University of Pittsburgh, Center for Regional Analysis. Pittsburgh.

Pounds, N. 1959. *The Geography of Iron and Steel*. London: Hutchinson.

Rodgers, A. 1952. "Industrial Inertia: A Major Factor in the Location of the Steel Industry in the United States." *Geographical Review* 42:56–66.

Scherrer, C. 1988. "Mini-Mills: A New Growth Path for the U.S. Steel Industry." *Journal of Economic Issues* 22:1179–1200.

Warren, K. 1973. *The American Steel Industry, 1850–1970: A Geographical Interpretation*. New York: Oxford University Press.

17

SERVICES

Ronald F. Abler

The number of workers employed in service industries located in Pennsylvania grew from 1.53 million in 1940 to 3.17 million in 1989, an increase of 1.64 million jobs.[1] Over the same half century, total employment in Pennsylvania grew from 3.23 to 4.54 million, a gain of 1.31 million jobs. Growth in the service industries has outpaced overall employment growth in Pennsylvania, as it has in the United States and many other parts of the world. The service sector has come to dominate employment in Pennsylvania; in 1989, more than 70 percent of Pennsylvania's employed labor force worked in service industry establishments.

Today's national and international economies have often been described as *service* or *postindustrial* economies. Pennsylvania now has a service economy in terms of the industries that dominate employment, as does the United States and much of the world. Dramatic as it undeniably is, the shift to a labor structure dominated by service industries and jobs is neither new nor revolutionary. As early as 1940, almost half of Pennsylvania's labor force was employed in service industries. Nor does the current dominance of service employment foretell the death of manufacturing and other kinds of work, in Pennsylvania or elsewhere. Pennsylvania's current service economy is a logical outgrowth of past economic conditions, and a prelude to further economic evolution in the future.

This chapter assesses the current status of service industry employment in Pennsylvania and suggests some possible scenarios for the future. Specifically, it:

- Describes the changes in Pennsylvania's service industry employment that have occurred since World War II

1. Unless otherwise noted, all data in this chapter, including map data and tabular statistics, are taken or derived from decennial U.S. census reports and from the U.S. Dept. of Commerce, *County Business Patterns, 1985,* and *County Business Patterns, 1989.*

- Places Pennsylvania's service employment in national and international contexts
- Maps variations in service employment within the state
- Discusses the implications of global, national, and Pennsylvania service employment for continued economic growth in Pennsylvania

The Nature of Service Industries

Service industries consist of enterprises that engage in transportation, wholesale and retail trade, finance, insurance, and real estate. Firms that provide temporary lodging, personal services, business services, and repair services, and that provide health care, legal advice, education, social services, and public administration, are also classified as service enterprises. The service industries exclude agriculture, forestry, fishing, mining, construction, and manufacturing. Service industries have three traits that strongly influence the ways they are provided and the kinds of locations chosen by those who purvey them.

In contrast to agricultural, construction, and manufacturing industries, service industries purvey *intangibles*. Farmers grow crops and raise livestock; fisherfolk capture fishes; construction workers build highways, houses, and skyscrapers; factory workers make everything from abrasives to zwieback. In contrast, service establishments create or alter the conditions or states of individuals or tangibles, and they purvey activities, events, and sensations. Services are performed by people for other people. A service worker repairs your broken heater or air conditioner because you want to be warm or cool—you prefer a certain condition and a service industry worker makes it possible. A hotel provides you with temporary shelter, and perhaps some pampering if it is a luxury hotel or resort. A health club provides equipment you can use for exercise. Television networks, the motion picture industry, and live theater companies furnish entertainment. Schools and universities proffer education. When service enterprises use or provide tangibles, they are usually incidental or secondary to the services performed. The equipment you use in a health club is secondary to the primary product offered—a place to exercise. The canisters of film produced by a motion picture studio are incidental to showing the films to audiences for a hefty profit. The food you eat in a restaurant accounts for a small fraction of the cost of your meal; most of the tab pays for the labor of preparing and serving the meal, and for the cost of providing a setting in which to consume it, even in fast-food outlets.

Compared with the outputs of agriculture and manufacturing, services are *perishable*. An ear of corn that was picked yesterday remains edible today, even if it is less delectable than yesterday. It may be sold a week from now at the same price it costs today. An automobile produced nine months ago still commands a new-car price. But most services deteriorate rapidly in value and in price, and many are gone forever if they are not consumed when offered. A restaurant meal that was ready to serve an hour ago may be less than appetizing now. An airplane seat that was vacant ten minutes ago is worthless if the plane backed away from the gate one minute ago, worthless to the airline company as well as to the passenger who will arrive in five minutes seeking transportation to the plane's destination. The need to use or lose capacity in many service sector industries results in complicated pricing schemes such as those that prevail in air transportation, as managers try to fill—but not overfill—planes whose unused capacity is lost forever each time they depart with empty seats. Another temporal consideration is that people do not like to wait for many services, even if they are not as perishable as airplane seats. If you hunger today, the promise of a meal tomorrow will do little to silence your growling stomach. A free haircut next week will be of little interest to a job applicant who must look good at a critical interview this afternoon.

A typical service enterprise must cope with inputs and outputs that are *less standardized* than the products of the typical agricultural or manufacturing enterprise. In fact, producers of certain products often strive for uniformity of product, whereas service producers usually strive to tailor their outputs for groups or individuals. Barbers and beauticians accommodate individual variations in head shape, hair quality, and customer preferences for a multiplicity of styles. Hotel and motel chains stress their abilities to satisfy travelers' individual needs and wants, at the same time that they offer sufficient degrees of standardization in room layout to provide a comforting sense of familiarity. And when competing services are identical or nearly so, as in fast-food retailing or health clubs or automobile rentals, each competitor strives mightily to convince customers that its offerings and its service are in some way unique, and of course therefore better than any others.

Service Employment Data

Despite discussions about service industries and the service economy that have been under way for decades, defining service employment unambiguously presents great difficulties. The statistical agencies of the U.S. government define *industries* by the end products that *establishments* produce. They define *occupations* on the basis of the tasks workers perform, without reference to industry. Establishments are physical locations where business is conducted or where services or industrial operations are performed. An establishment that produces a tangible manufactured product—or that is part of an organization whose end product is something tangible that is made by the organization—is by definition part of a manufacturing industry. Difficulties and ambiguities arise because, depending on what an establishment manufactures (say, widgets), only a small number (perhaps 20 percent) of the establishment's employees may be production workers engaged in actually making widgets. The other 80 percent may work at such tasks as cleaning, equipment maintenance, management, packing, shipping, sales, or supervision. But regardless of the work they actually do (their occupations), all the establishment's employees will be classified as manufacturing-industry employees. Conversely, a worker who fabricates tangibles—for example, a carpenter employed by a university—will be included as a service industry employee because the end product of his or her employing *establishment* is a service.

Difficulties arise in using either industry or occupational data as a basis for examining service employment. This chapter relies primarily on industry statistics because the data for industries are more detailed and more recent, and available at greater geographical specificity, than the data for occupations. A more comprehensive treatment of services, which is not possible here, would consider occupations as well as industries in order to track and map the rising proportions of service occupations within almost all manufacturing industries. Suffice it here to note that, in Pennsylvania in 1989, only about one of four employed individuals worked as craftspersons, operatives, or laborers—that is, in occupations devoted to growing or making things. More than three out of four employed Pennsylvanians have service *occupations,* which means that the industry statistics used in this chapter *understate* service employment because they camouflage the large numbers of nonproduction employees in agriculture, construction, and manufacturing industries.

Kinds of Services

The kinds of services used as examples above were those that analysts classify as *consumer services:* They are delivered or performed for individuals who consume them as ends in themselves. Retail sales, public transportation, and entertainment and leisure services are primarily offered to consumers. The creative or productive chain ends when they are sold to customers because they do not become inputs for further productive activities. *Producer services,* on the other hand, are those that are consumed or used by other establishments in the furtherance of their goals. Wholesale trade firms, consultants, public relations specialists, bookkeepers and accountants, and market research organizations provide producer services. Distinctions between consumer and producer services—like most definitions one encounters in the complicated world of labor force statistics—are far from precise. A seat on an airplane, bought by a salesperson traveling on business, is a producer service. The adjacent seat on the same plane, bought at the same price by a person journeying to a wedding, is a consumer service; the transportation produces no further economic benefit. For the salesperson who travels to call on a client, the trip may yield a multimillion-dollar manufacturing order for the widgets the salesperson sells. Though sometimes fuzzy, the distinction between consumer services and producer services often proves useful; the two kinds of services have different locational needs and grow at different rates.

Governments or quasi-governmental organizations often provide *public-sector services.* Activities such as education, health care, police protection, national defense, and postal service are intended to be available to all members of society. Public-sector services are therefore often supported in part or in whole by taxation, and normal market considerations are disregarded or relaxed in providing such services. The government agencies that support or provide such services usually try to "fill up" the spaces over which they have dominion by making their services equally available everywhere within those territories. Citizens may have rights to public-sector services regardless of where they live or how much of a public-sector service they require.

Some services—especially some kinds of producer services—can be provided to distant or dispersed customers; they can be exported to locations distant from those where establishments providing them are situated. Corporate headquarters, research facilities, universities, tourist attractions, and conference centers all serve customers or producers who may be at or who may come from distant locations. Increasingly, competition in exportable services has taken on national and even global dimensions. Universities and research facilities compete for talented teachers and investigators internationally, and the knowledge academic institutions produce is an international commodity. With declining international airfares, Europe competes directly with domestic locations for American vacation dollars. Global competition in some service specialties—computer software, for example—has become as commonplace as the global competition earlier evident in automobile manufacturing.

Geographic Aspects of Service Production and Consumption

Because service establishments provide few or no tangibles, service production can be said to be *dematerialized.* Other things being equal, dematerialized production can potentially be performed at many more locations than production based on access to or the assembly of heavy inputs, or production that yields heavy outputs as its final product. Agriculture, mining, and manufacturing often consume large quantities of bulky commodities such as fertilizer, wheat, coal, and iron, and they produce heavy outputs, such as railroad cars for grain or automobiles. Materials used in service industries are, as a rule, much smaller in volume and higher in value per kilogram or pound than agricultural or manufacturing inputs. Similarly, the outputs of the service industries are, again as a rule, lighter and more portable than manufactured goods. Services enterprises are therefore inherently more footloose than agricultural or manufacturing enterprises. Theoretically, they can locate at a much greater variety of places than manufacturing enterprises.

But the locational flexibility theoretically inherent in service industries may be more apparent than real, especially for consumer services. For most consumer services, people—usually the same people—are both the raw material and the finished product. When you enter a barbershop or beauty salon, you are the enterprise's raw material. When you emerge after undergoing the ministrations available there, you are its finished product. Because haircutting and hairstyling—and a host of other consumer services—cannot be standardized, because of variations in customer hair and preferences, immediate and direct contact between the service provider and the service consumer are mandatory. Above all, then, consumer service enterprises require access to large numbers of people in order to succeed. It is therefore no accident that consumer services cluster in cities and metropolitan areas. They have more access to more people at such locations than they would elsewhere.

Some producer services are tied to specific locations, but many are increasingly *deterritorialized.* They can be provided equally well from several or many locations because inputs are easily assembled and outputs are easily transported or transmitted. In the modern world, information services are especially likely to be free of locational or territorial constraints, owing to improved telecommunications technology and capacity and the low cost of using advanced telecommunications services. Telecommunications services have even been used to facilitate the export of routine information-processing tasks to low-wage regions within the United States and as far afield as Barbados and Singapore.

Important Issues

Several final caveats are in order before we embark on a detailed consideration of service industry employment in Pennsylvania. These emphasize the point that although some aspects of service industry employment cannot be treated in detail in this chapter, they should not be totally ignored.

First, goods and services complement each other. The complexity of modern agriculture, construction, goods production, and even mining requires considerable service expertise. Many service industries prosper by distributing, maintaining, and repairing tangible products produced or used in agriculture, mining, and manufacturing. Moreover, many service specialties, such as health care and national defense, generate major equipment manufacturing.

Second, achievement of a postindustrial or service economy does not mean that agriculture and manufacturing have vanished. Agriculture, mining, construction, and manufacturing together employed 42 percent of Pennsylvania's labor force in 1980. Despite many reports of their deaths, Pennsylvania's goods-producing industries are alive and well; Pennsylvania had 400,000 more manufacturing jobs in 1980 than it had in 1940. Manufacturing employment in Pennsylvania has grown absolutely, even though it has declined relative to services, and manufacturing remains an important component of Pennsylvania's economy.

Third, service productivity is difficult to measure. An automobile coming off an assembly line is easily counted and its value can readily be assessed because it is a standard product that is sold in a normal marketplace. If the same number of manufacturing workers produce more products this year than they made last year, productivity is clearly up. In contrast, breakthroughs in medical research are not standard products. Analysts have trouble determining how many minor breakthroughs equal one major advance, and whether more breakthroughs are being produced in relation to inputs this year than last year. Because of the intangible natures and lack of comparability among services, most states and nations are only just beginning to come to grips with how they should count (and tax) services. For these reasons, it is difficult to determine how efficiently service industries operate, and even more difficult to measure productivity decreases or increases in service industries. It is clear that Pennsylvanians, in common with consumers elsewhere in the country, are paying more for services. Whether they are getting more value, and proportionally more, is much more difficult to say.

Fourth, gender and other social issues should be inherent components of any comprehensive discussion of the provision of services. Women predominate in some service industries and jobs (nursing, for example), and they often dominate low-wage service jobs. Employment shifts among industries always have social implications, and because services are growing more rapidly than any other industrial sector, the social dimensions of industrial evolution will be debated largely in connection with the service industries in the years to come.

Changes in Service Employment Since 1940

Service industry jobs engaged almost half of Pennsylvania's labor force as early as 1940. By 1980 the number of service industry workers had more than doubled in absolute numbers, and constituted 64 percent of the state's employed workers (Table 17.1). Statewide, service industry employment grew at almost the same rate as the total labor force from 1940 through 1960, rising from 48 to 51 percent. After 1960, however, service industry employment spurted. More than a million service jobs were

Table 17.1. Major industry employment in Pennsylvania, by industrial sector, 1940–1990.

	Agriculture	Manufacturing	Services	Unclassified	Total
1940	194,263	1,441,511	1,532,845	56,448	3,225,067
%	6.0	44.7	47.5	1.8	100.0
1950	164,804	1,804,445	1,909,574	52,346	3,931,169
%	4.2	45.9	48.6	1.3	100.0
1960	110,509	1,773,618	2,092,300	150,781	4,127,208
%	2.7	43.0	50.7	3.7	100.0
1970	80,184	1,836,569	2,620,150	0	4,536,903
%	1.8	40.5	57.7	0.0	100.0
1980	80,472	1,716,644	3,164,385	0	4,961,501
%	1.6	34.6	63.8	0.0	100.0

SOURCE: U.S. Census Bureau Special Reports.

Table 17.2. Employment in Pennsylvania by detailed industry group, 1940–1980.

	1940	%	1950	%	1960	%
Agriculture	194,263	6.0	164,804	4.2	110,509	2.7
Mining	224,942	7.0	192,253	4.9	65,528	1.6
Construction	144,163	4.5	215,872	5.5	205,205	5.0
Manufacturing	1,072,406	33.3	1,396,320	35.5	1,502,885	36.4
Transportation	248,416	7.7	330,906	8.4	290,157	7.0
Wholesale trade	66,127	2.1	113,177	2.9	121,320	2.9
Retail trade	462,802	14.4	576,625	14.7	586,576	14.2
Finance/insur./real estate	94,512	2.9	116,400	3.0	146,318	3.5
Business services	59,591	1.8	88,836	2.3	90,612	2.2
Personal services	247,027	7.7	197,083	5.0	189,860	4.6
Entertainment services	23,138	0.7	29,946	0.8	26,965	0.7
Professional services	239,397	7.4	311,498	7.9	459,244	11.1
Public administration	91,835	2.8	145,103	3.7	181,268	4.4
Not elsewhere classified	56,448	1.8	52,346	1.3	150,781	3.7
Total	3,225,067	100.0	3,931,169	100.0	4,127,208	100.0

	1970	%	1980	%
Agriculture	80,184	1.8	80,472	1.6
Mining	41,708	0.9	55,645	1.1
Construction	247,055	5.4	240,162	4.8
Manufacturing	1,547,806	34.1	1,420,837	28.6
Transportation	288,425	6.4	347,197	7.0
Wholesale trade	161,107	3.6	194,512	3.9
Retail trade	693,257	15.3	778,164	15.7
Finance/insur./real estate	190,529	4.2	256,725	5.2
Business services	124,748	2.7	186,589	3.8
Personal services	163,289	3.6	124,371	2.5
Entertainment services	27,330	0.6	37,075	0.7
Professional services	757,281	16.7	1,011,813	20.4
Public administration	214,184	4.7	227,939	4.6
Not elsewhere classified	0	0.0	0	0.0
Total	1,536,903	100.0	4,961,501	100.0

Source: U.S. Census Bureau Special Reports.

created in Pennsylvania between 1960 and 1980, increasing the service industry's share of all employment by 7 percent in the 1960s, by another 6 percent in the 1970s, and by a similar increment in the 1980s. As of 1989 there were 2.31 service industry workers for every person employed in agriculture, mining, construction, and manufacturing.

In recent decades, growth in professional service industries has outpaced growth in all other industrial groups. Employment in medical services, education, and legal and engineering services increased from 240,000 jobs and 7 percent of the labor force in 1940 to more than 1,000,000 jobs and 20 percent of the labor force in 1980 (Table 17.2). Within the professional category, health-related industry employment grew most rapidly. There were 76,300 people employed in medical facilities and establishments in 1940, and by 1980 the number had risen to 402,250—an increase of 427 percent. Health service employment continued to grow rapidly in the 1980s to become a large and critical labor force sector.

Service industry growth had to come at the expense of agriculture and mining originally, and then at the expense of manufacturing in the last two decades (see Table 17.2). Agricultural employment declined from 6 percent to 2 percent of Pennsylvania's labor force, and mining dropped from 7 percent to 1 percent. Manufacturing employment maintained its share of the total Pennsylvania labor force until the 1970s, when it dropped from 34 percent to 29 percent of all jobs. Further declines in the 1980s left manufacturing with a much diminished share of employment in Pennsylvania in 1990. Employment in most other industry groups grew at about the same rate as the state's total labor force. Except for agriculture, mining, and manufacturing, no nonservice industry group's share of the 1980 labor force differed by more than 2 percent from its share of the 1940 labor force. Personal services are an exception to the growth or rapid growth characteristic of service industries generally. Almost 250,000 workers (8 percent of employed persons) provided personal services in 1940; some 144,000 were private household workers, such as maids. By 1980 the number of household service workers had declined to 24,000, accounting for much of the decline in the personal services group, which constituted only 3 percent of the labor force in 1980.

In summary, industry shares of Pennsylvania's total employment changed comparatively slowly from 1940 to 1960. Agriculture and mining declined during the period, but other industry groups employed stable shares of the state's labor force. Between 1960 and 1980, changes that were even more dramatic occurred, with service industries growing in all categories except personal services. Employment in professional services more than doubled from 1960 to 1980. (The occupational shifts that occurred from 1940 to 1980 document the growing share of nonproduction workers across all industries and underline the degree to which industry employment statistics understate service dominance. Farmers, craftspersons, and operatives constituted 45 percent of Pennsylvania's 1940 labor force, but by 1980 only 17 percent of employed individuals worked at such occupations. On an occupational basis, there are now roughly six service workers for every worker who makes or raises a tangible product.)

Service Employment Within Pennsylvania

The examination of service industry employment in Pennsylvania undertaken in the remainder of this chapter makes use of the source that provides the most detailed locational data: the industry establishment data contained in the Pennsylvania volume of *County Business Patterns, 1989,* the latest information available at the time this chapter was written. The discussion progresses from a consideration of the locations of all service industries (the tertiary sector) to a more detailed look at employment in selected consumer and producer service establishments.

Location of Service Industry Employment

Because most service enterprises cater to people or to the things that people keep around them, service entrepreneurs seek access to large numbers of people. Other things being equal, therefore, one would expect to find large numbers and high proportions of service industry jobs in Pennsylvania's metropolitan regions. And indeed, service industry jobs are concentrated in such locations. There were 3.2 million service industry jobs[2] in the state in 1989; some 2.8 million (89 percent) of them were located in the 29 of Pennsylvania's 66 counties that the U.S. Department of Commerce has classified as parts of the 15 metropolitan regions located in whole or in part in Pennsylvania. Some 58 percent of all tertiary or service jobs in Pennsylvania were located in the nine counties comprising Pennsylvania's two largest metropolitan areas—Philadelphia and Pittsburgh.

Throughout Pennsylvania, service industry employment generally dominates the labor force in and near metropolitan areas (see Fig. 17.1). Paradoxically, though, service employment is highest in nonmetropolitan Montour County, where a small employed labor force (10,272 workers), combined with the presence of a large medical center, results in 91 percent of the labor force working in service establishments. Philadelphia County ranks second in service employment in Pennsylvania, with 81 percent of its labor force devoted to providing services. Statewide, 70 percent of Pennsylvania's employed workers work in service establishments; only 15 counties had proportions in services exceeding 70 percent. Clearly, service employment is concentrated in metropolitan regions, and within those regions, in the core counties. In the Pittsburgh metropolitan area, for example, service employment

2. As used here, the service industries include the following census industrial groups: transportation; wholesale and retail trade; finance, insurance, and real estate; and services. This cluster of economic activities is also often referred to as the *tertiary industries*.

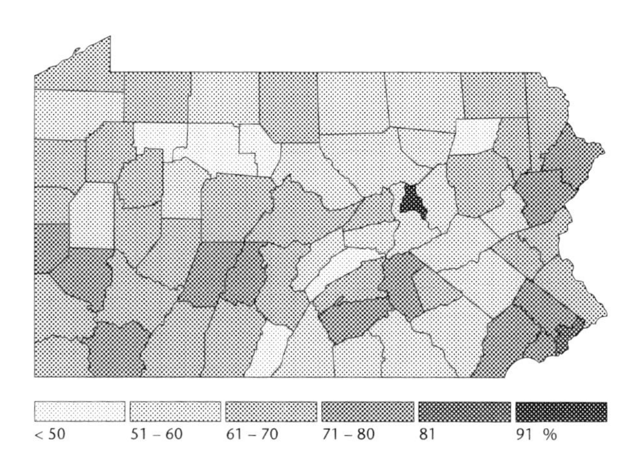

Fig. 17.1. Percentage of county employment in all service (tertiary) industries, Pennsylvania, 1989.

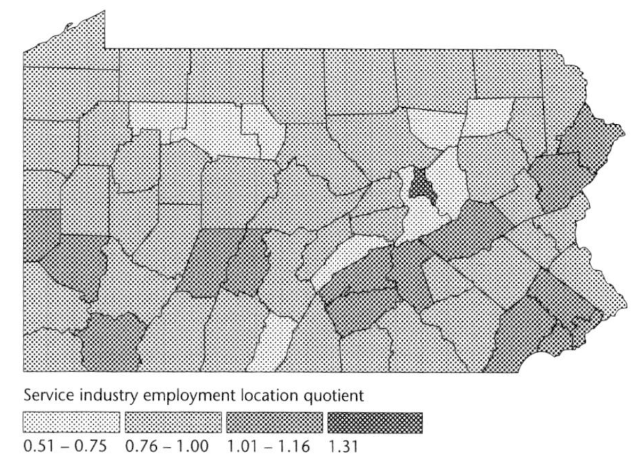

Fig. 17.2. Service industry employment location quotients by county, Pennsylvania, 1989.

exceeds the state proportion in Allegheny and Beaver counties but is less than the state ratio in Washington and Westmoreland counties.

Another way to compare service employment in different parts of the state is to use location quotients. Location quotients were originally suggested by Florence (1929). In this instance, they are obtained by dividing a county's share of employment in a particular industry by that county's share of total employment in the state, or, more generally:

$$LQ = (Ii/IT)/(Ei/ET)$$

Where,

Ii = Industry employment in the ith county
IT = Industry employment in the state
Ei = Total employment (all industries) in the ith county
and
ET = Total employment (all industries) in the state

In Montour County, for example, 9,377 people are employed in services, compared with 3,168,205 employed in services in all of Pennsylvania. Montour County's total employment is 10,272, and total employment in Pennsylvania is 4,539,619. Thus, (9,377 / 3,168,205) / (10,272 / 4,539,619) = 0.002959720 / 0.002262744 = 1.308. Location quotients enable analysts to make reasonable comparisons of individual areas (counties in this instance) with each other and with larger units, such as states. The location quotient of 1.31 for services in Montour County, for example, tells us that Montour County has 1.31 times as many service jobs as might be expected, given Montour County's share of all the state's jobs. Similarly, a county service location quotient of 1.0 says that that county has exactly the number of service jobs one could expect, given its share of the state's employment, and a quotient of 0.5 for a given industry in a given county implies that the county's employment in that industry is half what might be expected given its share of all employment in Pennsylvania.

The patterns of service industry employment in Pennsylvania produced by means of location quotients (Fig. 17.2) differ little from those evident when percentages of county employment are mapped (see Fig. 17.1), except that the location quotient maps mute some of the extremes on the higher and lower ends of the distributions. Service employment below what would be expected given their shares of state employment (location quotients less than 1.00) characterize most of Pennsylvania's counties. Only a few counties have service industry location quotients greater than 1.00, and with the exception of Montour County's 1.31 most service employment quotients do not greatly exceed 1.00. No Pennsylvania county has a quotient lower than 0.50. Even if service industries in total employ fewer people than expected, nowhere in Pennsylvania does the reduced number of service jobs amount to less than half of what might be expected. Service employment is a local necessity to the degree that although service jobs may be fewer than expected given the size of a county's labor force, no county is without a base level of service jobs that exceeds half the "normal" number the statewide relationship defines. Three predominantly rural counties—Fayette, Monroe, and Pike—have service industry location quotients that exceed 1.00. Fayette County's greater-than-expected service employment (1.10) most likely results from the fact that the northern part of the county lies within the Pittsburgh exurban region, while the central and southern parts of the county export services to the sparsely populated areas of West Virginia and Maryland to its south. Larger-than-usual numbers of service jobs in Monroe (1.06) and Pike (1.14) counties reflect the recreation-based economy of the Pocono Mountains recreation area.

Counties with three-fourths or fewer of the service jobs one might expect are usually sparsely populated regions. Having few people, and lacking large towns or cities that would be attractive to service enterprises, their service needs are partially met by firms located in adjacent areas. Being sparsely populated, some of these counties also have small labor forces. Of the nine counties with service industry location quotients ranging from 0.55 to 0.75 (Fig. 17.2), several had very small labor forces. In 1989 Cameron County employed only 1,747, and Sullivan County employed 1,261; a mere 690 were employed in Forest County. Small county labor forces create the possibility that a single enterprise employing several hundred people will skew county employment data in ways that cannot occur in counties with tens or hundreds of thousands of jobs.

In recent years, service industry employment in Pennsylvania has continued the patterns of growth it has exhibited for the last several decades. From 1985 to 1989, the number of people employed in service industry jobs increased from 2.68 million to 3.17 million, a growth of just under half a million jobs and 18.1 percent, during a period in which total employment increased 11.7 percent. Forest was the only county in Pennsylvania in which service employment declined (Fig. 17.3), from 339 jobs in 1985 to 318 in 1989. Because of Forest County's small employment base, the loss of 21 jobs produced a 6.2 percent decline. Philadelphia County's low percentage growth (5.1) would be expected, given the large percentage (77.8) of the county's labor force already employed in services in 1985. The rest of the Philadelphia metropolitan region grew at rates exceeding the state average, and Chester County's service employment grew especially rapidly; the area added more than 28,000 service jobs between 1985 and 1989, an increase of 35.4 percent for the period.

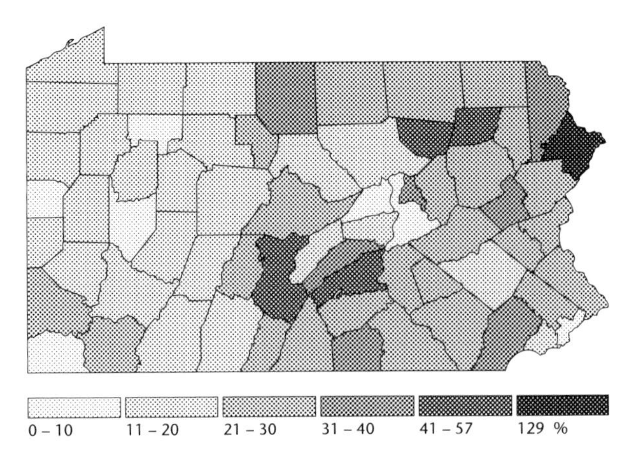

Fig. 17.3. Percentage change in service employment, Pennsylvania, 1985–1989.

The very high rates of service job increase in Huntingdon, Perry, Pike, Sullivan, and Wyoming are all built on small total employment bases, as in the case of Pike County's 129 percent increase from 2,050 service jobs in 1985 to 4,698 in 1989. Nonetheless, such gains are impressive, especially when they are built on somewhat larger 1985 employment bases. Clearly, some of the general shift to service employment occurring in Pennsylvania is finding its way into outlying areas.

Retail and Wholesale Trade

Among the several components of the service industries reviewed in Figs. 17.1, 17.2, and 17.3, retail trade, a consumer service, would be the activity that could be expected to most closely approximate the distribution of total population and total employment. On a daily basis, people need goods such as gasoline, milk, and basic foodstuffs, and merchants who purvey them strive to make them available at locations that are convenient to people. The retail trade industries include more than gas and food, of course, and on average consumers will have to travel farther to shop for large-ticket products purchased infrequently, such as furniture or automobiles (what merchants call *shopping goods,* as opposed to *convenience* items such as gas and milk). Because the economy of any area can support fewer furniture stores than supermarkets, for example, the average individual will have to travel considerably farther to buy a bedroom set than to buy a quart of milk.

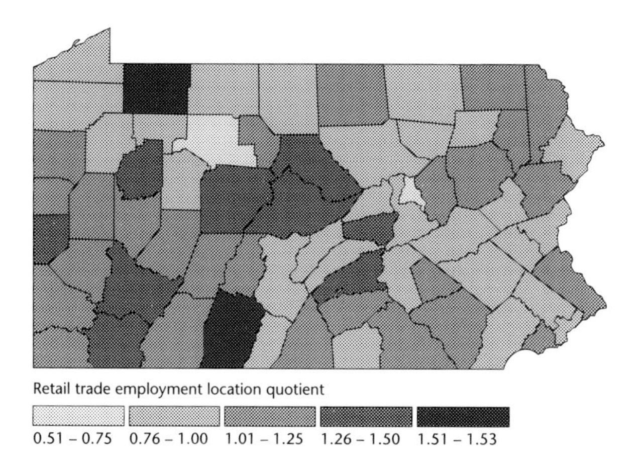

Fig. 17.4. Retail trade employment by county, Pennsylvania, 1989.

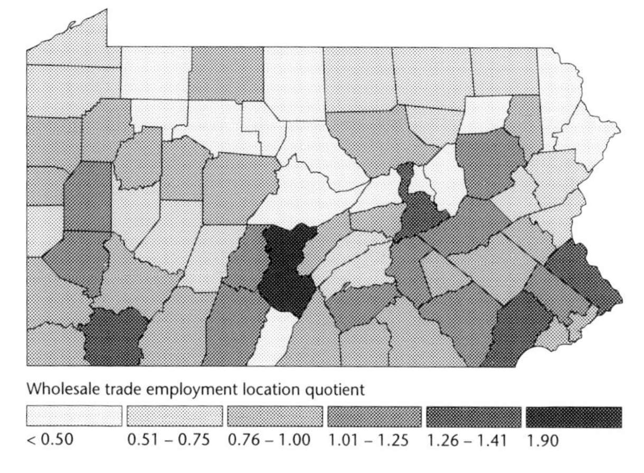

Fig. 17.5. Wholesale trade employment by county, Pennsylvania, 1989.

Retail trade establishments of all kinds—those purveying convenience goods through major shopping items such as automobiles—employed 925,718 people in Pennsylvania in 1989; they constituted 20.4 percent of the state's labor force. Most of Pennsylvania's counties had retail trade location quotients reasonably near 1.00 (Fig. 17.4), which is to be expected given the nature of the activity; the mean ratio for all Pennsylvania counties was 1.06, with a small standard deviation (0.20). No county had a retail trade location quotient below 0.50, and only Elk and Montour counties had quotients below 0.75. Elk County's labor force is heavily engaged in manufacturing (its manufacturing location quotient is 2.54), and it has no urban centers of consequence. It is likely that Elk County residents buy many of their nonconvenience goods in adjacent counties. As we previously noted, Montour County's employment is dominated by a large medical center (its health services location quotient is 6.80), and it too lacks a large urban center to host a major retailing complex.

Counties with unusually large shares of retail trade employment are scattered in the western half of the state with the exception of the cluster formed by Centre, Clearfield, and Clinton counties. To some degree, all three provide retail services for adjacent counties with below-average retail trade employment. Also, Interstate Highway 80 passes through all three counties (and Clearfield County as well), which enables establishments that serve the needs of travelers to capture mobile trade that moves along that important east-west transportation artery. Interstate 80 also traverses Jefferson County, but that county lacks urban centers comparable to Clearfield, DuBois, and Clarion to serve as nuclei for highway-oriented trade establishments. Bedford and Warren counties have over half again as many workers employed in retail trade as would be expected given their shares of Pennsylvania's labor force. Warren County provides retail services to populations in adjacent counties where service establishments are sparse. Similar to the counties traversed by Interstate 80, Bedford benefits from sales to the transients moving along the Pennsylvania Turnpike, especially the thousands that patronize the massive lodging and retail complex located at the Bedford interchange of the Pennsylvania Turnpike and U.S. Highway 220. Beaver and Westmoreland counties support larger-than-usual retail trade employment on the basis of Pennsylvania Turnpike trade and the locations within their borders of large retail shopping malls on the periphery of Pittsburgh. Fayette County enjoys the same advantage with respect to Pittsburgh, and also profits from access to underserved areas in Maryland and West Virginia to its south. Small labor forces and the absence of other options lead to the emphasis on retail trade in Perry and Snyder counties.

Other things being equal, wholesale trade—a producer service—should be more geographically variable than retail trade, because one wholesaler needs many retail establishments to remain in business. Wholesale trade employment is highly variable in Pennsylvania (see Fig. 17.5). Wholesale trade location quotients are below 1.00 in most of the counties in the northern half of the state, and counties with quotients over 1.00 are located largely in the southeastern quadrant, with outliers in the south-central and western parts of the state. Huntingdon County has almost twice the number of workers engaged in wholesale trade (1.90) as would be expected given its share of Pennsylvania's labor force. The number of workers employed in wholesale trade establishments (1,182) there is not large in absolute terms, and the county's modest-size labor force (10,193 in 1989) means that even a few larger-than-usual wholesaling establishments could produce the unusually high quotient. Higher-than-usual wholesale trade employment is also evident in several counties of the Philadelphia and Pittsburgh metropolitan areas and in Northumberland County. At one time, wholesaling was railroad-oriented and clustered primarily in the cores of cities and metropolitan areas. In recent decades, high land prices and truck transportation have made metropolitan peripheries the favored locations for wholesaling. The high location quotients for wholesaling in Bucks, Delaware, Fayette, and Northumberland counties reflect the shift to highway orientation on the part of wholesalers and the importance of access to major interstate routes.

A further examination of the relationship between retail and wholesale trade in each county offers insights into the problems that arise in using arbitrarily defined statistical units such as counties, in addition to insights into the locations of these critical services and the geographical differences between consumer and producer services. In 1989, some 277,528 people were employed in wholesale trade in Pennsylvania, and there were 925,718 retail trade workers. On a statewide average then, for all kinds of wholesale and retail trade, each retail trade worker was served by 0.30 of a wholesale trade worker (277,528 / 925,718). On a county-by-county basis, this wholesale/retail trade ratio varied from 0.68 wholesale trade employees for each retail trade employee in Huntingdon County, to 0.03 in Cameron County (Fig. 17.6), where there is virtually no wholesaling. As noted above, Huntingdon County has a modest labor force and an unusual concentration of wholesale trade establishments, resulting in a ratio beneficial to the county, because people can be employed there based on wholesale services that are provided to people outside the county. High wholesale/retail trade employment ratios are evident in the Philadelphia metropolitan area (with the exception of Delaware County), which contained 39 percent of the state's total wholesale employment in 1989. Butler and

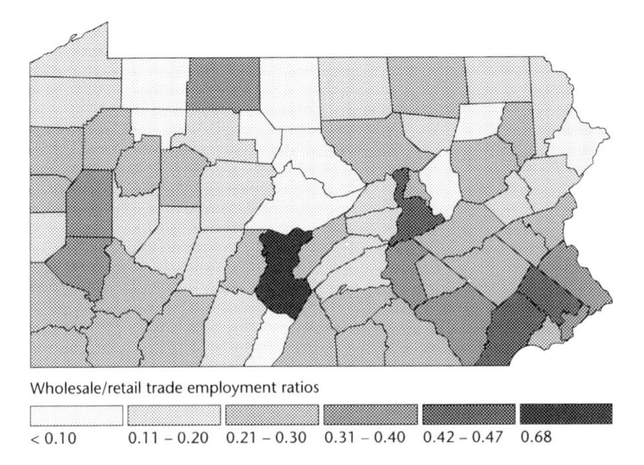

Fig. 17.6. Wholesale/retail trade employment ratios by county, Pennsylvania, 1989.

Allegheny counties also show wholesale strength. Northumberland County's wholesale employment is high with respect to its retail trade (0.43), although the number of wholesale jobs there in 1989 (2,309) was not outstanding in absolute terms.

The patterns of wholesale trade location quotients and those of wholesale/retail employment ratios demonstrate the affinity of producer services for metropolitan and suburban locations and their distaste for rural and sparsely populated areas. Butler and Northumberland counties are the only counties extending into the northern half of the state in which wholesale/retail ratios exceed the state relationship of 0.30 wholesale workers per retail worker—except for McKean County, where a slightly elevated number of wholesale jobs (812) in a small labor force (13,434) produces a ratio (0.32) slightly above the state ratio. And with the exceptions of Allegheny, Butler, Huntingdon, and McKean counties, no county outside the southeastern quadrant exceeds the state ratio. Throughout much of Pennsylvania, retail trade is serviced by wholesalers located in distant metropolitan counties, and a large share of the state's retailers rely on wholesalers located in the southeastern part of Pennsylvania.

Finance, Insurance, and Real Estate

Financial, insurance, and real estate (commonly abbreviated FIRE) services are mixed consumer and producer services; they are supplied to final consumers as well as to commercial and industrial enterprises that use them in furtherance of additional

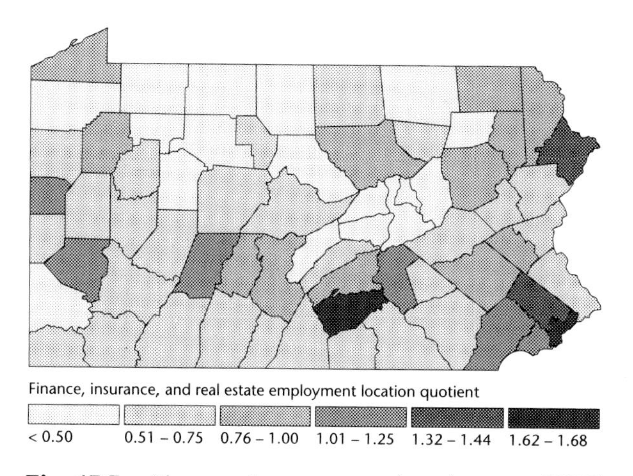

Fig. 17.7. Finance, insurance, and real estate (FIRE) employment by county, Pennsylvania, 1989.

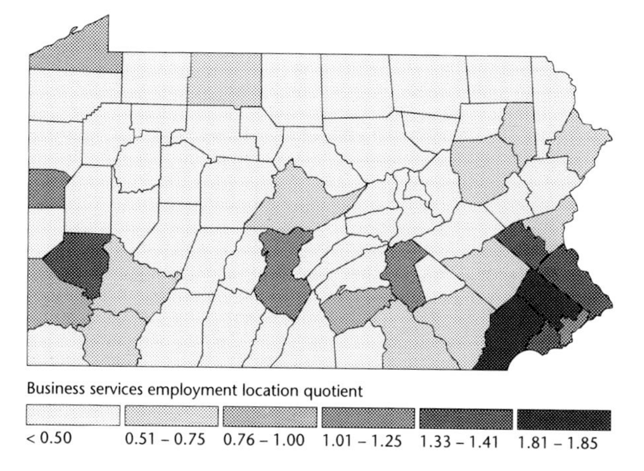

Fig. 17.8. Business services employment by county, Pennsylvania, 1989.

production. Employment in the FIRE industries has increased from 2.9 percent of the state's labor force in 1940 to 5.2 percent in 1980; more than 304,000 Pennsylvania workers now occupy themselves with providing FIRE services. Despite their retail components, FIRE services show a strongly metropolitan focus, compared with total employment (Fig. 17.7). Only 10 of Pennsylvania's 66 counties have FIRE location quotients higher than 1.00, and in most parts of the state FIRE employment is less than three-fourths, or even less than half of what would be expected, given county shares of total state employment. Cumberland and Philadelphia counties have the most salient concentrations of FIRE employees (location quotients of 1.68 and 1.62, respectively). The former reflects the influence of Harrisburg, in neighboring Dauphin County, which also has a high ratio (1.13). Philadelphia County's high quotient is based on the city's historical focus on financial and insurance services. Allegheny County (Pittsburgh) has a high FIRE location quotient (1.41) for similar reasons. Cambria and Lawrence counties' excess of FIRE employees is not as large (1.20 and 1.11, respectively); their central cities (Johnstown and New Castle, respectively) provide FIRE services for surrounding low FIRE employment areas to a degree sufficient to employ larger than expected FIRE work forces. Pike County's FIRE location quotient of 1.44 results from FIRE employment of 571 workers in a total county labor force of 5,921 in 1989. Most Pike County FIRE employees are engaged in real-estate sales associated with the area's Pocono Mountains recreational activities. Finance and insurance are *high order* functions, in which small numbers of large enterprises serve customers over large territories. One can therefore expect that they would cluster in metropolitan areas more than real-estate employment, which is a more geographically generalized service.

Business Services

Business services are by definition producer services. As counted by the census and as analyzed here, they consist of enterprises that furnish advertising, credit reporting, mailing and reproduction services, computer and data processing services, management and public relations, and protection services. Except for the firms that provide protection and services to buildings, they deal primarily in information or advice to other businesses that foster further production. In 1989 some 4.4 percent of Pennsylvania's employed labor force worked in business service enterprises. Because they are heavily information- and producer-oriented, the locations chosen by business service enterprises provide helpful keys to forecasting the geography of Pennsylvania's emerging information economy.

Most counties have few jobs in the business services industries (Fig. 17.8)—and in much of the state, business service employment is less than half of what it should be, given county shares of total state employment. Counties in which business service employment exceeds what would be expected given their shares of state employment number only nine; together they host 149,896 (74.7 percent) of Pennsylvania's 200,564 business services

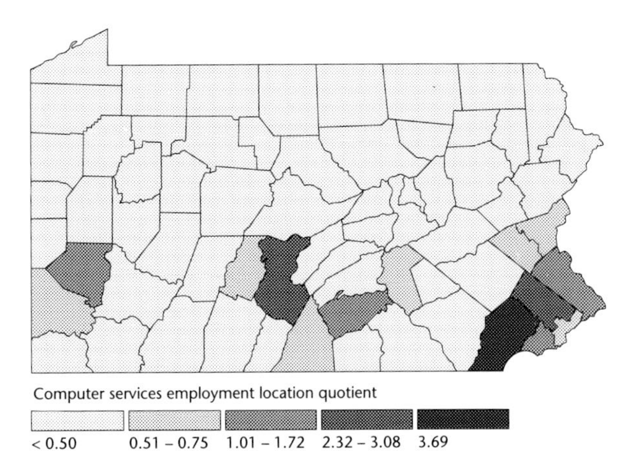

Fig. 17.9. Computer services employment by county, Pennsylvania, 1989.

jobs. Just three of those counties—Allegheny, Montgomery, and Philadelphia—contain 102,878 (51.3 percent) of the state's business service jobs. A significant number (513) of Huntingdon County's unusually high number of service workers (6,557) are employed in business service enterprises. Lawrence County's business service location quotient exceeds unity, but only by a small margin (1.16).

Computer and data processing services are a specialized subcategory of the general business services category of producer services, and the jobs they provide are even more highly concentrated than are business services (Fig. 17.9). In 1989, Chester County had the highest concentration of such jobs in the state, with a location quotient of 3.69; Montgomery (3.08) and Huntingdon (2.32) counties were next in quotient order, although Huntingdon County's computer services employment is far smaller in absolute terms than that of other counties with high quotients. Most computer services jobs in the state in 1989 were located in Montgomery (9,847), Allegheny (6,677), Chester (4,235), Philadelphia (2,703), Delaware (2,534), Bucks (1,634), and Cumberland (829) counties, which together hosted 84.7 percent of the state's total computer services work force of 33,615 workers.

In theory, computer and data processing services ought to be highly footloose. They deal almost entirely in digital information that is easily and inexpensively transmitted via telephone lines, and such equipment as is needed to perform those services is highly portable in many instances. In empirical fact, computer service enterprises restrict their locations to metropolitan areas in Pennsylvania, and to few of those. The metropolitan concentrations of enterprises such as the FIRE industries that purchase computer and other information services reinforce a tendency to select metropolitan locations because of the better supply of well-educated and trained employees in metropolitan counties and urban places. The high degree of metropolitan clustering evident in business and computer services to date bodes ill for those who view such industries as solutions to problems of few jobs, and fewer well-paying jobs, in Pennsylvania's outlying areas. Unless some major shift in locational behavior and criteria occurs, depressed and stagnating areas cannot rely on many kinds of service employment to replace lost agricultural, mining, and manufacturing employment.

Health Services in Pennsylvania

Health services are consumer services that are to some degree information services, although less directly so than those described above. Health services are currently of great concern to individual citizens as well as to policy makers at all levels of government because of their rapid growth and because of the rapidly growing shares of both personal and national income they absorb. In recent decades, health service employment has grown from 2.7 percent of Pennsylvania's 1960 labor force, to 4.2 percent in 1970, to 8.1 percent in 1980, and to 10.8 percent in 1989. In absolute terms, the number of health care jobs increased from 382,019 in 1985 to 488,098 in 1989, an increase of 106,079 jobs in just four years and a rate of increase of 6.9 percent a year. A population growing at 6.9 percent a year will double in just over 10 years, so it is little wonder that health care employment and its geographical distribution are of great interest to citizens and policy makers.

Ideally, health care should be readily available in all parts of the state. When people need health services, they usually need them quickly and conveniently. Economies of scale and the need for specialized equipment and facilities, on the other hand, dictate that health services, and especially high-order services, will be concentrated in some areas, leaving

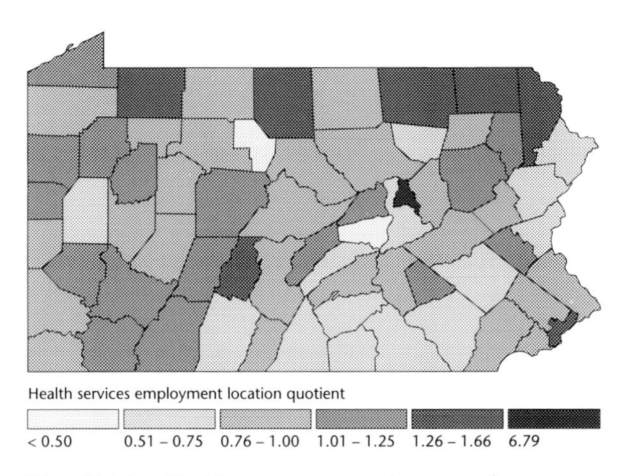

Fig. 17.10. Health services employment by county, Pennsylvania, 1989.

deficits in others (Fig. 17.10). Counties with low health quotients (less than 0.75) form no consistent pattern, either geographically or with respect to labor force structure. The highest health care industry location quotient occurs in Montour County, where about three-fourths of the labor force is employed in that single industry. Less-dramatic concentrations yield location quotients exceeding 1.25 in Potter (1.66), Bradford (1.59), Philadelphia (1.39), Warren (1.35), Wayne (1.34), Blair (1.29), and Susquehanna (1.26) counties. Where total labor forces are small or modest in size, as they are in five of those seven counties, the personnel needed to ensure minimal local health care—plus those employed in the occasional private or public regional health facility—form large enough shares of county employment to yield high location quotients. Blair (44,520) and Philadelphia (613,875) counties have substantial labor forces that are engaged differentially in health services for different reasons. A large state hospital in Blair County boosts its quotient, and the provision of high-order medical services for much of the state and surrounding areas raises the quotient for Philadelphia. It is worth noting that modest location quotients do not, of course, necessarily mean that few people are employed in an industry. Allegheny County, at the core of the Pittsburgh region, has only a few more people employed in the health industries than would be expected given its share of state employment, but its 1989 health services location quotient of 1.13 represents the employment of 75,547 people. The number so employed in Philadelphia County in 1989 was 91,826.

The Geography of Services in Pennsylvania

What general patterns emerge from the service industries examined in this chapter? First, service enterprises clearly prefer urban and metropolitan locations. Geographers and planners often refer to small towns and cities as *service centers,* but such locations hold few attractions for most service establishments other than those engaged in retail trade. And even retailing—the service whose geographic distribution should most closely approximate that of Pennsylvania's population—clusters in larger places because today's highly mobile consumers will travel considerable distances for ordinary retail goods, and even longer distances for higher-order services such as specialized medical care. Conversely, some service enterprises such as business services prefer to bring their offerings to dispersed customers from their urban and metropolitan locations.

The higher the order of the service—that is, the fewer locations from which it is provided in the state—the more likely it seems that enterprises offering that service will cluster in several or even a few metropolitan areas. The progression of specific services examined in this chapter—from retail trade and wholesale trade, through FIRE services, business services, and computer services (Figs. 17.4 through 17.9)—was a progression from those that were more dispersed to those that are highly concentrated in just a few metropolitan counties. Even information services, which might be expected to be the most flexible locationally because they are almost completely dematerialized and deterritorialized, have not dispersed, as many analysts thought they would. They are not located in the out-of-the-way places theoretical considerations might suggest, but are instead clustered densely in a surprisingly small number of places. Even health services increasingly cause consumers to move toward their restricted locations rather than being available locally.

Services—and especially information services—have often been viewed as solutions to development problems in regions that are lagging economically. Experience to date in Pennsylvania suggests that such hopes will be disappointed. Pennsylvania's service economy, as measured by the locations of its service employment, has not heretofore provided large numbers of jobs in rural and sparsely popu-

Table 17.3. Industrial sector employment in Pennsylvania, the U.S., and selected countries, 1989 (percent).

	Pa.	U.S.	Germany	Japan	United Kingdom
Agriculture	1.1	2.7	3.9	7.2	2.1
Manufacturing	29.1	26.6	40.1	34.4	29.2
Services	69.8	70.7	56.0	58.4	68.7

Source: Department of Economics and Statistics, Organization for Economic and Commercial Development, *Quarterly Labour Force Statistics, 1990,* no. 1 (Paris, 1990).

Table 17.4. Resource, transportation, and settlement eras.

Era	Resource	Movement	Settlement	Period
Agricultural	Land	Foot/wagon	Dispersed	to 1800
Industrial	Materials	Ship/rail	City	1800–1920
Service	People	Rail/auto/air	Metropolis/megalopolis	1920–80
Information	Data/knowledge	Air/telecommunication	Megalopolis/ecumenopolis	1980–?
?	?	?	?	?

Source: After John R. Borchert.

lated parts of the states. Service enterprises find urban and metropolitan locations more attractive and profitable. Although theory and logic may suggest that services should offer alternatives for declining employment in other economic sectors such as agriculture, mining, and manufacturing, past and current locational behavior argue to the contrary. Services in general, and information services in particular, offer little hope for economic development in Pennsylvania's economically lagging regions.

Despite widespread reports of its demise, Pennsylvania's economic development, including the evolution of its service economy, has kept pace with that of the nation and with other competitive countries (see Table 17.3). The proportion of Pennsylvania's labor force engaged in manufacturing exceeds the national percentage by only 2.5 percent, which is considerably less than popular accounts of the plight of the rust belt might suggest. The proportion of Pennsylvania's labor force in the broad services category (utilities, transportation, trade, FIRE, and services) is almost identical to the national share. Although there are clearly some differences between Pennsylvania's labor force structure and the structure of the work force in Germany, Japan, and the United Kingdom, the similarities are more striking than the differences. Pennsylvania's continued economic evolution parallels that of the nation and of other economically advanced nations of the world.

What Next?

Economic development is a general and continuous evolutionary process. The term *economic development* and concern about it suggests third world nations and grinding poverty to most people. It is important to remember that the world's most economically advanced nations develop also; they are not standing still, waiting for other nations to catch up with them. Service or postindustrial economies are but one phase in a series of industrial and occupational shifts that have occurred in the past and that will continue to evolve in the ongoing process of economic development. There have been predecessors to current industrial and labor force structures, and there will be successors, and they will have their own geographies. As John R. Borchert has argued in a number of lectures and publications, a few key variables govern the economic geography of any region at any given time (Table 17.4). The dominant or most valuable resource, and the dominant movement technology, combine to create *critical locations,* which become hot spots for creating wealth and therefore attract large numbers of people to settle in and near them. Each era is typified by the form of settlement organization appropriate to the conditions of resource use and transportation that prevail.

Pennsylvania contains relics or examples of settlements from all previous eras. Philadelphia and Lancaster are examples, respectively, of the seaports and inland trade centers that served dispersed rural hinterlands. Pittsburgh and Reading exemplify manufacturing cities based on inland water and rail transportation. Philadelphia and Pittsburgh have kept pace with economic change through the service era and into the current information era, with larger and smaller numbers of outliers around the state playing lesser or greater roles, depending on the kinds of services.

It is difficult to forecast how long the current services and information era will last, or what will supplant it. It is safe to say, however, that most of the economic action—in the remainder of this era or the next—will take place in a selected number of locations, most of which have historically been economic hot spots. Employment and the creation of wealth have never been ubiquitous, even during the agricultural era. In fact, they have become progressively more localized through the subsequent industrial, service, and information eras. That high degree of localization will likely persist in the future, in Pennsylvania and elsewhere in the country and the world.

Bibliography

Daniels, P. 1982. *Service Industries: Growth and Location.* Cambridge: Cambridge University Press.

———. 1988. "Some Perspectives on the Geography of Services." *Progress in Human Geography* 12(3):431.

Feketekuty, G. 1988. *International Trade in Services: An Overview and Blueprint for Negotiations.* Cambridge, Mass.: Ballinger.

Florence, P. S. 1929. *The Statistical Method in Economics.* London: Kegan Paul.

Gershuny, J. I., and I. D. Miles. 1983. *The New Service Economy: The Transformation of Employment in Industrial Societies.* New York: Praeger.

Giarini, O. 1987. *The Emerging Service Economy.* New York: Pergamon.

Grimmeau, J. P., C. Jodard, and M. Roelandts. 1985. "Les variations spatiales de structure des professions tertiares en belgique." *Acta geographica Lovaniensia* 26.

Kellerman, A. 1985. "The Evolution of Service Economies: A Geographical Perspective." *The Professional Geographer* 37(2):133–143.

Kirn, T. J. 1987. "Growth and Change in the Service Sector of the U.S.: A Spatial Perspective." *Annals of the Association of American Geographers* 77(3):353–372.

Noyelle, T. J. 1985. *New Technologies and Services: Impacts on Cities and Jobs.* College Park: University of Maryland Institute for Urban Studies.

———. 1987. *Beyond Industrial Dualism: Market and Job Segmentation in the New Economy.* Boulder, Colo.: Westview.

Nusbaumer, J. 1987. *The Services Economy: Lever to Growth.* Boston: Kluwer.

Ó hUallacháin B., and N. Ried. 1991. "The Location and Growth of Business and Professional Services in American Metropolitan Areas, 1976–1986." *Annals of the Association of American Geographers* 81(2):254–270.

Price, D. G., and A. M. Blair. 1989. *The Changing Geography of the Service Sector.* London: Belhaven.

Stanback, T. M., Jr. 1979. *Understanding the Service Economy: Employment, Productivity, Location.* Baltimore: Johns Hopkins University Press.

———. 1981. *Services: The New Economy.* Totowa, N.J.: Rowman & Allanheld.

Wheeler, J. O., and R. L. Mitchelson. 1989. "Information Flows Among Major Metropolitan Areas in the United States." *Annals of the Association of American Geographers* 79(4):523–543.

Independence Hall in Philadelphia

Point Park at the Golden Triangle in Pittsburgh

The State Capitol building in Harrisburg

PART FOUR

THE CITIES

Pennsylvania's cities have long been the focus of the state's economic, cultural, and political lifeblood. They represent some of the most impressive landscapes and institutions that men and women have built and modified to meet individual and societal needs. Their size and structure and character reflect changing fortunes and misfortunes—both in the physical environment and in the marketplace—and the indelible imprints of the rich history of different ethnic and racial groups, their livelihoods, and technologies for making and moving things.

Pennsylvania's 15 metropolitan areas now cover half the state's counties and are home to more than 92 percent of its population and an even larger share of its employment opportunities. The first chapter in Part Four examines the location and growth of the state's metropolitan places from the colonial period to the present. The spatial pattern of growth is woven from the threads of shifting economic structure—from agriculture to manufacturing to services—along with changing transportation, communications, and production technologies. Because Pennsylvania has traditionally provided key mineral and energy resources, labor, capital, and markets to fuel the engines of production, the state has been a microcosm of the effects of changes over more than two centuries of metropolitan growth throughout the nation.

The internal spatial structure of Pennsylvania's metropolitan areas also reflects the long history of development. Despite the obvious uniquenesses of different places, there are many similarities in the spatial patterning of Pennsylvania's metropolitan areas. The effects of transportation and communication technologies, certain types of architecture or city-building, and modes of industrial production created notable similarities of structure among places. All the metropolitan areas carry a certain amount of this "baggage" from earlier eras of development, depending on the relative amount of addition to the built environment at particular times and the destructive effects of later reuse and redevelopment of urban land. The second chapter in Part Four explores these similarities and differences in Pennsylvania's internal spatial arrangement and their driving processes across space and time.

For most Americans, the images of urban Pennsylvania logically turn to Philadelphia or Pittsburgh, or both. Representing the two largest population concentrations at opposite ends of the state, the rivalries are legendary and often overshadow the many interdependencies of these two nodes in the Pennsylvania economy. The Philadelphia and Pittsburgh chapters provide the reader with more in-depth urban geographies of these two areas, focusing on their spatial evolution and historical patterns of shifting populations, on suburbanization and changing industrial structures of their economic bases, on urban redevelopments, and on transitions of key urban neighborhoods. These chapters examine the processes of change that suggest future patterns of development in these two nationally and internationally prominent metropolises.

18

THE LOCATION AND GROWTH OF PENNSYLVANIA'S METROPOLITAN AREAS

Rodney A. Erickson

Pennsylvania's cities have always had an important place in the history, legend, and economic fortunes of the nation. For most Americans, Pennsylvania's cities evoke vivid images of important military battles, gritty steel mills, rowhouses, and river towns—images that are sometimes appropriate, but as often as not outdated or inaccurate. Throughout the past three centuries, the state's cities and towns have undergone constant change, waxing and waning with shifting economic conditions.

Today, Pennsylvania continues to rank as one of the most populous states in the nation, following only California, New York, Texas, and Florida. While Pennsylvania has a lower than (national) average share of its population living in urban places, its 15 county-based metropolitan statistical areas (MSAs), as defined by the U.S. Census Bureau, span 33 of the state's 67 counties (Fig. 18.1).[1] These metropolitan areas each comprise a densely settled core of 50,000 or more inhabitants and its surrounding urban and rural populations. The MSAs are home to nearly 85 percent of Pennsylvania's population and represent functional areas with a strong degree of internal economic integration, reflecting the concept of the "greater city." Because they are based on county boundaries, these census-defined areas also represent a useful way to track metropolitan population growth in relatively constant geographic units over time, a situation not possible with the frequent changes in corporate city boundaries due to annexation.

1. The definition of the Pittsburgh MSA was changed following the 1980 census. Fayette County was added to the Pittsburgh MSA, and Beaver County, previously part of the Pittsburgh MSA, was broken off to form a separate Beaver Valley MSA. Both the Pittsburgh and the Beaver Valley MSA are combined in the Pittsburgh Consolidated Metropolitan Statistical Area (CMSA). Because of the close historical ties of Beaver Valley to Pittsburgh, the current CMSA definition is reported as Pittsburgh throughout this chapter. As a result, there is information reported for 14 metropolitan areas.

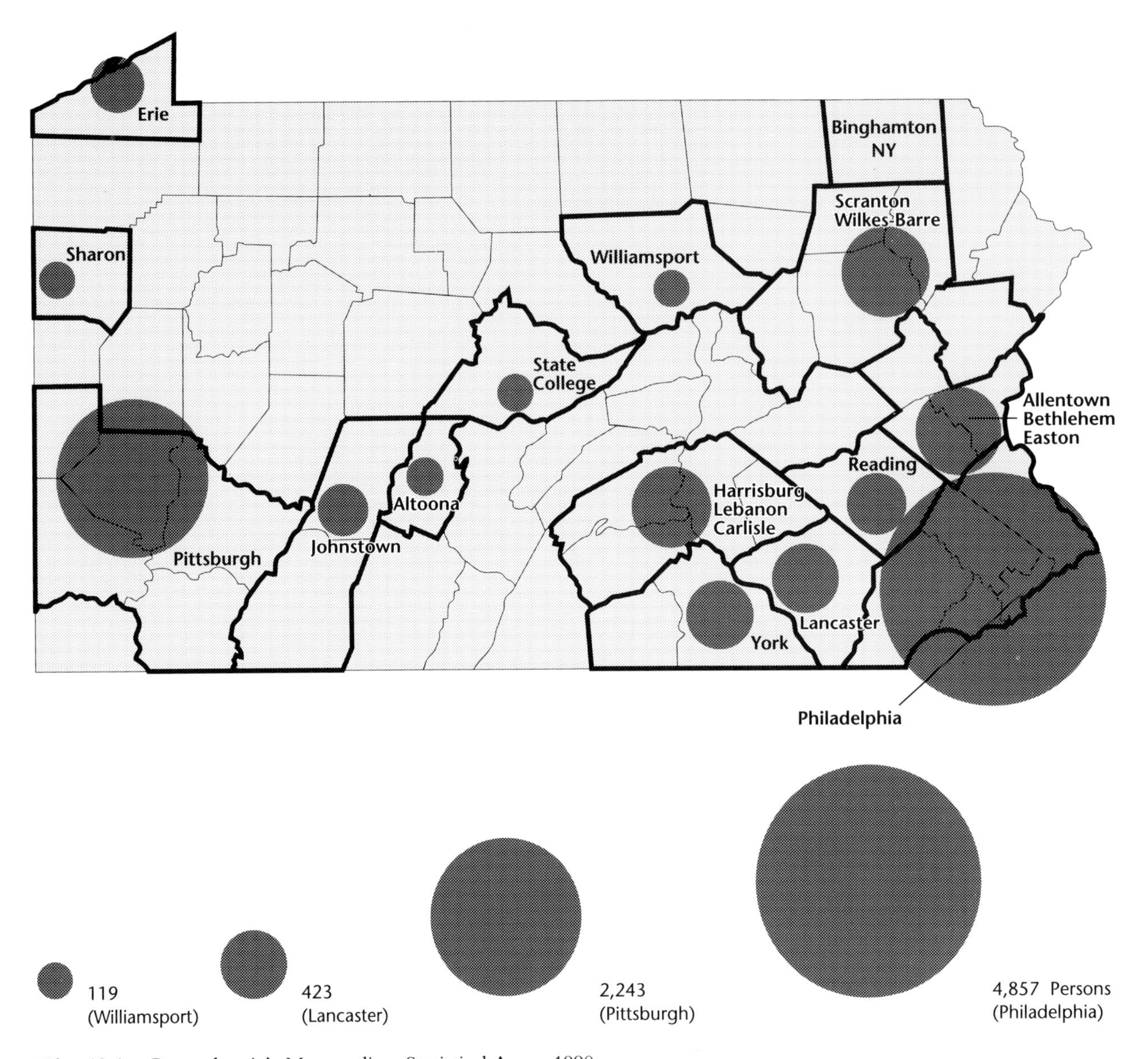

Fig. 18.1. Pennsylvania's Metropolitan Statistical Areas, 1990.

Pennsylvania's metropolitan population is by no means uniformly distributed across the state. Philadelphia, anchoring the state on the east with its 4.86 million residents (stretching into New Jersey), and Pittsburgh–Beaver Valley, anchoring the west with its 2.24 million residents, together account for nearly three-fifths of the state's total population.[2] In between, metropolitan places of widely varying size are located in and around old core cities. The metropolitan areas of Scranton–Wilkes-Barre, Harrisburg-Lebanon-Carlisle, and Allentown-Bethlehem-Easton each add well over half a million residents, and the Erie, Johnstown, Lancaster, York, and Reading MSAs all have at least a quarter of a million population. The remaining MSAs—Altoona, Williamsport, State College, and Sharon—are all relatively small, the latter two having attained metropolitan status only recently.

This chapter addresses the questions of how and why this pattern of metropolitan location and

2. In addition to the five Pennsylvania counties, the Philadelphia MSA includes the New Jersey counties of Burlington, Camden, and Gloucester.

growth evolved in Pennsylvania, and provides some clues as to where the state may be headed and some possible consequences of these growth patterns. The approach is based on the premise that the historical dimensions of this metropolitan geography are critically important to understanding the contemporary scene. Metropolitan growth is clearly a cumulative process of adding to the built environment and attracting more and more laborers and entrepreneurs. But this accumulation process typically works in fits and starts and builds on the strength and momentum of previous accomplishments. Before embarking on this journey through time and space, a framework that ties together some central concepts is provided.

Pennsylvania's Metropolitan Evolution: A Framework in Space and Time

Cities have always served several basic functions that have taken on differing relative importance at various times in their existence. These functions include economic, social, religious, political, administrative, transportation, and military activities. Some, such as military functions, were especially important factors in the early locations of settlements where defensible sites were necessary. Pittsburgh and Erie are ready examples, where the military function has no contemporary relevance but was a key to their initial establishment. Political and administrative functions have also been important, as the location and growth of Harrisburg and that of many county seat towns, later to become metropolitan areas, would suggest. But in the end, the overwhelmingly important function of cities is economic, for the sustained ability to collect, produce, trade, and transport raw materials, goods, and services is the lifeblood of cities. Their comparative advantage in these endeavors with respect to other cities and shifting patterns of market demand and local resource supplies is the principal factor underlying their growth or stagnation and decline.

The evolutionary nature of the metropolitan mosaic has been a topic of considerable interest among geographers for several decades. The most comprehensive framework for structuring such inquiry has been provided by Borchert (1967) in his classic research on American metropolitan evolution nearly three decades ago. The framework rests on the impacts of and adaptations to major innovations, paying particular attention to the effects of changes in the size and resource base of a metropolitan area's hinterland and the technology of transportation and energy for the manufacture of material resources. Borchert also argues that these two factors—resources and technology—are strongly interrelated, for the technology helps define the resource base and also affects the geographic extent and thus the resource base of the hinterland.

The innovations and changes of principal concern in this framework of metropolitan evolution include the steamboat and "iron horse," the use of steel rails, the application of electric power to industrial tasks, and the internal combustion engine, along with the shift to a service economy. Borchert identifies four major eras in American metropolitan evolution: (1) the Wagon-Sail Era, 1790–1830; (2) the Iron Horse Era, 1830–1870; (3) the Steel-Rail Era, 1870–1920; and (4) the Auto-Air-Amenity Era, 1920–1960.

Each of these eras is roughly defined by the growth and peaking of the major sources of technological change during the three eras since 1830. Before that time, most raw materials and goods were carried by ship in international or coastwise trade, and internal commerce movements were confined to crude turnpikes and canals that were built along the major waterways. The application of the steam engine to boats permitted a steady buildup of waterborne commerce on the inland lakes and rivers beginning in the 1830s. Likewise, rail mileage increased sharply as locomotives were improved and capable of pulling greater loads. The introduction of steam power created major inland transportation corridors for the first time. The decade of the 1870s saw railroad shipping overtake inland waterborne commerce, and the rapid expansion of the American steel industry following the introduction of the Bessemer process. The development of the steel industry was paralleled by the growth of bituminous coal production, the latter of which peaked about 1920. Railroads, steel, and coal all carried powerful consequences for the metropolitanization of the population in several parts of the nation. The railroads also made possible the large-volume haulage of coal for use in electrical power generation, a factor that helped to centralize manufacturing industry in large metropolitan agglomerations, free from the locational constraints imposed by waterpower sites. By 1920, motor vehicle and oil production, as well as the mileage of surfaced roads, began to increase rapidly. Commercial air transport also was initiated during the 1920s and has grown steadily since. Thus, Borchert found that by 1960 there was a relatively mature metropolitan system whose economies were typically based more on services production than industrial or agricultural output.

The years since 1960, when Borchert brought his study of metropolitan evolution to a close, have witnessed improvements in many of the innovations set into place during the 1920–1960 era. The Interstate Highway System was largely completed in the ensuing years and created a high-speed network that had further severe repercussions for the railroads and the places that had grown around their functions. Air traffic—both passengers and freight—increased dramatically. Both trends made the nation more dependent on petroleum resources that would add new dimensions of instability to the American economy and metropolitan evolution. Amenity locations have thrived to an even greater degree with earlier retirements, greater financial independence, and longevity of America's senior citizens and the more footloose character of many service-producing industries that currently account for more than 70 percent of total national employment.

But the most significant innovation for metropolitan evolution in the post-1960 period was the introduction and rapid adoption of computer and telecommunications technology. The technology has permitted the greater concentration of economic activities in large corporations with a centralization of management functions in a limited number of corporate "control points." Organized research and development, and the more demanding skills of many service and industrial design industries that depend on highly educated personnel, have spurred the growth of higher education—with growth repercussions for metropolitan places that specialized in these activities. The concomitant effects of the application of new computer-based technologies to agriculture, mining, and manufacturing industries has been to enhance their productivity while reducing the relative labor input. In addition, the increasingly global nature of trade and production has ushered in an era of much greater international competition for many products, especially manufactured products. This fifth era, from 1960 to the present, which can be referred to as the Computer-Telecommunications Era, has been accompanied by massive economic restructuring.

Borchert (1967) identified five size classes of metropolitan places that represented five levels in the American metropolitan hierarchy in 1960. First-order places had a population of more than 8 million; second-order places had 2.3 to 8 million; third-order places had 820,000 to 2.3 million; fourth-order places had 250,000 to 820,000; and fifth-order places had less than 250,000 population. The set of counties comprising each MSA in 1960 was used as a benchmark geographic area in order to retain a common set of boundaries—and comparability—over time. The requirement for central cities of MSAs to have a population of at least 50,000 was scaled backward and forward from 1960 for each of the era time breaks in proportion to the size of the national population at the time. The limits of the size orders in the hierarchy were also scaled back to be commensurate with the size of the national population in the earlier periods. While relatively few of the places identified in the early years of the analysis can be considered "metropolitan" by contemporary standards, the largest places were nonetheless the principal population clusters of the nation at that time.

In the analysis below, Borchert's methodology is used, the principal alteration being that 1990 MSA boundaries are used throughout the various eras. This change not only permits an updating of the information but also results in the inclusion of Pennsylvania's two most recent MSAs: State College and Sharon.

Before the location and growth of Pennsylvania's metropolitan places are analyzed in these eras, however, it is crucial first to explore the pattern of location and growth of colonial urban settlements in Pennsylvania. The patterns that were set into motion before nationhood—before regular population census data were even collected—would strongly influence the future course of metropolitan development.

The Colonial Period

Originally a proprietary town of the Penn family, Philadelphia was incorporated in 1701. From the outset, the city was envisioned as a commercial center. Within three years of William Penn's landing, shipping services had been established (Currier 1979). The shipping trade included routes to the Far East, the "triangle trade" routes of the West Indies/Iberia-England-Philadelphia and coastal trade stretching to

New England and the Carolinas. The Revolutionary War compelled the forging of new trading links, but the city regained its position as the nation's leading port after hostilities had ended, a status it did not lose until later in the nineteenth century. In addition, Philadelphia early became an important ship-building center, and its local entrepreneurs also developed iron furnaces, potteries, milling, and paper-making operations (Mason 1979). Service sector activities also revolved around trade, with insurance, finance, brokerage, and other related endeavors as important supporting enterprises.

During the colonial period, Philadelphia is described as a city of widespread economic opportunity, one of artisans and entrepreneurs. Warner (1968) suggests that most of its citizens shared the experience of entrepreneurship, but Philadelphia was also a growing metropolis by eighteenth-century standards and was increasingly acquiring contemporary urban traits. It had an evident and growing disparity between its richer and poorer citizens, with increasing numbers of the less fortunate (Clark 1975). Furthermore, pollution in the form of noise, traffic, and garbage was already being reported, and the needs for improved infrastructure were clearly apparent.

Settlement was confined to the Delaware Valley until about 1700. This situation occurred partly as a result of a desire not to antagonize local American Indian populations with whom relatively friendly relations existed, but also as a result of the absence of any significant population pressure (Nelson 1969). However, the spread of population thereafter was very rapid; the Mennonites' purchase of land adjoining Conestoga Creek in 1709 near what would later be the settlement of Lancaster marked an early expansion of settlement well inland. The advance of the frontier in Pennsylvania did not constitute a steady advance of settlement, but rather a complex process of striking out beyond the lines of existing settlement with subsequent infilling of areas initially passed over (Florin 1966).

Between the advancing frontier and the Delaware Valley, there developed a region known as the backcountry. This area was defined as lying between one day's journey from Philadelphia—that is, approximately 30 miles to the east—and the frontier to the west. The backcountry expanded therefore as the frontier moved westward in the nineteenth century. In 1710 the frontier was at the western edge of Philadelphia; in 1720 it was at the Schuylkill River; in 1730, at the east bank of the Susquehanna River; in 1750, about 150 miles from Philadelphia; and in 1774, at Pittsburgh (Nelson 1969). The spread of population and settlement into the backcountry can be seen in Fig. 18.2, showing the expansion of population and urbanization in the eighteenth century.

Lemon (1967 and 1972) has identified the pattern of eighteenth-century urbanization in Pennsylvania as falling into several periods of development: (1) "establishment" from 1681 to 1700, in which settlement was established in the Delaware Valley; (2) "stagnation" from 1700 to 1730, during which the frontier advanced westward but few towns or settlements of any consequence were founded; (3) "expansion" from 1730 to 1760, when several backcountry county seats and towns were founded; and (4) "consolidation, disruption, and re-establishment" from 1760 to 1800, in which the backcountry filled in, and the Revolutionary War temporarily disrupted the spread of urbanization. Viewed within this framework, the establishment and growth of Philadelphia originated in the first period. During the second period, the frontier moved westward, and the area within the counties of Philadelphia (including Montgomery County, which later became a separate county), Chester, and Bucks was settled. There was, however, very little urban development within this region, either of new settlements or of the existing towns of Chester, Bristol, and Newtown. This relative absence of growth in these towns surrounding Philadelphia was directly attributable to the hegemony of Philadelphia as the principal market and administrative center for the region, a situation that was not at odds with the desires of the Penn heirs to promote and protect the position of Philadelphia within the region. Beyond Philadelphia's immediate sphere of influence, the backcountry towns founded in the third period grew rapidly.

The county town of Lancaster was founded in 1729 and thereby broke a protracted period of stagnation in the growth of new urban settlements. It was established as the seat of Lancaster County, founded the following year. In contrast to successors, the reason for Lancaster's location appears to be entirely political—specifically, the political influence of its proprietor, Alexander Hamilton. Lancaster was the fourth county and was formed out of Chester County, one of the original three counties in the colony. Pressure for establishment of a new county rose from the farmers of the backcountry area who had settled beyond the watershed distance

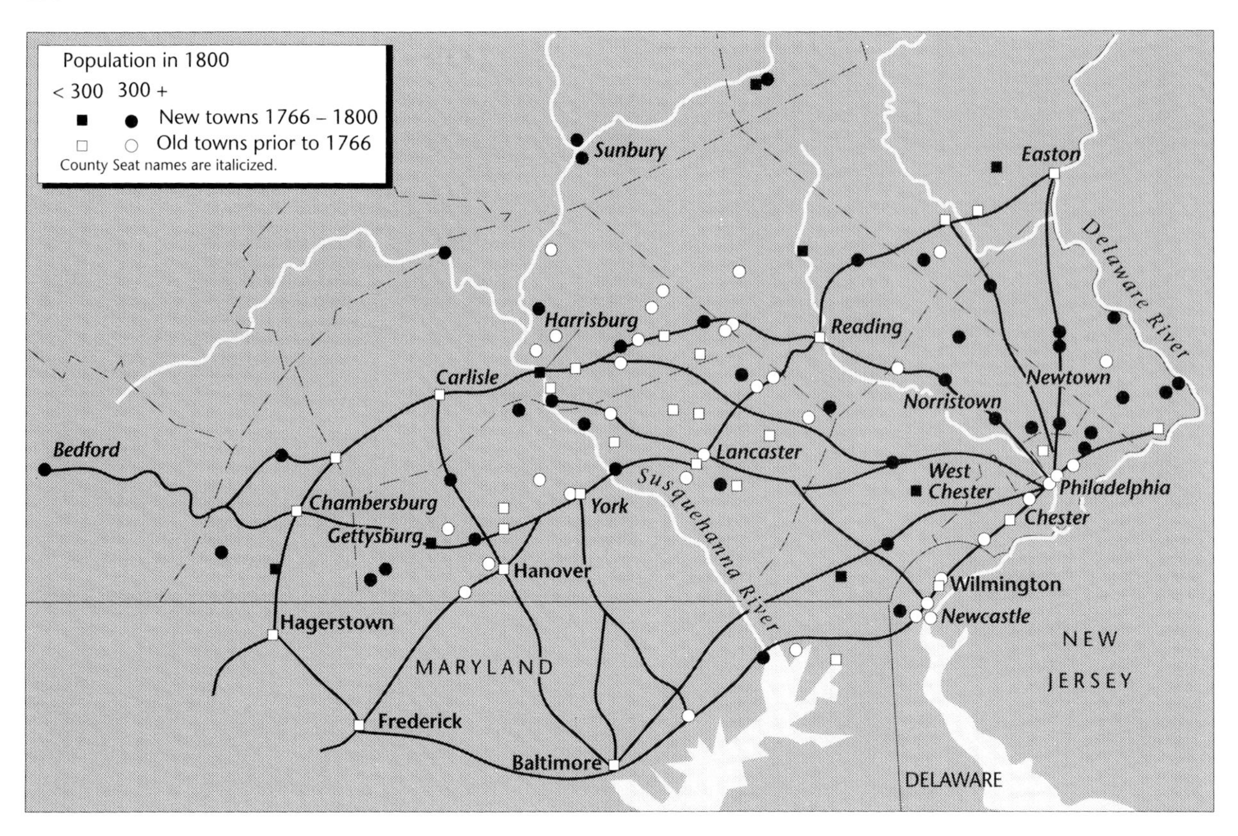

Fig. 18.2. Urbanization in southeastern Pennsylvania, 1766–1800.

of a day-long journey from Philadelphia and the county seat at West Chester.

Subsequently, as the backcountry spread across the Susquehanna River and northward to the Appalachian Mountains, similar demands for access were met with the formation of additional counties. York County, to the west of the Susquehanna River, became the fifth county in 1749, the county seat of the same name being situated centrally in the county. In the formation of many of these counties and their seats, the criteria for formation were clearly laid out by the Penn heirs: orientation to Philadelphia (that is, not so close as to pose undue competition for Philadelphia, while not so distant as to be beyond the range of feasible commercial linkages) and siting of the county seat at a central location within the county (Lemon 1967). Shortly after York, Cumberland County was founded in 1749 with Carlisle as its seat, and in 1752 Northampton and Berks counties were founded with their respective seats at Easton and Reading. The remaining counties of southeastern Pennsylvania formed during this third period of settlement were Dauphin County with Harrisburg, Bedford County with Bedford, and Northumberland County with Sunbury as their respective seats. In most of these cases—excepting only Lancaster and Harrisburg—these were proprietary towns of the Penn family. Sunbury was the last proprietary town, the practice of proprietorships ending with the new constitution adopted by Pennsylvania in 1789.

In addition to these county towns, other towns were founded during the period from 1740 to the Revolutionary War, a number of which went on to become an important part of the metropolitan fabric of Pennsylvania. These places included Bethlehem, Pottstown, Allentown, and Williamsport. Each of these was strategically located along the major inland river arteries that flowed to southeastern Pennsylvania or the Chesapeake Bay.

The transportation system of the colonial period consisted of a network of small roads around Philadelphia and utilization of the inland waterways (Rosenberger 1975). The transportation factor contributed significantly to the location of these settlements and their subsequent growth. Of the roads, the most significant was the King's Highway completed in 1741, laid out between Philadelphia and Lancaster and greatly enhancing the locational advantage of the latter (Landis 1918). Except where a county seat was located for the specific attribute of centrality, the other towns of this period were located on one of the major rivers of the region: Easton at the forks of the Delaware River, Allentown and Bethlehem on the Lehigh River, Pottstown and Reading on the Schuylkill River, and Harrisburg, Sunbury, and Williamsport on the Susquehanna River.

Transportation was critically important to the emerging central place functions and early industrial bases of county seat and other towns alike. The economy of colonial Pennsylvania was predominantly agricultural, and the efficient shipments of unprocessed or semi-processed commodities to the ports of Philadelphia and Baltimore was absolutely critical to their economic vitality. The towns also served important functions as local market centers for growing, albeit small, hinterlands surrounding them and as legal and, in the case of county seat towns, administrative centers. Throughout much of southeastern Pennsylvania, there were numerous iron plantations that were often small communities in and of themselves concentrated along the valleys of the Delaware, Schuylkill, and Susquehanna rivers (Billinger 1954; Murphy and Murphy 1937). Again, transportation access to the markets of the large mercantile and processing metropolises, particularly Philadelphia, was a primary determinant of the settlement's growth, stagnation, or decline.

By the late colonial period, frontier settlement was occurring both in western Pennsylvania and in the Wyoming Valley of the northeastern part of the state. The settlement of significant parts of western Pennsylvania occurred by way of Virginia and the Monongahela Valley, and northward to the forks of the Ohio River, the site of Pittsburgh (Buck and Buck 1939). Settlement in this region was not permanent until after 1768, in part because of the restrictions imposed by the Proprietary Government and the French and Indian War. The Wyoming Valley was settled in the 1750s by settlers from Connecticut, initiating a rivalry culminating in the Yankee-Pennamite wars of 1769 and 1775 and resolved only after the Revolutionary War. As early as 1776, however, the region had nearly 2,000 inhabitants as settlers flowed in following the river valley routes (Gallagher 1968).

The Wagon-Sail Era: 1790–1830

In 1790 Pennsylvania had 3 of the 39 places in the new nation that met the qualifications for metropolitan areas following Borchert's (1967) criteria: Philadelphia, Pittsburgh, and Lancaster (Fig. 18.3a). At that time Philadelphia had a population of nearly 172,000 (see Table 18.1) and was the largest metropolitan area in the nation; it was one of only three second-order places, just ahead of Boston and still surpassing New York by a significant margin. Pittsburgh grew very rapidly during the second half of the eighteenth century as a military and trading center and had achieved third-order status in the hierarchy with a population of more than 63,000. Lancaster, with about 36,000 residents, fell into the ranks of the fourth order in the national metropolitan hierarchy. Several other backcountry counties met the overall population-size criterion but represented predominantly dispersed agricultural settlement and did not have a central city of sufficient size to qualify as metropolitan areas.

By the turn of the nineteenth century, the backcountry region was a rapidly developing agricultural area, and with this production came the need for transportation improvements to reduce the costs of moving commodities to market. This necessity was felt both by the backcountry settlers who depended on trade and by the merchants of Philadelphia whose fortunes became increasingly entwined with those of the backcountry. Furthermore, relocation of the state capital to Lancaster in 1799, and then to Harrisburg in 1812, focused still more attention on transportation in the backcountry. Evidence of this development is found in the formation of the "Society for the Promotion and Improvement of Roads and Inland Navigation in the State of Pennsylvania" in 1790 (Cochran 1978).

This Society attended to the two transportation alternatives available at the time: roads and canals. The development of each of these transportation networks was intimately related to the growing settlement hierarchy—a relationship pervading this entire era and each of the subsequent ones brought about by successive transportation or communication innovations.

The first achievement of the Society was the securing of a charter for a turnpike between Philadelphia and Lancaster in 1792, the first of its kind in the

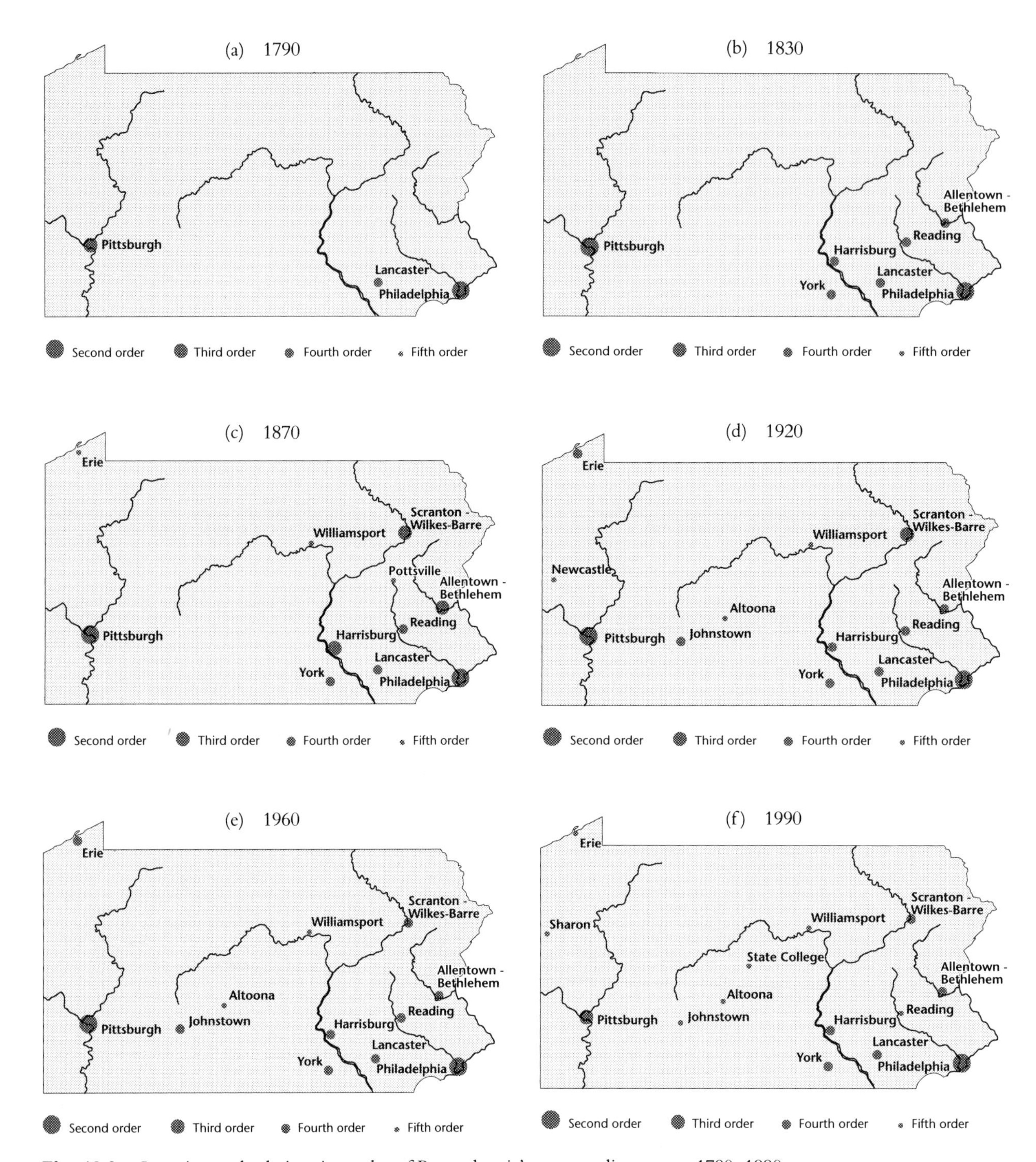

Fig. 18.3. Location and relative size-order of Pennsylvania's metropolitan areas, 1790–1990.

Table 18.1. Pennsylvania population and rank size order, selected years, 1790–1990 (in thousands).

Metropolitan Area	1790	1830	1870	1920	1960	1990
Philadelphia	172 (2)	402 (2)	1,059 (2)	2,714 (2)	4,343 (2)	4,857 (2)
Pittsburgh[a]	64 (3)	185 (2)	449 (2)	1,948 (2)	2,575 (2)	2,243 (3)
Scranton–Wilkes-Barre	—	—	223 (3)	764 (3)	691 (4)	734 (4)
Allentown-Bethlehem-Easton	—	80 (4)	181 (3)	409 (4)	545 (4)	687 (4)
Harrisburg-Lebanon-Carlisle	—	89 (4)	164 (3)	298 (4)	463 (4)	588 (4)
York	—	64 (4)	106 (4)	179 (4)	290 (4)	418 (4)
Lancaster	36 (4)	77 (4)	121 (4)	174 (4)	278 (4)	423 (4)
Reading	—	53 (4)	107 (4)	201 (4)	275 (4)	337 (5)
Johnstown	—	—	—	280 (4)	281 (4)	241 (5)
Erie	—	—	66 (5)	154 (4)	251 (4)	276 (5)
Altoona	—	—	—	128 (5)	137 (5)	131 (5)
Sharon	—	—	—	—	—	121 (5)
Williamsport	—	—	48 (5)	83 (5)	109 (5)	119 (5)
State College	—	—	—	—	—	124 (5)

SOURCE: Borchert 1967; U.S. Census Bureau.
NOTE: Numbers in parentheses are the rank order of metropolitan areas.
[a]See this chapter's footnote 1 for Pittsburgh definition used.

nation to make use of a macadamized surface (Cochran 1978; Landis 1918; Meyer 1948). The Turnpike Company was a wholly private concern, and although it was profitable its returns were modest. Most of the subsequent turnpikes in which the state became involved were public-private partnerships. The Lancaster-Philadelphia turnpike was an improvement over the earlier King's Highway and linked the nation's largest inland town with the leading eastern metropolis. In 1794 the turnpike was extended to Columbia, a key crossing point on the Susquehanna River and an important node on the future canal-turnpike network.

Following the success of the Lancaster-Philadelphia turnpike, "turnpike fever" gripped the state, and many local companies sprang up in the first decades of the nineteenth century. Early turnpikes were constructed between Erie and Waterford (New York); Wilkes-Barre and Easton; and York and Wrightsville (Livingood 1947). Following these initial improvements, turnpikes that proved to be important links were constructed between such emerging centers as Reading and Sunbury in 1812, and on to Williamsport. A turnpike between Reading and Harrisburg, and another for Lancaster, Middletown, and Harrisburg, opened in 1817, along with the Lackawanna Turnpike in northeastern Pennsylvania. A series of turnpikes made up the

Pennsylvania Road, linking Philadelphia and the East with Pittsburgh and the receding frontier of the West. Along the routes, York, Carlisle, Shippensburg, Chambersburg, Bedford, and Pittsburgh became increasingly significant centers, both for trading and shipping goods and for servicing the westward-bound migrants. The Pennsylvania Road continued to be a major thoroughfare even after the 1810 opening of the Cumberland Road, which linked the southwestern part of the state more directly with Maryland and its emerging port of Baltimore (Klein and Hoogenboom 1980).

Concomitant with the construction of turnpikes was the development of a network of canals and river transportation. The towns of southeastern Pennsylvania in the late eighteenth and early nineteenth centuries were processing and marketing centers and emerging transportation hubs. The canals that linked them conveyed agricultural products and manufactured goods to and from the frontier, and, later, anthracite coal from northeastern Pennsylvania. In addition, the canal boats conveyed passengers.

It was discovered early that the shallow, rocky rivers throughout most of Pennsylvania were not conducive to an efficient river transportation system. On both the Lehigh-Delaware and the Susquehanna rivers, arks were initially floated downstream (Livingood 1947). Clearly this system did not allow for upstream navigation and was unreliable at best for any usage. On each of the major rivers, the early improvements and canals that were constructed contributed significantly to the growth of towns such as Easton, Bethlehem, Allentown, and Mauch Chunk (later, Jim Thorpe) on the Lehigh River. These improvements made the shipping of anthracite coal a practical reality. Anthracite soon became the most important freight on the eastern canals. The Lackawaxen Canal linked Honesdale in the northeastern part of the state with the Delaware River and ultimately with New York City. The Schuylkill Canal opened in 1825 and made navigation possible along the route from Pottsville to Reading and on to Philadelphia (Klein and Hoogenboom 1980).

Navigation of the Susquehanna had been attempted by steamboat in 1825–1826 but proved unsuccessful and was followed by a number of canal projects. The Union Canal, opened in 1827, had more symbolic than economic success, but it linked the Susquehanna River (at Middletown) to the Schuylkill River (at Reading) and thence Philadelphia (Meyer 1948). Throughout the nineteenth century, the region bordering the Susquehanna was at the center of a trade rivalry between the Philadelphia merchants and those in Baltimore to the south, and the canal-building projects in this area were oriented to one or the other of these two metropolitan ports (Livingood 1947). The Susquehanna River became the main artery, particularly after both Lancaster and York were linked to it by canals. Tapping the Susquehanna River trade became vitally important for both Philadelphia and Baltimore, and the Union Canal reinforced Philadelphia's early lead. Philadelphia's uncanny ability to thwart its rival through control of state government concessions enabled it to overcome what would have otherwise been a serious locational disadvantage. The opening of the Main Line Canal in 1834 further enhanced Philadelphia's position, but the Susquehanna and Tidewater Canal, opened in 1840, from Columbia to Havre de Grace in Maryland tipped the balance toward Baltimore. Thereafter, the rivaly was renewed in competing railroad projects.

The Pennsylvania Main Line canal system (the State Works) project was begun in 1826 and was a concerted effort on the part of the state to link Philadelphia with Pittsburgh and the rapidly expanding West. The State Works resulted from the panic following the opening of New York State's Erie Canal in 1825, which threatened (and succeeded) in diverting much of the western trade from Philadelphia to New York City. Indeed, the threat was a double-edge sword, with the threat that Baltimore would steal the trade of western Pennsylvania. Although the Main Line was not financially successful, it was both an engineering and a political success.

The inadequacies of the Schuylkill and Union canals necessitated the construction of a railroad from Philadelphia, through Lancaster to Columbia on the Susquehanna. This railroad carried canal boats to Columbia, whereupon they were placed in the Main Line Canal, which ran northward and then westward along the Juniata River to the foot of the Allegheny Mountains at Hollidaysburg. A portage railroad hauled the canal boats over the mountains for 36 miles and down to a canal basin at Johnstown, from whence the canal proceeded along the Kiskiminetas and Allegheny rivers to Pittsburgh. The system was a significant factor in establishing settlements in central and western Pennsylvania, particularly those towns along the route, such as

Lewistown, Huntingdon, Hollidaysburg, and Johnstown. Political opposition to the scheme led to appeasement of the northern region in the form of the West and North Branch canals; these followed the West Branch of the Susquehanna River to Williamsport and Bellefonte, and the North Branch to Sunbury, Wilkes-Barre, and Athens (New York), respectively. Also, the State Works included the Erie Division canal from Pittsburgh to Erie and thus connected the city at the forks of the Ohio with the Great Lakes.

Following the Revolutionary War, Philadelphia had been compelled to realign its foreign trading links. From being an importer of goods from Britain and its colonies, it switched its attentions to South America, East Asia, and Russia and also began to export domestic goods, especially anthracite coal. By 1800 about 40 percent of the city's work force was engaged in manufacturing—a sector that became increasingly important as Philadelphia's preeminent trading position was lost to New York and Baltimore. Though overtaken by New York as the dominating metropolis of the national urban hierarchy during this period—a position never since regained—Philadelphia nonetheless retained its position as the leading industrial and commercial center of Pennsylvania during the Wagon-Sail Era. By 1830 Philadelphia had a population of 402,000 and was firmly established as a second-order metropolitan area (see Fig. 18.3b and Table 18.1).

In the old backcountry, towns grew up as nodes on the evolving canal-turnpike system or retained their marketing functions for hinterlands whose density of settlement was enhanced in the course of continued agricultural development. Also, the larger towns, especially Lancaster and Reading, became early manufacturing centers specializing in agricultural processing, machinery, and wagons. By the end of the Wagon-Sail Era the Allentown-Bethlehem-Easton, Harrisburg, Reading, and York metropolitan areas each had a population of more than 40,000 and moved into the fourth order of the hierarchy, bypassing the fifth order altogether. These emerging metropolitan areas joined Lancaster, which, although growing significantly to nearly 77,000 residents, remained in the fourth order to complete the metropolitan set in the backcountry of southeastern Pennsylvania (see Fig. 18.3b and Table 18.1).

During this period, in addition to those metropolises of southeastern Pennsylvania a number of small towns were established that were destined for very different fortunes in the ensuing eras. Throughout the region, agriculture and scattered iron plantations were the predominant economic base, and access to the transportation system was essential to economic success. Some, like Bedford and Hollidaysburg, lay on the east-west thoroughfares; others, such as Sunbury and Bellefonte, were incipient regional centers. In most cases, these smaller centers had initial advantages that were insufficient to carry them well forward into the succeeding eras as major focuses of population and economic growth. For the fortunate few others, the coming together of key features of energy resources, transportation, and production advantages propelled them into prominent places in the state's metropolitan hierarchy.

The Wyoming Valley remained predominantly agricultural throughout much of the Wagon-Sail Era, with a scattered population of about 5,000 in 1800. But throughout the region, small communities were emerging on the anthracite coal fields and the canal arteries that tied them into the eastern industrial markets; many of these places would eventually become part of the Scranton–Wilkes-Barre MSA to emerge in the subsequent era.

In western Pennsylvania, Pittsburgh rapidly ascended to regional hegemony, growing as a trading and outfitting node on the east-west trade in goods and migrants, at the terminus of the Pennsylvania Road and the Main Line Canal, and at the head of the Ohio-Mississippi river system (Baldwin 1937). Pittsburgh became a major port on the Ohio, as well as a shipbuilding center. Successively, flatboats, keelboats, sailboats, and steamboats were built at Pittsburgh. The larger part of the region's population in 1790 was located in the Monongahela Valley, where the major economic activities were agriculture and iron-making—the seeds of later industrial development (Bining 1973). By the end of the Wagon-Sail Era, Pittsburgh had moved rapidly up the urban hierarchy to second-order status, with a population of more than 185,000, at the same rank but still significantly smaller than Philadelphia (Fig. 18.3b and Table 18.1).

In the northwest, Erie was emerging as the state's door onto the Great Lakes and was developed enough to lobby for the north-west extension of the State Works canal system. By the time this canal was completed in 1844, it had been turned over to a private commission with a large part of the stock held by Erie residents. The small settlement was emerging as a milling and industrial center on the

Great Lakes, factors that would later propel the place into the ranks of Pennsylvania's metropolitan areas.

The Iron Horse Era: 1830–1870

This period saw the widespread development of the railroads, and with that a new transportation network on which towns would grow. In general, railroads came after the canals, but initially there was a considerable degree of overlap in routes. Indeed, there were heated debates concerning the design of the Main Line Canal system centered on the relative merits of canals versus railroads. At that time, canal technology was better understood and the canal lobby won the day, although there was something of a compromise in building the railroad sections on the canal system (Rubin 1961).

The earliest railroads were horsedrawn and utilized wooden rails, but these gave way to gravity designs, steam power, and iron rails. Before the Iron Horse Era, railroads were local and shortline projects usually associated with industrial users. For instance, an early gravity railroad was built at Mauch Chunk in 1827 for carrying anthracite coal to the Lehigh River (Billinger 1954), and from Carbondale to Honesdale a steam railroad also carrying coal was opened in 1829 (Swetnam 1964). Another early railroad was constructed between Philadelphia, Germantown, and Norristown in 1832 that carried both freight and passengers, also reflective of the geographic expansion of metropolitan Philadelphia.

The Main Line system, although primarily a canal network, demonstrated the feasibility of long-distance railroads in place of canals by virtue of its successful railroad sections over the Allegheny Mountains. Of course, it also sowed the seeds of its own subsequent demise in the process. The Main Line Canal did not face a serious railroad challenge until 1852, by which time a number of regional railroads had been built. These included the Lehigh & Susquehanna Railroad in 1837, which connected Wilkes-Barre on the Susquehanna with the Lehigh River. One of the most important regional railroads of the epoch was the Philadelphia & Reading in 1842, which linked Pottsville, Reading, Pottstown, and Philadelphia (Roberts 1980). This railroad was an important carrier of anthracite coal and effectively replaced the Schuylkill Canal. By mid-century a number of railroads were being constructed in the northern and eastern parts of the state, primarily for the shipment of anthracite. In the northern coal fields, the Delaware, Lackawanna & Western Railroad, completed in 1853, linked the emerging Scranton with New York City; in the western coal fields, a railroad completed between Sunbury and Mount Carmel in 1853 provided a shipping outlet down the Susquehanna River (Snyder 1972).

In the Susquehanna Valley, the Philadelphia-Baltimore rivalry switched to the new railroad transport technology, and again a railroad to the south from Pennsylvania was thwarted by the legislature until after 1838, when a line from Baltimore to York was completed (Livingood 1947). This line was subsequently extended to Wrightsville on the Susquehanna River in 1840 and to Bridgeport in 1851, where it connected with the Pennsylvania Railroad. Thereafter, lines extended northward in turn to Sunbury, Williamsport, and Wilkes-Barre. Harrisburg began to emerge as a node on the railroad network with lines to Carlisle and Chambersburg in 1837, to Lewistown in 1849, and linking with the Pennsylvania Railroad three years later.

The railroad connection between Philadelphia and Pittsburgh was made in 1852, when the Pennsylvania Railroad opened, traversing the state east to west. The company had been founded in 1846, but unlike the earlier State Works that built the Main Line Canal, the railroad received no funding from the state legislature (Rubin 1961). The competition with the Main Line was short-lived, however, and in 1857 the railroad company purchased the canal. A later, east-west railroad was completed in 1864 between Philadelphia and Erie via Sunbury (Rosenberger 1975). Perhaps the primary contributions of both these railroads, although the latter was ultimately not a financial success, was to open large parts of the still-undeveloped central and northern regions of the state to settlement.

In the rapidly developing western region, railroads emerged quickly. As early as 1851, the Ohio & Pennsylvania Railroad connected Pittsburgh with Cincinnati and Cleveland (Schusler 1960). By 1860, additional carriers included the Pennsylvania, the Pittsburgh & Connellsville, and the Allegheny Valley railroads. In the northwestern part of the state, Erie became a major terminus on the developing through-routes to the Great Lakes. Erie attained this strategic position during this period not only because of its location on the Great Lakes but also because of the fragmented nature of railroad companies which consisted of several small, local companies, often with different gauges. So Erie became a transfer node, at which point passengers and freight were transferred to trains running on different-gauge rails.

Philadelphia continued to develop as a manufacturing center during the Iron Horse Era. By 1840 the city had outgrown its corporate boundaries and was surrounded by a number of satellite towns, such as Germantown, Manayunk, and Frankford. Between the city and its satellites, there developed a variety of manufacturing industries, the most commonly reported being textiles, and the mills of which dotted the Schuylkill River in the vicinity of Philadelphia. Following the construction of the railroads, industries also stretched along the tracks radiating from the emerging central area terminals. Metropolitan Philadelphia's population had increased by about 600,000 during the Iron Horse Era to register 1.06 million residents by 1870, placing the growing metropolis firmly among the second-order places in the national hierarchy (Fig. 18.3c and Table 18.1).

In the former backcountry, most of the established urban settlements also became manufacturing centers. Allentown-Bethlehem developed into a burgeoning producer of processed minerals, iron, textiles, cigars, and leather goods, spurred on by the opening of the Lehigh Valley Railroad. By 1870 its population had grown to nearly 181,000, the fourth largest metropolitan area in the state, placing it among the ranks of the third order nationally, albeit only a single era that its rank was that high (Table 18.1). Reading grew on the trade in anthracite and developed an engineering sector based in part on the new railroad industry (Patton 1983; Proctor and Matuszeski 1978). While Reading did not grow as fast as the Allentown vicinity, its growth was nonetheless sufficient to place it firmly among fourth-order metropolitan areas (Fig. 18.3c).

Farther to the west in the Piedmont region, the Lancaster, Harrisburg, and York metropolitan areas also experienced significant population growth (Table 18.1). Harrisburg grew most rapidly and cracked the ranks of the third-order metropolitan areas by the end of the era, a status that was not maintained after 1870. Harrisburg's growth was in large part built around its role as a hub for inland river–canal and the emerging railroad traffic carrying manufactured goods, raw materials, and passengers. By 1870 it had reached a size of nearly 165,000 residents in the counties presently composing the metropolitan area. Lancaster and York did not fare quite so well. Lancaster developed significant textile and tobacco industries and was an early industrial leader expanding especially rapidly betweeen 1840 and 1850 (Klein and Hoogenboom 1980). Similarly, York grew nearly in pace with Lancaster, based on its diversified processing and manufacturing activities. However, both Lancaster and York were already experiencing a slowing of their rates of growth because they were not major river or rail hubs.

In the northeastern region of the state, a rapidly developing settlement system was taking shape. Scranton was founded and grew very quickly in this era to become a regional capital. Fundamental to its success was the Delaware, Lackawanna & Western Railroad and its shipments of anthracite and, later, iron. To the north, Carbondale enjoyed a brief spell as a regional center based on the same ingredients. In their wake a number of other towns, such as East Stroudsburg grew, created by the railroad in 1856. By 1870 Scranton–Wilkes-Barre had not only cracked the metropolitan ranks with 223,000 population, but skyrocketed to the third order (Table 18.1). It is interesting to note that Pottsville also entered the ranks of metropolitan places in 1870 (Fig. 18.3c) but did not diversify its traditional anthracite mining activities to any great degree. As a result, it achieved metropolitan status (albeit at a fifth-order rank) but subsequently dropped from the metropolitan ranks in the next era, never to regain its standing again.

In central and northern Pennsylvania, towns also grew up along the railroads. Altoona, Johnstown, Bellefonte, Lewistown, Tyrone, and others sprang up along the Pennsylvania Railroad, while such places as Renovo, Williamsport, Ridgway, Lock Haven, Warren, and Erie developed on the Philadelphia & Erie Railroad. Altoona was a creation of the railroad company in 1849 as an engineering center and a base for the extra engines needed for the climb up the Allegheny Front and the cross-mountains haul. By 1870, within the short period of 20 years, it had attained fifth-order metropolitan status (Fig. 18.3c). Williamsport, on the Philadelphia & Erie and the Philadelphia & Reading railroads, also established itself as a regional center for the assembly and break-in-bulk of transported goods and for the lumbering industry (Plankenhorn 1957). Williamsport grew rapidly during the last half of the Iron Horse Era, reaching 48,000 population and fifth-order status by 1870 (Table 18.1).

In western Pennsylvania, Pittsburgh continued to emerge as a manufacturing center during this era; its commercial activities, though growing and important, were increasingly overshadowed by coal mining, iron-making, and the glass and oil industries (Lubove 1976). Much of the output from the developing manufacturing industries was shipped westward on the Ohio and Mississippi river systems or eastward by railroad and canal. A contemporary description of Pittsburgh was provided by Parton in 1866 (quoted by Lubove 1976, 8): "It is chiefly at Pittsburgh that the products of the Pennsylvania hills and mountains are converted to wealth and distributed over the world." That the burgeoning young metropolis was becoming a veritable caldron of industry is evident from his further descriptions of it as "Hell with the lid off."

In the case of western Pennsylvania's metropolitan development, the important role of the major inland lake and river transportation routes must also be noted. The application of steam power technology to shipping meant that far larger vessels could ply the deeper inland waterways and effectively move heavily laden boats against the upstream currents in the case of river traffic. Indeed, during the early decades of the Iron Horse Era, the steamboat was a critically important means of transport on the western rivers and the inland lakes. Thus, places with good natural harbors and access to local resources were particularly favored. Erie, with its deep harbor on the Great Lakes, protected by Presque Isle, was destined to grow as a break-in-bulk and processing location. Pittsburgh, at the head of Ohio River navigation, not only had a strategic transportation location with access to the key routes to the Philadelphia and the East, but was also centrally located with respect to raw materials for the emerging materials industries.

It was soon clear that Pittsburgh would become the dominating metropolis of western Pennsylvania. By 1870 it had attained a population of 449,000, placing it firmly among the second order of the hierarchy and by far the largest place in Pennsylvania aside from Philadelphia (Table 18.1 and Fig. 18.3c). Erie, with nearly 66,000 population, attained metropolitan status for the first time within the fifth order of places.

The Steel-Rail Era: 1870–1920

The Steel-Rail Era was ushered in by two related industrial innovations: (1) the Bessemer process, an efficient steel-production process patented in Johnstown, and (2) the concomitant and widespread use of steel rails in railroad trackage (Berger 1985). Accompanying these changes was a continued expansion of the railroad network that linked Pennsylvania into a national system of transportation, and a consolidation into larger, regional companies that created corporate giants.

The changing spatial pattern of the iron and steel industry in the nineteenth century represents a broadly east-to-west movement, with a gradual shift in its location from the Delaware Valley to central Pennsylvania and, in the Steel-Rail Era, to western Pennsylvania (McLaughlin 1938). By way of exception in this period, the industry also migrated to the anthracite region of northeastern Pennsylvania. Thus, a qualification of this east-to-west thesis, and one that might better describe this process, lies in the frontier thesis put forward by Glasgow (1985). This position suggests that the movement of the iron and steel industry has broadly followed the movement of the frontier, and as a descriptive device it is useful for this analysis.

The Steel-Rail Era is perhaps best introduced by a consideration of the fortunes of two Pennsylvania industrial towns: Bellefonte and Scranton. At the beginning of this era, Bellefonte was an iron-making town riding high on the Civil War demand for iron and serviced by a branch of the Pennsylvania Railroad. Scranton was a newly forming iron-making town, succeeding in a shaky fashion on railroad rail contracts and astute entrepreneurship. The innovations of the era heralded new technological requirements: the Bessemer steel-making process technology, large-scale plant and equipment, and a locational orientation toward fuel—that is, coal—sources. Bellefonte could not meet the demands of the new production regime and fell into a protracted decline (Lewis 1972). In contrast, Scranton was blessed with abundant anthracite fuel and entrepreneurial foresight that led to its success, very much a product of its era (Folsom 1981). By 1876, Scranton had a large-scale steel-making plant and was producing steel rails for a booming market. These two experiences were played out in a number of other cases across the state, thereby affecting the pattern of metropolitan development greatly.

In western Pennsylvania, the Scranton experience occurred on an even larger scale. Between 1860 and 1914 some 33 percent of the nation's steel, 40 percent of its coal, and 30 percent of its railroad tonnage originated in this western area of Pennsylvania

(Miller 1981). Pittsburgh was at the heart of this growth and became the center of corporate empires, most notably that of Andrew Carnegie. Furthermore, Pittsburgh typified the "gritty cities" of the period, with its preeminence of industrial capital and a general lack of concern for environmental and human welfare. Considerable industrial and hence urban development also took place in the river valleys of western Pennsylvania. Monessen was a typical industrial boomtown of this period, growing from farmhouses to a town of more than 10,000 in just 17 years (Magda 1985). Johnstown also grew rapidly in this period, the industry there being an early adopter of the new steel-making technology. Like Scranton, an early stimulus was the production of steel rails, and the town's location on the Pennsylvania Railroad was enhanced after 1881 by the addition of the Baltimore & Ohio Railroad (Berger 1985). In northwestern Pennsylvania, the discovery of oil and the production of petroleum products in the latter half of the nineteenth century caused a number of boomtowns to appear (Michener and O'Malley 1984). Oil City, founded at this time, was one of the few towns to survive; Pithole City lasted only three years but at its peak had 15,000 residents.

In the anthracite region of northeastern Pennsylvania, the Scranton experience was repeated often as well. The growth in the demand for anthracite, especially in the earlier years of the era, and the extension of railroads into the region was accompanied by the development of a new settlement hierarchy. In the latter years of the nineteenth century, Pottsville was joined as a significant settlement by Shamokin, Shenandoah, and Hazleton (Karaska 1962). However, most of the population growth in the anthracite region had slowed or ceased before the end of the era.

Thus, the major factors shaping Pennsylvania's metropolitan evolution during the Steel-Rail Era were manufacturing concentrations made possible by the increasing density and carrying capacity of the developing railroad network. Coal and minerals could then effectively be shipped at relatively low costs to large and centralized production sites, preferably near major railroad hubs. Coal also served to create the steam in large plants for electricity generation, contributing to further industrial concentration. Concomitantly, areas with the resources necessary to fuel this industrial machine experienced significant increases in population. This resource orientation was especially salient in the case of places on or near bituminous coal, which rapidly replaced anthracite coal for industrial uses in the late nineteenth and early twentieth centuries.

The concentration of economic activities in the industrial metropolis is perhaps nowhere better demonstrated than in Pennsylvania's two largest metropolises, which continued to evolve as major rail transportation hubs. Philadelphia, already large, grew by 1.6 million persons over this half-century to 2.7 million. Philadelphia became the hub of the developing railroad network as local rail companies consolidated into large regional corporations. Similarly, Pittsburgh grew by 1.5 million persons to 1.95 million. Both retained their second-order rankings along with other burgeoning industrial metropolises such as Cleveland, Chicago, and St. Louis (Fig. 18.3d and Table 18.1).

In the Lehigh Valley, Allentown continued to diversify from its early iron-industrial base into engineering and textiles; Bethlehem, like Scranton, became a major steel-making center based on the new Bessemer technology. Indeed, Bethlehem was the only significant steel-making center of the Lehigh Valley to survive into the twentieth century. The Allentown-Bethlehem metropolitan area more than doubled its population, growing to 409,000 residents by the end of the era. Reading also nearly doubled its population, to retain its fourth-order ranking. Like the Allentown-Bethlehem area, it too functioned as a lesser rail transportation hub, developing a major railroad engineering function and diversifying into other manufacturing endeavors.

In the rest of the old backcountry of southeastern Pennsylvania, the metropolitan areas of Lancaster, York, and Harrisburg all benefited from the concentrating forces of industrialization. Harrisburg even developed a modest steel industry. However, they did not attain nearly as high a growth, either in absolute or relative terms, as metropolises that were the largest centers and transportation hubs going into the Steel-Rail Era. The growth rates of each of these places as manufacturing centers was lagging by the turn of the century. None developed the heavy manufacturing orientation of the large industrial metropolises of Pennsylvania—a fact that undoubtedly has eased their economic transitions throughout the twentieth century. Despite having a population of nearly 300,000 by 1920, Harrisburg slipped back in rank to the fourth order. Lancaster and York, both with less than 180,000 population by the end of the era, retained their fourth-order ranking (Fig. 18.3d and Table 18.1).

In addition to Pittsburgh, the metropolitan areas of western Pennsylvania also experienced significant population growth. Altoona more than quadrupled in population, thanks to the swelling demand for railroad repair and rolling-stock construction facilities. Johnstown also quadrupled in size, with its specialization in the coal-steel complex driving its population growth. Likewise, Erie had nearly tripled its 1870 size by the end of the era, having evolved as a center for metalworking and machine industries. Each of these smaller western Pennsylvania metropolitan areas was a fourth- or fifth-order place in 1920 (Fig. 18.3d). Like Pottsville in the preceding era, Newcastle also attained metropolitan status during this period, only to see it slip away after 1920.

In northeastern Pennsylvania, Scranton–Wilkes-Barre also grew very rapidly during the Steel-Rail Era, especially in the earlier years. The tremendous anthracite boom coupled with steel-making and metalworking industries and a transportation hub of some importance propelled this metropolitan area to 764,000 by 1920, sufficient to achieve third-order status in the national hierarchy of places. Williamsport, on the other hand, lagged in growth, increasing by only about 35,000 residents over the span of a half-century. Several reasons help to account for this, including its reliance on the lumbering industry, the depletion of local resource supplies, and the shifting of major timber harvest areas to the Upper Midwest. And a flood in 1889 swept away much of the area's timber resources that were awaiting sawing, which was a shock to Williamsport's economic base from which it did not recover for many years.

The Auto-Air-Amenity Era: 1920–1960

The most significant innovation heralding this era was undoubtedly the internal combustion engine and its subsequent applications to automobiles, trucks, and air transportation. Once again, the innovations instigated a new transportation network regime, and hence new locational relationships. The developing highway and, to a lesser degree, air travel patterns, in concert with the changing resource requirements, spelled growth (or decline) for urban areas well-disposed (or not) to the new requirements, or having local economies tied to the new or former resources and technologies.

As in previous eras, the pattern of metropolitan settlements changed as a result of the new transportation and industrial innovations. There were significant differences in the relative growth of places, but relatively little change in the membership list of metropolitan places. This is partly a reflection of the inertia created by the previous eras and of the flexibility of the automobile, truck, and aircraft systems to fit the established patterns of metropolitan areas. As Borchert (1967) has indicated, the contemporary pattern of metropolitan areas was largely set into place by 1920.

With the development of the automobile, and its widespread use after 1920, attention returned to the long-neglected road network. During the previous Iron Horse and Steel-Rail eras, roads had generally fallen into a progressive state of disrepair. This situation began to improve in 1903 with the formation of a State Highway Department and with state financial backing in 1905 and federal support in 1915. The Lincoln Highway—later U.S. Route 30—was constructed during this era as one of the earliest transcontinental highways, largely by resurfacing existing roads. In the late 1920s and early 1930s, the state, under Governor Pinchot, took over responsibility for many small rural roads and similarly resurfaced them. Interregional highways were further developed with the construction of the Pennsylvania Turnpike between 1937 and 1940. This road effectively became the first all-weather highway across the Allegheny Mountains, utilizing the abandoned bed of the incomplete South Pennsylvania Railroad. Later, as part of the developing national Interstate Highway System, the Keystone Shortway (Interstate 80) was begun in the waning years of this era and would eventually provide the most direct east-west route across the state—in the process bypassing all of Pennsylvania's largest metropolitan areas.

Before the Interstate Highway Act of 1956, the major highways constructed in Pennsylvania were designed to link the major urban areas (National Interregional Highway Committee 1944). The primary result of this thrust was to create a significant degree of inertia in the metropolitan and smaller urban hierarchy of Pennsylvania. In the later years of this era, the highways, while linking the metropolitan and urban areas, often bypassed or circled them.

In doing so, they helped to create a new locational pattern that spread the influence of each of the state's metropolitan areas across wider geographic areas.

The application of the internal combustion engine to aircraft also had a significant influence on the metropolitan hierarchy—in particular, it tended to cement the existing metropolitan order. Early development of aircraft was stimulated greatly by World War I, and by the early 1920s the Postal Service was transporting mail by air. The development of trimotor aircraft in the mid-1920s provided a greater measure of safety, and by 1930 air passenger travel in the United States was growing rapidly and commercial airlines were carrying mail to subsidize their passenger operations (Miller and Sawers 1968).

The introduction of larger planes took the air transport industry out of the grass field stage, in which almost any town with suitable topography could have a landing strip, and thrust it firmly into the realm of larger metropolitan areas. Only the larger metropolitan areas could generate the volume of passenger traffic necessary to support the commercial air carriers. In addition, only the largest metropolises could support the airport facilities, first developed by private interests and subsequently taken over by public entities. As the costs of equipping and maintaining these facilities mushroomed—along with the amount of land required for ever-longer runways, especially as the jet age dawned in the mid-1950s—the smaller metropolitan centers found themselves at a further disadvantage.

The result of these trends was that the high-order standing of Philadelphia and Pittsburgh in the metropolitan hierarchy of Pennsylvania and the nation was reinforced. But Philadelphia was increasingly overshadowed by New York to the north and Washington to the south in its quest to capture air passenger traffic. Like Baltimore, which also lay between these two dominating air traffic hubs, Philadelphia has never generated the air passenger traffic that might have been expected based on its size. However, Philadelphia did become a major center for airmail shipment, a distinction it carries to this day. Pittsburgh, by contrast, emerged as an early air traffic hub with less competition from any close and large metropolitan areas. Most of Pennsylvania's other metropolitan places had only a commuter-service and feeder role during the Auto-Air-Amenity Era. Thus, the nature of air transport technology and the evolving network accentuated the communications cost and the time advantages of progressively larger metropolitan places.

The tremendous growth in auto and air transportation industries produced some very differential effects on the fortunes of Pennsylvania's metropolitan places. Although the demand for steel and other metals, such as aluminum, used in transportation equipment grew rapidly, the locus of the metal industries shifted to other regions. More of the nation's steel production shifted to the Great Lakes and tidewater locations (Markusen 1985). Pittsburgh and Allentown-Bethlehem remained more competitive than most other areas in the state, but the depressing effects on the more marginal production areas, such as Scranton–Wilkes-Barre, were startling.

The shift to petroleum-based energy sources benefited mainly the largest metropolitan area transportation hubs, where the oil was brought in by rail and water—oil production areas having long since moved to the more productive regions to the south and west. At the most important hubs, Philadelphia and Pittsburgh became the centers for refining and petrochemicals production in the state.

Concomitantly, the shift to motor freight transportation, especially for the lower-bulk higher-value shipments, coupled with the loss of substantial passenger traffic, sent the railroad industry into a downward spiral from which it has never recovered in Pennsylvania. Coal-mining areas fell to a similar fate. The MSAs most severely impacted were those that had remained both relatively specialized in railroading activities and small in size. Altoona and Johnstown were particularly hard hit, the latter's problems compounded by a second devastating flood. Metropolitan areas most specialized in manufacturing also bore the brunt of the Great Depression. The smaller and more diversified metropolitan areas, such as Lancaster and York, fared relatively better.

By 1920 the age of the industrial metropolis was waning, to be replaced by the emerging service-economy metropolis. Manufacturing employment, as a share of national employment, was near its peak—further gains were mitigated by the significant increases in productivity brought about by improved technology. As personal incomes rose, the demand for services rose more than proportionately, as did the demands for greater leisure-related activities and amenities.

While Pennsylvania's "gritty cities" had limited

amenities, the largest ones did represent huge markets for services. The growth of service sector employment produced further demand stimulation that engendered a strong cumulative growth factor in these largest places—much more than sufficient to offset the declines in certain manufacturing and transportation industries. The demand for higher-education services also accompanied the shift to the predominantly white-collar occupations of the service industries, and the growth of the education establishment provided significant positive stimulus in several of Pennsylvania's metropolitan areas, particularly after World War II. Likewise, the demands for a greater public-sector role in the affairs of the state promoted the growth of the Harrisburg-Lebanon-Carlisle MSA and encouraged the further development of ancillary service activities there.

The effects of these changes on Pennsylvania's metropolitan evolution during the Auto-Air-Amenity Era are seen in Fig. 18.3e and Table 18.1. No new places in Pennsylvania became metropolitan in status during this era. Only one, Scranton–Wilkes-Barre, changed ranking, and that was a fall from third to fourth order—a result of its significant absolute population loss of more than 70,000 during this era.

Pittsburgh and particularly Philadelphia were the big gainers in the era of the service economy. Philadelphia gained more than 1.6 million residents, while Pittsburgh, despite its concentration on steel and other manufacturing, gained more than 625,000 in population. The old backcountry metropolitan areas of southeastern Pennsylvania, along with Erie, recorded moderate gains, while the rest of the western metropolitan areas stagnated.

The Computer-Telecommunications Era: 1960–Present

It is tempting to assume that patterns established during the Auto-Air-Amenity Era have simply been perpetuated during the years since 1960. After all, the technologies and trends set into motion during that period are still applicable today. However, the large-scale introduction of computers and telecommunications technology after 1960 not only reinforced some of the trends that typified the previous era, but also helped create some significant changes in the metropolitan hierarchy and growth patterns—all of which affected Pennsylvania.

The past three decades have witnessed a tremendous increase in merger and acquisition activity among America's corporate community. Today, a small proportion of all corporations control a significant majority of the assets, sales, and employment of the nation's businesses. The scale of this phenomenon would simply have been impossible if it were not for the communications revolution and the computer capabilities that now permit management of such large enterprises.

The result of these processes has been an increasing centralization of certain high-order management functions and a decentralization of production activities over the past three decades. The ability to manage branch and subsidiary units at many distant locations—whether manufacturing, trade, or services—has permitted multiunit enterprises to operate nationally or internationally. Inputs can be purchased from and products sold more easily in other parts of the world. Labor productivity, especially in manufacturing, has also been improved dramatically by the introduction of computer-based automation. Industry has also increased its purchases of services as the needs for more specialized ancillary activities has grown, and those services can often be purchased at lower cost than if they were provided in-house (Noyelle and Stanback 1984).

In many instances, this technology has made it possible to redistribute industrial activities to smaller and often satellite metropolitan areas and the smaller towns of nonmetropolitan America. This trend was particularly evident during the 1960s and early 1970s as population "turnaround" became an important trend affecting most developed industrial nations. This trend, coupled with increasing foreign competition from both developed nations and the newly industrializing Third World countries, spelled big declines in industrial employment for most MSAs, especially the largest ones. It also set in motion an era of slow metropolitan growth, which was further slowed by a very low birth rate in the United States by the late 1960s (Phillips and Brunn 1978).

Not only did manufacturing activities continue to decline as a share of national employment, and to decentralize to less-populated areas, but there is also evidence of an increasing decentralization of corporate headquarters functions out of their previous strongholds of the largest northeastern United States metropolitan areas. Their destinations were most often the smaller satellite metropolitan areas and the major metropolitan areas of the sun belt, such as Atlanta, Dallas, Houston, Los Angeles, and San Francisco. The big northeastern metropolises still retain a high proportion of all corporate-headquarters functions, but their share has been steadily eroded over the past 25 years (Borchert 1978). While some highest-order management functions are still best transacted in New York or Chicago, the new communications-computer technologies mean that the major metropolitan areas of the sun belt have to give up very little. In addition, the continuing influence of amenities, especially the climate of the sun belt, put Pennsylvania's MSAs, and other surrounding rust belt states, at a clear disadvantage.

The largest metropolitan areas of Pennsylvania and other northeastern states were generally better able to make the transition to the new service and deindustrialized economy of the post-1960s era than many of the smaller ones that had very specialized industrial economies. The large population bases of the former group still represented large markets for services, and while rapid population increase was a thing of the past, they were still generally able to maintain their population bases (Borchert 1972). The completion of the Interstate Highway System provided high levels of access for all the major metropolitan areas in Pennsylvania. This, coupled with the increased air-travel needs of the service economy, also helped to maintain the position and regional influence of these major metropolitan areas in the state.

The energy crisis in the decade following 1973 also worked against the metropolitan areas of Pennsylvania and other northeastern states. The energy-rich states of the Southwest experienced economic boom conditions, and those who depended on oil experienced sharp production-cost increases, further contributing to the rust-belt to sun-belt shifts. While the demand for coal increased during this period, the temporary rejuvenation of the industry did little to restore the economies of the Pennsylvania MSAs most affected by coal production. With the cost of conversion to coal, the problems of burning high-sulfur coal, the enhanced capital intensity of modern coal production, and its nonmetropolitan production sites, coal could contribute relatively little to metropolitan growth in Pennsylvania.

The changes during this era introduced a period of relatively slow aggregate metropolitan growth in Pennsylvania, much slower than the national average. The Pittsburgh metropolitan area was the first major MSA in the country to lose population in the period 1960–1970 and since. Further losses were sustained in the 1970s and 1980s. The growth of service sector jobs, spurred on by Pittsburgh's role as host to the headquarters of many large corporations, was unable to compensate for the huge losses in manufacturing, more than 80,000 of which occurred in the steel industry alone from 1960 to 1985. As a result, Pittsburgh dropped to a third-order place in the national hierarchy of places (see Fig. 18.3f and Table 18.1).

Philadelphia, traditionally less dependent on manufacturing and more reliant on services, has been able to grow over the past 25 years, but at a rate far lower than that of the nation. Its 1990 population had increased to 4.86 million, and it was able to retain its second-order ranking in a solid fashion.

Most of the old backcountry metropolitan places of southeastern Pennsylvania fared relatively well. York and Lancaster (the only Pennsylvania MSAs to exceed the national growth rate from 1960 to 1990) grew most rapidly, followed by Harrisburg-Lebanon-Carlisle and Allentown-Bethlehem, all of which retained their fourth-order status. Harrisburg benefited greatly from its role as state capital in this era because of the substantial growth in state government until the late 1970s, following the typical trend in states across the nation. The Reading metropolitan area grew most slowly among this set, slowed by its antiquated industrial structure of textiles, railroad transportation, and metalworking. Indeed, it slipped to the fifth order by 1990. State College also joined the ranks of metropolitan places for the first time in 1980 as a fifth-order place, its growth buoyed by the rapid growth of the educational establishment and related trade and service industries.

In northeastern Pennsylvania, Scranton–Wilkes-Barre grew slowly, a victim of its aging industrial structure, although its previous growth (predominantly during the Steel-Rail Era) was sufficient to maintain its position in the fourth order. Williamsport also grew slowly, and remained among the fifth-order places.

Of western Pennsylvania's smaller metropolitan places, Johnstown, Altoona, and Erie all experienced population losses between 1960 and 1990. The problems in the steel, transportation, and metalworking industries noted earlier in the discussion of this period all have plagued development in these areas.

The losses in Johnstown were particularly pronounced, and it fell to the fifth-order rank by 1990. Although Erie experienced a growth of about 25,000 persons over the 1960–1990 period, its increase was not sufficient to maintain its previous place as a fourth-order metropolis, and it joined the fifth order along with Johnstown. Sharon, which like State College entered the metropolitan ranks by dint of change in statistical definition, was basically stagnant during the entire 1960–1990 period.

The Future Course of Metropolitan Evolution

Recent population changes and prospects for metropolitan development in Pennsylvania suggest a future scenario of modest eastern Pennsylvania growth and western Pennsylvania stagnation—at least throughout the remainder of the twentieth century. Pennsylvania's relatively old population structure will further dampen its metropolitan area growth, which is already suffering from a lack of net in-migration.

In the eastern portion of the state, the smaller and more diversified metropolitan areas (including State College), especially those of the southeastern quadrant, will continue to grow most rapidly. They benefit from economies that have made the great transformation to economic health, and they will continue to benefit as business and industry decentralize from the major metropolitan centers, such as New York and Philadelphia. Higher-technology-based manufacturing industries and segments of the service industries that are growing most rapidly will fuel these advances, although at rates less than the national average. These characteristics also typify the Philadelphia situation, although growth in percentage terms will obviously be much slower than in the rest of southeastern Pennsylvania. In the northeastern part of the state, Scranton–Wilkes-Barre and Williamsport appear to have sustained their greatest economic losses, and modest future growth—more modest than in southeastern Pennsylvania—should be expected.

In western Pennsylvania all but Erie experienced losses during the 1980s. Pittsburgh continues to lose population, although the most severe declines have probably taken place already. Its industrial structure today looks very different from what it did just 20 years ago—it is much more heavily represented by high-order service activities. The back-to-back national economic recessions during the early 1980s, and a sluggish economy in recent years, have also had a devastating effect on many of the smaller, more specialized industrial economies, such as Johnstown, Altoona, and Sharon, where protracted periods of negative or very slow growth can be expected for some time to come.

This study of Pennsylvania's metropolitan evolution has taken us through three centuries of Pennsylvanians carving out an urban experience—metropolitan areas waxing and waning, influenced by forces of changing technologies and national economic conditions beyond their abilities to control. The result is a most interesting metropolitan landscape, with its inertia and vestiges of the past, its reflections of the present, and its hopes and prospects for the future.

Bibliography

Baldwin, L. D. 1937. *Pittsburgh: The Story of a City, 1750–1865.* Pittsburgh: University of Pittsburgh Press.

Berger, K., ed. 1985. *Johnstown: The Story of a Unique Valley.* Second edition. Johnstown, Pa.: Johnstown Flood Museum.

Billinger, R. D. 1954. *Pennsylvania's Coal Industry.* Pennsylvania History Studies No. 6. Gettysburg: Pennsylvania Historical Association.

Bining, A. C. 1973. *Pennsylvania Iron Manufacture in the Eighteenth Century.* Second edition. Harrisburg: Pennsylvania Historical Commission.

Borchert, J. R. 1967. "American Metropolitan Evolution." *Geographical Review* 57:301–332.

———. 1972. "America's Changing Metropolitan Regions." *Annals of the Association of American Geographers* 62:352–373.

———. 1978. "Major Control Points in American Economic Geography." *Annals of the Association of American Geographers* 68:214–232.

Buck, S. J., and E. M. Buck. 1939. *The Planting of Civilization in Western Pennsylvania.* Pittsburgh: University of Pittsburgh Press.

Clark, D., ed. 1975. *Philadelphia, 1776–2076: A Three Hundred Year View.* Port Washington, N.Y.: Kennikat Press.

Cochran, R. C. 1978. *Pennsylvania: A Bicentennial History.* New York: Norton.

Currier, W. R. 1979. "The Ports of Philadelphia." In *The Philadelphia Region: Selected Essays and Field Trips,* ed. R. Cybriwsky. Washington, D.C.: Association of American Geographers.

Florin, J. 1966. "The Advance of Frontier Settlement in

Pennsylvania, 1683–1850: A Geographic Interpretation." M.S. thesis, Department of Geography, The Pennsylvania State University.

Folsom, B. W. 1981. *Urban Capitalists: Entrepreneurs and City Growth in Pennsylvania's Lackawanna and Lehigh Regions, 1800–1920*. Baltimore: Johns Hopkins University Press.

Gallagher, J. P. 1968. *A Century of History: The Diocese of Scranton, 1868–1968*. Scranton, Pa.: Diocese of Scranton.

Glasgow, J. 1985. "Innovation on the Frontier of the American Manufacturing Belt." *Pennsylvania History* 52:1–21.

Karaska, G. J. 1962. "Patterns of Settlement in the Southern and Middle Anthracite Region of Pennsylvania." Ph.D. dissertation, Department of Geography, The Pennsylvania State University.

Klein, P. S., and A. Hoogenboom. 1980. *A History of Pennsylvania*. Second edition. University Park: The Pennsylvania State University Press.

Landis, C. I. 1918. "History of the Philadelphia and Lancaster Turnpike." *Pennsylvania Magazine of History and Biography* 42:1–28, 127–140.

Lemon, J. T. 1967. "Urbanization and the Development of Eighteenth-Century Southeastern Pennsylvania and Adjacent Delaware." *William and Mary Quarterly,* 3d ser., 24:501–542.

———. 1972. *The Best Poorman's Country*. Baltimore: Johns Hopkins University Press.

Lewis, P. F. 1972. "Small Town in Pennsylvania." *Annals of the Association of American Geographers* 62:323–351.

Livingood, J. W. 1947. *The Philadelphia–Baltimore Trade Rivalry, 1780–1860*. Harrisburg: Pennsylvania Historical and Museum Commission.

Lubove, R. 1976. *Pittsburgh*. New York: New Viewpoints.

Magda, M. S. 1985. "Life in an Industrial Boomtown: Monesson, 1898–1923." *Pennsylvania Heritage* 10 (Winter): 14–19.

Markusen, A. E. 1985. *Profit Cycles, Oligopoly, and Regional Development*. Cambridge: MIT Press.

Mason, H. 1979. "Early Industrial Geography of the Delaware Valley." In *The Pennsylvania Region: Selected Essays and Field Trips*, ed. R. Cybriwsky. Washington, D.C.: Association of American Geographers.

McLaughlin, G. E. 1938. *Growth of American Manufacturing Areas*. Pittsburgh: University of Pittsburgh, Bureau of Business Research.

Meyer, B. H. 1948. *History of Transportation in the United States Before 1860*. Washington, D.C.: Carnegie Institute.

Michener, C. K., and M. J. O'Malley. 1984. "The Last Frontier: Venango County." *Pennsylvania Heritage* 10 (Spring): 32.

Miller, E. W. 1981. "Pittsburgh: Patterns of Evolution." *Pennsylvania Geographer* 14 (October): 6–20.

Miller, R., and D. Sawers. 1968. *The Technical Development of Modern Aviation*. New York: Praeger.

Murphy, R. E., and M. Murphy. 1937. *Pennsylvania: A Regional Geography*. Harrisburg, Pa.: Telegraph Press.

National Interregional Highway Committee. 1944. *Interregional Highways*. Report of the National Interregional Highway Committee. Washington, D.C.: Government Printing Office.

Nelson, R. 1969. "Backcountry Pennsylvania, 1704–1774: The Ideals of William Penn in Practice." Ph.D. dissertation, University of Wisconsin.

Noyelle, T., and T. Stanback. 1984. *The Economic Transformation of American Cities*. Totowa, N.J.: Rowman & Allanheld.

Patton, S. G. 1983. "Comparative Advantage and Urban Industrialization: Reading, Allentown, and Lancaster in the Nineteenth Century." *Pennsylvania History* 50:148–169.

Phillips, P. D., and S. Brunn. 1978. "Slow Growth: A New Epoch in American Metropolitan Evolution." *Geographical Review* 68:274–292.

Plankenhorn, W. 1957. "A Geographic Study of the Growth of Williamsport." Ph.D. dissertation, Department of Geography, The Pennsylvania State University.

Procter, M., and B. Matuszeski. 1978. *Gritty Cities*. Philadelphia: Temple University Press.

Roberts, J. P. 1980. "Railroads and the Downtown: Philadelphia, 1830–1900." In *The Divided Metropolis: Social and Spatial Dimensions of Philadelphia, 1800–1975,* ed. W. Cutler and H. Gillette. Westport, Conn.: Greenwood.

Rosenberger, H. T. 1975. *The Philadelphia and Erie Railroad: Its Place in American Economic History*. Potomac, Md.: Fox Hills Press.

Rubin, J. 1961. "An Imitative Public Improvement: The Pennsylvania Mainline." In *Canals and American Economic Development,* ed. C. Goodrich. New York: Columbia University Press.

Schusler, W. K. 1960. "The Railroad Comes to Pittsburgh." *Western Pennsylvania Historical Magazine* 43: 203–238.

Snyder, M. P. 1972. *City of Independence: Views of Philadelphia Before 1800*. New York: Praeger.

Swetnam, G. 1964. *Pennsylvania Transportation*. Pennsylvania History Studies No. 7. Gettysburg: Pennsylvania Historical Association.

Warner, S. B., Jr. 1968. *The Private City: Philadelphia in Three Periods of Its Growth*. Philadelphia: University of Pennsylvania Press.

19

THE INTERNAL SPATIAL STRUCTURE OF PENNSYLVANIA'S METROPOLITAN AREAS

Rodney A. Erickson

Introduction: Finding Order in the Mosaic

Pennsylvania's metropolitan areas are fascinating places, rich in variety and visual appeal, a product of their long and sometimes tumultuous histories. To visitors and residents alike, these places often appear to have very different internal geographic patterns. The many land uses, building types, road layouts, densities of development, and residential districts present a mosaic that often seems to defy order, each the product of unique circumstances. Yet when the geographer probes deeper into the internal spatial structure of cities, be they in Pennsylvania or elsewhere in North America, there are many similarities along with the differences. One can often trace a particular way of city-building, a certain building style or architecture, or a type of transportation structure, to common origins, whereby similar processes result in a significant uniformity of spatial pattern.

A metropolitan area can be viewed as an evolving system (Bourne 1982). Individuals, activities, and the places of the natural and built environment are all elements in the system. The energy to keep the system moving comes from the continual investment and reinvestment of capital in plants, stores, housing, and the social and economic overhead comprised of schools, streets and roads, government facilities, and the like. The skills and entrepreneurial talents of laborers and managers are also critically important to the continuing development of metropolitan places. Transportation and communications serve as the vital conduits that link the various elements together and make it possible for the metropolis to function. As in any system, feedback is present; there are few, if any, instances in which changes in one dimension of the urban process do not create repercussions in the spatial patterns of many elements.

The purpose of this chapter is to examine the internal spatial structure of Pennsylvania's metropolitan places. The historical dimension has considerable significance for the analysis, given that many additions to the built environment of cities last, if not indefinitely, for many decades or centuries. We have also seen in the preceding chapter that Pennsylvania's metropolitan areas have experienced very different growth trajectories during each of the developmental eras. Areas with large additions to the built environment in any particular era will carry with them relatively more of the "baggage" from those eras. Finally, transportation technology—the linkage mechanism in the internal metropolitan system—has played a most crucial role in shaping the spatial form and functioning of our metropolitan places.

Stages in the Evolution of Metropolitan Form

The development of metropolitan form in North America has many obvious similarities with that of greater cities around the world. Yet there are also certain deeply ingrained factors that have given some special characteristics to the metropolitan areas of Pennsylvania and other places across our nation. Many of these differences relate to the particular circumstances of culture, natural resources, and development timing, factors that are essential to an understanding of internal spatial structure.

The cultural foundations of American metropolitan development are built on an attitude of a rural ideal and an accompanying antiurban bias. This attitude is rooted in a Jeffersonian view of democracy based on the virtues of a rural society with a strong and independent freeholder agricultural system. As the industrial revolution unfolded in the midst of the nineteenth century, Americans brought their rural ideals with them to the emerging urban-industrial centers and sought to make these centers into noncities (Muller 1981). Given the unsanitary and generally unhealthful conditions of early industrial cities in Pennsylvania and elsewhere across the manufacturing belt, the attempts of the more affluent members of society to distance themselves from the most urban parts of metropolitan places were not unexpected. One result has been a century and a half of developing and adapting technologies to permit living in more rural-like environments while simultaneously taking one's income from urban-industrial pursuits. Another result has been repeated attempts to impose a more rural environment onto the city by setting aside and improving land for parks, tree planting, parkway and boulevard construction, and the preservation of open space.

Pennsylvania's metropolitan places, like their counterparts elsewhere in the nation, have long been influenced by the American perception of limitless land resources. The continuing spread to the rural fringes and development at very low densities, often at the expense of prime agricultural land, has been a readily apparent feature of our metropolitan landscapes (Lewis 1983). This pattern stands in sharp contrast to most other urban-industrial societies outside North America where limited arable land and a strong mandate for land as a societal resource have led to more stringent public land use controls, higher urban densities, and a more orderly pattern of development. The American reverence for the rights of private property holders to use land as they alone see fit has added further complexity to the mosaic of internal spatial patterns (Warner 1968).

The timing of metropolitan developments has also been a critical factor in the look of the city and suburbs. Pennsylvanians, like other colonists, brought with them ideas of city-building largely drawn from whence they came: predominantly northwestern Europe in the age of the mercantile city. This was a time in which land was increasingly viewed as a marketable commodity to be sold in lots and blocks. In a setting where there was no urban built environment already in place, there were no major constraints—save the local topography—to developing cities afresh and to prevailing ideas of the time as to how those cities should look and function.

Pennsylvania's urban areas, like any that have been around for some time, have undergone significant building booms and busts. It is apparent from the preceding chapter, on the state's metropolitan population, that their growth was anything but smooth or continuous. As population was added to these evolving metropolitan places, additions to the built landscape increased, but surely not apace. The construction industry in national aggregate has always been characterized by both short-term and long-wave fluctuations, and so also have those cycles of individual metropolitan areas. There is an "accelerator" phenomenon that, once set into motion, produces a strong cumulative effect as earnings from the construction sector are plowed back into the metropolitan economy and speculative building increases. While there may be several smaller short-term cycles within a longer building boom, the latter have always been followed by some degree of bust. The bust is typically softened in areas that continue to experience healthy population growth, while few additions, mostly of a replacement nature, are made to the built environment in places where population growth has stagnated or decline has set in.

The importance of building booms is not only that they occur in different intensities at various times but also that they occur at times in which certain transportation technologies are most prevalent. Adams (1970) has developed a model that relates the evolution of metropolitan form in a framework that incorporates both building cycles and transportation technologies. While the model is most relevant for residential development patterns, there is widespread applicability to the evolution of metropolitan form for many types of land uses.

Each growth stage in the model is dominated by a particular transportation technology whose network strongly influences the spatial pattern of urban expansion and metropolitan organization (Fig. 19.1). Four periods adapted from the Adams model are identified: (1) the Pedestrian-Omnibus Era (up to the 1860s); (2) the Railway Era (1860s to World War I); (3) the Early Auto Era (1920s to 1945); and (4) the Freeway Era (post–World War II).

While the eras relevant to internal spatial structure do not correspond precisely to those of Borchert (1967) governing the overall population growth of the system of metropolitan places, there is a strong overlap with respect to transportation technologies. The Railway Era here corresponds to Borchert's late Iron Horse and the Steel-Rail Era, the period from about 1880 onward characterized by the application of electrification to many tasks. Adams's two periods since 1920 correspond in time to Borchert's single Auto-Air-Amenity Era that evolved in response to the technology of the internal combustion engine.

The Pedestrian-Omnibus Era is dominated by a "walking" mode of transportation throughout the period considered. During the evolution of metropolitan form in Pennsylvania and other American cities, the journey-to-work has been effectively limited, for the most part, by about a 45-minute time constraint. The limitations of time therefore imposed distance constraints such that most people

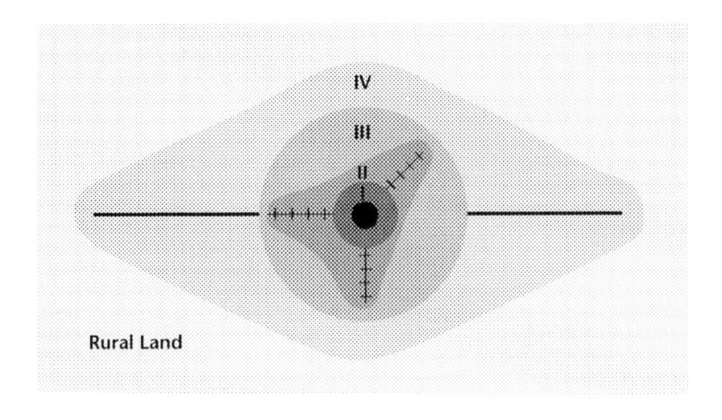

Fig. 19.1. Intra-urban transportation eras and metropolitan growth patterns: (I) Walking-Horsecar Era, (II) Electric Streetcar Era, (III) Recreational Auto Era, (IV) Freeway Era.

had to live within two or at most three miles of the inner city, where jobs were concentrated. The result was a relatively compact city, one of considerable heterogeneity of land uses and even social structures. Only the most affluent could afford private horse-and-carriage transportation; the introduction of the omnibuses late in this era had some modest effect in extending the range of feasible commuting, but the costs remained well beyond the means of most urban-dwellers.

The introduction of the horsecar in many of the larger cities of the Northeast after mid-century represented a modest improvement in internal transportation, but its average speed of movement was only about half again as fast as walking. More important, the rapid introduction of railroad technology by the 1860s promoted the development of major railway use for passenger carriage. The inter-urban railroad permitted far more rapid travel, and for the first time commuter movements on a significant scale were possible at relatively long distances from the city center. Accordingly, sizable and affluent commuter residential communities grew up around stations along the route of the railroads. The introduction of cable cars expanded the density of the networks, but it was the electric streetcar made possible by the centralized generation and distribution of power that had an even greater impact on urban spatial patterns. At speeds that doubled those of the typical horsecar, the geographic area over which the industrial metropolis could effectively be organized was increased substantially along the routes of the streetcar lines.

The result of the inter-urban railroad and the

streetcar was a star-shaped development pattern, the points of which reflected these lines poking progressively farther out into areas often beyond the corporate boundaries of the central city into the suburban fringe. This star-shaped configuration was also reinforced by the tendency of manufacturing industries in search of larger tracts of cheaper land to spread out along the railroad at sidings farther and farther from the burgeoning city center. Meanwhile, the core of the city was developing (and redeveloping) as a very high-density and dominating commercial focus with many types of businesses crowded into the loop to gain maximum urban accessibility.

The large-scale introduction of the private automobile after about 1920 had the effect of promoting development in many of the interstitial areas between the spokes of the rail lines. Road-building in the suburbs proceeded slowly in the early years and accelerated as the federal and state governments became heavily involved in construction and maintenance. For much of this period, however, automobile use remained largely "recreational" in character, and although the growth of passenger traffic on streetcar and rail lines slowed, the auto by no means replaced this latter movement for the journey-to-work until after World War II. Trade and service activities were initially slow to suburbanize until the market was well established. The application of the internal combustion engine to truck freight carriage also provided greater overall accessibility for manufacturers and other businesses, although most preferred the security of proximity to rail sidings. This era also witnessed the final encircling of most larger corporate cities by a fragmented set of suburban local governments.

Following World War II, the freeway and improved highway building, coupled with automotive technological advances and safety improvements, provided a new set of spokes along which development could be organized. Most freeway segments were built along inter-urban stretches of existing federal or state highways, most being built on the cheaper land of the suburbs where huge additions to the housing stock were under construction. The result was a major decentralization of both population and economic activities. Typical speeds of postwar automobile travel were two to four times faster than the streetcar and significantly faster than those of the Early Auto Era. Of course, a doubling of average speed represented a quadrupling of geographic area that could be effectively brought into the metropolitan realm (Ullman 1962). The suburban planned industrial park and shopping mall were also very much a part of this decentralization process.

By 1960 the massive construction program of the Interstate Highway System was well under way in most larger metropolitan areas, including those of Pennsylvania. Under the guise of redevelopment, some freeway links reached into and through the central business districts. Others skirted the corporate city almost entirely, instead bypassing major activity concentrations of the metropolitan areas; others encircled part or all of the metropolis as a beltway, typically cut by major inter-urban arterials or freeway links. Although areas have continued to develop farther out along the inter-urban freeway corridors, average speeds of movement have increased relatively little since the early 1960s, and a large share of development has occurred along the beltways and bypasses, where a high level of accessibility is assured. The result has been continued selective decentralization along with infilling of areas previously passed over in the development process. The energy crisis of the 1970s also served to discourage some far-distant development on the suburban fringe and encouraged development on sites that were already served by transportation and public utilities and in closer proximity to more densely settled labor markets.

Given this general framework of technology and timing in a spatial context, it is now appropriate to explore in greater detail the shaping of internal patterns in Pennsylvania's metropolitan areas. These five eras presented above are still visible to varying degrees, the most recent ones obviously having obliterated many traces of former eras, especially where rapid growth and intense pressures for development and redevelopment have occurred.

The Pedestrian-Omnibus Era: To 1860

Pennsylvania's Pedestrian Era cities were very compact settlements. While the areas that constitute the present-day boundaries of the state's metropolitan places cover a county or more in size, the corporate cities that formed the cores of these early settlements were very small. The (corporate) city of Philadelphia, for example, had less than 25,000 residents in 1776, covering an area of two miles in length and seven blocks wide (Warner 1968, 5; Alexander 1975, 9). Even by the middle of the nineteenth century the city remained relatively compact, covering only about six square miles. Other emerging Pennsylvania metropolises were much smaller.

The pedestrian city produced a relatively high density of population clustered around and behind the waterfront, the major crossroads, or the water-power sites, where the bulk of the economic and transportation activities were concentrated. These sites were also frequently the points at which ferry traffic crossed the larger, unbridged rivers. Early settlements typically consisted of a mixture of industrial, commercial, and residential land uses on adjacent parcels, although manufacturing activities were quite limited, and those that existed were typically oriented to trade and shipping pursuits (Swauger 1978). Early portraits of Philadelphia, Pittsburgh, and Lancaster reveal small houses next to large ones, banks next to liveries, and the usual residential quarters situated above the shops and cottage industries of the first-floor premises. In 1850 the average distance traveled to work in Philadelphia was a mere one-half mile (Hershberg et al. 1981, 138).

According to Warner (1968, 80), "social and economic heterogeneity was the hallmark of the age." Such traditional spatial arrangements of work and housing created many miles of new streets and alleys of resident workers and shopkeepers. To the extent that limited segregation existed in Pennsylvania cities, the patterns were generally the opposite of those we find today. Early in the pedestrian city period, areas of the core were more likely to be inhabited by professional and business persons living close to their places of work, while the periphery was more likely to be inhabited by blue-collar workers and laborers (Blumin 1973; Swauger 1978).

Because most Pennsylvania cities were located on waterways, the streets and roads were typically laid out parallel and perpendicular to the waterfront (or the crossroads, in the case of landlocked settlements), the waterfront being more important than strict adherence to cardinal points. Philadelphia, Pittsburgh, Allentown, Harrisburg, Erie, and Johnstown are all examples of this orientation (Fig. 19.2). The grid pattern also became an important feature of the layout of the pedestrian city. William Penn originally platted Philadelphia's rigid grid layout and square blocks in 1682. As part of a speculator's colony, the city's land was laid out with all the rationalization of a merchant apportioning his land for market. This rigid grid pattern stood in sharp contrast to the organic, beachfront pattern of New England and the baroque geometry of many of the early cities of the South built on a model of rural aristocracy.

The grid pattern proved to be not only an efficient pattern for moving goods and people through the limited spatial expanse of the city, but also a boon to the orderly sale and transfer of legal rights to the property and an ideal system for the orderly expansion of the urban area (Taylor 1966). Indeed, a considerable amount of land was sold to buyers both abroad and in other colonies even before their arrival in Pennsylvania, based largely on the maps and property descriptions used to market the land outside Pennsylvania.

The grid-pattern idea of city-building was carried west to Lancaster and the other backcountry county seats, centers that were also speculators' towns. City center squares were often included at the major crossing of the grid roads; some—such as Wilkes-Barre, platted in 1804—centered on a diamond surrounded by a grid, a slight but common variation (Harvey 1909, 1755). But even in cities that were not traditional speculators' towns the grid became the firmly established pattern of physical layout. Many early Pennsylvania cities expanded out of their original grid plan almost immediately, especially along the rivers, but they remained grid-like nonetheless. Extensions to the original plats were typically gridded during the nineteenth century, even though the block dimensions and directional orientation often changed and old and new streets did not always intersect smoothly. These ideas of the grid pattern city were subsequently carried beyond Pennsylvania to areas farther west in an almost universal fashion; in that sense, Pennsylvania was a model for the nation (Warner 1968).

Within the structure of the grid system, the larger blocks lent themselves well to buildings that fronted on the streets for maximum accessibility. Interior spaces served as limited gardens or yards and contained the immediate common spaces as might be available. Alleyways and interior streets were cut into many of these larger blocks as the areas developed, frequently with secondary or even primary frontage on the small interior alleys. Elfreth's Alley in contemporary downtown Philadelphia pro-

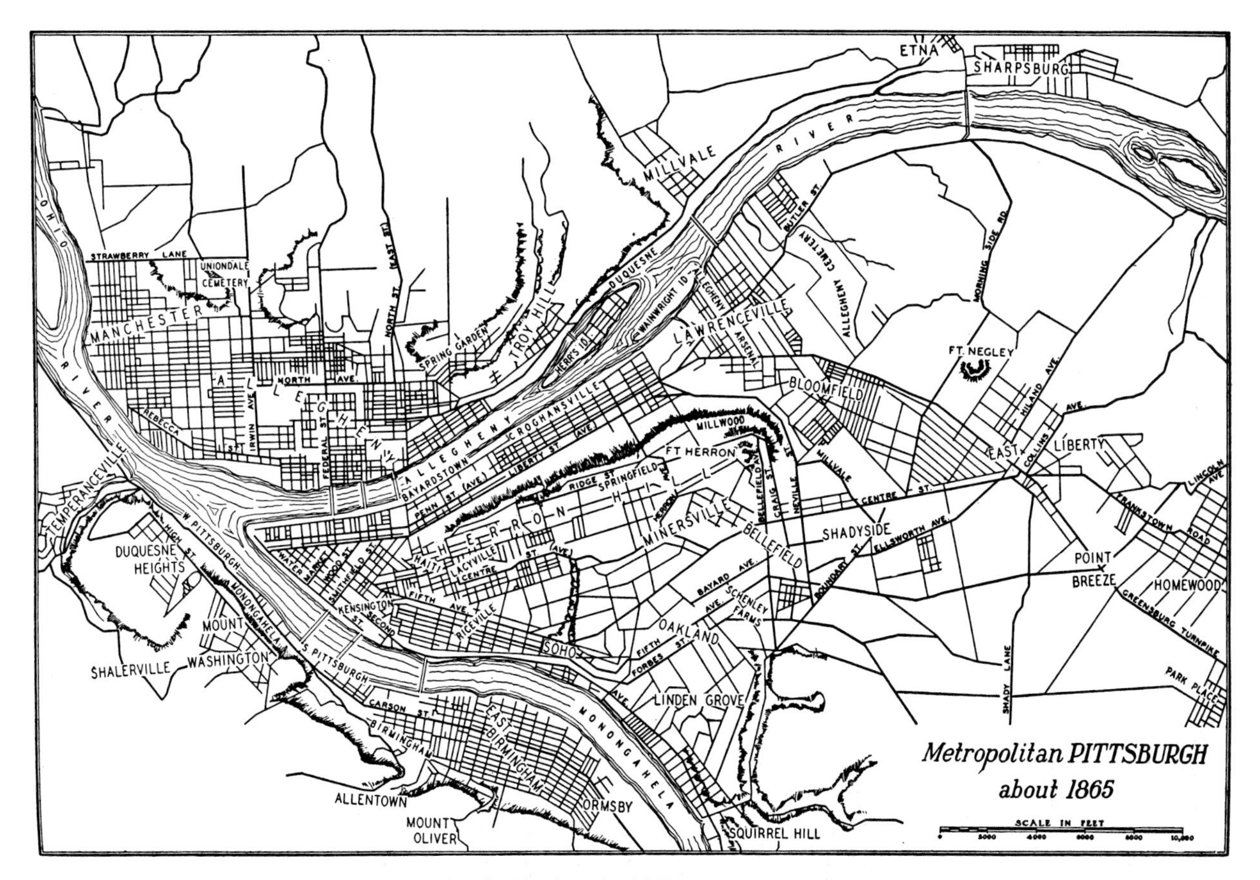

Fig. 19.2. Street-to-river orientation of early Pittsburgh, 1865.

vides a good indication of the crowding typical of the pedestrian era city. A few of these areas remain today much as they did in the eighteenth and early nineteenth centuries in parts of Pennsylvania's oldest cities—areas such as Society Hill in Philadelphia or the oldest parts of Lancaster or the other back-country county seats.

Access to customer traffic at the street front was obviously important for shopkeepers and artisans alike, and common-wall-building characterized the relatively high-density and compact pedestrian city. Fire was a constant threat, and prohibitions against building with materials other than brick were instituted early in these common-wall areas of cities. The continuous-front rowhouse developments came to characterize Philadelphia, Lancaster, and other early Pennsylvania cities, and this style of building can be seen in its fullest expression in the southeastern quadrant of the state, with lesser amounts in other areas (Fig. 19.3). The style remained particularly popular throughout the nineteenth century, and even to this day, to a greater degree in Pennsylvania cities than elsewhere.

The building technology of the pedestrian city also placed severe limits on the heights of buildings. Lacking a strong internal superstructure, buildings of more than five or six stories were rare, and three or four stories were the norm. Such heights as the latter could also be accommodated more easily without elevator facilities, the technology for which was as yet not well developed.

By the 1840s in Philadelphia and Pittsburgh, the omnibus became the first regularized system of intra-urban transportation. The omnibuses followed the grid layouts within the densely settled parts of the burgeoning Pennsylvania cities (Fig. 19.4), some of which doubled in population size during the course of a decade, and to some of the freestanding small towns that comprised the fringe areas of the metropolis. The omnibus was a relatively costly and quite inefficient and slow transportation system, used only by a small number of the wealthier

Fig. 19.3. Row-housing built in early Lancaster, still in use.

citizens. While the more affluent had tended to cluster increasingly in higher-income and more prestigious districts of the core city during the earlier years of this era, the omnibuses helped to accelerate the outmovement of the wealthier from inner-city areas (Swauger 1978; Tarr 1978).

The metropolitan areas of Pennsylvania in the Pedestrian-Omnibus Era were probably most accurately portrayed as a set of small, compact settlements dominated by a focal city, such as Philadelphia, Pittsburgh, Harrisburg, or Reading. It was, in that sense, a multinucleated metropolitan region, but with relatively limited interaction between these settlements, each of the surrounding ones carving out its own specialty, such as agricultural or other processing, and often built on local waterpower sources. Around Pittsburgh, the mill towns began to appear along the three rivers. Manayunk, Frankford, and Germantown were thriving smaller processing centers, and Darby and Fairmount were early upper-class suburbs in the metropolitan shadow of Philadelphia, which consolidated and captured many of these settlements within the city in 1854 (Murphy and Murphy 1937; Taylor 1966; Gillette 1980). This experience was typical of all the evolving metropolitan areas of Pennsylvania.

Thus, the Pedestrian-Omnibus Era left some significant marks on the internal spatial structure of Pennsylvania's metropolitan places through such contributions as the grid pattern, town squares, and rowhouses and their alleyways and courtyards. But much of this era's heritage in the built environment was lost as redevelopment of subsequent eras occurred. Many of the oldest areas of settlement, such as early waterfront districts and what is now the central business district, were obliterated, as subsequent means of transport and industry and commerce cleared these parts of the city for redevelopment. Those that remain, however, represent a rich heritage and a constant reminder of the important role they have played in structuring urban space in the Pennsylvania metropolis.

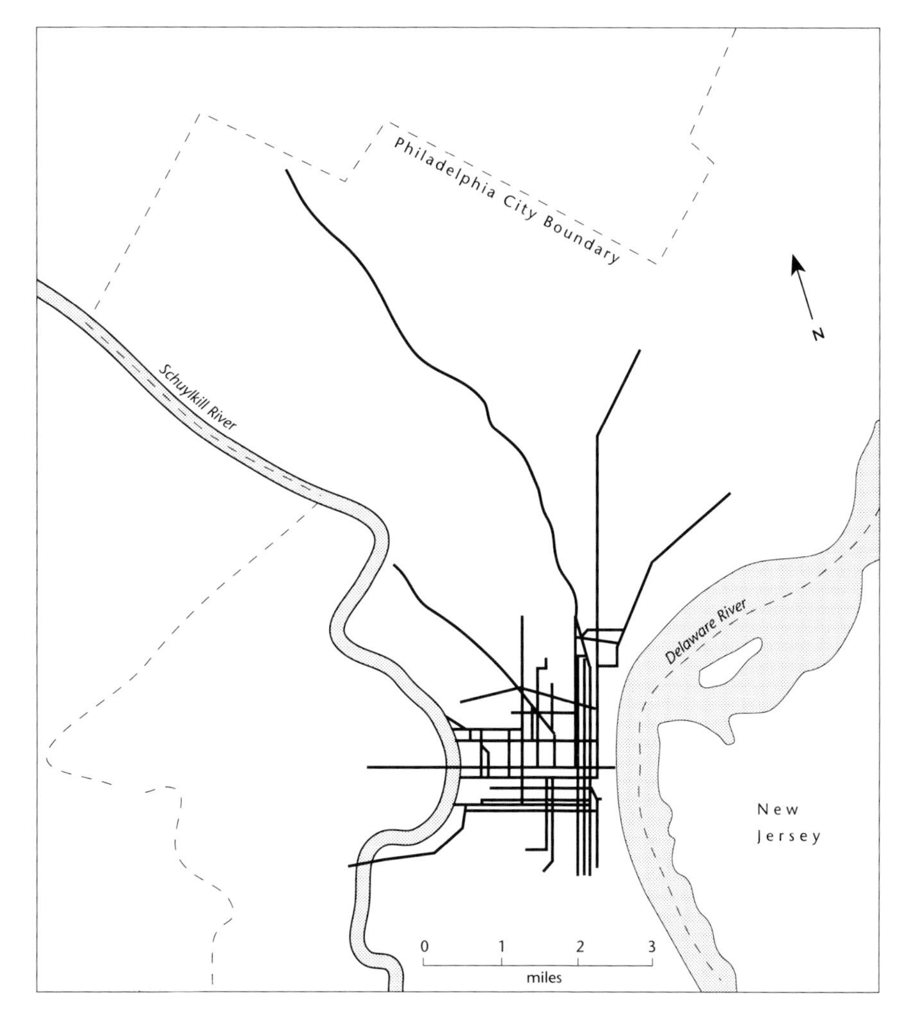

Fig. 19.4. Philadelphia omnibus lines reach out from the core, 1854.

The Railway Era: 1860–1920

From about 1860 the metropolitan areas of Pennsylvania expanded spatially at a far more rapid pace than before. This was partly a result of a population increase—particularly with large numbers of immigrants arriving in the late nineteenth and early twentieth centuries—but it was also a function of the development of better intra-urban transportation networks. For the first time, reasonably efficient systems of intra-urban transport appeared, and these facilitated the expansion of the urban area by enabling people and businesses to locate in a wider range of potential locations. Decentralization was thus not only possible but also urged by social reformers as a way of averting a possible revolution in the dirty and crowded industrializing cities of Pennsylvania (Sammartino-Marsh 1977).

The previously small cities with their peripheral settlements at the fringes underwent a metamorphosis to become the urban-industrial, core-dominated metropolises of this era. Specifically, center city areas became the commercial hearts of cities, and industry chose to relocate—or was pushed by market forces—to more peripheral parts of the urban region. The new transportation systems affected segregation at a number of levels, separating businesses and residential activities from each other, and within these shifts segregating different businesses and different classes of residences. Thus, specialized and relatively homogeneous areas developed—particularly in the previous undeveloped suburbs—

and as a result the average journey-to-work increased for all classes of workers. Furthermore, while facilitating urban expansion and enabling it to be variegated in form, the transportation systems also channeled the direction and often controlled even the rate of this growth.

The transportation system referred to thus far was in effect a series of overlapping competing and complementary transportation innovations. The first significant innovation in Pennsylvania was the railroad, and this usually entailed the application of a shortline, freight, or regional inter-urban railroad line for local passenger traffic. Subsequently, street railroads, or streetcars, developed. First horsedrawn, then electric, they extended the urban area still farther and also caused much infilling between stations. Streetcars were the major innovation of this period, but they were complemented by other, less-common systems including cable cars, inclines, and horse-drawn and electric trolleys. Until the early twentieth century, intra-urban transportation was characterized by public systems that increasingly replaced the earlier, private buggies, but this era was also a period in which intra-urban transportation became commonplace and an integral part of the urban scene.

The effects of this population growth, coupled with earlier railroad-based improvements in intra-urban transportation, can be seen in the expansion of the built-up area of Philadelphia, which by 1880 had grown to 129 square miles (Sammartino-Marsh 1977, 103). Horsecars appeared in the larger metropolitan areas of Philadelphia and Pittsburgh in the mid-1850s. Other areas in the eastern half of Pennsylvania, such as Allentown, Reading, Scranton–Wilkes-Barre, and Williamsport, all had horsecar service by the 1860s or early 1870s, while the western Pennsylvania metropolitan areas such as Altoona and Johnstown had service by the early 1880s.

Wedges of development began to appear in Philadelphia, Pittsburgh, and other metropolitan areas as the various forms of street railways pulled commerical land uses and higher-density housing to locations along their routes, and especially near the streetcar stops. There was a particularly large growth in population in those areas within one hour's commuting distance from the city center, and considerable residential construction within a 10-minute walk from the line. In Pittsburgh, for example, the horsecars hauled 4 million passengers a year by 1863, and 8 million by 1871 (Tarr 1978, 7).

The horsecar developments actually began as a complement to the railroads. These earliest railroads often did not enter the most densely built-up areas of the city. In Philadelphia, for example, the railroad terminated at the edge of the city and looped around the city, thereby creating early fringe area industrial corridors (Roberts 1980). Horsecars linked the downtown to this terminus, allowing longer distance commuting and further suburban development (Roberts 1980). By the 1860s, some 13 commuter railroad lines spawned suburban development in Philadelphia (Loeben 1979, 178). The Philadelphia, Germantown & Norristown Railroad became a major corridor to the northwest, attracting an almost continuous wedge of development along its route (Taylor 1966). Chestnut Hill, Mount Airy, and numerous other railroad suburbs emerged, while the Main Line of the Pennsylvania Railroad evolved early as a series of housing developments near the stations, which eventually pulled a vast wedge of higher-income residential and commercial development to the commuter suburbs in its wake.

Similarly, by 1851 Pittsburgh had railroads running to the east and west of the city, with commuter rail service in the mid-1850s (Lowry 1963, 132). By 1857 there were six trains a day to East Liberty, and additional lines were added in subsequent years (Jucha 1979, 302). Pittsburgh was able to annex some of these early peripheral suburbs, much the same as Philadelphia was. To the east, where topography presented lesser problems than other directions for Pittsburgh, a series of early residential settlements sprang up, including Oakland, Wilkinsburg, Shadyside, and Squirrel Hill (Toker 1986). Other lines linked Pittsburgh with towns such as McKeesport, which were already established industrializing centers (Baldwin 1937).

The steep bluffs of the Pittsburgh and Johnstown areas, where the settlements were nestled primarily along the river valleys, created unique problems for the spread of development. The inclined planes provided a ready transportation solution in which a pair of cars counterbalanced on two sets of tracks hauled passengers up and down the hill in large numbers, effectively opening up areas south of the Monongahela and Ohio rivers to development and later annexation by the city of Pittsburgh (Lowry 1963). The Mount Washington incline began operation in 1870, and others were opened later in that decade through the early 1890s (Lawther 1981; Lowry 1963; Berger 1985).

The cable car represented a brief but significant addition to the intra-urban transportation network. Cars, propelled by gripping moving cables run through a network of underground pulleys and powered by coal-fired steam engines at a central plant, plied their routes in Pittsburgh and some other Pennsylvania cities in the late 1880s and early 1890s (Tarr 1978). The technology was particularly well adapted to steep slopes where either horsecars or conventional railroad equipment was ill-suited for service, particularly in winter conditions. The difficulties of route flexibility and extensions, however, also severely limited its usefulness.

The electric streetcar, along with the inter-urban commuter railroad, proved to be the most lasting of the railway technologies during the 1860–1920 period. The streetcars, powered by overhead electrical wires, were relatively easy to construct, compared with cable technology. Tracks were laid down the center of streets shared with horse and wagon or carriage, and wires were strung between the poles used to carry electricity to businesses and households beginning in the 1880s. Diffusion of the new technology was very rapid, and by about 1890 all Pennsylvania's metropolitan areas of that time had electric streetcar service. The growing urban-industrial metropolitan areas increased both the number of lines and the lengths of the routes by the end of the century; the streetcars reinforced and repeated the patterns of commercial and high-density development along the routes and street-corner stops initiated by the horsecars some decades earlier.

The streetcar and inter-urban railroad opened up the suburbs and, for the first time on a large scale, promoted widespread residential segregation in the cities. Whereas the journey-to-work in 1880 was typically about a mile in length, the streetcar and inter-urban railroad made commuting from several miles to the center city both feasible and relatively inexpensive for the time. The "nickel ride" on the streetcar was a tradition that lasted well into the early years of the twentieth century. White-collar urban Pennsylvanians could increasingly escape not only the undesirable conditions of the core city, but even the higher population density, noise, and dirt of the cities' streetcar corridors for the more pastoral settings of suburban housing clusters. In addition to middle-class commuting, working-class recreation was also provided in the form of amusement parks, zoos, and other facilities frequently incorporated into the early streetcar network, another example of taking the urbanite to the periphery to experience the virtues of rural living (Muller 1981).

By 1900 the streetcar lines were producing radiating networks surrounding both large and small metropolitan cores in Pennsylvania. Uneven development frequently occurred due to the location of lines and developers building at selected places along the lines, for the streetcars were the critical mechanism dictating financial success for new housing or commercial areas. By the end of the nineteenth century, half the municipalities in Delaware County to the southwest of Philadelphia were founded on the trolley lines (Webster 1982). Once-peripheral areas, such as West Philadelphia, initially typified by poor housing and industry, became decidedly upper middle class by the 1880s (Warner 1968). Similar effects of the streetcar prevailed in cities across the state, even in those of less than metropolitan stature.

The patterns of the freight railroad routes reinforced the wedges and corridors of industrial development that occurred during the late nineteenth and early twentieth centuries. Freight railroads normally followed the shorelines or river valleys, particularly where situations of severe local relief dictated that the tracks follow the relatively modest grades along the waterways. Because most of Pennsylvania's nineteenth-century metropolitan areas were already established before the heyday of railroad shipping from the 1870s onward, the freight terminals were often located at the edge of the densely developed parts of the city core. Belt lines or interlines that served as connecting links between the railroads coming into the city from different directions served as powerful attractors for industrial development in their vicinity. Freight classification yards, in which cars are sorted and switched and train loads are assembled, are among the most space-consuming railway facilities. Storage for perishable commodities, railcar repair facilities, and round-houses for engines also required considerable space. But because there were neither sufficient expanses of flat topography nor land at prices the railroads could afford to pay, the classification yards were usually located much farther out in the city than the core area freight terminals. For extensive land-using industries that likewise could ill afford the high land rents near the core, location in proximity to both belt lines and classification yards produced strong agglomerative locational effects.

Passenger terminals near the emerging central

business districts of the core were increasingly in demand both for intercity rail traffic and for local passenger lines. Passenger-coach yards were usually not located near the central passenger terminals, again a result of the high costs of land, although they were generally located not so far away as the freight classification and repair yards. In Pennsylvania's larger metropolitan areas, which were served by multiple railroads, trunk lines and shared passenger terminals accommodated the movements into the relatively scarce land areas available at the very edge of downtown.

The large downtown passenger terminal of the late nineteenth century became a major symbol of the metropolis itself. Because it was the visitor's first exposure to the city, costs were not spared to impress the new arrival with the grandeur and a reflection of the headiness of the growing industrial metropolis. Some of Pennsylvania's finest and grandest buildings are still its passenger railroad stations. Philadelphia's 30th Street Station, the Pennsylvania stations in Pittsburgh and Harrisburg, and Scranton's Delaware, Lackawanna & Western Railroad Station (now an attractive hotel following redevelopment) are all vivid examples of the urban symbolism and architectural styles of their time. Unfortunately, others, such as Altoona's Victorian era station, were razed in the early years of urban renewal. Many ancillary types of activities focused on transportation and commuter services clustered near these passenger terminals, and some continue to do so to this day.

While the metropolis of the railway era was expanding outward at a rapid rate, significant changes were also under way in the core. The high costs of moving goods within the city relative to the costs of moving goods between cities or moving people within cities contributed to the concentration of the core (Fales and Moses 1972). Costs per ton mile on the intercity trunk railroad lines fell sharply in real terms during this period. Similarly, walking and the "nickel ride" on the streetcar provided relatively inexpensive movement of people to jobs. In contrast, most goods coming into central freight or passenger terminals were moved by carters driving teams of horses pulling small wagons, a relatively expensive means of intracity freight movement. Thus, many types of businesses preferred to be located very near both central terminals, where pedestrian traffic was highest for commercial businesses oriented toward the mass market, and where costs of moving goods were lowest for industries that relied on these movements. Wholesaling or middleman activities in goods distribution came to replace traditional warehousing by the 1890s, and warehouse districts formed at the edge of downtown in Pennsylvania's core cities, often characterized by large, multifloored brick and arch-windowed designs characteristic of the Richardson architectural style or smaller loft-type buildings.

In addition, the rise of big corporate capitalism in the late nineteenth century resulted in an early division of labor within the firm. Processing activities were steadily shifted to the cheaper land farther from the core, while corporate office facilities that used land very intensively remained downtown. The core as office center was reinforced by the advantages of face-to-face communications at a time when communications technologies such as the telephone or teletype were not yet in widespread use.

The result for Pennsylvania and other North American core cities was a major transformation of the downtown. Less-intensive space users such as heavy industries were pushed by land market forces to more peripheral locations along the rails and/or waterways. The improvements in elevators and the steel-girder building technologies permitted vertical expansion in the central business district that was not previously possible. That meant that earlier buildings were not as intensive land users as the taller buildings and hence did not lead to as high rates of return. Accordingly, a major wave of downtown redevelopment occurred in the late nineteenth century and proceeded throughout this period as both taller and larger-based buildings, sometimes covering entire blocks, began to appear on the skyline. Small hills such as those of downtown Pittsburgh were leveled or smoothed for the rebuilding, and shorelines were filled and smoothed as core-area land became more valuable. A significant share of present-day downtown buildings in Pennsylvania's large metropolitan cores remains from the latter years of construction and reconstruction during this era. Land and housing in the path of the advancing downtown was often held for speculation, and the rundown housing and multifamily conversions became home to ethnic and racial ghettos, some of which remain to this day in Pennsylvania's larger cities.

Considerable amounts of parks and open space were also created or retained during the latter years of the nineteenth century. The parks movement—itself a product of the antiurban bias inherent in rural idealizations—took root in many Northeast indus-

Fig. 19.5. Pittsburgh's bridges, 1902. Bridges played an important role in the spatial form of Pennsylvania's metropolitan areas.

trial cities where green space and trees had been sorely lacking. Commons areas and public squares and land donated by corporate benefactors and social reformers were planted with trees and lawns, especially as the City Beautiful movement spread at the turn of the century. Boulevards, ostensibly to help relieve the severe congestion of the central business district, were also planted with greenery, and many of the major public buildings, such as opera houses, civic centers, and art museums, were built during this era. Pennsylvania's larger metropolitan areas generally were participants in these trends.

A final transportation factor of the railway era that must be noted is the changes that were brought about by technological improvements in bridge construction. Because many of Pennsylvania's nineteenth-century metropolitan areas were located along rivers, bridge crossings were essential for improved interregional trade. But they were also important for the growth stimulation of the downtown areas that were often trapped by the waterways on at least one side. Bridges provided access from across the river to downtown businesses, transport terminals, and the heart of the communications network (Fig. 19.5). Early arch-span bridges were sufficient for the wide but shallow rivers in such places as Harrisburg, and many of these classic bridges are used to this day. Other narrower and deeper rivers, such as those of western Pennsylvania, were better spanned by the iron and later steel bridges of which Pittsburgh and most of the state's other metropolitan areas located on waterways have been so well endowed. The suspension bridge for spanning the widest and deepest waterways was well-established technology by the end of the nineteenth century and also was a vast improvement over the ferryboats that carried passengers and freight across the rivers. The Benjamin Franklin Bridge built across the Delaware River in the early 1900s was a major contributor to the continued growth of both downtown Philadelphia and downtown Camden.

The Early Auto Era: 1920–1945

By 1920 suburbanization in Pennsylvania's metropolitan areas was already well established, and both

Table 19.1. Population of Pennsylvania central cities and share of metropolitan population, 1920, 1940, 1960, 1990 (population in thousands).

Central City or Cities[a]	Population and Share (%) by Year 1920	1940	1960	1990
Allentown-Bethlehem-Easton	116 (28.5%)	189 (41.2%)	216 (39.6%)	203 (29.5%)
Altoona	60 (47.0%)	80 (57.1%)	69 (50.6%)	52 (39.7%)
Erie	93 (60.8%)	117 (64.7%)	138 (55.2%)	109 (39.5%)
Harrisburg-Lebanon-Carlisle	111 (37.4%)	125 (35.9%)	126 (27.3%)	96 (16.3%)
Johnstown	67 (24.0%)	67 (22.3%)	54 (19.2%)	28 (11.7%)
Lancaster	53 (30.1%)	61 (28.9%)	61 (21.9%)	56 (13.1%)
Scranton–Wilkes-Barre–Hazleton	244 (31.9%)	265 (31.5%)	207 (29.9%)	154 (21.0%)
Philadelphia	1,972 (72.7%)	2,087 (65.2%)	2,159 (49.7%)	1,704 (35.1%)
Pittsburgh	635 (32.6%)	727 (31.8%)	650 (25.2%)	395 (19.2%)
Reading	108 (53.7%)	111 (45.7%)	98 (35.6%)	78 (23.3%)
Williamsport	36 (43.6%)	44 (47.4%)	42 (38.4%)	32 (26.9%)
York	48 (26.5%)	57 (26.1%)	55 (18.8%)	42 (10.1%)

SOURCE: U.S. Census Bureau.
[a]Includes the individual or multiple cities designated as central cities in the 1990 census.

Pittsburgh and Philadelphia were among the "metropolitan districts" that the U.S. Census Bureau first identified as "greater cities" in 1910. Using 1990 census county boundaries backcast to 1920 for uniformity, the data in Table 19.1 indicate that even by 1920 all of Pennsylvania's older metropolitan areas had a sizable share of their populations living beyond the central cities. This suburbanized population was far greater in number than that accounted for by the older, outlying settlements on the fringe, and represented the effects of early streetcar and automobile decentralization. This dispersal trend was to become a hallmark of twentieth-century metropolitan development in Pennsylvania (and elsewhere throughout the industrialized nations).

The shift from mass rail transportation to private automobiles and motor buses was rapid in Pennsylvania metropolitan areas, as it was in the nation. Automobile registrations in Pennsylvania's metropolitan areas grew exponentially during the 1920s and 1930s, despite the economic difficulties of the Great Depression years (Tarr 1978). Likewise, by the 1920s motor buses were being used in Pittsburgh, Philadelphia, Erie, and other places in Pennsylvania to add to the existing route structure and in many cases to replace portions of the trolley networks.

However, automobile usage remained largely "recreational" during this first auto era, for mass movements to jobs, still predominantly located in the central city or densely developed old industrial suburbs, continued to be served by the inter-urban and street railway systems. Early highway-building had tended to concentrate on rural segments linking cities and linking farms to city markets, rather than on large-scale urban street-building programs. Early parkways, such as the East and West River Drives along the Schuylkill River in Philadelphia's Fairmount Park, exemplify this type of construction pattern (Muller 1981).

The congestion of downtown areas also discouraged automobile commuting, for few public parking facilities were available at that time. As early as 1910, more than 100,000 commuters came into Pittsburgh's central business district every day (Tarr 1978, 26). The congestion there and in other larger Pennsylvania cities, made somewhat more bearable by public transit, became increasingly problematic as more and more automobiles began to invade the central areas in the 1930s. Early parking lots and garages appeared in larger numbers to accommodate this shift to automobile commuting.

Outside the central business district, street patterns began to change in response to the automobile. Hard-surface road-building accelerated beyond the early parkway phase, and new thoroughfares and highways were built in most of the larger Pennsylvania metropolises (Loeben 1979). Building new bridges was also necessary to accommodate the key auto arterials of this era (Muller 1981). These arterials radiated from the central business district and, where they did not follow the older trolley streets, were often routed through the less densely built-up areas of the interstices, which provided fewer impediments to wider streets and highways, and lower land costs (Tarr 1978).

The effect of these arterials was to open up the interstitial areas between the old trolley and interurban railroad lines for suburban development. Private developers no longer found it necessary to subsidize the streetcar lines in order to make their peripheral settlements financially viable. Instead, they could purchase large tracts of relatively cheap land, build model homes along the roads, and market them to motorists passing by. The rise of large-scale Pennsylvania suburbia was on, and the demise of many mass transit companies was one concomitant of this development pattern.

Successive waves of suburban residential building began to characterize all the larger Pennsylvania metropolises. The Philadelphia suburbs expanded particularly to the north and northwest (Warner 1968), while areas of Pittsburgh expanded especially to the south and east (Lowry 1963; Tarr 1978). The clusters of middle- and upper-income buyers soon found it to their advantage to incorporate into municipalities separate from the central city, thereby providing limited fiscal support for the infrastructure-intensive core city while still taking their economic livelihood from it. Johnstown, for example, created an area-wide planning commission in 1916 to encompass the city and its many boroughs, an effort in urban amalgamation that was not typical of most other larger Pennsylvania urbanized areas before or during this era (Berger 1985).

The expansion of industrial districts usually followed the intercity freight lines outward and spilled across the corporate city boundaries, increasingly by the early automobile era. Motor freight shipping initially reduced the costs per ton mile of intra-urban shipping by more than half (Moses and Williamson 1967) and had a strongly parallel effect of putting the carters out of business over a relatively short period of time. While many manufacturers and distributors preferred the security of access to rail sidings, motor freight shipping provided more freedom to choose larger tracts of cheaper land with roadway access at "greenfield" sites. By the 1930s, many of the problems inherent in an industry dominated by small, independent motor-freight transportation companies had been ameliorated through stronger public regulation. In most Pennsylvania metropolitan areas, one can see evidence of this early motor-vehicle era decentralization characterized by its single-story plant layouts.

Retail and commercial functions typically lagged behind the decentralization of manufacturing activities and, before 1920, most suburban commercial establishments were located in small, isolated clusters at streetcar junctions or stops and in the older mill towns surrounding the core city. The building of the urban arterials provided the higher accessibility locations that commercial businesses required. The arterials began to fill, slowly at first, from the edge of the central business district outward, with businesses catering to the automobile traveler including automobile dealers and gasoline stations.

Although zoning had been upheld by the courts shortly before the start of the era, early zoning practices in Pennsylvania municipalities and elsewhere actually promoted strip development by setting aside land along the urban arterials and thoroughfares for commercial purposes. Thus, strips of relatively intense business development came to dominate the area to a depth of a half or full block behind the road, with housing development located behind. To this day Pennsylvania's metropolitan areas reflect some of the least aesthetic aspects of this strip development begun predominantly during the Early Auto Era. Proudfoot's (1937) classic study of city retail patterns in Philadelphia during the 1930s provides strong evidence of this outward arterial expansion pattern of the retail sector. In addition, early "shopping centers"—such as Philadelphia's 69th Street area—began to appear during the closing years of the 1930s (Muller 1981).

Despite early attempts to improve the flow of transportation in downtown areas, congestion was only increased by the greater density of building that had been initiated during the later years of the Railway Era. The first of the "skyscrapers" appeared in Pittsburgh and Philadelphia during the Early Auto Era, the latter diminished by height conventions, although neither approached the heights of New York and some of the other large industrial cities of the Northeast. Although buildings in Pennsylvania's smaller metropolitan areas were not of the scale found in Philadelphia or Pittsburgh, they nonetheless helped to change the skylines significantly during this period. As land rents rose, banks, hotels, and offices forced out more and more of the warehouses and manufacturers that had not been driven out before the 1920s. Philadelphia's central business district was described as a "monument of the middle class, the theater of its work and shopping" (Warner 1968, 188).

Meanwhile, the middle- and upper-class populations were pulled to the central-city fringes and suburbs by the lure of large lots, single-family houses, and improved Auto Era accessibility. Neighborhood enclaves that were homogeneous in their socioeconomic structure, bolstered by zoning and building code requirements, created vast tracts of similar housing quality and occupant characteristics. In Philadelphia the Main Line and Chestnut Hill areas became upper class, and Northeast Philadelphia toward the city was working-class and mill-town residents, while away from the city it was inhabited by higher-income white residents (Warner 1968). In contrast, inner-city areas such as North and South Philadelphia increasingly became lower-income black and ethnic ghetto areas, respectively, as the flight to the suburbs was aided by the Early Auto Era developments. This type of scenario was being played out across Pennsylvania's metropolitan areas, be it Pittsburgh, Wilkes-Barre, Reading, or other of its larger places.

The Freeway Era: 1945 to the Present

By the mid-1940s the monopoly quality of inner suburban sites with superior transportation accessibility was being seriously weakened. The population of Pennsylvania's metropolitan areas continued to suburbanize rapidly (see Table 19.1). In short, the dispersal of population that had begun in earnest some decades earlier was accelerated following World War II. This deconcentration of population was aided significantly by postwar government policies that subsidized housing costs for veterans and others, encouraged the building of single-family detached housing (the variety most common to the suburbs), and plowed billions of dollars into new highway construction, much of it in the fringes of the metropolitan areas.

Aside from the Pennsylvania Turnpike, opened in the late 1930s, most of the early post–World War II freeway building occurred outside the central business districts and often outside the central cities, such as expressways that served as limited links between major federal highways or feeder routes. The Interstate Highway program, enacted in 1956, in many ways merely accelerated the building of many freeway segments that had previously been started or planned. The early freeway segments built first tended to be in the suburban areas just outside the perimeter of the central city, where congestion brought about in large part by commuting was most severe. These segments were usually part of the intermetropolitan links. Because these segments were built in what was initially rather sparsely settled parts of the metropolitan area, construction proceeded relatively faster than it would on later segments that went through or near downtown or other more densely built-up areas. This pattern of building extended the feasible commuter shed out very quickly to 40 miles or more and precipitated dense development along many stretches of these intermetropolitan corridors closest to the central city. The areas incorporated into these wider commuter regions have become known as "urban fields," which when taken together permit metropolitan Pennsylvania's labor markets to cover nearly all of the state's population (Friedmann and Miller 1965).

Manufacturing plants increasingly opted for the greenfield sites of the more rural parts of the suburban fringe, many newly freed from the locational monopoly of sites with railroad sidings. The growth in the number of planned industrial parks following World War II was indeed phenomenal, and many of Pennsylvania's suburban communities zoned land for industrial purposes far in excess of realistic demands (Erickson and Wollover 1987). In the Pennsylvania portion of the Philadelphia metropolitan area, there were already 32 suburban industrial parks by 1960, nearly all of which were located along or very near major U.S. highways. By 1981 there were 153 industrial parks in the same area. The large growth in the supply of suburban sites has been fueled by speculators who regard control over large-land-parcel developments as a means to ensure greater profits.

Manufacturing growth in Pennsylvania's metropolitan areas continued, especially in the suburban places, through the first three decades of the postwar period. During the 1980s, however, manufacturing in the central cities declined sharply, while suburban areas have been fortunate to have remained steady. Many of the older central city (and early suburban mill town) industrial plants are no longer profitable to operate, and in many of the state's larger metropolitan areas these relics—like the steel mills along the Monongahela River in Pittsburgh—stand as industrial dinosaurs of the nineteenth and early twentieth centuries.

The rapid suburbanization of population in Pennsylvania's metropolitan areas during the postwar era created for the first time a mass market in the suburbs. The higher-income characteristics of the middle- and upper-class suburbanites attracted retailers on a large scale beginning in the 1950s after the riskiness of suburban sites was no longer in question. The suburban shopping mall was one of the retailers' responses to that demand. In suburban Pennsylvania the mall concept took off during the 1950s and has continued to the present time, although in recent years the rate of new mall construction has slowed and there is currently renewed investor interest in the redevelopment of older malls, such as the Camp Hill Mall in suburban Harrisburg, and even the building of center city malls, such as the Gallery in Philadelphia.

Shopping malls have generally acquired sites at or near major freeway interchange areas so as to maximize the number of potential buyers with access to the facilities. Lancaster's Park City Mall, Reading's Wyomissing Mall, Harrisburg's East Mall, and Philadelphia's King of Prussia Mall are only a fraction of the major shopping malls in Pennsylvania's metropolitan areas that illustrate this location principle. The suburban shopping mall has often become a central attraction for many other types of service sector economic activities or even light industrial/office parks.

The almost exclusively single-family detached character of housing that characterized Pennsylvania's suburban areas in the 1940s and 1950s was first broken by the so-called "suburban apartment building boom" that occurred in the 1960s. Builders responded both to the demand for housing by the large number of baby boomers in the population and to the fact that more and more jobs were being located in the suburbs. Indeed, by the early 1970s there were often more jobs in the suburbs of Pennsylvania's metropolitan areas than there were in the core central cities.

In Pennsylvania's larger metropolitan areas, perhaps the most notable trend since the 1960s has been the coalescence of economic activities around a few major suburban nodes. This pattern has intensified during the 1980s as new office and other high-order service activities have clustered at these key sites. This trend has been spurred onward by the construction of the remaining links of the Interstate Highway System, both around and through or near the central business districts of major metropolitan areas. Pennsylvania's metropolitan areas have not been subject to the same degree of beltway-type interstate highways or ring-road freeways as many other parts of the nation, especially those whose topography is more hospitable. This factor has tended to string out development along the bypass segments and intermetropolitan corridors more than one might expect from the traditional circular beltway construction pattern.

King of Prussia in northwest suburban Philadelphia near the interchange of the Pennsylvania Turnpike and the Schuylkill Expressway, and Monroeville in east suburban Pittsburgh near the Pennsylvania Turnpike, I-376, and U.S. 22 interchange are both classic examples of these high-density nucleations of economic activity and surrounding residential development (see Fig. 19.6). The King of Prussia nucleation includes not only a super-regional shopping mall (one of the nation's largest) but also corporate offices, light industry, research and development, hotels and motels, a convention center, restaurants, and many other assorted activities (Fig. 19.7). Around the commercial core, residential areas tend to follow a classic pattern of declining density with distance from the interchange area. Areas nearest the commercial development have a far greater share of apartment and other multifamily housing, which generally tails off into traditional single-family detached housing within a few blocks, these gradients being stretched out along the roads with access approaching the area. With growth has come increasing congestion, a problem whose magnitude may someday approach the congestion in the central business district of some of our largest core cities.

Nucleations on different scales have been forming on the suburban periphery in all of Pennsylvania's metropolitan areas during the 1970s and 1980s. Even State College, which ranks as the smallest of Pennsylvania's metropolitan areas, is at present evolving with two clusters of economic activities at key locations along its limited-access bypass route.

Within Pennsylvania's inner cities, the effects of the dispersal of population and economic activities have also been felt. The flight of residents and businesses alike has meant declining opportunities

Fig. 19.6. Nucleations of high-density development outside the central city: the Monroeville area of east suburban Pittsburgh.

for employment for those persons left behind in the transition. The inner city is increasingly home to the vast majority of racial and ethnic minorities, especially in the larger metropolitan areas. In these parts of the city, housing decay and abandonment, and deteriorating public services, continue to be pressing problems.

One of the more limited but notable changes that has been occurring over the past 15 years has been the gentrification and revitalization of some inner-city neighborhoods. Areas of Philadelphia, Harrisburg, Lancaster, and virtually all of Pennsylvania's other inner cities have undergone some restoration and revitalization as young urban professionals have taken control of certain of these neighborhoods. The result has been sometimes remarkable changes in the physical and socioeconomic character of neighborhoods. In its wake, however, has been the displacement of the existing residents, most or all of whom are unable to afford to live in the neighborhood as gentrification proceeds.

The Future Spatial Structure of Pennsylvania's Metropolitan Areas

In the absence of any significant new transportation technologies, we are likely to witness a continuation of recent trends in the internal spatial structures of Pennsylvania's metropolitan areas. The "galactic metropolis" that Lewis (1983) so eloquently described, with its Auto Era satellite city nucleations, will continue to extend the geographic area over which the metropolis is effectively organized. The concomitant of this trend is the relative decline of the central city accompanied by greater and greater reverse commuting as workers find an increasing share of the job opportunities located in Pennsylvania's suburbs.

For Pennsylvania's metropolitan core central cities, the prognosis is mixed. In an era of increasing service sector dominance, the downtowns of large

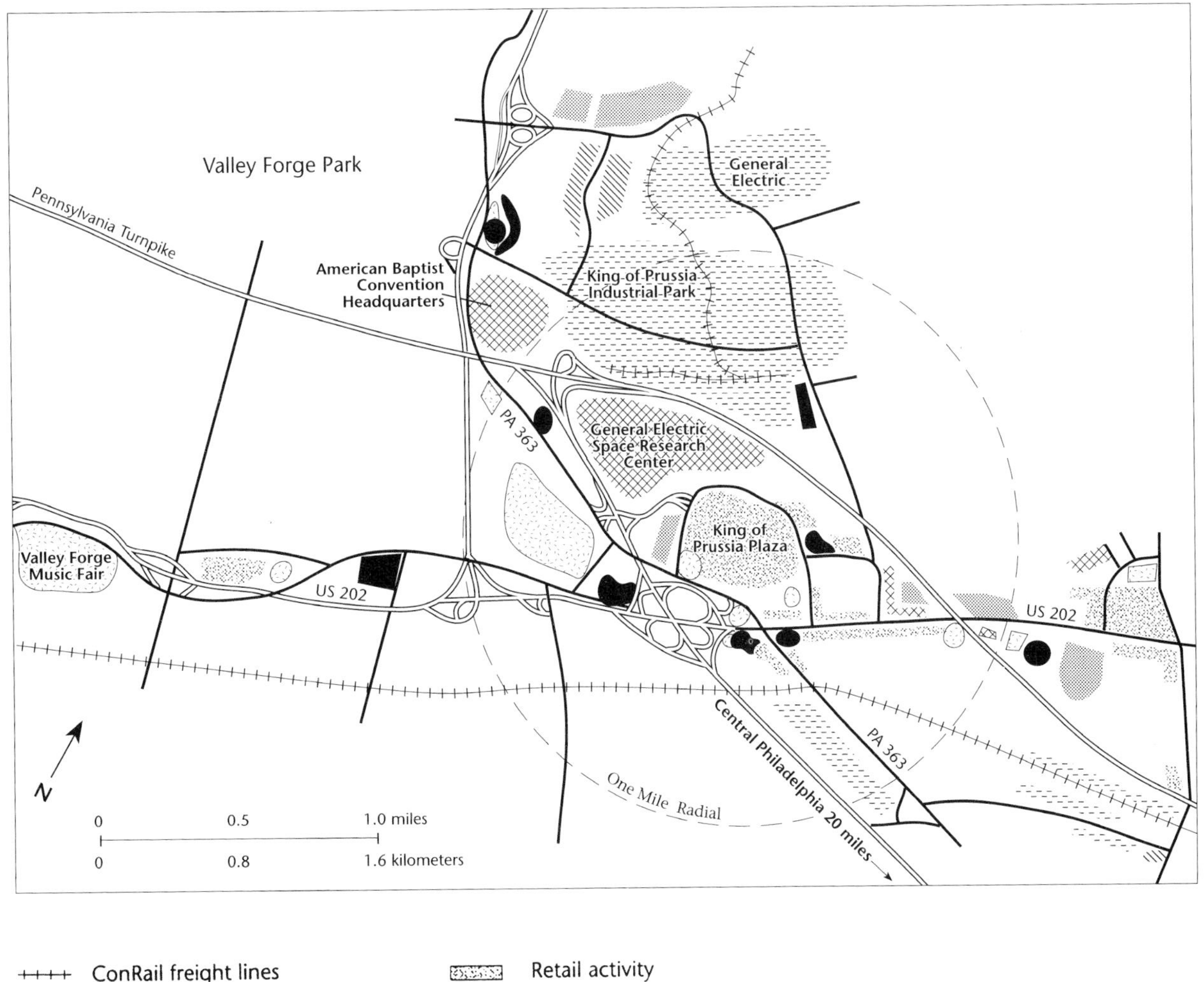

Fig. 19.7. The internal spatial structure of the King of Prussia suburban nucleation, mid-1970s.

central cities remain attractive sites for office industry, especially the types of activities that rely on face-to-face contacts and flows of specialized information. We are likely to witness the continued "Manhattanization" of these downtowns, especially in Pittsburgh and Philadelphia, as unique buildings (such as the Pittsburgh Plate Glass "glass castle"), often occupying entire blocks, represent the prevailing corporate images and office functions of the central business district. Retailing and other service functions cater to the classes of professionals and other workers that make up an increasing share of the central business district daytime population. In the wake of this redevelopment, more and more of the former skylines of the central business districts will disappear. Even the central business districts of smaller metropolitan areas will also see more of this effect as this area of metropolis becomes a place of increasingly specialized economic activities.

For those inner-city neighborhoods with older houses of unique historical or architectural qualities, continued gentrification seems probable. For the segments of the population desiring different lifestyles and access to professional jobs and the amenities of large downtowns, certain inner-city neighborhoods remain attractive. The increase in this group of residents, however, will not be sufficient to counterbalance the continued population losses from other areas of the central cities.

For much of the rest of the central cities in Pennsylvania's largest metropolitan areas, further economic and demographic deterioration is likely to continue. The absence of large-scale new investment or reinvestment in these areas suggests continued blight and deterioration, particularly in the absence of any significant public-sector involvement in the renewal of these areas. The fact that most of Pennsylvania's racial minorities and poorer residents are confined to these areas as a result of labor and housing market inequalities, combined with widespread public indifference, suggests that these areas of the central cities will continue to suffer from a lack of new capital investment and little change in spatial structure or built environment.

Older industrial districts, be they located in the central cities or in the suburbs, are likely to undergo further decline due to their uncompetitive economic situations; in and near these districts, commercial and residential blight will undoubtedly continue to plague these areas. Many of these older industrial activities will never reopen as industrial restructuring proceeds. Nonetheless, some of these areas, particularly those along the riverfronts and floodplains, represent potentially prime sites for redevelopment in Pennsylvania's older metropolises.

Changes in internal spatial structure will obviously be greatest where the most new capital investment is plowed into the built environment. Those areas continue to be the suburban portions of the metropolitan statistical areas that account for all of the net population growth in nearly all of Pennsylvania's metropolitan places.

The completion of beltway and feeder freeways to relieve the growing congestion along existing routes represents only temporary solutions, for the infrastructure investments that help to create the nucleations in the "galactic metropolis" serve to encourage further growth as they alter the accessibility surface of the metropolitan area. The results of these trends in the state's larger metropolitan areas will be more intensive land uses and greater specialization of economic activities and residential communities. The hearts of these nucleations will increasingly look like the "new downtowns" that they are, complete with higher-rise buildings, vehicular and pedestrian traffic congestion, and more higher-density residential building on the relatively expensive land.

The evolving internal spatial structure of Pennsylvania's metropolitan places will be interesting to watch as it unfolds. While this structure creates or perpetuates many metropolitan problems, it also creates new opportunities and challenges as Pennsylvanians continually try to cope with and shape their future metropolitan environments.

Bibliography

Adams, J. S. 1970. "Residential Structure of Midwestern Cities." *Annals of the Association of American Geographers* 60:37–60.

Alberts, R. C. 1980. *The Shaping of the Point: Pittsburgh's Renaissance Park.* Pittsburgh: University of Pittsburgh Press.

Alexander, J. K. 1975. "A Year . . . Famed in the Annals of History: Philadelphia in 1776." In *Philadelphia, 1776–2076: A Three Hundred Year View,* ed. D. Clark. Port Washington, N.Y.: Kennikat Press.

Baldwin, L. D. 1937. *Pittsburgh: The Story of a City, 1750–1865.* Pittsburgh: University of Pittsburgh Press.

Berger, K., ed. 1985. *Johnstown: The Story of a Unique Valley.* Second edition. Johnstown, Pa.: Johnstown Flood Museum.

Blumin, S. 1973. "Residential Mobility Within the Nineteenth-Century City." In *The Peoples of Philadelphia: A History of Ethnic Group and Lower-Class Life, 1790–1940,* ed. A. F. Davis and M. H. Haller. Philadelphia: Temple University Press.

Borchert, J. R. 1967. "American Metropolitan Evolution." *Geographical Review* 57:301–332.

Bourne, L. S. 1982. "Urban Spatial Structure: An Introductory Essay on Concepts and Criteria." In *Internal Structure of the City,* 2d ed., ed. L. S. Bourne. New York: Oxford University Press.

Erickson, R. A., and D. R. Wollover. 1987. "Local Tax Burdens and the Supply of Business Sites in Suburban Municipalities." *Journal of Regional Science* 27:25–37.

Fales, R. L., and L. N. Moses. 1972. "Land-Use Theory and the Spatial Structure of the Nineteenth Century City." *Papers of the Regional Science Association* 28:49–80.

Friedmann, J., and J. Miller. 1965. "The Urban Field." *Journal of the American Institute of Planners* 21:312–319.

Gillette, H. 1980. "The Emergence of the Modern Metropolis: Philadelphia in the Age of Its Consolidation." In *The Divided Metropolis: Social and Spatial Dimensions of Philadelphia, 1800–1975,* ed. W. Cutler and H. Gillette. Westport, Conn.: Greenwood Press.

Harvey, O. J. 1909. *A History of Wilkes-Barre, Luzerne County, Pennsylvania.* Volume 3. Wilkes-Barre, Pa.: Raeder Press.

Hershberg, T., H. Cox, D. Light Jr., and R. R. Greenfield. 1981. "The 'Journey to Work': An Empirical Investigation of Work, Residence, and Transportation, Philadelphia, 1850 and 1880." In *Philadelphia: Work, Space, Family, and Group Experience in the Nineteenth Century: Essays Toward an Interdisciplinary History of the City,* ed. T. Hershberg. New York: Oxford University Press.

Jucha, R. J. 1979. "The Anatomy of a Streetcar Suburb: A Development History of Shadyside, 1852–1916." *Western Pennsylvania Historical Magazine* 62:301–19.

Lawther, D. E. 1981. "Mount Washington: A Demographic Study of the Influence of Changing Technology, 1870–1910." *Western Pennsylvania Historical Magazine* 64:47–72.

Lewis, P. F. 1983. "The Galactic Metropolis." In *Beyond the Urban Fringe: Land Use Issues of Nonmetropolitan America,* ed. R. H. Platt and G. Macinko. Minneapolis: University of Minnesota Press.

Loeben, A. F. 1979. "Philadelphia Suburbs." In *The Philadelphia Region: Selected Essays and Field Trips,* ed. R. Cybriwsky. Washington, D.C.: Association of American Geographers.

Lowry, I. S. 1963. *Portrait of a Region.* Volume 2 of the Pittsburgh Regional Planning Association Economic Study of the Pittsburgh Region. Pittsburgh: University of Pittsburgh Press.

Miller, R. 1982. "Household Activity Patterns in Nineteenth-Century Suburbs: A Time-Geographic Exploration." *Annals of the Association of American Geographers* 72:355–371.

Moses, L. N., and H. R. Williamson Jr. 1967. "The Location of Economic Activity in Cities." *American Economic Review Proceedings* 57:211–222.

Muller, P. O. 1981. *Contemporary Suburban America.* Englewood Cliffs, N.J.: Prentice Hall.

———. 1986. "Transportation and Urban Form: Stages in the Spatial Evolution of the American Metropolis." In *The Geography of Urban Transportation,* ed. S. Hanson. New York: Guilford Press.

Murphy, R. E., and M. Murphy. 1937. *Pennsylvania: A Regional Geography.* Harrisburg, Pa.: Telegraph Press.

Proudfoot, M. 1937. "City Retail Structure." *Economic Geography* 13:18–34.

Roberts, J. P. 1980. "Railroads and the Downtown: Philadelphia, 1830–1900." In *The Divided Metropolis: Social and Spatial Dimensions of Philadelphia, 1800–1975,* ed. W. Cutler and H. Gillette. Westport, Conn.: Greenwood Press.

Sammartino-Marsh, M. 1977. "Suburbanization and the Search for Community: Residential Decentralization in Philadelphia, 1880–1900." *Pennsylvania History* 44:99–116.

Swauger, J. 1978. "Pittsburgh's Residential Pattern in 1815." *Annals of the Association of American Geographers* 68:265–277.

Tarr, J. A. 1978. *Transportation Innovation and Changing Spatial Patterns in Pittsburgh, 1850–1934.* Chicago: Public Works Historical Society.

Taylor, G. R. 1966. "The Beginnings of Mass Transportation in Urban America: Part I." *Smithsonian Journal of History* 1:35–50.

Toker, F. 1986. *Pittsburgh: An Urban Portrait.* University Park: The Pennsylvania State University Press.

Ullman, E. L. 1962. "The Nature of Cities Reconsidered." *Papers of the Regional Science Association* 9:7–23.

Warner, S. B., Jr. 1968. *The Private City: Philadelphia in Three Periods of Its Growth.* Philadelphia: University of Pennsylvania Press.

Webster, N. 1982. "Delaware County: Where Pennsylvania Began." *Pennsylvania Heritage* 8 (Fall): 2–7.

20

POSTINDUSTRIAL PHILADELPHIA

Roman A. Cybriwsky

The geography of Philadelphia has changed profoundly a number of times since the city was founded by William Penn in 1682. During its first years, Philadelphia was a compact commercial and administrative center crowded along the sides of a small cove off the Delaware River. However, through a series of steps tied closely to patterns of national development and economic progress, as well as to countless technological innovations that created and fueled industry, enhanced transportation, and facilitated city-building, Philadelphia grew to become a manufacturing center of preeminent proportions by the latter part of the nineteenth century. The population exceeded 1,000,000 by 1890, by which time the city could boast of far-reaching influence on all aspects of American life.

The population of Philadelphia is now approximately 1,558,577 (1990 census count), the fifth largest total in the United States. Moreover, the city is the hub of a huge metropolitan area that sprawls over a wide section of the Delaware Valley and includes an additional 3.3 million persons. All told, the Philadelphia Standard Metropolitan Statistical Area (SMSA), which includes Philadelphia and four "suburban" counties in Pennsylvania and three counties in New Jersey, numbers more than 4.8 million inhabitants and ranks fourth in size in the nation.

The full story of Philadelphia's growth to prominence is long, complex, and rich in detail. The most important aspects of this development can be summarized under three convenient headings: preindustrial city; industrial city; and postindustrial city. These categories are defined by three fundamentally different bases for economic activity in the city and by contrasting patterns of urban form and spatial dimensions. They also represent different periods in history. While exact dates are difficult to pin down because adjacent periods overlap, for our purposes we can use the following time frames:

1. Preindustrial city: 1682 to early 19th century
2. Industrial city: early 19th century to early 20th century

3. Postindustrial city: evolving since the 1920s, but especially since the end of World War II

The emphasis of this chapter is on the last category—on Philadelphia as a postindustrial city. As the term implies, the manufacturing economy that had sustained the city before has been winding down and a new economic structure is taking shape. This is the so-called "services sector," which emphasizes, among other jobs, various white-collar professional occupations. Moreover, there is a concomitant shift taking place in land-use patterns within the city, as well as in the socioeconomic characteristics of a broad range of residential neighborhoods. Our purpose, then, is to illustrate this emergent new geography as it applies to Philadelphia, and to discuss some of its implications. Because present patterns are cast on foundations that were set in the past, we begin with a review of Philadelphia's geography during preindustrial and industrial times.

Historical Patterns

Philadelphia was founded to be the capital city of Pennsylvania, a "holy experiment" by William Penn in social and religious tolerance. According to Penn, the goal of his colony was "to afford an asylum to the good and oppressed of every nation." This was a timely invitation that appealed to numerous maltreated groups in Europe, notably Penn's own English Quakers, and that set off successive waves of immigration. Other groups in the early city included Friends from Wales and Ireland, large numbers of German Mennonites, and transplants from other British colonies in the New World.

However, the promise of social tolerance was only one factor in the city's early development. More important, certainly over the longer term, was that Philadelphia was also intended to be a good place for business. Penn was a practical entrepreneur as well as a liberal visionary, and he sold land in Pennsylvania even before he himself arrived. He offered for sale a creative package deal that combined large parcels in rural areas with city building lots, and that attracted numerous business-minded investors. This, plus the city's intrinsic geographical advantages of a fine harbor, two large rivers for trade (the Delaware and the Schuylkill), ample water, and flat, fertile farm lands, quickly propelled Penn's venture into a prosperous commercial city. Within 20 years after founding, Philadelphia (present boundaries) had more than 4,000 inhabitants and ranked as one of the largest cities in British North America. By the time of the Revolution, the population had increased to approximately 40,000. By then Philadelphia was well established as first in population rank, a distinction the city held until the more rapid rise of New York in the second decade of the nineteenth century.

During the colonial period, as well as approximately a half-century beyond, urban development in Philadelphia was concentrated along the Delaware River not far from the original nucleus. Even as population grew to the tens of thousands, the city expanded inland only slowly (see Fig. 20.2). As late as the Revolution, for instance, the built-up area extended inland only about half a mile, to about Seventh Street, and this only along the Market Street axis. This contrasted with a dimension of about one and a half miles north-south along the bank of the river. Urban expansion was fastest here, because the economy of the city was dominated by waterborne trade and related occupations. Businesses gravitated as close as possible to dockside locations, and when there was no room spilled over to Northern Liberties and Southwark, riverfront "suburbs" immediately north and south of the city limits. The only significant urban growth away from the river in the preindustrial period was at such centers as Germantown and Newtown. Both are now well-integrated into the fabric of Philadelphia, the former as a close-in urban neighborhood, the other as a residential suburb in Bucks County, but they were originally established as farmers' towns at the Pennsylvania frontier.

The core of preindustrial Philadelphia was extremely crowded, not only because businesses and residents scrambled to be near the river, but also because the spatial extent of the city was constrained by limited transportation. Moreover, there was little pattern to the arrangement of land uses. Residences, shops, industries, churches, meetinghouses, marketplaces, taverns, burial grounds, and other uses were jammed together into a compact, pedestrian-scale form. Nearly every street combined a mixture of some of these, and many buildings combined residential and commercial functions. Merchants, for example, commonly lived above or behind their shops, as did most craftspeople and other artisans. Only along the waterfront itself, where shipbuilding and dock

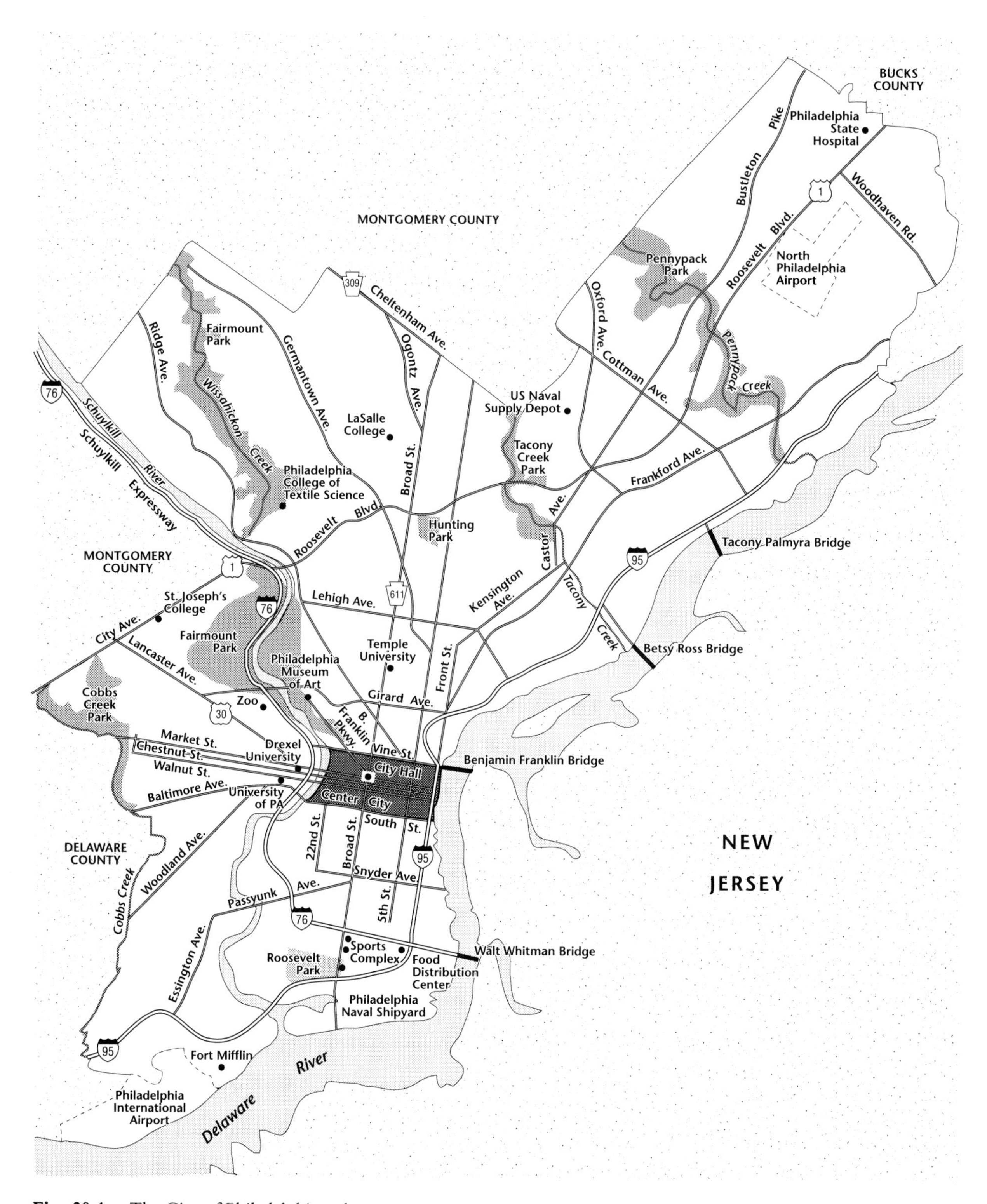

Fig. 20.1. The City of Philadelphia today.

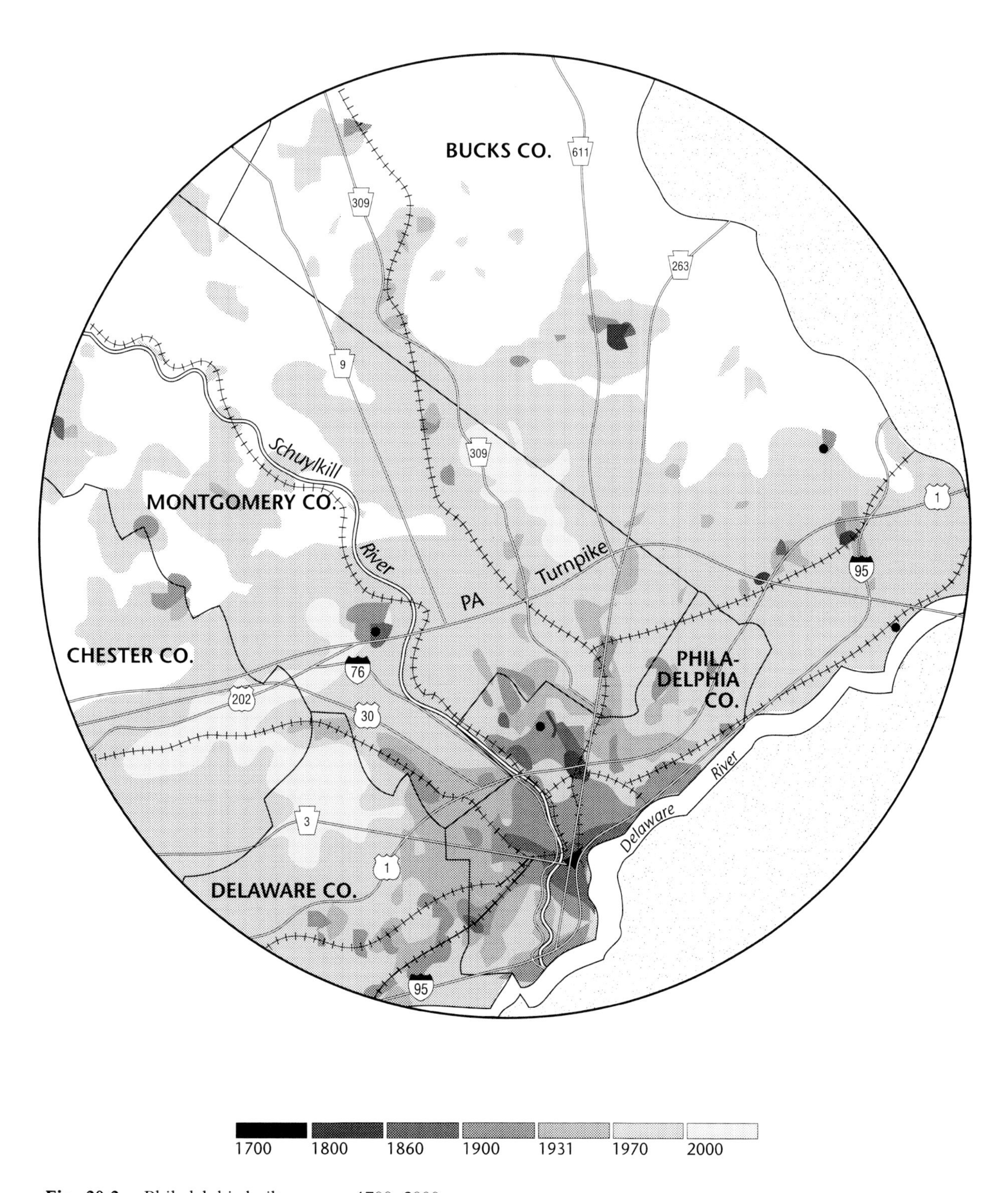

Fig. 20.2. Philadelphia built-up area, 1700–2000.

facilities were preponderant, was there any significant land-use specialization.

Such patterns began to change in the nineteenth century, as Philadelphia's geography was transformed by industrialization. The new landscape emerged first in the 1820s at Manayunk, a site near the falls of the Schuylkill River, several miles upstream from the core of Philadelphia, where waterpower was plentiful. This advantage, plus the improvements to transportation that followed construction of a canal in 1823, changed the riverfront at Manayunk to a specialized factory district. It was lined with large mills for spinning cotton and weaving wool, as well as with dyeing establishments and other related industries. By the middle of the nineteenth century, as coal replaced falling water as the chief source of power for industry, more factories with different technologies were established in other precincts. For example, Fishtown, Kensington, Brewerytown, and other areas north of Philadelphia were developed into industrial districts, as were most other parts of what we now call North Philadelphia. So too industry expanded in select sections of South and West Philadelphia, in Germantown and East Falls in the northwest, and in such districts as Frankford, Bridesburg, and Port Richmond in the Northeast. The largest of these concentrations was the North Philadelphia–Kensington area, particularly along rail corridors, where coal was moved conveniently and where rivers could be used for transportation and waste disposal.

Philadelphia's industrial growth was spectacular. By the early 1900s the city had a diversified industrial base that comprised more than 8,000 manufacturing establishments and more than 250,000 employees. This put Philadelphia high among the leaders in the world in manufacturing production. The largest industrial sector was textiles and garments, but Philadelphia was also noted for the making of carpets and rugs, oil cloth and linoleum, athletic equipment, saws and other hardware, surgical equipment and artificial limbs, leather goods, umbrellas, fertilizers, the refining of petroleum and sugar, and many other products. Most factories were small to medium-size, employing 200 or fewer workers. However, a select few companies, consisting of Baldwin Locomotives and the Stetson Hat Company in North Philadelphia, William Cramp's shipyards in Fishtown, and the Bromley textile mills in Kensington, grew to be truly large, with more than 5,000 workers each.

The labor force for such expansion came largely from immigration. In the early part of the nineteenth century most arrivals were from England, Scotland, or Wales. By the 1840s great waves of Irish Catholics were added to the population. Immigration from Germany peaked a short time later (1850s–1870s). The turn of the century (1880s–1910s) brought immigration levels to an all-time high, as a great mix of peoples arrived from Italy, Poland, Hungary, Russia, and numerous other countries in southern and eastern Europe. Racial heterogeneity increased as well, as small groups of Chinese came from the West Coast in the 1870s, and as growing numbers of African-American migrants, mostly from the U.S. South, supplemented the small number of African-Americans who had been established in the city since colonial times.

Thus, the character of Philadelphia changed dramatically over the period of industrialization. The population skyrocketed from about 81,000 in 1800 to 409,000 by mid-century, and then to more than 1,000,000 by 1890. By 1920, when the industrial period was at its peak, Philadelphia had more than 1.8 million inhabitants. Concomitantly, the city broke loose from its historic nucleus and expanded laterally. By 1854 the old boundaries of Philadelphia, unchanged since 1682 when they were established by the plan for the city drawn by Thomas Holme, had to be abolished. New limits that encompassed all of Philadelphia County were set instead. This changed the size of Philadelphia from 2 square miles to about 130 square miles.

The physical expansion of Philadelphia was facilitated by improvements to transportation. Rail lines were especially important, because in addition to forming the industrial corridors they also facilitated expansion of the city's residential land. This was especially the case with respect to districts that were built up as commuter suburbs (either within or outside city limits) for the wealthiest classes. Examples include a part of Germantown that was developed as an elite enclave as early as the 1850s; Chestnut Hill, located at high elevation in Philadelphia's northwest and developed starting in the 1880s at the terminus of two different commuter lines; and such famous "Main Line" suburbs in Montgomery County as Haverford, Bryn Mawr, and Villanova. Other new residential areas grew up in conjunction with different transportation systems—namely, trolley lines and electric streetcars. The so-called "streetcar suburbs" of the 1880s–1890s were most notable, especially in North, West, and South Philadelphia.

These are areas of tightly packed row houses, often for owner-occupancy, built to accommodate the burgeoning numbers of middle-income working people who crowded the inner city at the time of Philadelphia's most rapid growth. Examples of such places are Fairmount and Strawberry Mansion in North Philadelphia, and Mill Creek and Haddington in West Philadelphia (see Fig. 20.3).

Such neighborhoods contrasted greatly with the older parts of the city, such as Southwark and Northern Liberties. Many of these places had become overcrowded immigrant quarters for the newest arrivals in Philadelphia, as well as home for many of the city's poorest African-American residents. Housing conditions, sanitary conditions, social services, and other aspects of the urban environment were below accepted standards. Construction of the new neighborhoods, then, often provided housing relief for immigrants and their descendants, who moved to the better areas as soon as incomes allowed. But most African-Americans continued to be confined by discrimination to overcrowded, older neighborhoods.

Meanwhile, other parts of the historic core, particularly the middle ground between the Delaware and Schuylkill rivers, were transformed into yet another new type of land use—the Central Business District (CBD). In Philadelphia this is now called "Center City" and is a specialized zone of major banks and other financial institutions, company offices, government buildings, large hotels, department stores, significant cultural attractions, and other important functions. Such businesses emerged in this area because Center City was the focus of all transportation systems in Philadelphia, including the train and streetcar lines, and the subway system (constructed in the early twentieth century). Thus, the CBD served all parts of Philadelphia and was unrivaled during the industrial period as the nerve center of the entire metropolitan region.

Postindustrial Patterns

The geography of Philadelphia continues to evolve and change since the industrial period. We now speak of Philadelphia as a postindustrial city, different in a number of important respects from urban patterns displayed earlier. The most significant shifts include (1) a steep decline of population in Philadelphia relative to a rapidly expanding suburban ring, and associated shifts in urban social geography; (2) suburbanization of much of the business activity of Philadelphia; (3) a major shift in the economic base of the Philadelphia area; and (4) a redefined role for the Central Business District and adjacent neighborhoods.

Population Shifts

Perhaps the most conspicuous characteristic of postindustrial Philadelphia is the huge drop in population that has taken place in the decades after World War II. The total peaked with the 1950 census count at 2,071,605 and has subsequently slipped with every decennial tally to 1,558,577 in 1990 (Table 20.1). Estimates suggest that the next census (2000) will show a total that has fallen below the 1.5 million level.

The loss of population is most striking in many of the neighborhoods that were developed at the time of most rapid industrialization. For example, North Philadelphia (defined generally as the area between Spring Garden and Wingohocking streets and between Front Street and Fairmount Park) lost more than half its population between 1950 and 1990 because of out-migration from more than 500,000 inhabitants to about 243,000. Population losses were also recorded for Kensington, South Philadelphia, and most other close-in Philadelphia neighborhoods. In many of these areas, the decline is seen in the landscape in the form of empty and abandoned houses and apartments, as well as in empty lots where residences once stood. All told, Philadelphia has some 30,000 abandoned housing units. About half this total is in North Philadelphia, where the population decline has been steepest. Between 1970 and 1980 alone this area lost 11,642 housing units (14 percent of the total) due to abandonment and/or demolition.

In contrast to the city itself, most suburban areas around Philadelphia gained in population in recent decades. For example, the combined populations of the seven counties in Pennsylvania and New Jersey that comprise the suburban ring of the Philadelphia SMSA more than doubled between 1950 and 1990, from 1.6 million to just over 3.3 million. Thus, suburban population exceeds that of Philadelphia by more than 1.7 million. This is in marked contrast to 1950, when Philadelphia's total was 25 percent larger

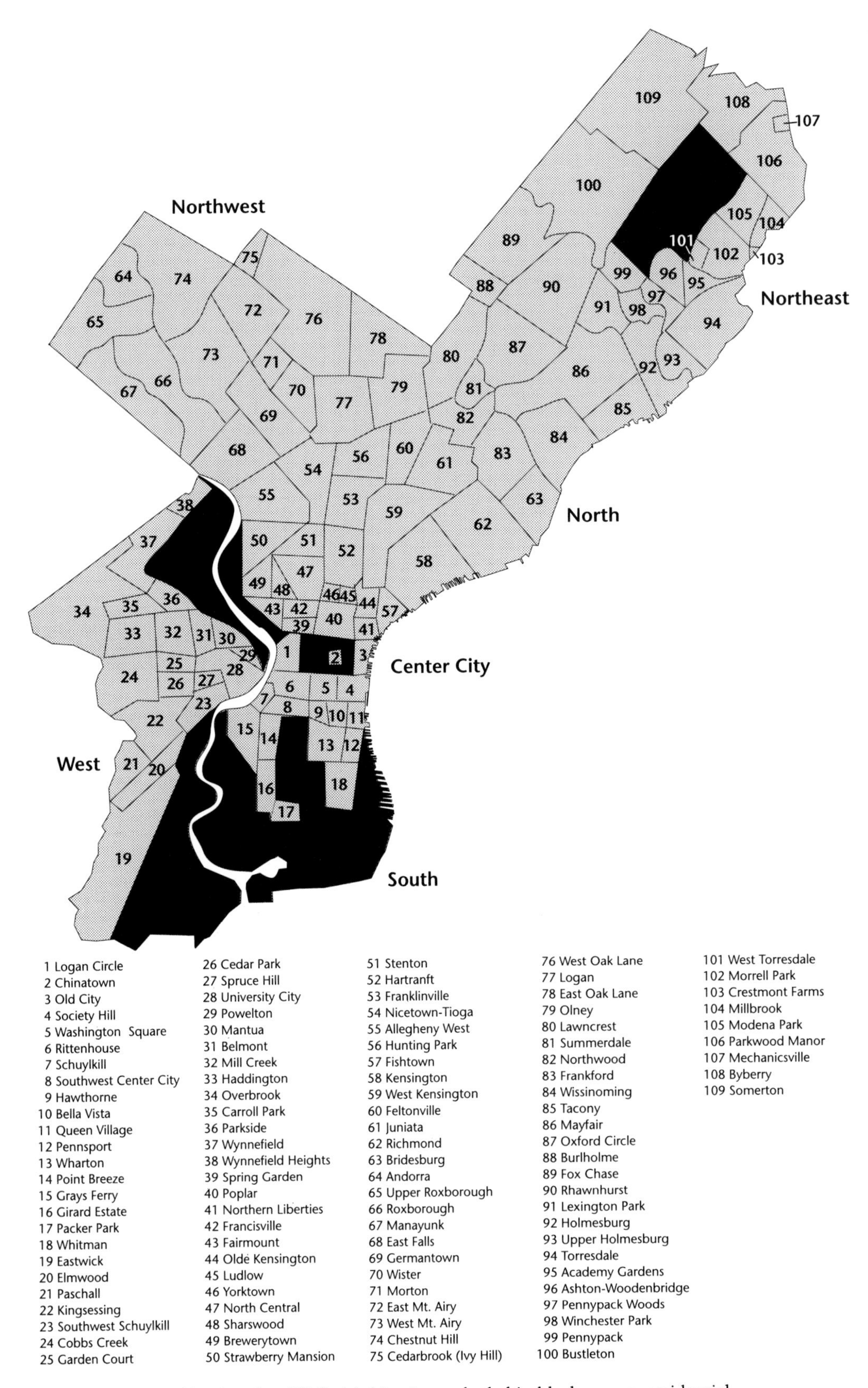

Fig. 20.3. Neighborhoods of Philadelphia. Areas shaded in black are nonresidential.

Table 20.1. Philadelphia population, 1790–1991.

Census Year	Population (in thousands)[a]	% Change	National Rank[b]
1790	54	—	1
1800	81	48.9	1
1810	111	37.3	2
1820	137	23.3	2
1830	189	37.8	2
1840	258	36.6	2
1850	409	58.4	2
1860	566	38.4	2
1870	674	19.2	2
1880	847	25.7	2
1890	1,047	23.6	3
1900	1,294	23.6	3
1910	1,549	19.7	3
1920	1,824	11.8	3
1930	1,951	7.0	3
1940	1,931	− 1.0	3
1950	2,072	7.3	3
1960	2,003	− 3.5	4
1970	1,949	− 2.8	4
1980	1,688	− 15.4	4
1990	1,559	− 8.3	5

Source: U.S. Census Bureau.

[a]All population totals are for Philadelphia's present boundaries (Philadelphia County).

[b]The cities that have surpassed Philadelphia in population are, in order, New York City, Chicago, Los Angeles, and Houston.

than the suburban population, or in 1900 when the city total was more than double the suburban total. The last census count, however, recorded losses (or at least very slow growth) in many of the older, inner-ring suburbs. So too, there have been some other pockets of population loss in the suburbs, such as at older industrial centers like Norristown and Bristol (Bucks County). On the whole, however, the suburbs continue to grow, especially in some of the SMSA's more remote townships, where suburban land development is just beginning to take off.

The growth of suburban population has been fueled by a combination of factors, including the "push" forces of rising city problems, and such "pulls" as the lure of lower-density living, increased opportunities for home ownership, and the attraction of new public facilities (schools and community centers). Land for suburban development was opened up by construction of highways to serve a public that, increasingly, had come to rely on private automobiles for transportation. The completion of the Schuylkill Expressway in 1955, for example, helped to open up much previously inaccessible land for residential development in the Valley Forge–King of Prussia area, in Montgomery County approximately 15 miles from the core of Philadelphia. Similarly, the rapid growth during the 1960s of what is called the Far Northeast, a suburb-like section at a distant extremity of the city, is tied closely to highway improvement projects, most notably the widening of U.S. 1—Roosevelt Boulevard—to cut commuting time between new residential neighborhoods and the center of the city.

As has been the case in most other cities in the United States, suburbanization in the Philadelphia area has been selective according to race, economic status, and other socioeconomic variables—that is, the majority of movers to the suburbs were white, middle- and upper-income households, most typically those with children. Their exodus coincided, not entirely by chance, with stepped-up migration to Philadelphia during the post–World War II decades by low-income African-Americans, mostly from rural areas in Virginia, the Carolinas, and other parts of the South, as well as by low-income Hispanics. The majority of the latter are from Puerto Rico. This shift is seen most clearly in any of more than a dozen older, inner-city neighborhoods that changed from nearly all-white to mostly African-American or Puerto Rican over fairly short periods.

The changing racial geography is seen in the fact that Philadelphia's population has shifted from 82 percent white and 18 percent nonwhite in 1950, to 53.5 percent white and 46.5 percent nonwhite in 1990. The African-American population increased from 18.2 percent of the total in 1950 to 39.9 percent in 1990. It is heavily concentrated in North, South, and West Philadelphia and includes many types of neighborhood conditions and housing environments (see Fig. 20.4). Similarly, the Hispanic population has increased and now comprises approximately 5.6 percent of the total. The principal Hispanic neighborhood is the West Kensington section in North Philadelphia (Fig. 20.5). By contrast, the nonwhite proportion in the suburbs is small. According to the 1990 census, only a little more than 10 percent of the more than 3.3 million residents in the seven suburban counties were African-American or Hispanic. Moreover, the majority of these persons resided in smaller central cities such as Camden (New Jersey) and Chester and in segregated suburbs. Many of the latter are located adjacent to predominately African-American neighborhoods in the city.

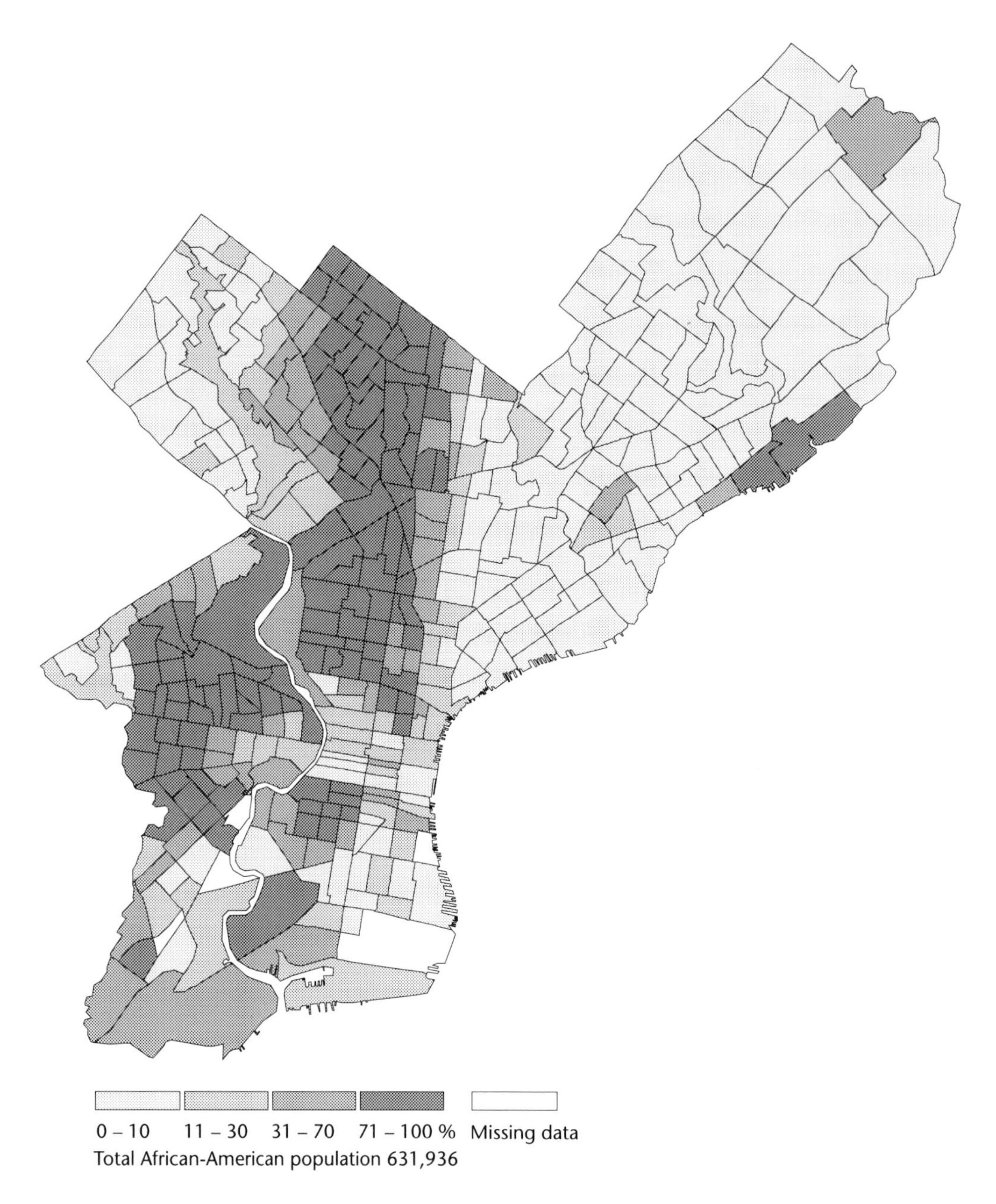

Fig. 20.4. African-American population in Philadelphia by census tract, 1990.

We also see contrasts between the central city and suburbs in economic status, measured by income levels. For example, according to the 1990 census, the median annual income for households in the city of Philadelphia was $24,603, only about two-thirds of the suburban average of $36,023. So too we see that the city has a much larger fraction of families with income below the poverty line (16.2 percent, compared with 4.0 percent for the suburban counties), more households dependent on public assistance payments (13.9 percent versus 4.4 percent), and more households receiving Social Security income (31.5 percent versus 26.1 percent). In all these indicators, the majority of low-income persons in the suburbs are heavily concentrated in the smaller central cities and other pockets of poverty. Most

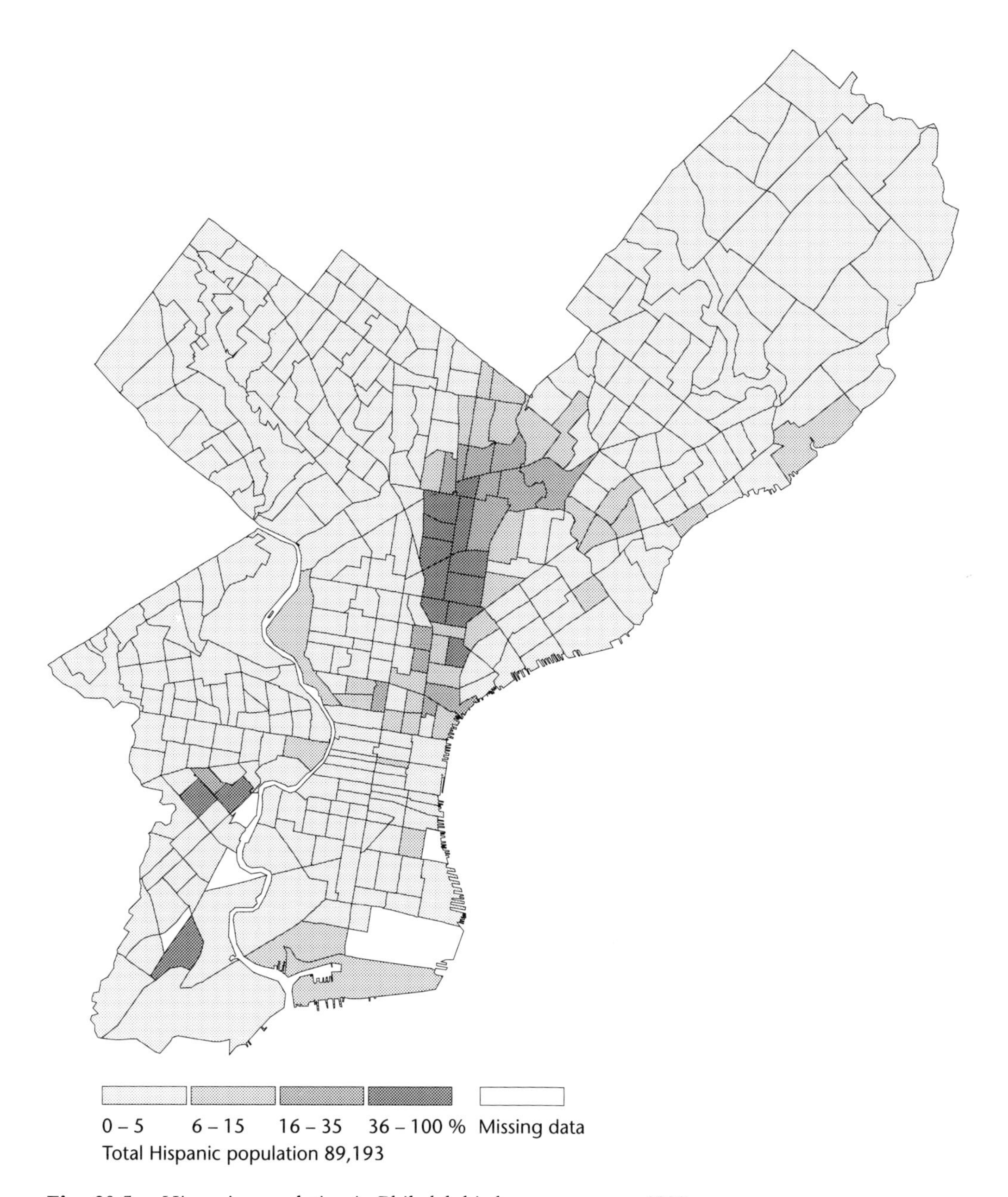

Fig. 20.5. Hispanic population in Philadelphia by census tract, 1990.

suburban jurisdictions have very low, perhaps even negligible, proportions of their populations with poverty-level incomes.

However, it would be wrong to think that postindustrial Philadelphia is nothing more than low-income ghettos formed after white flight to the suburbs. While this has come to be an all-too-familiar type of neighborhood in the city, Philadelphia retains many other types of districts as well, even in some of the older sections (see Fig. 20.6). Manayunk, for example, has been a stable, comparatively unchanged neighborhood and retains much of the same mixture of European ethnic groups that came over in the nineteenth century to labor in the mills. Strong ethnic ties have also contributed to stable neighborhoods among Italians in parts of

Fig. 20.6. View across North Philadelphia. A landscape of century-old rowhouse neighborhoods, aging industrial plants, and public housing projects from the post–World War II urban renewal era. The photograph also shows Girard College, a private educational institution built before the industrialization in North Philadelphia, when the land was open space.

South Philadelphia, among Poles in Port Richmond, and among Chinese in a small Chinatown just north of the central business district, as well as among other groups elsewhere in the city. There are also several solid middle-class African-American neighborhoods. Examples include parts of Germantown and Mount Airy in the northwest, West Oak Lane at the northern reaches of the city, and parts of such areas in West Philadelphia as Wynnefield and Haddington. The city also retains an opulent (overwhelmingly white) district in Chestnut Hill, the elite enclave of large estates mentioned earlier in connection with Philadelphia's railroad-industrial era. There are also some racially integrated neighborhoods, such as East Oak Lane (north) and parts of both East and West Mount Airy (northwest). However, districts where African-Americans and whites are finely intermixed house by house, and where there is long-term numerical balance, are very few. As most examples show, the great majority of neighborhoods are strongly segregated by race.

In addition to the types of neighborhoods given above, there are a number of notable new sections of Philadelphia. These developed in tandem with the postwar rise of suburbia and have socioeconomic profiles that typify middle-class suburban communities. The Far Northeast, for example, defined as the area "above" Pennypack Park, was still largely rural in the 1950s but during the 1960s developed quickly as a "suburb within the city." The population rose from 27,563 in 1950 to 69,627 to 1960 to more than 160,000 in 1990. Most new residents were young, white families with children who moved out of older, inner-ring neighborhoods as Philadelphia's minority population increased. Other examples of sections within Philadelphia that gained population during the postwar decades are Roxborough in the city's Far Northwest and Eastwick in the far

Southwest. Overall, however, increases in such neighborhoods have been much smaller than population losses for the city as a whole. Suburbanization is far and away the dominant trend.

Thus, Philadelphia is a city of many kinds of neighborhoods, as well as many racial and ethnic groups and different categories of economic status. The social mix continues to evolve, as new immigrants arrive in the city and establish niches of their own in the Philadelphia mosaic. For example, in just the past ten years distinctive new communities have formed among Vietnamese in South Philadelphia, Cambodians and other Asians in Logan (in the north of the city), Koreans in Olney (also north), West Indians in parts of Germantown, and Russians in certain subsections of the Northeast. Moreover, as we shall see later, numerous other neighborhoods, particularly those in the vicinity of the central business district, are changing in yet another way and becoming quite fashionable upper-income districts with mostly new residents.

Suburbanization of Business

In addition to redistribution of population, there has also been a shift to the suburbs of much of the business activity that was once centered in Philadelphia. This applies to such mainstays of the employment base as manufacturing and company offices, as well as to almost all types of retailing, various categories of personal and professional services, and other businesses. In part, the shift reflects that business has moved to be near to the growing consumer markets in the suburbs and to the suburban labor force. However, the relocation of business is catalytic as well, as suburbanization of population proceeds, in part, because so many of the SMSA's business establishments are in the suburbs.

The greatest growth in business has been at such developing "regional centers" as King of Prussia in Montgomery County, Oxford Valley in Bucks County, and Cherry Hill in New Jersey. Each of these places is highly accessible to a large population by automobile (e.g., by being at the intersection of expressways) and features a sprawling enclosed shopping mall and other retailing along major access roads. In addition, each has numerous new office buildings and other employment centers, new entertainment facilities, hotels, restaurants, and other economic activity, as well as manufacturing in modern industrial parks. Peter Muller, an urban geographer who has studied the economic development of suburbia, has referred to such centers as "mini cities." By this, he means that the great clustering of commercial activity within major suburban centers suggests a high degree of independence from the central city. And indeed, studies have shown that residents of suburban areas increasingly work and shop outside the central city, and travel to the urban core only infrequently on special occasions. In fact, such centers have grown to be so large and important that they are no longer "mini." The new term for them is "edge cities," taken from the title of a popular book by Joel Garreau.

The decline of the central city relative to the suburbs is documented most clearly for retailing (see Table 20.2). As recently as 1958 half of all sales in retailing in the eight-county SMSA were in the city of Philadelphia, but this has declined steadily to the point where by 1987 only about 22 percent of the SMSA's sales volume was in the city. Over the same period, the number of retailing establishments in Philadelphia declined by more than two-thirds, from

Table 20.2. Changes in retailing: Philadelphia, suburban counties, and Philadelphia CBD, 1958–1982 (as measured by number of stores and volume of sales).

	1958	1963	1967	1972	1977	1982
No. of retail establishments						
Center City	$ 2,621	$ 2,293	$ 1,946	$ 1,786	$ 1,887	$ 1,705
Philadelphia	22,934	18,980	17,952	15,131	13,175	11,543
Suburban ring	18,733	18,341	19,443	21,243	22,392	24,064
Retail sales (× $1,000)						
Center City	$ 603,615	$ 563,248	$ 573,070	$ 649,344	$ 838,184	$ 1,029,244
Philadelphia	2,305,467	2,361,996	2,748,382	3,378,337	4,165,555	5,391,039
Suburban ring	2,357,092	3,175,497	4,338,501	6,769,683	10,206,646	15,512,548

Source: U.S. Census of Retail Trade.

Note: The dollar figures are not adjusted for inflation.

Table 20.3. Retail sales in Center City as percentage of Philadelphia and SMSA retail sales, 1958–1982.

	1958	1963	1967	1972	1977	1982
Central city as % of Philadelphia	26.2	23.9	20.9	19.2	20.1	19.1
Central city as % of SMSA	12.9	10.2	8.1	6.4	5.8	4.9

SOURCE: U.S. Census of Retail Trade.

22,934 to 8,388. By contrast, the suburban counties experienced an increase in the number of establishments. Furthermore, sales volume in the suburban ring increased more than ten-fold between 1958 and 1987, from $2.4 billion annually to about $24.1 billion. In the city, however, the increase was by a factor of only 3, from $2.3 billion to just under $7.0 billion, a gain that failed to keep pace with inflation.

Similar statistics illustrate the relative decline of the central business district (CBD). Once the dominant shopping center for the entire Philadelphia area, Center City Philadelphia comprises less than 5 percent of total sales volume in the SMSA (Table 20.3). It continues to be the single biggest retailing center, but not nearly as, say, a generation ago. For example, in 1963 downtown Philadelphia transacted nearly 11 times the retailing dollars of the second-ranking center, the Roosevelt Boulevard–Cottman Avenue area in Northeast Philadelphia. By 1972 the CBD's dominance over the second-ranking center (still the Roosevelt-Cottman area) slipped to a ratio of 5.8 to 1. By 1982, by which time numerous suburban shopping malls had been built, the Oxford Valley Mall in Bucks County emerged as the second-ranking center, and the ratio dropped from 2.1 to 1.

Economic Shifts

As the term "postindustrial" suggests, there has been a decline in manufacturing in the Philadelphia area. This drop has been steep and relatively steady for at least a half century. Consider, for example, that the number of manufacturing establishments in the city declined from a peak of more than 9,000 shortly after World War I, to 3,329 in 1972, and then to fewer than 1,900 in the latest (1987) census count. Similarly, we see that the number of workers engaged in manufacturing has declined too. The most recent total (95,000 in 1987) is much less than half the manufacturing work force at the turn of the century, and less still compared with the 1947 total of approximately 329,000. Even in the worst year of the Great Depression, 1933, there were more workers in manufacturing in Philadelphia than there are now. The loss of manufacturing jobs has been especially precipitous in recent years. Since the 1970s alone well over 100,000 such positions have been lost to the Philadelphia economy.

This decline reflects a more general deindustrialization that has taken place in recent decades across the United States, but most especially in the aging industrial cities of the Northeast. It has affected nearly all the major categories of manufacturing established by the U.S. Census, including the aforementioned examples of Philadelphia industries. There are many reasons for this, notably automation of manufacturing processes where machines replace human labor, declining overall demand for particular classes of manufactured goods, and strong competition in manufacturing from companies in foreign countries where labor and other production costs are cheaper. Furthermore, manufacturing in a city like Philadelphia is also at a disadvantage because of such problems as aging, inefficient factory buildings, crowded, narrow streets that are inadequate for today's transportation demands, an unfavorable local tax structure, and high crime and poor neighborhood conditions.

Consequently, over the years, hundreds of factories have closed altogether or moved from the city to relocate elsewhere. Many have been drawn to modern industrial parks that are better suited by design and setting for the demands of contemporary business. Some of these places are in newer areas of Philadelphia itself, such as planned industrial parks in the Far Northeast section of the city or at Eastwick, in the southwest near the airport. Especially in the last two decades, however, most new factory sites have been in the suburban counties that surround the city. The advantages to employers of such locations often include ample relatively inexpensive land, convenient access to expressways and turnpikes, nearness to a work force that resides in the suburbs, and avoidance of a stiff (and immensely unpopular) tax on wages imposed by Philadelphia

Fig. 20.7. The deindustrialization of Philadelphia. Lehigh Avenue in the Kensington neighborhood, showing the ruins after a fire of what was once one of the city's biggest manufacturing employers, Bromley Textile Mills. The building had been vacant for years before it was burned by vandals.

on employees of companies within city limits. Still other manufacturing plants, including many from older industrial centers near Philadelphia like Chester (Delaware County) and Norristown (Montgomery County), have left the metropolitan area for new sites in such sun-belt states as Georgia or Texas. Such places often offer substantially lower labor, energy, and tax costs.

Losses in manufacturing have been balanced, at least in part, by growth of the services sector. This too reflects a nation-wide economic trend. This sector comprises a broad range of occupations, including such categories as financial, legal, and medical services, managerial positions in company offices, teaching and professorial jobs, many government jobs, and much of the professional employment in computer-oriented high-technology industries. It also includes lower-wage service jobs, such as in hotels and restaurants, clerical work of various kinds, and many of the lower-rung positions in public and private sector office settings. While such growth has been fastest in the sun belt, the Philadelphia area has received a substantial share of the new industry. This is in part because of the advantage of central location within the Boston-to-Washington megalopolis. We can also point to Philadelphia's long history as a city with services-sector-type employment. For instance, several of Philadelphia's most prominent services-sector institutions of today can be traced back directly for more than 200 years. The University of Pennsylvania, Pennsylvania Hospital, and the Insurance Company of North America are examples. Today's services sector is also well rooted in the industrial era, when Philadelphia prospered as headquarters for manufacturing firms and railroad companies, large banks, insurance companies, brokerage houses, and other institutions.

As with manufacturing, growth in the services sector has been fastest in the suburbs. As a result, this area has had a substantial overall increase in recent years in total employment. Often, the newest businesses are found in modern office parks (or "office campuses") amid green surroundings and accessible to major highways. A rapidly developing corridor of high-technology companies along U.S. 202 in Chester and Montgomery counties is a notable example. By contrast, the city of Philadelphia has not gained enough in the new economy to offset losses in manufacturing and other declining sectors. Thus, for the city as a whole there was a net loss of 160,000 jobs between 1970 and 1984. This represents a drop of about 16 percent of all jobs in the city in less than 15 years, which has resulted in a chronic problem of unemployment in Philadelphia and is in itself one of the major causes of the high levels of poverty and dependence on public assistance that characterize many of its older neighborhoods.

Center City Revitalization

Downtown Philadelphia has changed greatly over the postindustrial period. We have already seen that its role as the SMSA's dominant retailing center has been eroded significantly by competition from expanding regional centers in the suburbs. Furthermore, there has been a long-term decline in total employment in Center City, most notably in manufacturing jobs in a broad zone of aging industrial loft buildings at the northern edge of the central business district (CBD). Between 1970 and 1984 alone there was a net loss of more than 25,000 jobs, representing 7 percent of the total, in the Center City area. Thus, the downtown of Philadelphia is in many ways part and parcel of the economic problems of the city as a whole and reflects faithfully the types of transitions that have taken place in the city over the past decades.

However, in recent years there has also been considerable revitalization in this area. In part, this is attributed to a creative comprehensive plan for Center City put into effect in the 1950s and 1960s, and that has guided most aspects of the area's redevelopment ever since. The brainchild of Edmund Bacon, Philadelphia's famous master planner during this period, the plan called for the removal by urban renewal of the most noxious blighting influences, and for construction by public-private sector cooperation of spectacular new projects designed to modernize such categories of land use as offices, retailing, transportation, and spaces for tourism and recreation. The plan also provided for elaborate historic preservation efforts in older neighborhoods, particularly for the preindustrial urban fabric of the area called Society Hill near the Delaware River. Such projects coincided extremely well with broad structural shifts in the economy to favor services and have helped to establish Center City as a rapidly developing center for newest employment in such fields as company offices, banking and finance, legal services, medicine, and other professions. As was the case with services in the suburbs, there are also gains in such lower-rung jobs as office clerks and restaurant help. All told, service sector employment in Center City has increased by 18,000 jobs since 1977, and in most years has more than offset losses in the downtown in other categories of employment.

Another approach to downtown revitalization in Philadelphia has been the establishment since early 1991 of what is formally called the "Center City District." This is a special taxation district that encompasses most of Center City, including many of the retailing streets and the major office buildings subsection within the downtown, and that applies funds that are raised from a specially authorized surcharge on real-estate taxes directly to improvements in the local area. Specifically, all landowners within the district (except churches and others who are exempt), ranging from the owner of the smallest shop to the owners of the biggest high-rises, pay an added charge to their annual real-estate tax bill (the amount of which depends on the assessed value of the property) to support daily cleanup of local streets and sidewalks, additional security patrols, and various promotions designed to improve the public's image of Center City. Early indications are that this strategy, which can be summarized as "If you clean it they will come," is being successful at enhancing the overall reputation of the downtown area, and that it can in fact support local economic growth.

The revitalization of the downtown is seen clearly in the landscape. In addition to cleaner streets, the most striking aspect is the newly constructed office towers that now dominate the skyline. Since 1970 more than 10 million square feet of new office space have been added to the Center City inventory. Between 1980 and 1985 alone, 12 new buildings totaling more than 4.3 million square feet were under construction in 1985. The most notable new building is One Liberty Place, a prime-site office tower that rises to 940 feet (60 stories) and exceeds by some 75 percent the height of City Hall, the long-

Fig. 20.8. Philadelphia skyline as seen from Fairmount Park. Liberty Place is in the center, surrounded by other new high-rises in the city's fast-developing office towers district in the central business district. City Hall, with a statue of William Penn at the top, is left of center.

established tallest building. There are also several other new office and hotel buildings in the downtown, an attractive new convention center, and various significant improvements to retailing districts. The most notable change is Market Street East, a "ground-up" rebuilding of the CBD's largest shopping area. It features "The Gallery," a huge, multilevel, suburban-mall-like shopping center with dozens of shops and restaurants and three department stores as anchors. This project also has considerable new parking, direct access to an attractive new passenger rail station, and convenient location to some of the city's largest office-building developments. In addition to all of this, the tourism industry in Philadelphia has been expanded by the establishment of Independence National Historic Park at Independence Hall, and the Liberty Bell, and by such improvements to the Delaware waterfront as the new park and recreation-oriented commercial district called Penn's Landing.

There has also been considerable revitalization of residential neighborhoods in the area that surrounds the central business district. This began with the Society Hill urban renewal project in the 1960s that transformed a decaying, low-income neighborhood into a showpiece of historical preservation and one of the SMSA's wealthiest and most expensive districts. It has since spread to cover an ever-widening ring of older, low, and moderate-income neighborhoods, including such places as Queen

Village, south of Society Hill, the Northern Liberties area, and Fairmount and Brewerytown in the industrial-era landscapes of lower North Philadelphia. This revitalization is now one of the characteristic elements of the social geography of the postindustrial city. Its physical manifestations include significant renovation of older housing, adaptive reuse of vacant buildings, considerable construction of new clusters of townhouses and other residences, and various types of rejuvenation in neighborhood commercial districts. All this is oriented for a middle- and upper-income market, particularly the so-called "young professionals" who are attracted to employment in the expanding services sector of the downtown core. However, this type of neighborhood change, called gentrification, typically causes displacement of existing residents. This is done by market forces and is most harmful to low-income renters, elderly residents on fixed incomes, and others for whom decent and affordable replacement housing is almost impossible to find. Consequently, this revitalization is a controversial type of neighborhood change, and its benefits to the city in the form of physical improvements to buildings and enhanced tax base have to be balanced against the hardships it imposes on those who are pushed from their homes and communities.

The geography of Philadelphia has changed greatly over the city's history, and it continues to evolve. At present, the city is in the throes of change from industrial to postindustrial patterns of economy and land use. The manufacturing economy that gave contemporary Philadelphia its basic form is ever less a dominant feature. Instead, in Philadelphia, as indeed across much of the United States, the most rapidly growing sector is the provision of services, the processing of information, and the exchange of goods. As seen, such work is arranged spatially with concentrations in the central business district and in widely scattered "regional centers" in the suburbs.

The key trends outlined in this chapter will probably continue into the near future. For example, it is likely that the city will continue to lose population for another decade or two. Furthermore, the metropolitan region as a whole may also begin to show declines, with substantial losses yet to come in the inner suburbs. Much of this decline will reflect continued economic problems. Not even the most optimistic forecasts would predict that a reversal of manufacturing shutdowns is imminent. However, it is expected that the downtown will continue to thrive as the services economy expands, and that the residential neighborhoods in the vicinity of the central business district will see continued upgrading and even increases in numbers of households. With this, however, there will be a widening gap in the quality of life between the well-off—the professionals, managers, and others who have the requisite skills to participate in the modern growth economy—and those persons who in one way or another are left behind. This suggests a continuation and even aggravation of social ills in the city, and intensification of segregation in living and work areas.

The transition to postindustrial patterns has not been smooth. It has brought obsolescence to many parts of the city and hastened their decline into slums. Many neighborhoods, such as large parts of North Philadelphia and sections of both West and South Philadelphia, are now defined by a landscape of dilapidated housing, vacant storefronts and empty lots, abandoned factories, dangerous public housing projects, graffiti, vandalism, and neglected streets, school buildings and playgrounds, and other ills. They have been written off by private capital, which turns its attention to expanding suburbs and a booming Center City, and depend instead on support from government-funded programs and the determination of hardworking but poorly capitalized civic groups. Yet the future of the city depends on the direction that such neighborhoods are able to take in the years to come, as well as on the extent to which change provides direct benefits for their poor residents. Fortunately, there are enough examples in which grassroots efforts have succeeded amid adverse conditions to have at least some optimism. Even in the worst districts, one finds revitalized shopping streets, successful housing programs for the poor, and social programs that make a positive difference in people's lives. These efforts are testimony to the strength of the city's ultimate resources—the creativity and the great energy of dedicated citizens.

Bibliography

Adams, C., D. Bartelt, D. Elesh, I. Goldstein, N. Kleniewski, and W. Yancey. 1991. *Philadelphia: Neighborhoods, Division, and Conflict in a Postindustrial City.* Philadelphia: Temple University Press.

Cuff, D. J., et al. 1989. *The Atlas of Pennsylvania.* Philadelphia: Temple University Press.

Cutler, W. W., and H. Gillette Jr. 1980. *The Divided Metropolis: Social and Spatial Dimensions of Philadelphia, 1800–1975.* Westport, Conn.: Greenwood Press.

Cybriwsky, R. A. 1968. "Social Aspects of Neighborhood Change." *Annals of the Association of American Geographers* 68:17–33.

Cybriwsky, R. A., and T. A. Reiner. 1982. "Philadelphia in Transition." *Focus* 33:1–16.

Cybriwsky, R. A., and J. Western. 1982. "Revitalizing Downtowns: By Whom and for Whom?" In *Geography and the Urban Environment,* ed. D. T. Herbert and R. J. Johnson. Chichester, U.K.: John Wiley & Sons.

Golab, C. 1977. *Immigrant Destinations.* Philadelphia: Temple University Press.

Hershberg, T., ed. 1981. *Philadelphia: Work, Space, Family, and Group Experience in the Nineteenth Century.* Oxford: Oxford University Press.

Levy, P. R. 1978. *Queen Village: The Eclipse of Community.* Philadelphia: Institute for the Study of Civic Values.

Ley, D. 1974. *The Black Inner City as Frontier Outpost: Images and Behavior of a Philadelphia Neighborhood.* Washington, D.C.: Association of American Geographers.

Lowe, J. R. 1967. *Cities in a Race with Time: Progress and Poverty in America's Renewing Cities.* New York: Random House.

Muller, P. O., K. C. Meyer, and R. Cybriwsky. 1976. *Metropolitan Philadelphia: A Study of Conflicts and Social Cleavages.* Cambridge, Mass.: Ballinger.

Philadelphia City Planning Commission. 1976. *Philadelphia: A City of Neighborhoods.* Philadelphia: Planning Commission.

———. 1988. *The Plan for Center City.* Philadelphia: Planning Commission.

Warner, S. B., Jr. 1968. *The Private City: Philadelphia in Three Periods of Its Growth.* Philadelphia: University of Pennsylvania Press.

Weigley, R. F., ed. 1982. *Philadelphia: A 300-Year History.* New York: Norton.

21

PITTSBURGH: AN URBAN REGION IN TRANSITION

E. Willard Miller

Pittsburgh at the fork of the Monongahela, Allegheny, and Ohio rivers has one of the most strategic geographical positions in the world. Since the first settlement, the city has evolved from a military base to a trading post, and in the nineteenth and early twentieth centuries the city was the apex of one of the world's largest heavy-industrial regions. With the decline of the heavy-industrial economy, the city is experiencing a transition to a service economy. In each of these stages of development, the city has been the center of activity. The river valleys have provided the lowlands along which the activities have spread.

Frontier Settlement

The pioneer settlement of Pittsburgh began in 1750 when a group of Virginians, who had formed the Ohio Company in 1747 to engage in land speculation and Indian trade, secured a grant of 250,000 acres in southwestern Pennsylvania. Pittsburgh began as a fortress settlement when in March 1754 the British built Fort Pitt at the fork of the three rivers. Pittsburgh became a point of conflict between the French and English in the French and Indian War. The French recognized the importance of this location and forced the British to evacuate the fort in July 1754, holding it as Fort Duquesne until 1758, when the British once again gained control. With the withdrawal of the French from their western outposts in 1759, the British once again took possession and enlarged the fort, renaming it Fort Pitt.

In 1763, with the outbreak of Indian attacks in the Pittsburgh area, all white settlers were ordered into the fort from the region. The refugees totaled 630. In the Indian attacks the many settlements outside the fort were destroyed.

As the Indian attacks subsided, settlement of the region progressed, and by the late 1760s Pittsburgh was changing from a military outpost to a trading

Fig. 21.1. The old and the new in downtown Pittsburgh.

post, assuming a position as the commercial gateway to the West. Pittsburgh possessed the most strategic location for the distribution of merchandise from the east to new settlers moving westward along the Ohio River valley. The goods came by wagon train from such eastern cities as Philadelphia and Baltimore to Pittsburgh. Pittsburgh became a transshipment center of warehouses, and boat-building emerged as one of the earliest industries. Pittsburgh flourished as a trade center, with population increasing from about 300 in 1780 to nearly 5,000 in 1810.

Gateway to the West

By the early nineteenth century the demand for goods locally and in the western frontier market had increased so rapidly that industry was developing in Pittsburgh. By the time Pittsburgh was incorporated in 1816, its economy no longer rested solely on commerce. Growth in manufacturing had given the town a new role in the productive system of the region. Its iron industry was well established. The first iron foundry had been built in 1793, and the industry had advanced so rapidly that the first rolling mill was built in 1812. In addition, there were glass factories, breweries, potteries, cotton and wool factories, printing offices, a steam engine factory, and metal fabrication plants producing such common articles as nails, shovels, scythes, hoes, axes, and frying pans. Between 1812 and 1826 some 48 steamships were built in Pittsburgh for the western river trade. In addition, barges, flatboats, and keelboats were constructed. During the Mexican War, and later the Civil War, Pittsburgh became a leading center for the output of guns, cannon, and ammunition. The city became a major importer of iron, lumber, glass sand, cotton, and wool to support the industrial growth.

From its industrial origin Pittsburgh became known for its dirty industrial environment. In 1828 a visitor wrote: "The city lay beneath me, enveloped in smoke—the clang of hammers resounded from the numerous manufactories." And when Charles Dickens visited Pittsburgh in 1842, he wrote: "Pittsburgh is like Birmingham in England, at least its townspeople say so." By the middle of the nineteenth century, a diversified industrial economy had evolved in Pittsburgh.

As a response to economic development, Pittsburgh was the largest city in the interior. Its population grew rapidly from about 375 in 1790 to about 6,000 in 1815. By 1830 the city had 12,500 people, with an additional 10,000 in the surrounding area. In 1860 the population reached 50,000, with about 100,000 more in Allegheny County. Migration was primarily responsible for the growth. The newcomers came not only from the eastern United States but also from Europe. At least one-third of the inhabitants in 1860 were European immigrants, largely from Ireland and Germany, followed by English, Scots, and French.

Hearth of the Nation

The evolution of Pittsburgh from a commercial and diversified industrial center to the nation's principal iron and steel and heavy-industry center in the last half of the nineteenth century was due to a number of unique advantages the area possessed. Before 1860 the iron industry was localized in eastern Pennsylvania, utilizing anthracite as a fuel. Widely distributed iron furnaces throughout the state were using small, local bog iron ore deposits and charcoal. The catalyst that resulted in the centralization of the iron and steel industry in the Pittsburgh region was a technological change in iron-making and steel-making from the small anthracite furnaces to the larger Bessemer converter, which required coke to support the heavy furnace load of iron ore and limestone. Pittsburgh was particularly favored, because the Pittsburgh coal seam, one of the few Appalachian coals that could be coked in the beehive oven process, had massive reserves in southwestern Pennsylvania in Fayette and Westmoreland counties. Although western Pennsylvania had sufficient iron ore deposits for the small charcoal furnaces, the nearest source of large iron ore deposits was in the Lake Superior fields. The iron ore was transported by lake ore vessels to ports on Lake Erie like Ashtabula and Conneaut, and then carried by railroads to Pittsburgh's iron and steel centers. The industry was fuel-oriented, for it required about two tons of coal to smelt one ton of Lake Superior iron ore. The integrated iron and steel industry thus developed at the place of greatest cost advantage in transportation for assembly of fuel and raw materials and the marketing of the finished products. This location was found to be in the Pittsburgh area, particularly in the Monongahela Valley and to a lesser extent in the Ohio, Allegheny, and other river valleys.

In the 1870s and 1880s Pittsburgh became the heavy-industry heart of the nation. As the district's iron and steel industry grew in importance, new heavy industries that depended on steel migrated to the area, primarily to save freight charges on the transportation of iron and steel. Heavy industry occupied most of the valleys. Scores of blast furnaces, rolling mills, and factories belched smoke into the congested valleys. Pittsburgh became known not only as the "Steel City" but also as the "Smoky City."

The period from 1860 to 1914 was one of tremendous economic expansion. At the turn of the century the Pittsburgh area produced one-third of the nation's steel, 40 percent of the nation's coal, and about 50 percent of glass production, and accounted for 30 percent of the nation's rail tonnage. Pittsburgh's dominance in the heavy metallurgical industry is well illustrated by the control the industrialists were able to impose on freight rates. They established the freight rate system known as "Pittsburgh Plus." Its primary purpose was to stabilize the industry in the region by discouraging competition. The freight rate policy operated in this manner: If a Kansas City, Missouri, buyer purchased steel from a Gary, Indiana, plant, the purchaser paid the f.o.b. Pittsburgh price and also the freight rate that would have been assessed against shipment from a Pittsburgh mill (the "plus"). In 1900 the cost of producing iron and steel was lower in the Pittsburgh region than anywhere else in the nation. Shortly after 1900, however, the cost of producing iron and steel in a number of places, notably the Chicago region, fell below that of Pittsburgh. As a result of this system of freight charges, however, Pittsburgh possessed a tremendous competitive advantage over all other regions. The Pittsburgh industrialists were able to maintain this freight rate system from the 1880s to 1924, when it was declared unconstitutional by the U.S. Supreme Court.

A number of other factors were important in making the region a dominant steel center. The transportation system in which the Pennsylvania Railroad occupied a key role was one of the best developed in the nation and gave the region access to a national market. The concentration of railroad lines in the valleys centering on Pittsburgh made the city a railroad center as inevitably as it was an inland port of an earlier period. The industry of the region was directed by a group of aggressive industrial and financial giants. Of them, the Pittsburgh steel industry owes most to the organizing genius of Andrew Carnegie. He entered the iron and steel business in 1863 and within a few years merged a number of small companies into the Carnegie Steel Company. In the succeeding decades Carnegie built blast furnaces to supply his steelworks, gained control of the largest coke companies in the Connellsville region, and controlled a large share of the Lake Superior ores to ensure a steady flow of ore to his mills. He also owned or controlled steel-fabricating plants to secure a guaranteed market for his basic steel. When Carnegie retired in 1903, his company became the nucleus for the U.S. Steel Corporation. For many years he dominated the entire economic structure of Pittsburgh and was respected as the steel magnate throughout the world.

Besides the great iron and steel companies, Pittsburgh developed a number of other major industrial empires in the nineteenth and early twentieth centuries. Of these, the Westinghouse Electric Corporation was one of the first major electrical machinery companies in the world. George Westinghouse, through the development of the alternating current, made possible the transmission of electricity. No single person contributed more through invention and technology to the evolution of the electrical machinery industry.

Pittsburgh's industrial economy spawned a number of great financial leaders. The creation of numerous fianncial empires was a direct result of the genius of such men as Andrew W. Mellon. The Mellon bank provided the financial backing for the creation of some of the largest corporations in the nation. One of the earliest financial endeavors was the backing of the infant aluminum industry in Pittsburgh, which became the Aluminum Company of America (ALCOA). In 1896 Mellon became a partner with Edward G. Acheson to form the Carborundum Company. In 1901, when the prolific oil strike was made at Signal Hill in Texas, Andrew Mellon was the principal financier in the development of the Gulf Oil Corporation, the first major oil company to provide competition to the oil empire of John D. Rockefeller. With two young engineers, Mellon formed a construction company that built the Panama Canal locks, the George Washington Bridge, and the Waldorf-Astoria Hotel. He controlled the Pittsburgh Coal Company, and in 1918, when he saw the value of Henry Kopper's by-product coke ovens (the ovens recovered vast quantities of hydrocarbon materials that had previously been lost in the beehive coking process to produce a wide variety of chemical products) he and three associates purchased the American assets of the German-controlled company. In all, Andrew Mellon held directorships in more than 62 corporations. He was certainly among the most important financiers in the United States between 1880 and the 1920s. He exemplified the archtypical capitalist that evolved out of the industrial revolution of the nineteenth century.

As the economy expanded, the population of Pittsburgh grew rapidly. By 1880 the population of the city was 156,000, a three-fold increase from 1860, and the metropolitan region in the 20-year period had grown from 178,000 to about 250,000. The growth in the political city was due not only to immigrants and natural increases but also to the annexation of outlying towns of the East End (Oakland, East Liberty, and Lawrenceville) in 1867 and of the southside boroughs on the left bank of the Monongahela in 1872. With these acquisitions the area of the city increased more than 15 times between 1866 and 1872.

The 1880s experienced the maximum rate of increase in population when the annual growth was about 4 percent. By 1910 the population of Pittsburgh was 533,905, an increase of 377,516 since 1880, and during the same period the population of Allegheny County increased from 355,869 to 1,018,463. In the decade 1910–20, population slowed greatly, and in 1920 the city's population was 588,343.

An Age of Stagnation

By the 1920s, Pittsburgh was an old industrial region. There had been little urban planning, so the city sprawled across the hills. The streets were congested and the rivers were polluted. Floods inundated the streets at the confluence of the river systems. In his *Prejudices,* Henry L. Mencken described the industrial landscape of Pittsburgh:

> Here was the very heart of industrial America, the center of its most lucrative and characteristic activity, the boast and pride of the richest and grandest nation ever seen on earth—and here was a scene so dreadfully hideous, so intolerably bleak and forlorn that it reduced the whole aspiration of man to a macabre and depressing joke. Here was wealth beyond computation, almost beyond imagination—and here were human habitations so abominable that they would have disgraced a race of alley cats.
>
> I am not speaking of mere filth. One expects steel towns to be dirty. What I allude to is the unbroken and agonizing ugliness, the sheer revolting monstrousness, of every house in sight.

The economy was dominated by heavy industry. These major forms of industrial activity were all capital-goods industries closely tied to the stage of industrial development in the nation as a whole. As the national economy expanded, new centers of production were developing elsewhere. Consequently, the rate of growth in the iron and steel industry was near zero, and the coal and coke production exhibited declining trends. The district was so dominated by coal, coke, and steel production that generally favorable rates of growth in glass, aluminum fabrication, and electrical equipment were of limited importance in raising industrial output.

In the period from 1920 to 1940, Pittsburgh's industry did not keep pace with a changing national economy. In the nineteenth century the metallurgical industry expanded, with demands for steel in construction and railroad transportation. In the twentieth century, however, these demands declined, for the new expanding industries required a different type of steel. Unfortunately, there was little development of such twentieth-century industries, such as motor vehicles, in Pittsburgh. Because coal and allied coke production was directly related to the output of the heavy industries, their production had a corresponding drop, which was heightened by increasing technological efficiency.

With decline, the area began to experience industrial structural changes. These included the marked slowing of manufactural output, rapid increase in service activities, a marked geographical shift of manufacturing from the city of Pittsburgh to the smaller industrial cities, especially along the Ohio, Monongahela, and Allegheny rivers, and the centralization of the service functions in Pittsburgh.

A major effect of economic maturity was not only the slowing of capital investment but also the migration of capital to other industrial centers. As local investment opportunities declined, the Pittsburgh industrialists sought "economic colonies" in other regions. In part, migrating investments took the form of branch plants of Pittsburgh-controlled concerns. When capital was invested in Pittsburgh, it did not attract new industries but was used primarily to further mechanization and rationalization of existing industries.

As industrial expansion slowed, unemployment problems began to appear in the 1920s. Even in the peak economic year of 1929, unemployment was 5 to 10 percent in many towns of the region. From 1929 to 1932 the decline in production and employment was drastic, with only a modest recovery before World War II. As a consequence, population growth during this period was modest. Between 1920 and 1940 population in the city of Pittsburgh increased slowly, from 588,343 to 671,659, and in Allegheny County from 1,185,808 to 1,411,539.

Out of the great industrial growth of the nineteenth century, two classes of people dominated the scene. On the top was a very small group of wealthy industrialists, businessmen, and financiers, and at the bottom the overwhelming number of industrial workers. The wealthy lived in ornate mansions that cost millions to build and were furnished with elegant and imported tapestries and silks for wall hangings. For example, the Fifth Avenue residence of R. B. Mellon had 65 rooms and 11 baths; Henry C. Frick's home at Penn and Homewood had 30 rooms; and the house of Michael Benedum, "the great wildcatter," on Woodland Road (now Chatham College) had marble pillars and statuary, walls paneled in oak, a dining and living room in red mahogany, and a billiard room in limed oak. The McCook mansion had its own chapel. When the wealthy entertained, they did so with formal parties, dinners, or cotillions at such exclusive clubs as the Pittsburgh Club or the Pittsburgh Golf Club. Many homes had ballrooms and even bowling alleys.

In contrast to the well-to-do, the working people lived in overcrowded tenements or simple houses without sanitary facilities. As late as 1934, 3 out of 4 houses on the lower North Side or in Woods Run had neither hot water nor bathrooms, and 9 out of 10 were without furnace heat. The workers spent their leisure time in bars and saloons, at sports events, or at the movies. The immigrants from foreign countries located in specific districts, where the Nationality Halls—for example, the National Slovak Society, the Greek Catholic Union, the Polish National Alliance, the German Turn, and Gesangverlins—became the focus of activities. The Russian-Americans, the Serbian-Americans, the Hungarians, the Czechs, the Croats, the Slovenians, the Italians, and the Irish all had national organizations. In their halls they spoke in their native tongue and could find newspapers from their country of origin.

The Renaissance Era

During World War II, Pittsburgh once again became the "forge of the nation." In four years of war, 70 million tons of steel ingot, valued at $19 billion, were produced. But the war had brought no new industries, and doubts again grew about the economic future of Pittsburgh and its region. In addition, the city was older, grimier, and more unlovely than ever. Although Pittsburgh had some buildings of distinction, such as the Mellon Bank (1923), the Cathedral of Learning of the University of Pittsburgh (1926), the Koppers Building (1929), the Gulf Building (1932), and the East Liberty Presbyterian Church (1935), the business and financial heart of the city reflected the congestion of its nineteenth-century origin. Pittsburgh's housing problems persisted, with 40 percent of the area's dwellings in the late 1940s considered inadequate for habitation. Furthermore, most of the mill towns in the suburban region struggled with vast slums—a heritage of lack of planning in the past.

Catalysts for Change

It became abundantly clear in the 1940s that Pittsburgh's image as a mill town, a coal town, a railroad town, and a river town, with its grit and filth, did not have a future. The civic, business, and industrial leaders saw the need for a more hospitable environment. They wanted an environment where a person could go to the office downtown and put in a day's work without bringing along a second shirt to change during the noon hour because by noon the morning shirt had been so soiled by soot and grime from the city's mills and factories that it was no longer presentable.

Two men are credited with being the catalysts that transformed the "Smoky City" into America's "Renaissance City." The political forces were led by Mayor David L. Lawrence, and the financial and industrial forces were led by Richard King Mellon, head of the Mellon interests. These two men had little in common except power and the knowledge of how to use it. Mellon, from an elite patrician family, was a conservative Republican, a philanthropist, and a Protestant. Lawrence came from an Irish worker's family. He was a liberal Democrat politician, a pragmatist, a Catholic, and an iron-fisted product of the ward wars. At an earlier time these two men would probably never have met, but at this time of urban crisis they gave direction to what was to become known as "the Renaissance." They commanded a force of planners, dreamers, and visionaries and molded them into a powerful team of doers and achievers in both the private and the public sectors.

Implementation

In the mid-1940s the Allegheny Conference on Community Development was established to formulate the revitalization procedures. The resulting plan was unique for Pittsburgh and may have been unique for the nation. The basic long-range goal of this group was a complete inventory of the resources, industries, people who were part of the labor force, housing, and transportation to determine the needs and potentials for growth. In order to focus attention on the revitalization, the first major project was rehabilitation of the Golden Triangle—a 59-acre tract bounded by Stanwix Street and the Allegheny and Monongahela rivers. It had become a warren of decayed buildings, dilapidated warehouses, parking lots, and ugly railroad yards.

While the rehabilitation of the Golden Triangle gained the attention of the average Pittsburgh resident and provided the catalyst for change, the first test of the conference's power to revitalize Pittsburgh came when smoke-abatement legislation was proposed. Although the legislation was quickly passed, a major struggle began when it was to be implemented. The coal companies and many industries wanted implementation postponed indefinitely. It was only when Richard K. Mellon appealed to the opposing factions that the battle ended. As a result, the effective date for enforcement of the legislation for industry and commercial buildings was set for fall 1946, and for home furnace conversion on October 1, 1947. In 1949 smoke abatement control was extended to all of Allegheny County. By 1954 the Weather Bureau reported that Pittsburgh was receiving 89 percent more sunshine than before smoke control. The hours of heavy smoke had been reduced by more than 94.4 percent. Dieselization of the railroads, and conversion to oil and natural gas for home heating, practically solved the problem of railroad and household smoke. "If we could clear the smoke away, if we could make the sun visible, we can do anything" was the spirit of the times.

In its next phase the Allegheny Conference began plans to develop a comprehensive revitalization program. In order to accomplish this, aid from the state government, in addition to local legislation, was necessary. In 1947 the state assembly spent most of its session passing enabling legislation known as the "Pittsburgh Package."

As a response to these actions, the program included (1) a highway program to relieve congestion by building expressways to the city center; (2) the construction of flood-control dams, such as the Conemaugh and Kinzua, to control floods, of which Pittsburgh had 64 between 1900 and 1950; (3) the development of a County Sanitary Authority to ban dumping of raw sewage into streams and rivers; (4) the construction of parking garages and off-street parking lots; (5) a new development corporation to prepare plant and industrial sites; (6) a program to improve blighted housing in the area; and (7) a plan to develop better labor-management relations.

Many of these projects saw immediate results. A number of major parkways were completed in the 1950s. Eight major flood-control dams were constructed on the Allegheny and Monongahela river systems. More than 60 communities in Allegheny County entered into an agreement to develop a system of sewer collectors and disposal plants. A 330-acre tract in the Golden Triangle, which came to be known as the Gateway Center, was revitalized. Large parking garages were strategically located in the central city. The development of the housing program lagged and is probably the least successful of the early revitalization programs.

The original Renaissance spirit lasted from the mid-1940s to the late 1960s, but sustained rejuvenation cannot last forever. Why did the high cycle of creativity and productivity decline? Could it be that the death of David L. Lawrence in 1966 and Richard K. Mellon in 1970 marked the end of a period of dynamic leadership? Could it be that the living conditions of the average citizen were little changed by the Renaissance? Could it be that the traditional economy continued its decline in spite of the Renaissance?

Although the 1970s were not filled with the crusade spirit of the earlier time, it was not without change. The financial center of the city grew with the construction of such large structures as the U.S. Steel, Pittsburgh National Bank, and Westinghouse buildings. The magnificent Heinz Hall, where the world-renowned Pittsburgh Symphony performs, was refurbished. Three Rivers Stadium, where the football Steelers and the baseball Pirates play, was erected, and new art galleries and museums included the Scaife Gallery and the Old Post Office Museum.

In the late 1970s and early 1980s there was a rekindling of the earlier spirit of revitalization. This movement has come to be known as "Renaissance II." This new development is centered on Market Square, consisting of some 20 blocks in the heart of the Golden Triangle. Renaissance II, a $4 billion-plus program, has not only built several skyscrapers but also is revitalizing a number of city neighborhoods. Between 1980 and 1986 more than 6 million square feet of prime office space was built, along with 800,000 square feet of renovated space. This construction recognizes that if Pittsburgh is to flourish it must become a major service center for corporations.

Neighborhoods of Pittsburgh

There are more than 40 neighborhoods in the city of Pittsburgh and its suburban areas (see Fig. 21.3). Of these, three—the Golden Triangle, The Hill district, and Oakland—have played special roles in the city's development. Because of the city's rugged topography, with deeply incised valleys, steep slopes, and rolling uplands, nearby neighborhoods are isolated. Many of these areas became the home of distinct ethnic groups when they migrated to Pittsburgh, and the topographic isolation has preserved their

Fig. 21.2. Pittsburgh skyline about 1970.

cultural origin. There is a substantial German population on Troy Hill and Spring Hill; Polish on Polish Hill; Jews and Episcopalians in Squirrel Hill; Slavs, Ukrainians, and Russians on the South Side; Italians in Mount Washington, Bloomfield, and Junction Hollow; Blacks on The Hill, in Homewood, and in Manchester; Arabs in Oakland; and the founding Scots-Irish in the suburb of Fox Chapel and Sewickley Heights.

The Golden Triangle

The Golden Triangle (Fig. 21.4) is a small area where no two points are more than a 15-minute walk apart. It has always been strategic, whether in military, industrial, or corporate prestige. In 1784 surveyor George Wood established the basic street pattern of Pittsburgh by creating a triangular town, with one street parallel to the Allegheny, another parallel to the Monongahela, each converging at the Point where the rivers meet. Until the 1830s, Pittsburgh's growth was limited to the Triangle, but by the 1880s newer suburbs had grown up around it and the Triangle lost its industry and most of its population and churches. By this time the Triangle was assuming its present function by becoming the corporate trading and financial center of the region.

By the early twentieth century, a few developers controlled most of the land in the Triangle. The railroads held the areas along the waterfronts for tracks and warehouses. Of the builders, Henry Frick erected four skyscrapers on Grant Street, Henry Oliver developed a block along Sixth and Oliver avenues, and a third partner erected five more office buildings on Sixth Street and the Allegheny riverfront. The Mellons, who owned more land than anyone else, built a half-dozen buildings.

While the Triangle was being developed internally, it also became the focus for all transportation in Allegheny County. The railroads were the first major means of land transportation. Shortly after 1900, four major carriage roads (Schenley Drive and Bigelow, Beechwood, and Washington boulevards)

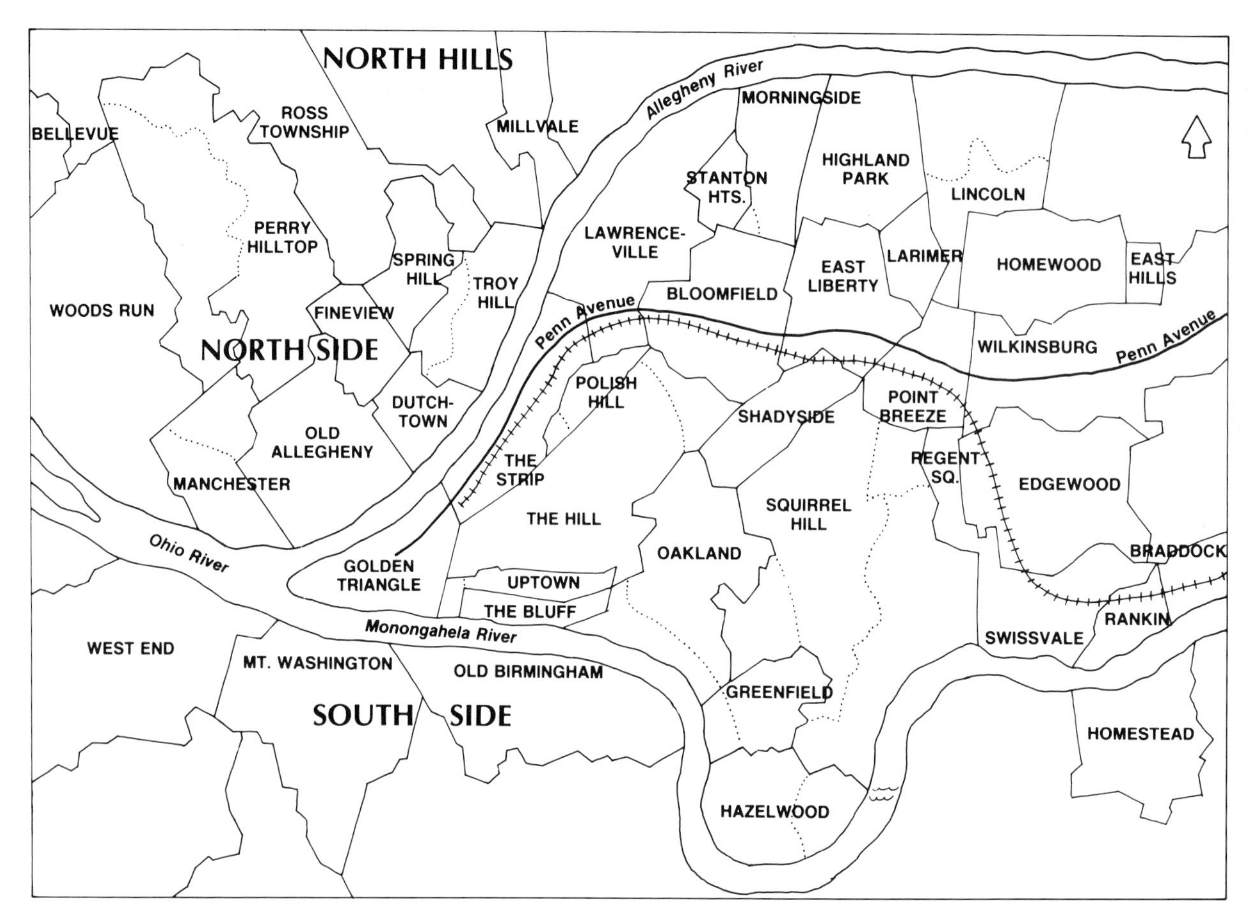

Fig. 21.3. Neighborhoods of Pittsburgh.

were extended into the residential districts of Oakland, Squirrel Hill, Shadyside, Point Breeze, and Highland Park, providing access to the city. Between 1922 and 1927 the Boulevard of the Allies was built along the Monongahela River. This was followed by the Liberty Bridge and Liberty Tunnel, providing a connection to the suburbs south of the Monongahela. Finally, in the 1950s the Penn-Lincoln Parkway was built, connecting Pittsburgh to the Pennsylvania Turnpike on the east and Pittsburgh Greater Airport on the west.

In the 1950s the Golden Triangle was a maze of old buildings and, due to lack of planning, experiencing great traffic congestion. A basic goal of Renaissance II was revitalization of the Golden Triangle. This progressed by removing the industrial activities at the Point and creating Point State Park. The next stage was building Gateway Center, located between Commonwealth Place and Stanwix Street.

In the development of the Point State Park, it was first proposed in 1946 to use the space for a historic park and office complex, but fortunately the idea of an office complex was abandoned. Instead, in the 1950s two bastions of Fort Pitt were rebuilt using archaeological drawings for accuracy. The reconstructed fort became the Fort Pitt Museum, with dioramas on the military and early industrial history of Pittsburgh. The pentagonal Blockhouse, many times modified, is located opposite the Fort with much of the original structure intact. At the very top of the Point, a magnificent fountain has been built. This fountain has a geyser that shoots water 150–300 feet in the air, spewing 6,000 gallons of water a minute from a 100,000-gallon hidden reservoir. It provides a spectacular sight at night and has become the symbol for the revitalized city.

Gateway Center was the first postwar investment to revitalize Pittsburgh as a modern city. The Equitable Life Assurance Society provided the investment funds to build the Gateway Plaza between 1950 and 1953 and Equitable Plaza between 1955 and 1968. Before the buildings were erected, such

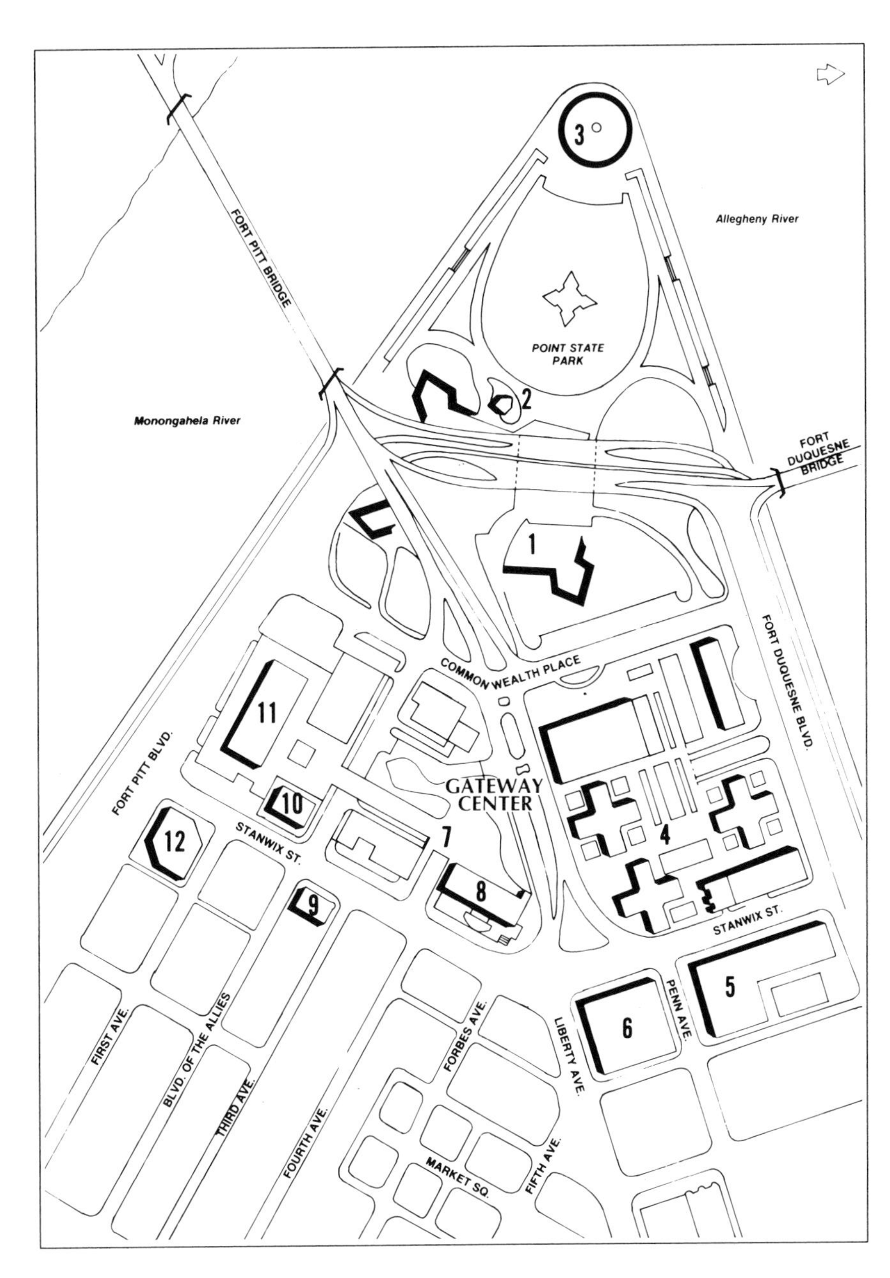

1 Fort Pitt
2 The Blockhouse
3 Point State Park Fountain
4 Gateway Plaza
5 Horne's
6 Fifth Avenue Place
7 Equitable Plaza
8 Four Gateway
9 St. Mary of Mercy Church
10 United Steelworkers
11 Westinghouse
12 Riverfront Center

Fig. 21.4. The Golden Triangle: The Point and Gateway Center.

Fig. 21.5. Early Renaissance of the Triangle, about 1960.

Pittsburgh corporations as J&L Steel, PPG Industries, People's Gas, and Westinghouse signed leases to occupy Gateway Center. These buildings demonstrate the importance of modern architecture in creating not only an attractive environment but also a functional city area. The open space around the building is used by thousands of office workers during the noon break, and in June the Three Rivers Arts Festival takes over the checkerboard terraces and vine-covered arbors for craft exhibits and performances that attract half a million visitors during its fortnight of existence.

Renaissance II concentrated on the centralization of 20 blocks in the heart of the Golden Triangle from Stanwix to Smithfield streets. The construction of the world headquarters for PPG industries from 1979 to 1984 was of fundamental importance to the continued urban renewal of downtown Pittsburgh. The PPG Place, the core of the development, occupies five-and-a-half acres in a complex of eight elements and six city blocks. The central element is a half-acre green plaza with a 40-story glass tower in front of it and five low-rise structures around it. All the buildings are sheathed in uniform mirror glass—20,000 pieces of silver PPG Solarban in the tower alone. Besides the spectacular PPG Place, Renaissance II has been responsible for a number of changes. These include the rebuilding of Grant Street, the Lawrence Convention Center, Liberty Center, Oxford Centre, Mellon Bank Center, Chatham Center II, Fifth Avenue Place, and the Consolidated Natural Gas Tower in the Golden Triangle, with spin-off developments in such adjacent areas as The Strip, Firstside, the South Side, and the North Side. Since the late 1970s, there has been about five times the building in the Golden Triangle as in the original Renaissance I. A subway system is in its initial stage of development. Renaissance II represents a technocratic process rather than the autocratic process that implemented the earlier development. The forces behind the present development have been the late Mayor Richard Caligiuri, the Allegheny Conference on Community Development, and the Urban Redevelopment Authority.

The Hill

The two miles between the Golden Triangle and Oakland is a poor residential area known as The Hill. Its rugged topography was the initial home for almost all the ethnic groups who came to Pittsburgh. These included Germans, Italians, Russians, Slovaks, Armenians, Syrians, Lebanese, Greeks, and Chinese, and it remains a home area for two groups who have lived there for generations. It is the center of the Black community of about 100,000, and of the 33,000 Jewish community.

The Hill has never played a significant role in the industrial development of the region, nor has it assumed a role in commercial relations. Its greatest contribution has been as a source of labor for economic activities outside the region. Over time it degenerated into one of the poorest residential areas in the city, and it has been noted for its crime.

In Renaissance I the reconstruction of the Lower Hill, nearest the Golden Triangle, became a priority. The project began in 1955 with a $17 million federal grant. Within the next few years in an area of 100 acres, 1,300 buildings including 413 small businesses and 8,000 residents were removed. This had a devastating impact on the social structure of the area. The program was, however, flawed from its beginning. Because the Crosstown Expressway separated the Lower Hill from the Golden Triangle, the two regions remained separated. Initially it was planned that the new concert hall, supported by the Heinz Foundation, would play a key role in providing a cultural center in the region, but instead the Heinz Foundation decided to locate its hall in the Triangle rather than The Hill. Furthermore, one of the major developers, William Zeckendorf, went into bankruptcy and many plans did not materialize. The great failure, however, was the animosity between the developers and the Black community, which came to a head in 1968 when rare riots occurred. Development ceased for many years.

The major development in the area is the Civic Arena. It began as a project in a cable-supported opera tent in the late 1940s. After long delays it evolved into an opera hall with a retractable dome for summer performances in the open air. In 1962 it opened with a stainless-steel dome 415 feet in diameter composed of eight sections. Each of the sections weighs 220 tons, and six sections can be opened in a little over two minutes. The seating capacity of the auditorium is 18,000. It is now used mainly for sports events and popular concerts.

Most of the Lower Hill is now taken up by the ubiquitous parking lots and a half-dozen new buildings in Chatham Center. Duquesne University lies on the bluffs above the Monongahela River. The only original building left in the lower Hill district is Epiphany Church. The small Beth Hamedrash Hagodol Synagogue, built in the 1960s, amalgamated several of the two dozen Jewish congregations scattered over The Hill.

Until about 1880 the Middle and Upper Hill were well-to-do residential areas with views of both the Allegheny and the Monongahela rivers. By the 1840s it had a dozen estates with Greek Revival mansions. By the 1860s there were hundreds of townhouses and it had become the home of most of the professional people of Pittsburgh. With the great influx of Central and Eastern Europeans, the residential area deteriorated. In recent times the renovated Italianate homes on Webster Street and the Victorian Gothic homes on Bedford Avenue reflect the grandeur of the past. The most striking evidence of the past, however, remains in the splendid churches. Of these, Saint Benedict the Moor Roman Catholic Church, Saint Michael's Russian Orthodox Church, the Miller Street Baptist, Mount Rose Baptist, and Zion Hill Baptist are the best examples. For generations the commercial and social center of this area has been on Centre Avenue between Roberts and Kirkpatrick.

The Middle Hill has one of the largest of the pioneer federal housing projects in Terrace Village. The project began in 1937, with President Roosevelt approving part of it on a visit to Pittsburgh in 1940. The development extended into the 1960s, when 3,727 units in 191 buildings had been completed. To build Terrace Village, three hills—Gazzam's, Goat, and Ruck's hills—were leveled by bulldozers and the valleys filled in. While this created a level area, it separated Terrace Village from the remainder of The Hill district. It became a separate and distinct enclave that has persisted to the present time as a period piece of a bygone day.

Oakland

While the Golden Triangle became the commercial center of Pittsburgh, Oakland evolved as the educa-

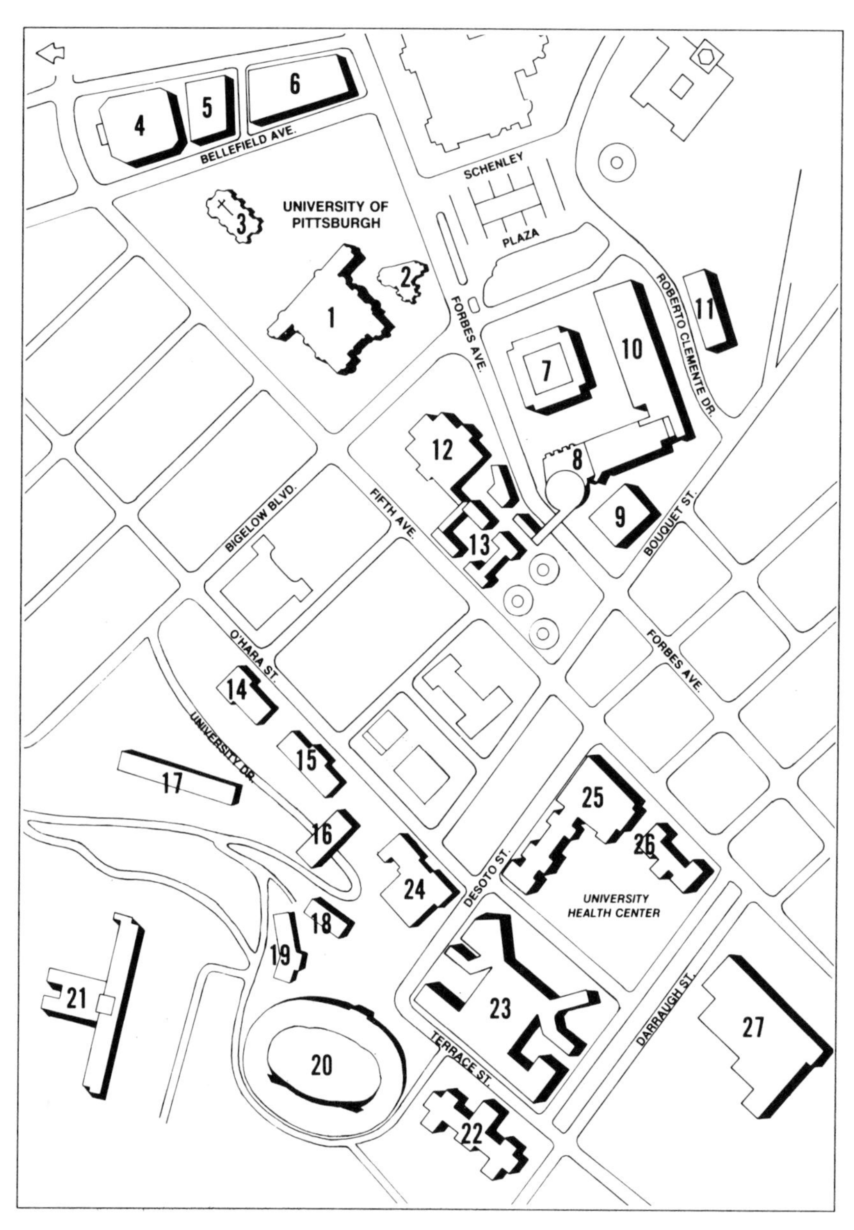

1 Cathedral of Learning
2 Stephen Foster Memorial
3 Heinz Chapel
4 Mellon Institute
5 315 S. Bellefield
6 Board of Public Education
7 Hillman Library
8 David L. Lawrence Hall
9 Law Bldg.
10 Forbes Quadrangle
11 Mervis Hall
12 William Pitt Union
13 Schenley Apts.
14 Thaw Hall
15 Allen Hall
16 LRDC
17 Alumni Hall
18 Mineral Industries Bldg.
19 Pennsylvania Hall
20 Pitt Stadium
21 Veterans' Administration
22 Salk Hall
23 Presbyterian-Univ. Hospital
24 Western Psych. Institute
25 Children's Hospital
26 Falk Clinic
27 Montefiore Hospital

Fig. 21.6. Chancellor Bowman's Oakland: The Cathedral of Learning, University of Pittsburgh.

tional and cultural center of the region (Fig. 21.6). Oakland is sharply detached from the rest of the Pittsburgh region by the valleys of Junction and Panther hollows to the east, the Monongahela to the south, and the rugged topography of The Hill to the north and west. For most of the nineteenth century, Oakland developed as a residential and industrial area. It was in 1889 that the character of the region changed, when Mary Schenley gave 400 acres of farmland to create Schenley Park on the eastern border of Oakland. To develop the area, Andrew Carnegie secured 20 acres of the park to construct a huge library, museum, and music hall. Carnegie quadrupled his complex in 1907.

Carnegie's patronage was the sign for other industrial leaders to enter the area. In 1893 Henry Phipps created a huge flower conservatory. Many of the builders attempted to re-create the buildings of the World Columbian Exposition held in Chicago in 1893. The new campus of the Carnegie Technical School (later the Carnegie Institute of Technology and now Carnegie Mellon University) was built between 1903 and 1922 in the general shape of the Chicago exposition's midway.

A major developer of the residential area of Oakland was Franklin Nicola. In 1898 Nicola built the Schenley Hotel, a stock venture supported by Carnegie, Frick, A. W. Mellon, Westinghouse, and Heinz. The hotel remains today as the University of Pittsburgh's student union. In 1905 Nicola's Schenley Land Company acquired the remainder of the Schenley estate in the heart of Oakland and planned to build a model city with a separate character for each of its four quarters. The two residential quarters were planned to have 135 houses; the educational quarter, planned in 1908, was to be a vast Acropolis for the University of Pittsburgh; and the fourth quarter was to house the soldiers' and sailors' Memorial Hall, the Masonic Temple, and other civic functions, as well as a half-dozen private clubs. There was also Forbes Field for professional baseball, which has since moved to Three Rivers Stadium on the Northside in Old Allegheny. The basic concept of an educational and cultural center was completed by 1910.

Since 1910 the Cathedral of Learning has been built, replacing the concept of the Acropolis. In 1920 John Bowman, chancellor of the university, planned to build the most spectacular university building in the world in order to raise the importance of education in the dismal industrial environment of the time. The Cathedral of modified Gothic design was begun in 1925 and completed in 1937 and became a symbol of education. For thousands of working-class people, education means entering the University of Pittsburgh. In addition to the Cathedral, the University has functions in more than 50 buildings on 125 acres of land in Oakland. The region has also become a major health center, with several nationally known hospitals where the Salk polio vaccine and synthetic insulin were developed and where pioneer work in heart and liver transplants has been done.

Changing Geographic Patterns of Economic Activities

The political city of Pittsburgh has long been losing its manufacturing to the suburban areas. Between 1950 and 1989 the number of workers in manufacturing within the city was reduced by about two-thirds. The migration of industry to the outlying areas is due to a number of reasons. With the revitalization of the city, particularly the Golden Triangle, the old industrial areas have been removed and replaced by office buildings or parks. Expressways built in the older industrial areas have removed vast areas from industrial development. The use of trucks to transport raw materials and finished products has lessened the concentration of industry along the railroad lines and river valleys. In the process of planning, there is a perception that industry and service functions should be separated. Finally, the suburban areas frequently present a more attractive economic and political environment.

To a considerable degree, the intensity and industrial mix of industry in the Pittsburgh region reflects the time of its origin. The traditional heavy industry continues to occupy the valley bottoms, with the Monongahela Valley having the heaviest concentration, followed by the Ohio and Allegheny valleys. Lighter manufacturing is typically found in the towns away from the rivers. The pattern thus appears as linear along the valleys and as clusters away from the valleys. The heavy industry located in the valleys is in decline, while manufacturing exhibits some growth in the nonriver towns.

There is also a changing pattern in many of the tertiary activities. The central business district of the industrial towns has declined in importance. Vacant

Table 21.1. Allegheny County employees by occupation, 1959–1989.

	1989	1979	1969	1959
Agricultural services	1,482	872	554	219
Mining	3,673	594	2,898	2,764
Construction	34,656	48,681	33,288	20,414
Manufacturing	82,802	155,718	200,288	181,843
Transportation and public utilities	39,968	39,725	36,429	32,114
Wholesale trade	41,366	42,425	41,775	36,424
Retail trade	127,241	110,249	98,796	78,553
Finance, insurance, real estate	48,387	43,387	32,785	27,660
Services	238,077	167,250	109,740	80,180
Total employees	621,871	615,745	557,257	460,116

SOURCE: U.S. Dept. of Commerce, *County Business Patterns*.

buildings characterize large areas of these once-thriving "downtowns." In contrast, large shopping malls have been built in the uplands where parking space is available and land costs less. The automobile has changed the shopping patterns of the people.

Changing Economic Structure

The occupational structure of the Pittsburgh region is in transition (Table 21.1). In 1953 manufacturing in Allegheny County employed 44 percent of the work force. By 1978 the role of manufacturing had declined greatly, with only 26.6 percent of employment. This decline has continued, and by 1990 manufacturing employed only 13 percent of the workers. In addition to the decline in employment, there has been a major shift in the type of manufacturing. The primary metals have experienced the greatest decline, decreasing from 80,000 workers in 1953 to about 10,630 by 1990. Other industries with major declines included fabricated metals, machinery, and electric and electronic equipment.

Like many cities in the Midwest and Northeast, Pittsburgh has been identified as a low-tech area. There is now an attempt to change this image with emphasis on selected aspects of high technology. Pittsburgh has become a center for the development and manufacture of industrial robotics. Carnegie Mellon University has conducted extensive research in robotics technology, and the findings have been applied in the manufacture of robots by Pittsburgh firms.

With the decline of the secondary activities in Pittsburgh, the tertiary activities have continually risen in importance and in 1990 employed about 80 percent of the labor force. Essentially every aspect of the service industries has increased in importance. Of these, a number of services stand out in importance. Pittsburgh is a major financial center in the nation. It remains the third largest city for corporate headquarters in the United States, outranked only by New York and Chicago. Fifteen of *Fortune*'s 500 industrial corporations have headquarters in the city. Pittsburgh is second only to New York in the amount of investment capital controlled by resident corporations. Banking in Pittsburgh is a $50 billion industry, and the city ranks as the fourth richest financial center in the nation. Its two largest banks are the largest in the state. In Allegheny County, financial activities employ about 48,000 people.

Health care has become a major industry. Health services in Allegheny County employed some 75,500 people by 1990—more than a 60 percent increase in the past decade. There are more than 50 hospitals in the region, supported by a $12 billion infrastructure. The University Health Center of Pittsburgh, affiliated with the University of Pittsburgh and its School of Medicine, has earned international acclaim as a medical treatment, research, and teaching organization. The world's first combined heart and liver transplant, for example, was performed at Presbyterian University Hospital, one of the Health Center's member institutions. Almost 4,000 students are in training in the Health Center and Pitt's health professional schools.

The industrial corporations of Pittsburgh now recognize that, if they are to survive, a vast modernization program is a necessity. This can only occur through the application of advanced technology. In the 1990s Pittsburgh-based firms are spend-

Table 21.2. Population of Pittsburgh and its metropolitan area by county, 1950–1990.

	Pittsburgh	Allegheny	Beaver	Butler	Washington	Westmoreland
1950	676,806	1,515,237	175,192	97,320	209,628	313,179
1960	604,332	1,628,517	206,948	114,639	217,271	352,629
1970	520,089	1,605,133	205,418	127,941	210,876	376,935
1980	423,960	1,450,196	204,441	147,912	217,074	392,184
1990	369,879	1,336,449	186,093	152,013	204,584	370,321

SOURCE: U.S. Census Bureau.

ing annually more than $1 billion on research and development (R&D). There are over 20,000 scientists, engineers, and support personnel employed in nearly 170 research laboratories, including 62 major facilities. It is believed that Pittsburgh is the third largest research and development center in the nation. Pittsburgh claims one of every 50 American scientists, engineers, and supporting technical personnel engaged in research and development activities in the nation. Pittsburgh has long been a leader in materials research, but modern demands have required that R&D become diversified.

A number of unique research developments are occurring in the region's laboratories. For example, the U.S. Bureau of Mines, at its 250-acre complex in South Park Township, is conducting fire and explosive testing in its underground mines. The Pittsburgh Energy Technology Center is experimenting with the liquification of coal. Kopper's research and development unit is experimenting with carbon in forms that will be more useful in the biomedical field. The Mobay Chemical Corporation's polyurethane laboratory is engaged in research to produce materials to satisfy the emphasis on lighter cars, safety, and fuel conservation. These are a few examples of the diversity of R&D programs in the Pittsburgh region.

The University of Pittsburgh and Carnegie Mellon University are increasing their roles in research and development. On a site along the Monongahela River, a steel mill was removed for the modern Pittsburgh Technology Center. Carnegie Mellon has created a Center for Advanced Manufacturing and Software and Engineering. The University of Pittsburgh is developing a Center for Biotechnology and Bioengineering. Leaders from both universities believe the new programs will attract industry eager to benefit from access to the universities' research.

The role of government has grown rapidly from about 8 percent of the labor force in 1955 to about 15.5 percent in 1992. This increase in government workers occurred at all levels of the goverment—local, state, and federal.

The wholesale and retail trade establishments provide 168,000 jobs in Allegheny County, 75 percent of which are in retail trade. Of the retail categories, eating and drinking places employ about 39,000 followed by general merchandise stores, food stores, and automotive dealers and service stations.

Population Trends

The population trends of Pittsburgh and its metropolitan area (Table 21.2) reflect the economic conditions of the region. As job opportunities declined, so did population. The city of Pittsburgh reached its maximum population in 1950 with a total of 676,806. Since then the population has steadily declined—to 604,332 in 1960, to 520,089 in 1970, to 432,960 in 1980, and to 369,879 in 1990. This is a loss in 30 years of 234,453 people. The present population is only 61 percent of that of 1960.

Although the population of Pittsburgh has experienced a long decline, the five counties of Allegheny, including Pittsburgh, and the four immediately adjacent counties, Beaver, Butler, Westmoreland, and Washington, have experienced diverse trends (see Table 21.2). Allegheny County reached its maximum population in 1960 with 1,628,517 people. Since then, population declined each decade to 1,336,449 in 1990 and is predicted to decline to 1,290,453 in the year 2000. Beaver County has also experienced a decline in population from a peak of 205,418 in 1970 to 186,093 in 1990 and a predicted continued decline to 179,213 in the year 2000. The population of Washington and Westmoreland counties has declined slightly. Washington County reached its peak in 1980 with a population of 217,074, but by 1990 the population declined to 204,584. It is predicted, however, that the popula-

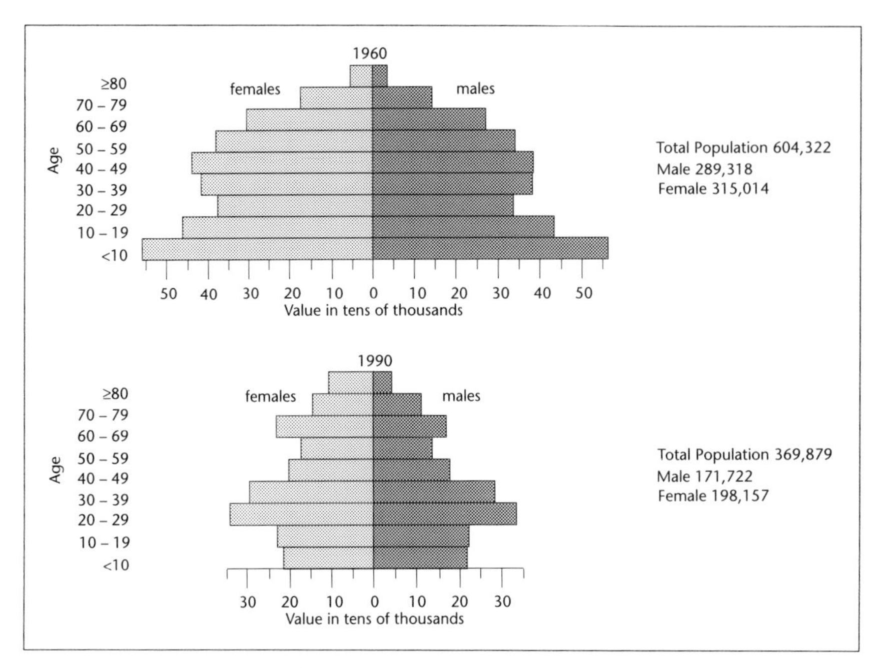

Fig. 21.7. Age pyramid of Pittsburgh, 1960 and 1990.

tion will rise to 214,394 by the year 2000. The same trends occurred in Westmoreland County, with a peak of 392,184 in 1980 and a decline to 370,321 in 1990, and it is forecast that the population will continue to decline to 365,276 in the year 2000. In contrast to the other counties, Butler County population has risen steadily from 114,639 in 1960 to a peak of 152,013 in 1990 and a projected rise to 170,117 in the year 2000.

The changing population pattern is a reflection of an economy in transition. Plant closings, primarily iron and steel, in Allegheny and Beaver counties has provided an economic environment for a decline in population. In Westmoreland County, residential suburbanization offsets industrial shutdowns. Nearly all of the more than 300 municipalities in the five counties have declined or shown only slow growth since 1960. The negative impact of industrial restructuring is most evident in Pittsburgh and the industrial towns along the three major rivers of the region.

Growth in population in a few places reflects economic focuses and continued outward redistribution of residences. The Pittsburgh International Airport has continued to expand, with the most recent addition in 1992, and has acted as a magnet for new businesses and residential development. The high-speed expressways have extended the commuter range and created additional growth poles with shopping malls, industrial parks, and interstate intersections. As a response, an outer ring of newer, growing communities has gradually evolved, but topographic complexity prevents a complete spatial integration.

Population Structure

The population structure of metropolitan Pittsburgh has exhibited a general aging since 1940 as a response to slow growth and the out-migration of the younger working-class adults (Fig. 21.7). In 1940 only 9 percent of the population was over 65, but by 1990 this number had increased to 18 percent. In 1940 the population structure reflected an industrial region with a balanced sex ratio. Although the Pittsburgh region experienced the baby boom of the 1940s and 1950s with a substantial increase in young people, the out-migration of working adults and the low depression-era birth rate reduced the number in young and middle-age groups. In addition, declining fertility rates in the 1970s, combined with out-migration, reduced the number of young people. Because males dominate out-migration and women have a long life-span, there are nearly 5 percent more women than men in the region.

Within the city of Pittsburgh, one district may

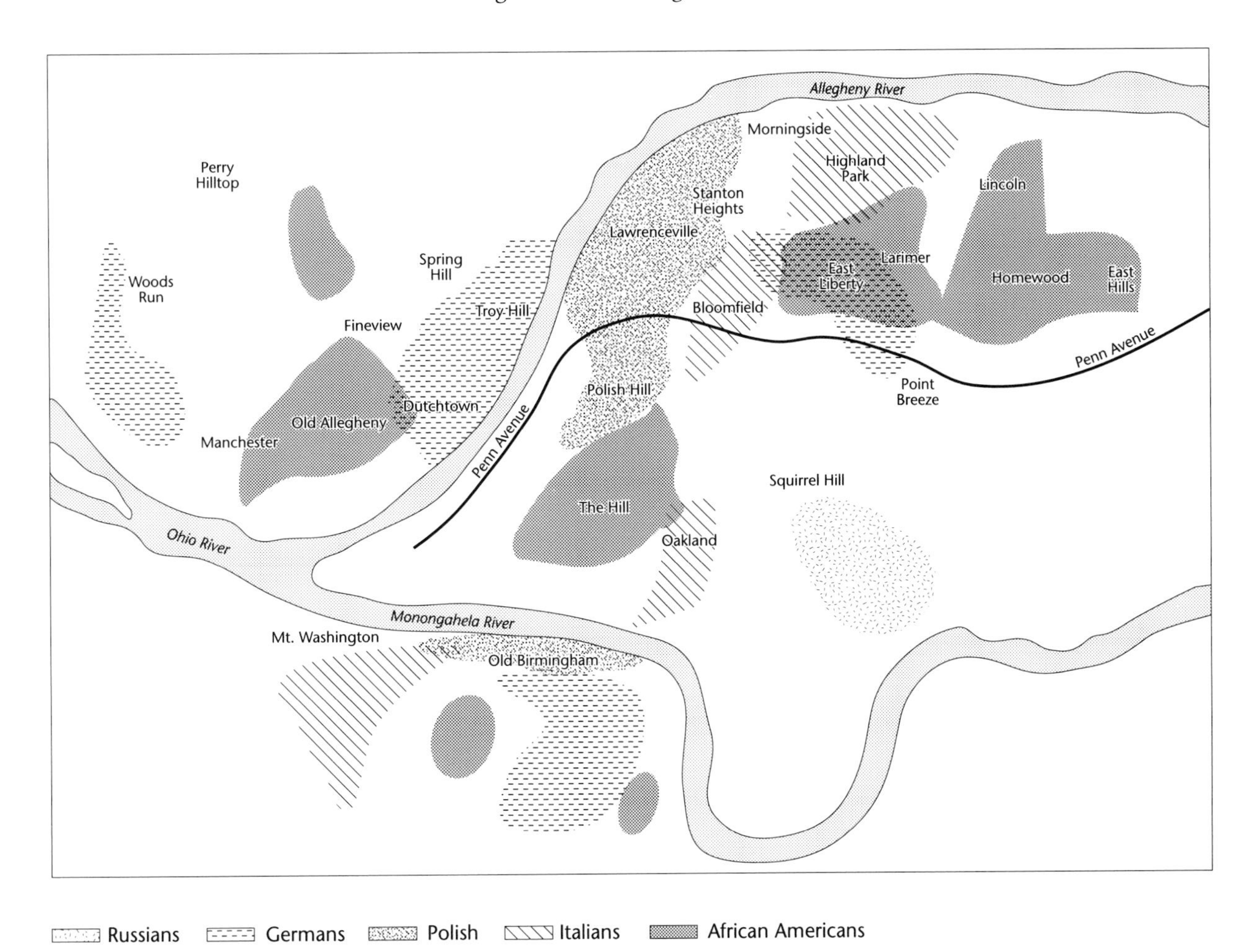

Fig. 21.8. Ethnic centers in Pittsburgh.

have a markedly different population structure than a nearby area. This is due to the concentration of special populations at a given place. For example, the number of older people may be very high in a specific place because that is where public and senior-citizen housing is located. There is also a tendency for older people to remain in the city. In contrast, younger people do not hesitate to move relatively often—to regions where the economy is more prosperous.

Ethnic Diversity

The Pittsburgh metropolitan area has long been characterized by its ethnic diversity (Fig. 21.8), and three distinct periods have evolved. The first, about 1730 to 1830, dates from the first settlers in the area. The British gained control of the region from the French in the 1750s, and the limited French influence disappeared. In the period from the 1750s to about 1780, the British and American colonists drove the Indians from the area. By the 1780s, the English, Scottish, and Scots-Irish provided Pittsburgh with a cultural identity. All these early settlers formed an elite that established the tradition for the future. Besides the dominant groups, small numbers of Germans, Swiss, Welsh, and French brought cultural diversity to the early settlements.

The second period of great European migration began about 1830 and continued until the federal restrictions on immigration were imposed in the 1920s. With a growing industrial economy, the Pittsburgh region was a magnet for large numbers of immigrants. From 1830 to about 1890 the Irish, Germans, and British were dominant in the flow of immigrants. In the 1840s the potato famine in

Ireland brought a great increase in the number of Irish immigrants, who were a distinct Catholic minority group with strong ethnic enclaves in the Strip District, on the Lower Hill, and around the mills on the north side in Allegheny City.

In 1890 the Germans were the largest ethnic group in the city. They created a number of neighborhoods, including Dutchtown in Allegheny City, the Strip District, the South Side, Uptown, and Lawrenceville. The diversity of the Germans was shown by the eight different church denominations they supported.

After 1890 the immigrants came largely from southern and eastern Europe. The region's population was diversified by the large number of migrants from such areas as Russia, Ukraine, Poland, Bulgaria, Serbia, Croatia, Lithuania, and Italy. Each of these groups created ethnic enclaves—for example, the Polish immigrants created communities in Lawrenceville, the South Side, and Polish Hill, and Italians concentrated in East Liberty, Larimer, and Bloomfield.

Although a few Blacks have been present since the city's early days, the population remained small until World War I. Then the demands for labor witnessed an unprecedented migration of Blacks to Pittsburgh from the South. By 1930 there were 55,000 Blacks in the city, many of whom worked in the least desirable jobs. The largest concentration of Blacks was in the Hill district. During the 1930s economic depression, the Black population remained stable, but with the industrial expansion of World War II it grew once again. By 1960 the Black population had grown to more than 100,000, and by 1990 it had increased to 174,551 in the five metropolitan counties, and 149,550 lived in Allegheny County. Although there has been some success at integration, such areas as the Hill district, Homewood-Brushton, and East Liberty are still highly segregated. With the slum clearance in the Hill district, about 1,500 Black families were scattered to various other neighborhoods.

Since the 1980s, a small number of Spanish-speaking and Asian groups have settled in the Pittsburgh region. Of the Asian groups, Indian, Chinese, and Korean are the largest. They are concentrated in Allegheny County in several East End communities from Oakland to Greenfield. The Spanish-speaking group is less than 1 percent of the population and reside mostly in the Oakland neighborhoods.

Although it is now more than 60 years since most of the European migrants came to Pittsburgh, ethnic clusters persist, although the locations of the groups are gradually changing. Germans are situated on the North Side and the eastern half of the South Hills, near their original localization. Italians are still found in South Oakland and Bloomfield, but they have left East Liberty and Larimer for the city's northeastern neighborhoods—Penn Hills and the western half of the South Hills. The Poles are widely distributed, but concentrations persist in the South Side, Polish Hill, and Lawrenceville as well as recent migration to the South Hills. The Ukrainians are concentrated in the South Side Flats along with the Poles. Russians are located in Squirrel Hill along with Hungarians and Jews. The Irish are widely distributed with a small concentration in the western part of South Hills.

Pittsburgh: Most Livable City

In March 1985 the Standard Metropolitan Area of Pittsburgh was rated as the most livable city out of 329 metropolitan areas that were evaluated in Rand McNally's *Place Rated Almanac*. For a nineteenth-century industrial city that was known as the Smoky City and sometimes called the Steel Capital of the World to be revitalized into the most livable city in the nation was a monumental task. Pittsburgh was followed by Boston, Raleigh-Durham, San Francisco, Philadelphia, and Nassau-Suffolk.

The livability of a region was determined by a number of criteria, including climate and environment, housing, health care, crime, transportation, education, the arts, recreation, and the economy. For the individual criteria Pittsburgh's scores ranged from 7 to 186 among the 329 metropolitan areas.

The city ranked highest in education, placing it number 7 in the nation. As the new city arose, there was a heightened awareness about the value of quality education in the public and parochial schools. Quality was improved through such initiatives as the city's magnet schools program in the middle-school areas. Magnet schools were established to emphasize particular subjects or academic disciplines—such as computer science, fine arts, or classical studies. These schools are always at capacity. Parochial schools have always been important in Pittsburgh because of the large Catholic population, but support for these schools has waned somewhat as operating costs have increased. Beyond the secondary schools, economic pressures have made people well aware of the value of higher education. Pittsburgh has 9 private and 10 public two-year colleges, and 11 private and 2 public four-year colleges and universities. These schools provide the widest range of educational programs.

The second highest position for the Pittsburgh metropolitan area was in the arts, where the region was ranked number 12. Although Pittsburgh's prominence as a cultural and arts center was long overshadowed by its industrial reputation, many of its nineteenth-century leaders provided support to these endeavors. Andrew Carnegie provided libraries throughout the region, and the first American library he endowed still exists in Braddock, in the shadow of a USX Corporation steel mill. Other family contributions are recognized in the names of Heinz Hall for the Performing Arts, the Frick Art Museum, the Buhl Science Center, the Scaife Galleries, the Phipps Conservatory, and the Mellon Institute, among others.

In health care the Pittsburgh region ranked 14th among the 329 cities. Pittsburgh has become a major medical center. Fifty hospitals in the nine-county area of southwestern Pennsylvania treat more than 750,000 patients a year. The area has 213 physicians per 100,000 people. Pittsburgh earned a rating of excellent for care affordability in the Places Rated Survey. Most of the region's premier health-care facilities, as well as the principal medical research and teaching facilities, are located in the city of Pittsburgh.

The development of transportation has always been a major problem in Pittsburgh, but improvements have occurred and the region was ranked 76 among the metropolitan regions. The rugged topography presents a challenge to development of a unified transportation system. The triangular shape of the central business district funnels traffic to a point. The shape of the Golden Triangle virtually dictates against establishment of an orderly street pattern. To develop a transportation system for the region, the Port Authority Transit (PAT) was organized in 1964, when 33 private and public bus and trolley companies were consolidated. This consolidation has improved the movement of people. Nevertheless, because of the slowness of the system and some delays, a large percentage of commuters continue to drive to work rather than use the mass transit. It has been estimated that close to 80 percent of PAT riders have no other means of transportation.

The highway system of the Pittsburgh region was long neglected, with regard to its maintenance and to development of modern highways. Modernization of the road system began in the 1970s, but lack of funds left many projects incomplete. To illustrate, there were unfinished segments of the East Street Valley Expressway between Downtown and the junction of Interstate 79, and ramps on the north end of the Fort Duquesne Bridge over the Allegheny River and the Crosstown Boulevard near the Civic Arena left hanging in air. In the 1980s new road-building occurred and most of the gaps in the system are now complete.

Crime was another factor considered in the evaluation of the metropolitan area. The Pittsburgh area ranked 78th in the nation, and only Nassau-Suffolk, New York, was rated more free from crime among the 13 largest metropolitan areas. Furthermore, the crime rate of Pittsburgh has improved since the survey was made. In 1985 the Federal Bureau of Investigation reported that in cities comparable to Pittsburgh in size the crime rate rose by 7 percent, while Pittsburgh experienced a decline of more than 12 percent. The FBI considers Pittsburgh the safest large city in the nation.

The climatic environment was also a factor in considering metropolitan livability. In this category Pittsburgh ranked 87th, ahead of such sun-and-fun spots as Fort Lauderdale, Daytona Beach, Las Vegas, Phoenix, and Colorado Springs. Pittsburgh was described as having a long summer with few excessively hot or muggy days and a mild autumn. A major climatic advantage was its location south of the Great Lakes snow belt, which still provided a cold, crisp winter.

The recreation facilities of the Pittsburgh region are considered good-to-excellent, and a wide variety of activities are available. The area supports 134 golf courses—more courses per capita than any other metropolitan area in the United States. In recent years, the state, counties, and municipalities have established a large number of parks. The river systems provide waterways for recreation. Traditionally, the number of pleasure-boat licenses issued each year in Allegheny County is among the highest for any county in the nation. Pittsburgh has three professional sport teams—the baseball Pirates, the football Steelers, and the hockey Penguins—and a large number of college and university sport teams. As a response to these facilities, the *Places Rated Almanac* researchers ranked Pittsburgh 90th in the nation and 9th among the 13 largest metropolitan areas.

In the economy category, Pittsburgh stood in 185th place. The economy rating is based on average household income (adjusted for taxes and cost of living), rate of income growth, and rate of job expansion. This rating is remarkbly high considering the decline in employment in the basic industries of the metropolitan region. The average household income in 1985 was almost $33,000, just below the national average. This figure rose almost 45 percent in the five years preceding the study. In the job category, Pittsburgh had more blue-collar jobs, despite the rapid growth of service jobs, than the national average. The unemployment threat was only moderate, indicating that the great job losses in the industries were in the past. In 1982 the *Places Rated Almanac* placed Pittsburgh in 232nd position among 277 metropolitan areas. With a 1985 standing of 185 out of 329 metropolitan areas, using the same standards in both years, Pittsburgh's economic position had improved remarkably in a few years.

Pittsburgh placed 186th in housing, the lowest rating of the individual categories. During the rapid growth in the late nineteenth century, much of the housing was built for those who worked in the heavy industries. As a result, more than 43 percent of all houses in the metropolitan area were built before 1940 and the average price of a house in 1985 was $60,500. Because much of the housing is old, housing costs are quite low. Only 1 of the top 10 metropolitan areas—Louisville—had a lower average cost than Pittsburgh. Although federal housing assistance has aided between one-third and one-half of the housing units built since 1965, a number of blighted areas remain.

Since the *Places Rated Almanac* named Pittsburgh as the most livable city in the nation in 1985, there is some evidence that the city is being viewed in a new light. The Greater Pittsburgh Convention and Visitors Bureau reports a significant increase in interest from out-of-towners who want to hold meetings and conventions in Pittsburgh. The region's old image of an ugly industrial region is changing. It is now seen as an area that provides the amenities needed for a high quality of life. This can only be a good omen for the future.

Bibliography

Baldwin, L. D. 1970. *Pittsburgh: The Story of a City, 1750–1865*. Pittsburgh: University of Pittsburgh Press.

Berger, K., ed. 1985. *Johnstown: The Story of a Unique Valley*. Second edition. Johnstown, Pa.: Johnstown Flood Museum.

Billinger, R. D. 1954. *Pennsylvania's Coal Industry*. Pennsylvania History Studies No. 6. Gettysburg: Pennsylvania Historical Association.

Bining, A. C. 1973. *Pennsylvania Iron Manufacture in the Eighteenth Century*. Second edition. Harrisburg: Pennsylvania Historical Commission.

Borchert, J. R. 1967. "American Metropolitan Evolution." *Geographical Review* 57:301–332.

———. 1972. "America's Changing Metropolitan Regions." *Annals of the Association of American Geographers* 62:352–373.

———. 1978. "Major Control Points in American Economic Geography." *Annals of the Association of American Geographers* 68:214–232.

Buck, S. J., and E. M. Buck. 1939. *The Planting of Civilization in Western Pennsylvania*. Pittsburgh: University of Pittsburgh Press.

Clark, D., ed. 1975. *Philadelphia, 1776–2076: A Three Hundred Year View*. Port Washington, N.Y.: Kennikat Press.

Cochran, R. C. 1978. *Pennsylvania: A Bicentennial History*. New York: Norton.

Currier, Wade R. 1979. "The Parts of Philadelphia." In *The Philadelphia Region: Selected Essays and Field Trips*, ed. R. Cybriwsky. Washington, D.C.: Association of American Geographers.

Florin, J. 1966. "The Advance of Frontier Settlement in Pennsylvania, 1683–1850: A Geographic Interpretation." M.S. thesis, Department of Geography, The Pennsylvania State University.

Folsom, B. W. 1981. *Urban Capitalists: Entrepreneurs and City Growth in Pennsylvania's Lackawanna and Lehigh Regions, 1800–1920*. Baltimore: Johns Hopkins University Press.

Gallagher, J. P. 1968. *A Century of History: The Diocese of Scranton, 1868–1968*. Scranton, Pa.: Diocese of Scranton.

Glasgow, J. 1985. "Innovation on the Frontier of the American Manufacturing Belt." *Pennsylvania History* 52:1–21.

Karaska, G. J. 1962. "Patterns of Settlement in the Southern and Middle Anthracite Region of Pennsylvania." Ph.D. dissertation, Department of Geography, The Pennsylvania State University.

Klein, P. S., and A. Hoogenboom. 1980. *A History of Pennsylvania*. University Park: The Pennsylvania State University Press.

Landis, C. I. 1918. "History of the Philadelphia and Lancaster Turnpike." *Pennsylvania Magazine of History and Biography* 42:1–28, 127–140.

Lemon, J. T. 1967. "Urbanization and the Development of Eighteenth-Century Southeastern Pennsylvania and Adjacent Delaware." *William and Mary Quarterly*, 3d ser., 24:501–542.

———. 1972. *The Best Poorman's Country.* Baltimore: Johns Hopkins University Press.

Lewis, P. F. 1972. "Small Towns in Pennsylvania." *Annals of the Association of American Geographers* 62:323–351.

Livingood, J. W. 1970. *The Philadelphia-Baltimore Trade Rivalry, 1780–1860.* New York: Arno Press.

Lubove, R. 1976. *Pittsburgh.* New York: New Viewpoints.

Magda, M. S. 1985. "Life in an Industrial Boomtown: Monesson, 1898–1923." *Pennsylvania Heritage* 10 (Winter): 14–19.

Markusen, A. E. 1985. *Profit Cycles, Oligopoly, and Regional Development.* Cambridge: MIT Press.

Mason, H. 1979. "Early Industrial Geography of the Delaware Valley." In *The Philadelphia Region: Selected Essays and Field Trips,* ed. R. Cybriwsky. Washington, D.C.: Association of American Geographers.

McLaughlin, G. E. 1938. *Growth of American Manufacturing Areas.* Pittsburgh: University of Pittsburgh, Bureau of Business Research.

Meyer, B. H. 1948. *History and Transportation in the United States Before 1860.* Washington, D.C.: Carnegie Institute.

Michener, C. K., and M. J. O'Malley. 1984. "The Last Frontier: Venango County." *Pennsylvania Heritage* 10 (Spring): 32.

Miller, E. W. 1981. "Pittsburgh: Patterns of Evolution." *Pennsylvania Geographer* 14 (October): 6–20.

Miller, R., and D. Sawers. 1968. *The Technical Development of Modern Aviation.* New York: Praeger.

Murphy, R. E., and M. Murphy. 1937. *Pennsylvania: A Regional Geography.* Harrisburg, Pa.: Telegraph Press.

National Interregional Highway Committee. 1944. *Interregional Highways.* Report of the National Interregional Highway Committee. Washington, D.C.: U.S. Government Printing Office.

Nelson, R. 1969. *Backcountry Pennsylvania, 1704–1774: The Ideals of William Penn in Practice.* Ann Arbor, Mich.: University Microfilms.

Noyelle, T., and T. Stanback. 1983. *The Economic Transformation of American Cities.* Totowa, N.J.: Rowman & Allanheld.

Patton, S. G. 1938. "Comparative Advantage and Urban Industrialization: Reading, Allentown, and Lancaster in the Nineteenth Century." *Pennsylvania History* 50:148–169.

Phillips, P. D., and S. Brunn. 1978. "Slow Growth: A New Epoch in American Metropolitan Evolution." *Geographical Review* 68:274–292.

Plankenhorn, W. 1957. "A Geographic Study of the Growth of Williamsport." Ph.D. dissertation, Department of Geography, The Pennsylvania State University.

Proctor, M., and B. Matuszeski. 1978. *Gritty Cities.* Philadelphia: Temple University Press.

Rosenberger, H. T. 1975. *The Philadelphia and Erie Railroad: Its Place in American Economic History.* Potomac, Md.: Fox Hills Press.

Rubin, J. 1961. "An Imitative Public Improvement: The Pennsylvania Mainline." In *Canals and American Economic Development,* ed. C. Goodrich. New York: Columbia University Press.

Schusler, W. K. 1960. "The Railroad Comes to Pittsburgh." *Western Pennsylvania Historical Magazine* 43: 203–238.

Snyder, M. P. 1975. *City of Independence: Views of Philadelphia Before 1800.* New York: Praeger.

Swetnam, G. 1964. *Pennsylvania Transportation.* Pennsylvania History Studies No. 7. Gettysburg: Pennsylvania Historical Association.

Warner, S. B., Jr. 1968. *The Private City: Philadelphia in Three Periods of Its Growth.* Philadelphia: University of Pennsylvania Press.

INDEX